Handbook for Vigilance in the Chemical and Biohazard Environments

Assessment and management of chemical and biohazard risk requires an understanding of risks in the laboratory, the environment, and the community. This handbook provides guidance on best management practices to help prevent costly oversights and dangers in practice, in industrial and scientific settings. It offers tools to actively avoid ill effects by addressing complacency, promoting good laboratory ethics, and encouraging proactive strategies to minimize risk. This handbook examines the safety culture of chemicals in a laboratory setting, simultaneously serving both as a proactive management exercise and as a means of preventing costly oversights and dangers in practice.

Handbook for Vigilance in the Chemical and Biohazard Environments

Edited by
Apollo Odhiambo Maima

CRC Press
Taylor & Francis Group
Boca Raton London New York

CRC Press is an imprint of the
Taylor & Francis Group, an **informa** business

Designed cover image: Shutterstock

First edition published 2026
by CRC Press
2385 NW Executive Center Drive, Suite 320, Boca Raton FL 33431

and by CRC Press
4 Park Square, Milton Park, Abingdon, Oxon, OX14 4RN

CRC Press is an imprint of Taylor & Francis Group, LLC

ISBN: 9781041048527 (hbk)
ISBN: 9781041048534 (pbk)
ISBN: 9781003630272 (ebk)

DOI: 10.1201/9781003630272

Typeset in Times
by codeMantra

This book is dedicated to:

The two most consequential people in my existence, each at various phases of my journey, and who have consistently placed significant faith in my abilities…

My adorable mother, the late Juanita, who shaped and gave me the gift of dreams and aspirations, and the capacity to actualize them. Regrettably, she departed us too soon; however, her memory will endure in our collective consciousness.

My beautiful wife, my confidante, my muse, my love and best friend, Susan, who supported me through the long journey to deliver these works. She has conspired with my present to support my growth and instill unparalleled tranquility within me, all while enabling and steering me toward an even greater future!

Contents

List of Figures

List of Tables

Foreword

The *Handbook for Vigilance in the Chemical and Biohazard Environments* provides an authoritative reference on safety procedures, the handling, disposal, and security of chemicals and biohazards at both laboratory and community levels. This book offers prudent guidelines to laboratory users for the development of regulatory policy for academic and nonacademic institutions. It stresses that all accidental chemical and biohazard injuries can be minimized if systems and attitudes are in place to prevent them, and that to fully materialize a safety culture, even with improved technology, everyone should be mindful of maintaining a safe environment.

Dr. Apollo O. Maima, the author, is Associate Professor and a Dean at Maseno University in Kenya. I met Dr. Maima 15 years ago when I was the Dean of the School of Arts and Sciences at the United States International University-Africa (USIU-A), and he was Assistant Professor in charge of the new Bachelor of Pharmacy program in my School. He later went on to become the first Dean of the School of Pharmacy and Health Sciences at USIU-A. We have since maintained a very cordial and collaborative academic relationship.

Dr. Maima is an avowed educationist and a respected pharmacist specialized in pharmaceutical chemistry and analysis, community health, epidemiology, health education, and pharmaceutical microbiology. He has designed a university-level curriculum for and taught laboratory safety, chemical safety and security, and biohazard environment for a long time. He has served in several professional and institutional boards and is the current Editor-in-Chief of the *Pharmaceutical Journal of Kenya.*

This handbook is, therefore, a meticulous intellectual product from someone who understands firsthand the importance of chemical safety and security after many years of teaching, practical, and community experience. It is meant to help readers in understanding the principles of chemical safety and security, and to provide guidance on the creation of a culture-shift toward chemical security, management of hazardous waste, and assessment and minimization of risks related to chemicals.

This handbook will assist faculty, researchers, staff, and students to provide a safe, secure, and healthy laboratory environment to teach, learn, and conduct research, and provide information that will assist them in meeting their goals in an environmentally sound and safe manner. Information has been included concerning the use and storage of chemicals, the use of personal protective equipment, safety practices, and the proper methods of waste disposal. This information is intended to minimize physical and health hazards to researchers and the university community while preventing pollution of our environment. As a microbiologist, biotechnologist, environmentalist, and laboratory-based researcher, I strongly advise that every university faculty member, researcher, and laboratory user should read this handbook, be familiar with and practice its contents, and keep it handy for reference.

Prof. James H.P. Kahindi, PhD, FRSB
Professor of Microbiology, and
Vice-Chancellor, Pwani University
P.O. Box 195-80108
Kilifi, Kenya.
Email: jkahindi@pu.ac.ke, vc@pu.ac.ke

Preface

Whereas chemicals, biohazards, and wastes are an important part of our lives, they also pose health and environmental risks, depending on their inherent hazardous nature and how they are used. This book was conceptualized on noting that hazardous substances can be found in foods, medicines, usable products, and, generally, in the biotic and abiotic environments. The use and/or distribution of such substances can jeopardize the long-term use of land and water resources, making them unfit for human consumption. This, in turn, can negatively impact the development and economies of countries. Exposure to hazardous substances can occur at work, home, and indirectly through the environment, potentially causing deaths, acute or chronic health effects.

Assessing and managing chemical and biohazard risks require an understanding of the risks in the laboratory, environment, and in the community. Threats to human and environmental health are on the rise, making it necessary to be vigilant and to prepare accessible responses and planning. Awareness of potential danger is challenging, especially if it is pernicious.

PURPOSE/NEEDS ASSESSMENT

The purpose of the *Handbook for Vigilance in the Chemical and Biohazard Environments* is to serve as a comprehensive resource for concepts, issues, and competencies in the management of laboratory design and safety, chemical security, and biohazard security for

1. academics with teaching, unit/site coordination, supervisory, and/or managerial responsibilities;
2. researchers and workers in chemical, pharmaceutical, and biotechnology settings;
3. government agencies and, especially, Regulating Authorities; managers of public and environmental health, especially in low- and medium-income countries (LMICs);
4. laboratory and industrial technical staff, and hygienists;
5. capital works & facilities staff responsible for overseeing the maintenance of laboratories;
6. school, college, undergraduate, master's, and PhD students in chemistry, biochemistry, biotechnology, medical laboratory science, and health sciences;
7. contractors engaged by the institution to undertake work within a laboratory.

Protection has become increasingly important in chemical and biohazard environments, whether in academia, government, or private sectors, requiring greater vigilance among personnel. This handbook highlights the meaningful contributions that bench-level, research, and regulatory personnel can make. It offers theoretical and practical insights on chemical and biohazard management drawn from experience in laboratory practice. For graduate and undergraduate students in the many laboratory-based programs in (bio)chemistry, biotechnology, microbiology, medical laboratory science, and other health sciences around the world, this will be a textbook. For managers, regulators, laboratory technologists, researchers, and university teaching staff, it will be a desk reference.

Hazardous substances in the environment include toxic chemicals, clinical agents, housekeeping poisons, irritants used in pest control, biological toxicants, and infectious agents. Contamination of food, water, or premises poses a potential hazard. Risk assessment is, therefore, crucial for evaluating exposure levels, consequences, and reduction methods. Threats to human and environmental health are on the rise, making it necessary to be vigilant and to prepare accessible responses and planning. Awareness of potential danger is challenging, especially if it is pernicious.

Assessment and management of chemical and biohazard risks require an understanding of the risks in the laboratory, environment, and community. This book provides guidance on best management practices to help prevent costly oversights and dangers in practice, in industrial and scientific settings, and offers tools to actively avoid ill effects by addressing complacency, good laboratory ethics, and proactive strategies to minimize risk. It examines the safety culture of chemicals in a laboratory setting and serves as a proactive management exercise for workers and users.

Unfortunately, there aren't very many good resources to teach chemical and biohazard management independently, in environmental science or even within other relevant laboratory-based sciences.

APPROACH

This book provides views and guidance on chemical and biohazard management practices through several strategic but focused and compounded lenses. *First*, the principles of the WHO/TDR's (Tropical Diseases Research) *Good Laboratory Practice Handbook*[1] form the foundation for this book. *Second*, to help prevent costly oversights and dangers in practice, in industrial and scientific settings, this book offers tools to actively avoid ill effects by addressing complacency, good laboratory ethics, and proactive strategies to minimize risk.

Third, it examines the safety culture of chemicals and other consumables in a laboratory setting. Hence, it serves as a proactive management key for workers and users, noting that the assessment and management of chemical and biological hazards require an understanding of the risks in the laboratory, environment, and the community.

Fourth, this book highlights the meaningful contributions that vigilant bench-level, research and regulatory personnel can make, as regulation and quality improvement is integrated into the text. Operational defects are curable through training and empowerment of internal users to comprehensively understand and apply the concept of quality control systems.[2]

Fifth, this book re-examines laboratory design, physical and materials management and system designs[3] to minimize hazardous consequences. Careful attitudes, training, knowledge of emergency procedures, and medical management play major roles.

The handbook is divided into four parts and has a total of 16 chapters as outlined below:

Part One: Laboratories as Major Chemical and Biohazard Environment (four chapters)
Part Two: Chemical and Biohazard Security (eight chapters)
Part Three: Design and Construction of Affordable Quality Laboratories (two chapters)
Part Four: Cost-Saving Laboratory Use (two chapters)

AIM

The aim of the *Handbook for Vigilance in the Chemical and Biohazard Environments* is to teach concepts, practices, skills, and competencies that will help eliminate chemical and biohazard risks in the laboratory, environment, and the community. This handbook could have a big impact.

Chemical Weapons Convention (CWC), in conjunction with the Organization for the Prohibition of Chemical Weapons (OPCW) programs, has held several regional workshops to teach concepts included in this handbook to their staff and those in the country programs.[4] There are efforts globally to improve education on toxic chemical hazards as the chemical and pharmaceutical industries grow. By promoting awareness and providing targeted training, individuals and communities can be educated and empowered to mitigate the potential hazards.[5] If this training were made widely available, it would improve the management of chemical, biological, and toxic hazards in laboratories, resulting in safer and more secure laboratories, environments, and communities, improved use of resources, and the fostering of positive change.

METHODOLOGY

A collaborative team approach has been used to develop the book chapters. All co-authors have contributed to chapters based on their fields of specialization and interest. At the core of this handbook is the interest of being able to share especially LMIC perspectives and to bridge the existing training/educational gap with developed nations.

The specific methodology used depends on the chapter, but largely, the authors exhaustively review relevant literature in a bid to determine key concepts to convey to the reader, or publish research findings on topical issues. While reviewing the existing literature, the authors look at international practices and write case studies to better illustrate the issues. Where training or technical teaching is requested or implied, our authors use GLP, recognized standards and publicly available resources that have been found scattered in the literature. Our unitary approach is conceptualized on the consolidation of the available literature material, mostly used in conducting short courses. Throughout the text, every effort has been made to illustrate a global approach to issues as well as cultural nuances.

Contributors

Simon Adede
Department of Chemistry and Material Science
Technical University of Kenya
Nairobi, Kenya

Austin O. Aluoch
Department of Chemistry and Material Science
Technical University of Kenya
Nairobi, Kenya

Collins O. Asweto
Department of Community Health
School of Nursing, University of Embu
Embu, Kenya

Yasodha K. Janapati
School of Pharmacy and Health Sciences
United States International University-Africa (USIU-A)
Nairobi, Kenya

Lincoln L. Were Munyendo
Department of Pharmaceutical Chemistry, Pharmacology & Pharmacognosy
School of Pharmacy and Health Sciences, United States International University-Africa (USIU-A)
Nairobi, Kenya

Stanley N. Ndwigah
Department of Pharmaceutical Chemistry, Faculty of Health Sciences
University of Nairobi
Nairobi, Kenya

Joab O. Onyango
Department of Chemistry and Material Science
Technical University of Kenya
Nairobi, Kenya

Patrick O. Onyango
School of Graduate Studies
Maseno University
Maseno, Kenya

Philip Mark Othigo
School of Architecture
Technical University of Kenya
Nairobi, Kenya

Geoffrey Otieno
Department of Chemistry and Material Science
Technical University of Kenya
Nairobi, Kenya

Sarah A. Otieno
Department of Chemistry
Binghamton University
Binghamton, New York

James J. Owuor
Department of Chemistry and Material Science
Technical University of Kenya
Nairobi, Kenya

Patrick M. Shem
Department of Chemistry and Material Science
Technical University of Kenya
Nairobi, Kenya

Acknowledgment

I sincerely wish to appreciate my family for providing me immense emotional and social support that enabled me to focus on this book-writing project: my father, late Francis, who supported this project but wouldn't live to see it completed; my wife Susan, and children Roselinder, Adelaide, Joan Nancy, Christabel, John Junior, Jaramogi, Georgina, Juanita, Stanley, Christine, and Amos Kyle; last, but not least, Roselyne Mama Chris. Without you guys giving me so much love, peace, material and moral support, this book would surely have been a flop.

I wish to express my gratitude to all authors for their essential contributions to this compilation. The expertise, professionalism, and diligence demonstrated by each author were instrumental in achieving the project's objectives. Furthermore, I acknowledge the reviewers, whose thoughtful critiques greatly enhanced the quality of the final chapters. Similarly, I wish to thank my friend and mentor, Prof. Kahindi, for his kindness in writing the foreword for this book.

The contributions of the Taylor & Francis Group production team in finalizing this edition warrant recognition. Notably, Senior Editor Hilary LaFoe played a pivotal role through her decision-making support and guidance during the initial stages of manuscript development. Meanwhile, Editor Adam Mlodzinski supported with the management of the editorial review, design, typesetting, and indexing tasks.

The editor invites constructive comments, criticism, and submissions on all or any of the sections of this book. Diverse perspectives from the contents, style, and analytical viewpoints are highly welcome and valued.

About the Editor

Author: Apollo Odhiambo Maima

School of Pharmacy, Maseno University, Kenya

Email: apollomaima@gmail.com/amaima@maseno.ac.ke

Author ORCID account: 0000-0002-9135-3930

Key Qualifications

Dr. Maima is Associate Professor of Pharmacy and Dean of the School of Pharmacy, Maseno University, Kenya. He holds a PhD in Community Health and Development and Master of Pharmacy in Pharmaceutical Chemistry and Analysis. He is a former Dean and Program Leader of the School of Pharmacy & Health Sciences at the United States International University-Africa. He is a fellow of the Centre for Effective Global Action (CEGA) of the University of California, Berkeley, and a member of the American Society of Tropical Medicine and Hygiene (ASTMH) and the International Society for Infectious Diseases (ISID). He is also a trained PhD supervisor.

Maima is a pharmaceutical chemist and analyst, public health specialist and epidemiologist, implementation research scientist, registered pharmacist in Kenya, and Good Standing Member of the Pharmaceutical Society of Kenya. His other current and past memberships include *International Pharmaceutical Federation (FIP)*, the *Experts/Specialization Committee*, the *Training and Assessment Technical Committee of the Pharmacy and Poisons Board of Kenya*, the Board of Management of the *National Quality Control Laboratories,* and the Ministerial Committee that prepared the *Kenya National Pharmaceutical Policy (KNPP)* of 2010. He is also the current President of the Association of Pharmacy Schools of Kenya (APSK) and the Editor-in-Chief of the *Pharmaceutical Journal of Kenya.*

SELECTED CERTIFIED MANAGEMENT COURSES

1. Resource person for "Workshop on Curriculum Development for Pharmaceutical and Agricultural Training Institutions to Mainstream Chemical Security (6th Oct 2022)".
2. The KENAS Assessor Training Course on ISO/IEC 17025:2017 (Competence of Testing and Calibration Laboratories) and ISO 19011:2018 (Guidelines for Auditing Management Systems).
3. The DIES/CREST Course for Doctoral Supervisors at Stellenbosch University, RSA (2019).
4. Post-Doctoral Fellow in Implementation Research at University of California, Berkeley (2018).
5. Workshop on Security and Counterterrorism Awareness, on 24/08/2015 at USIU-A, Nairobi.
6. "Good Manufacturing Practice" – World Health Organization (2004).
7. Certificate of the International Standards Organization, ISO 9000 Certification (2007), KBS.
8. "WHO/TDR-Workshop on Good Laboratory Practice" course (WHO/TDR and UON) – April 2004.

GRADUATE SUPERVISION AND EXAMINATION

Eighteen Masters Supervision & Examination (Pharmacy, Biotechnology or Public Health-related) and 12 PhD Supervision & Examination (Pharmacy, Biotechnology or Public Health-related).

BOOKS CO-AUTHORED

1. *Policy Guidelines on Pharmacy Specialties and Recognition* (2014). First Edition. Pharmacy and Poisons Board of Kenya.

2. *Guidelines for Evaluation and Assessment for Enrolment of a Pharmaceutical Technologist* (2013). Second Edition. Pharmacy and Poisons Board of Kenya.
3. *Guidelines for Evaluation and Assessment for Registration of Pharmacists* (2014). Second Edition. Pharmacy and Poisons Board of Kenya.

SELECTED RESEARCH PUBLICATIONS

1. Maima, A. (2024). Challenges in the pharmaceutical supply chain in Kenya. *The Pharmaceutical Journal of Kenya*, 28(1), 2024.
2. Balemba, B. L., Maima, A. O., & Wafula, C. (2023). Association of importance of knowledge and good practices in the proper use of llins in relation to malaria morbidity in households with children under 5 in the Karisimbi Health zone of Goma in North KIVU/DRC. *International Journal of Development Research*, 13(09), 63821–63830. https://doi.org/10.37118/ijdr.27227.09.2023
3. Judson, S., Torimiro, J., Pigott, D., Maima, A., Mostafa, A., Samy, A., . . . Njabo, K. (2022). COVID-19 data reporting systems in Africa reveal insights for future pandemics. *Epidemiology and Infection*, 54, 1–22. https://doi.org/10.1017/S0950268822001054
4. Nderitu, A., Karimi, P., Guantai, E., & Maima, A. (2021). Determinants of adherence to tuberculosis chemotherapy in Nairobi City County, Kenya. *The East and Central African Journal of Pharmaceutical Sciences*, 23(3), 100–103. http://uonjournals.uonbi.ac.ke/ojs/index.php/ecajps/article/view/717
5. Maima, A. O. & Were L. L. M. (2018). Contemporary advances and precincts of biopharming as drugs' production systems. *Journal of Natural Sciences Research*, 8(4), 921. www.iiste.org
6. Nassali, H., Karimi, P., Nyamu, D., Guantai, E. M., Maima, A. O., Kajungu, D., & Kirui, J. (2015). Evaluation of prescribing practices in Kenyatta National Hospital. *The Pharmaceutical Journal of Kenya*, 23(1), 2015.
7. Maima, A. O., Thoithi, G. N., Ndwigah, S. N., Kamau, F. N., & Kibwage, I. O. (2008). Phytosterols from the stem bark of *Combretum fragrans* F. Hoffm. *The East and Central African Journal of Pharmaceutical Sciences*, 11(2), 52–54.

SELECTED CONFERENCE PRESENTATIONS

1. **Maima, A. O.**, Okalebo, F. A., Othero, D. A., Osano, E. N., Orlendo, C., & Mugambi, W. (2024). The switch from 10 to 5-dose vial of Measles Rubella Vaccine in Kenya: A Logical Evaluation Model and Theory of Change. The Africa Evidence Summit in Dar es Salaam on July 10 and 11, 2024.
2. **Apollo O. Maima** (2022). "Felicitous Research and Workforce Training for a 21st Century Healthcare System" presented at the Health Workforce Conference 2022 held at Whitesands Hotel, Mombasa, from February 7th to 9th, 2022.
3. **Apollo O. Maima** (2018). "Contemporary environmental and health-related factors influencing human growth and development" presented at the 2018 Pharmaceutical Society of Kenya Annual Scientific Conference, May 30 – June 3, in Mombasa, Kenya.
4. Were L.L. Munyendo & **Apollo O. Maima** (2018). *Pleurotus ostreatus* mycelium powder aptitudes in formulations of fortified foods as patient specific nutraceuticals averting product distributorship. Pharmaceutical Society of Kenya 38th Annual Scientific Conference, Mombasa, Kenya, May 30th – June 3rd, 2018.
5. Poster Presentation Titled "Trends in facility caseloads of adult pneumonia in Nairobi", Presented at the International Pharmaceutical Federation (FIP) Congress 2015, September 29 – October 3, in Dusseldorf, Germany.

6. Poster Presentation Titled "Antimicrobial properties of some medicinal Plants of the Luo community of Kenya", presented at the International Pharmaceutical Federation (FIP) Congress 2014, August 30 – September 4, in Bangkok, Thailand.

SPECIAL PROJECTS UNDERTAKEN

1. Development of the Bachelor of Pharmacy curriculum for three Kenyan Universities (Maseno, USIU-A, Kenyatta). In each case, there was set up of relevant functional laboratories.
2. Development of the Bachelor of Technology in Analytical Chemistry curriculum for USIU-A.
3. Development of Core Curriculum for B.Pharm. degree for Regulator, Pharmacy and Poisons Board, Nairobi.
4. The development and writing of a technical proposal for financing the purchase of equipment for laboratories for the USIU Pharmacy Training program. This was a US$1,975,000 budget program.
5. He wrote and won a grant from Seeding Labs of Boston, USA (2018/2019) for supply of laboratory equipment worth US$110,000 for the United States International University (USIU)-Africa, Kenya.
6. Dr. Maima teaches two courses (Laboratory Design & Safety; Chemical Safety & Security) to undergraduate students of Bachelor of Pharmacy, B.Sc. Chemistry and B.Sc. Medical Laboratory Science at Maseno University.

Abbreviations & Acronyms

Abbreviation/Acronym	Meaning
AAN	Aristolochic Acid Nephropathy
ADA	American Disability Act
ADR	Adverse Drug Reactions
AGP	Aerosol Generating Procedures
AI	Artificial Intelligence
AIDS	Acquired Immunodeficiency Syndrome
ANSI	American National Standards Institute
AOPs	Advanced Oxidation Processes
APIs	Active Pharmaceutical Ingredients
ART	Assisted Reproductive Technology
ASEAN	Association of Southeast Asian Nations
ASEI	Activity, Student, Experiment Improvisation
ASHRAE 62-2001	American Society of Refrigerating and Air-Conditioning Engineers Standard 62.1
ASM	American Society for Microbiology
ATCC	*American Type Culture Collection*
AU	African Union
BARC	Bhabha Atomic Research Centre
BARDA	Biomedical Advanced Research and Development Authority
BCS	Biopharmaceutics Classification System
BIS	Bureau of Indian Standards
BMBL	*Biosafety in Microbiological and Biomedical Laboratories*
BOD	Biochemical Oxygen Demand
BPA	Bisphenol A
BREEAM	Building Research Establishment Environmental Assessment Model
BSC	Biological Safety Cabinet/Biosafety Cabinet
BSL	Biosafety (or Biological Safety) Levels
BSL1	Biosafety Level 1
BSL2	Biosafety Level 2
BSL3	Biosafety Level 3
BSL4	Biosafety Level 4
BWC	Biological Weapons Convention
CAGR	Compound Annual Growth Rate
CAM	Complementary and Alternative Medicine
CAPS	Chemicals Accident Prevention System
CARIS	Chemical Accidents Response Information System
Cas	Collaborative Arrangements
CAT	Centro Académico de Talentos
CBRN	Chemical, Biological, Radiological and Nuclear
CCMB	Centre for Cellular and Molecular Biology

Abbreviation/Acronym	Meaning
CCSE	Cooperation Centre for Scientific Expertise
CCTV	Closed-Circuit Television
CDC	Centers for Disease Control and Prevention
CDRI	Central Drug Research Institute
CDSCO	Central Drugs Standard Control Organization
CERCLA	Comprehensive Environmental Response, Compensation, and Liability Act
CERT	Company Emergency Response Team
CFATS	Chemical Facility Anti-Terrorism Standards
CFSLs	Central Forensic Science Laboratories
CH	Community Hospital
CHM	Chinese Herbal Medicine
CHoRFA	Chemical Hazard and Risk Framework Assessment
CITES	Convention on International Trade in Endangered Species (of Wild Flora and Fauna)
CLIA	Clinical Laboratory Improvement Amendments
CLP	Classification, Labeling and Packaging
CLRI	Central Leather Research Institute
CMHCC	Community Mental Health Care Center
CMS	Centers for Medicare & Medicaid Services
CMSR	Chemical Management and Safety Rules
COD	Chemical Oxygen Demand
COSHH	Control of Substances Hazardous to Health
CounterACT	Countermeasures Against Chemical Threats
COVID-19	Coronavirus Disease of 2019
CPCB	Central Pollution Control Board
CPR	Cardiopulmonary Resuscitation
CPS	Cyber-Physical Systems
CRMP	Cybersecurity Risk Management Plan
CSD	Central Sanitary Department
CSIA	Chemical Security Improvement Act
CSIO	Central Scientific Instruments Organization
CSIR	Council of Scientific & Industrial Research
CSP	Chemical Security Program (USA)
CSRA	Chemical Security Risk Assessment
CSSP	Council of Scientific Society Presidents
CSTR	Continuous Stirred Tank Reactor
CT	Computed Tomography
CVSS	Common Vulnerability Scoring System
CWC	Chemical Weapons Convention
DAL	Drug Administration Law
DCP	Department of Chemicals and Petrochemicals
DDA	Disability Discrimination Act
DDE	Dichloro Diphenyl Ethane
DDT	Dichloro-diphenyl-trichloroethane

Abbreviation/Acronym	Meaning
DGFASLI	Directorate General of Factory Advice Service & Labour Institutes
DHS	Department of Homeland Security
DISH	Directorate of Industrial Safety and Health
DNA	Deoxyribonucleic Acid
DRDO	Defense Research & Development Organization
DWWW	Darvill Waste Water Works
DXM	Data eXchange Message
E2P	Environmental Emergency Plans
ECs	Emerging Contaminants
ED	Explosive Device
EDCs	Endocrine-Disrupting Chemicals
EDs	Endocrine Disruptors
EID	Emerging Infectious Disease
ELNs	Electronic Laboratory Notebooks
ELT	Experiential Learning Theory
EMA	European Medicines Agency
EML	Essential Medicines List
EMT	Emergency Management Team
EN	European Norm (Equivalent to ISO)
EPA	Environmental Protection Agency
EPDM	Ethylene Propylene Diene Monomer
ERP	Emergency Response Plan
ES	Environmental Service
EU	European Union
FAHP	Fuzzy Analytic Hierarchy Process
FAO	Food and Agriculture Organization
FBI	Federal Bureau of Investigation
FCs	Fume Cupboards
FDA	Food and Drug Administration
FGD	Flue Gas Desulfurization
FGDs	Focus Group Discussions
FICCI	Federation of Indian Chambers of Commerce and Industry
FMEA	Failure Modes and Effects Analysis
FSL	Facility Security Level
FSSAI	Food Safety and Standards Authority of India
FTC	Federal Trade Commission
FTE	Full-Time Equivalent
G-8	The Group of Eight
GABA	Gamma-Aminobutyric Acid
GACP	Good Agricultural and Collection Practice
GAP	Good Agricultural Practices
GCP	Good Clinical Practice
GHS	Globally Harmonized system of Classification and labelling of Chemicals
GLP	Good Laboratory Practice

Abbreviation/Acronym	Meaning
GMMs	Genetically Modified Microorganisms
GMOs	Genetically Modified Organisms
GMP	Good Manufacturing Practices
GOARN	Global Outbreak Alert and Response Network
GRACE	Gravity Recovery and Climate Experiment
GSP	Good Storage Practices
HBs	Hepatitis B Surface
HCS	Hazard Communication Standard
HDB	*High-Dose Buprenorphine*
HDI	Herb–Drug Interactions
HEPA	High-Efficiency Particulate Air
HFACS	Human Factors Analysis and Classification System
HIPAA	Health Insurance Portability and Accountability Act
HIV	Human Immunodeficiency Virus
HOD	Head of Department
HPA	Hazardous Products Act
HPLC	High-Pressure Liquid Chromatography
HRF	High-Risk Facility
HVAC	Heating, Ventilation and Air Conditioning
IARI	Indian Agricultural Research Institute
IBC	Institutional Biosafety Committee
IBEF	India Brand Equity Foundation
ICMR-NIV	Indian Council of Medical Research – National Institute of Virology
ICMSR	India's Chemicals Management and Safety Rules
ICS	Industrial Control Systems
ICS	Indian Chemical Society
ICT	Information and Communication Technology
IDEM	Integrated Disaster and Emergency Management
IEC	International Electrotechnical Commission
IEDs	Improvised Explosive Devices
IHR	International Health Regulations
IICB	Indian Institute of Chemical Biology
IICT	Indian Institute of Chemical Technology
IIP	Indian Institute of Petroleum
IISc	Indian Institute of Science
IITs	Indian Institutes of Technology
INCB	International Narcotics Control Board
INTERPOL	International Criminal Police Organization
IoT	Internet of Things
IP	Implementation Protocols
IP	Intellectual Property
IPC	Indian Pharmacopoeia Commission
IPRs	Intellectual Property Rights
IQ	Intelligence Quotient
IRS	Incident Response Strategy

Abbreviation/Acronym	Meaning
ISI	Indian Standards Institution
ISO	International Standards Organization
ITB	Invitation to Bids
IUCN	International Union for Conservation of Nature
KCS	Kenya Chemical Society
KPA	Kenya Pharmaceutical Association
KPIs	Key Performance Indicators
KYC	Know-Your-Customer
L3	Long Life, Loose Fit, Low Energy
LAI	Lab-Associated Infections
LBG	Laboratory Biosafety Guidelines
LCSSs	Laboratory Safety Summaries
LED	Light-Emitting Diode
LEPC	Local Emergency Planning Committee
LIMS	Laboratory Information Management Systems
LIS	Laboratory Information Systems
LMICs	Low- and Medium-Income Countries
LRM	Legal Risk Management
MCM	Medical Countermeasures
MHOR	Manual Handling Operations Regulations
MIBBI	Minimum Information for Biological and Biomedical Investigations
ML	Machine Learning
MoEFCC	Ministry of Environment, Forest and Climate Change
MRI	Magnetic Resonance Image
MSC	Microbiological Safety Cabinet
MSDS	Material Safety Data Sheets
NABL	National Accreditation Board for Testing and Calibration Laboratories
NAFAC	National Agency for Food and Drug Administration and Control
NAM	New Approach Methodologies
NATCM	National Administration of Traditional Chinese Medicine
NATECH	Natural Hazard Triggered Technological Accident
NATO	North Atlantic Treaty Organization
NBA	National Biosafety Authority
NCB	Narcotics Control Bureau
NCL	National Chemical Laboratory
NDMA	National Disaster Management Authority
NDPS	Narcotic Drugs and Psychotropic Substances
NEERI	National Environmental Engineering Research Institute
NEQS	National Environmental Quality Standards
NFPA	National Fire Protection Association
NGOs	Nongovernmental Organizations
NGRA	Next-Generation Risk Assessment
NIH	National Institutes of Health
NIPER	National Institute of Pharmaceutical Education & Research
NMPA	National Medical Products Administration

Abbreviation/Acronym	Meaning
NMR	Nuclear Magnetic Resonance
NPP	Nuclear Power Plant
NPPA	National Pharmaceutical Pricing Authority
NQCL	National Quality Control Laboratories
NRAs	National Regulatory Authorities
NSAID	Nonsteroidal Anti-inflammatory Drug
NSTA	National Science Teachers Association
NTC	Federal Trade Commission
OCA	Occupational Chemical Accident (in Korea)
OCME	Office of Chief Medical Examiner
OECD	Organization for Economic Co-operation and Development
OPCW	Organization for the Prohibition of Chemical Weapons
OSHA	Occupational Safety and Health Administration
OT	Operational Technology
OTC	Over-the-counter
PAHs	Polycyclic Aromatic Hydrocarbons
PAPRs	Powered Air-Purifying Respirators
PBA	Pharmaceutical-Based Agent
PCBs	Polychlorinated Biphenyls
PDS	Proactive Defense Strategy
PDSI	Plan, Do, See, Improve
PEP	Post-exposure Prophylaxis
PFAS	Perfluoroalkyl and Polyfluoroalkyl Substances
PI	Principal Investigators
PIN	Personal Identification Number
PLC	Programmable Logic Controller
PM	Particulate Matter
PMMA	Polymethyl Methacrylate
PNC	Polymeric Nanocomposite
POCA	Preparation of Off-Site Consequence Analyses
POM	Pharmacy-Only Medicine
POP	Persistent Organic Pollutants
PP	Polypropylene
PPB	Pharmacy and Poisons Board (Kenya)
PPCP	Pharmaceuticals and Personal Care Products
PPE	Personal Protective Equipment
PQP	Pre-qualification Program
PRC	People's Republic of China
PSK	Pharmaceutical Society of Kenya
PSM	Process Safety Management
PT	Proficiency Test
PUI	Person Under Investigation
PVC	Polyvinyl Chloride
QC	Quality Control
QCI	Quality Council of India

Abbreviation/Acronym	Meaning
QIVIVE	Quantitative In Vitro–In Vivo Extrapolation
QMT	Quality Management Team
qPCR	Quantitative Polymerase Chain Reaction
R&D	Research and Development
RAM	Risk Analysis and Management
RAMP	Reporting and Management Plans
REACH	Registration, Evaluation, Authorization and Restriction of Chemicals
RFP	Request for Proposal
RFQ	Request For Quotation
RG-4	Risk Group-4
RIA	Radio-immunoassay
RMP	Risk Management Program
RO	Reverse Osmosis
ROW	Rest of World
RSIs	Repetitive Strain Injuries
RTOs	Research and Technology Organizations
SAARC	South Asian Association for Regional Cooperation
SAMR	State Administration for Market Regulation
SAR	Supplied Air Respirator
SARS-CoV2	Severe Acute Respiratory Syndrome Coronavirus 2
SCADA	Supervisory Control and Data Acquisition
SCBAs	Self-contained Breathing Apparatus
SCS	Somali Chemical Society
SDA	State Drug Administration
SDF	Strategic Defense Framework
SDG	Sustainable Development Goals
SDSs	Safety Data Sheets
SF	Strategic Framework
SII	Serum Institute of India
SITRA	South India Textile Research Association
SMASSE	Strengthening of Mathematics and Sciences in Secondary Education
SMT	Security Management Team
SNL	Sandia National Laboratories
SOP	Standard Operating Procedure
SOPHA	Somaliland Pharmacy Association
SPA	Somali Pharmacists Association
SPC	Security Planning Committees
SPCBs	State Pollution Control Boards
SSA	Sub-Saharan Africa
STEM	Science, Technology, Engineering and Mathematics
STMP	Strategic Threat Mitigation Plan
TBO	Total Benefits of Ownership
TCM	Traditional Chinese Medicine
TCO	Total Cost of Ownership
TDR	Tropical Diseases Research
TDS	Total Dissolved Solids

Abbreviation/Acronym	Meaning
TIC	Toxic Industrial Chemicals
TIFR	Tata Institute of Fundamental Research
TIM	Toxic Industrial Material
TLC	Thin-Layer Chromatography
TOC	Total Organic Carbon
TRIPS	Trade-Related Aspects of Intellectual Property Rights
TSCA	Toxic Substances Control Act
TSS	Total Suspended Solids
UCD	User-Centered Design
UNDESA	United Nations Department of Economic and Social Affairs
UNEP	United Nations Environmental Programme
UNIDO	United Nations Industrial Development Organization
UNODC	United Nations Office on Drugs and Crime
UNSC	United Nations Security Council
UPVC	Unplasticized Polyvinyl Chloride
USAMRIID	United States Army Medical Research Institute of Infectious Diseases
USEPA	US Environmental Protection Agency
USFDA	US Food and Drug Administration
VOC	Volatile Organic Compound
VTG	Vitellogenin
WHMIS	Workplace Hazardous Materials Information System
WHO	World Health Organization
WIPO	World Intellectual Property Organization
WMDS	Waste Management Data Sheets
WMDs	Weapons of Mass Destruction
WTD	Working Time Directive
WTO	World Trade Organization
WWTP	Wastewater Treatment Plant
ZLD	Zero-Liquid Discharge

NOTES

1 UNDP/World Bank/WHO Special Programme for Research and Training in Tropical Diseases. (2001). *Handbook: Good Laboratory Practice (GLP)/UNDP/World Bank/WHO Special Programme for Research and Training in Tropical Diseases.* World Health Organization. https://iris.who.int/handle/10665/66894

2 Elzagheid, M. (2023). *Chemical Technicians: Good Laboratory Practice and Laboratory Information Management Systems.* De Gruyter, Berlin, Boston. https://doi.org/10.1515/9783111191492

3 Teotia, A. K., Chauhan, A. S., Thota, P., Kumari, B., Verma, O., Phathak, C., … & Raghuvanshi, R. S. (2024). Engineering advances in safe laboratory design for pharmaceutical microbiology and assessment of their biohazard risk measures. *Safety Science*, 174, 106458.

4 OPCW (2024a). OPCW strengthens legislative and regulatory framework on chemical security in Africa. Accessed on 14 November, 2024 from: https://www.opcw.org/media-centre/news/2024/05/opcw-strengthens-legislative-and-regulatory-framework-chemical-security

5 Shem, P. M., Otieno, G., Onyango, J. O., Owuor, J. J., Mbugua, P. W., & Aluoch, A. O. (2021). Chemical Security in Kenya: Efforts, Challenges, and Opportunities in Academic Institutions and the Chemical Industry. *ACS Chemical Health & Safety*, 28(2), 112–118. acs.org

Part 1

Laboratories as Major Chemical and Biohazard Environments

1 Overview of the Chemical and Biohazard Environment

Apollo Odhiambo Maima

1.1 INTRODUCTION

This chapter has two primary foci. First, it elaborately introduces the book on the crucial matter of vigilance in a laboratory setting and elaborates on the importance of maintaining a high level of alertness and conscientiousness while handling or working with chemicals and biohazards. Additionally, it provides an overview of the significance of preventing potentially harmful incidents even if they only happen occasionally.

However, the chapter begins to raise awareness about the profound and long-lasting consequences that arise when, despite precautions, accidents do happen. It highlights the collective interest in comprehending the inevitability of unpredictable situations and underscores the need to constantly update our knowledge, enhance our skills, and intensify our commitment to analyzing and minimizing potential risks. This serves as a constant and continuous reminder for all individuals involved in laboratory activities to remain consistently attentive, extremely cautious, and proactive in their approach to safety, always prioritizing the well-being of themselves, their colleagues, and the surrounding environment (Chipangura, Abdussamad, & Lewis, 2024; Subedi et al., 2025).

Second, the chapter focuses on the complex and vital subject matter of sound chemical management. Highly hazardous chemicals must be strictly necessary to be stocked, and demand for meticulous control mechanisms. On the other hand, some chemicals that require relatively less stringent controls carry significant stewardship responsibilities that cannot be disregarded. Managing these substances is costly, as it involves oversight procedures aimed at controlling the impact at local, institutional, and even national levels. These challenges occasionally happen, calling for proactive measures to mitigate them. To tackle these complexities, handlers must strategize their approaches, considering careful planning, judicious utilization, and the implementation of effective inspection procedures for chemicals to identify any signs of aging and decay. Simultaneously, it is crucial to ensure the availability of chemicals from an energy efficiency, dust removal, and administrative standpoint. By meticulously addressing all of these factors, laboratories can foster a safe and efficient environment for chemical management.

This chapter introduces the indispensable guidance provided by the book, which serves as a compass for chemists, microbiologists, biotechnologists, biochemists, and individuals working with biohazards in managing an absolutely safe and secure laboratory and community environment through astute observations and invaluable suggestions. By diligently following the guidelines, protocols, and counsel outlined herein, laboratory and community environments will be shielded by a powerful armor of safety, protecting users from potential hazards, and safeguarding their health throughout their scientific journey (Ménard & Trant, 2020; Mohzana et al., 2023).

1.2 LABORATORY WORK IN SCIENCE EDUCATION

Laboratories are important components of science lectures and are indispensable pedagogically. Due to constraints in time and large classes, laboratory work is often neglected and given less emphasis than lectures or textbooks can command. Even though laboratory work has its own limitations and constraints, due importance should be given to it, as nowhere else can practical applications of, or the limitations to, theory be brought home to the students, if not by first-hand observation and trial. When properly planned, organized, executed, and thoroughly interpreted, laboratory work serves as a powerful conduit through which knowledge is transferred and solidified (Gericke, Högström, & Wallin, 2022; Kennepohl, 2022).

It is widely acknowledged that laboratory work is indispensable in the teaching of science. Experience, however, calls for some equilibrium between the theory and practice components so as to ensure the overall effectiveness of laboratory instruction. A study has investigated the pedagogical benefit of laboratory manual-supported hands-on and virtual-based teaching over traditional lecturing in college-level science courses. In this study, the traditional group members attended only the three lecture hours per week for the course, while the other groups attended two laboratory sessions in addition to the three lecture hours (Halasa et al., 2020; Jia et al., 2022).

The laboratory sessions entailed the students using laboratory manuals to teach the different procedures and to perform the practicals themselves using different teaching aids before a summative test was administered to evaluate mastery of the course content. The main goal for the use of a laboratory manual, however, is to shrink the divide between conceptualization and its application – bridging the gap between the abstract and the concrete phases. The laboratory manual ensures the connection of theory to the practical aspects of knowledge. A student is hence able to not just have an overview of science subject matter but to expand the world of scientific information by executing the facts in the laboratory (Chan et al., 2021; Sharma et al., 2022a).

Success in the preparation and subsequent use of laboratory manuals is complicated by the expectation that students would obtain knowledge from them, in addition to practicing laboratory operations. These manuals are thus expected to influence a student's ideation, perception, comprehension, and retention of course content. It is, however, noteworthy that modern-era students approach their sources of information differently, and thus the laboratory manual should be prepared to suit their nature (Balakrishnan, Deepika, & Srinivasan, 2024; Gorıa & Konstantınıdıs, 2023). Textbooks have become a second choice to internet sources because students access information quickly using the internet (Baron, 2021; Szymkowiak et al., 2021). Therefore, prepared laboratory manuals should not be a heavy and complicated textual source of information; neither should they be a step-by-step explanation of how experiments are performed, but rather be a brief but informative source of knowledge that can push students to think for themselves.

1.3 LABORATORY PRACTICES AND SAFETY

Laboratory safety is of utmost importance and significance. In order to address the ever-expanding knowledge base of intricate chemical and biological manipulations, the current laboratory environs should continuously evolve, intricately considering the learning curve as well as the magnitude of experiments, while simultaneously adapting to the ever-changing needs of laboratory workers (Ménard & Trant, 2020; Lippi & Plebani, 2020). Enhanced safety training, meticulous chemical management, and continuous lifelong learning in this domain are absolutely indispensable for fostering a secure and highly efficient research atmosphere (Mohzana et al., 2023).

There is serious concern that many investigators and laboratory users lack fundamental skills in basic safety issues, including management and disposal of chemical and biological hazards (Lippi et al., 2020; Abumalloh et al., 2021). Recent trends in reduced safety instruction are potentially dangerous, especially considering the current advantage of decades of safety-based regulation at the local, national, regional, and global levels. Knowledge of such best practices should be dictating safety at appropriate levels, even in regulated research environments that often need to speed up project completion for patent extensions and drug delivery special access requests (Ménard & Trant, 2020). Considering the scenario, the development and implementation of comprehensive safety training programs need prioritization across all teaching, research, and regulatory institutions. Investigators and laboratory users need comprehensive education not only in the rudimentary principles of safety but also in the nuanced intricacies of managing and disposing of chemical and biological hazards. Possession of these essential skills can mitigate potential risks and ensure a safer working environment for all (Mohzana et al., 2023).

Institutions should allocate resources to develop comprehensive training modules that cover all aspects of safety, from hazard identification to proper handling and disposal techniques. Furthermore, close collaboration between teaching and research institutions, regulatory bodies, and industry stakeholders should be encouraged to exchange best practices, share expertise, and collectively establish higher standards of safety (Ménard & Trant, 2020; Mohzana et al., 2023).

Strengthening safety measures is not only a moral imperative but a strategic move to uphold the integrity and credibility of scientific research (Elkhalifa et al., 2024; Gonyora & Ventura-Medina, 2024). With concerted efforts and a renewed commitment to safety education, a culture can be fostered where every investigator is equipped with the necessary skills and knowledge to navigate potential hazards safely and responsibly. Only then can the potential of scientific progress be maximized without compromising the well-being of individuals or the broader community. Laboratory safety is paramount in the chemical and biohazard environs (Hall, 2018). Proper practices and protocols must be followed at all times to ensure the well-being of individuals and the community (Abedsoltan & Shiflett, 2024; Abdul Wahab, Isa, & Ahmad Rahiza, 2022; AlShammari, Alhussain, & Rizk, 2021).

1.4 SECURITY AND REGULATION OF PHARMACEUTICAL AGENTS

1.4.1 Background

A pharmaceutical agent is a chemical compound used for diagnosing, treating, or preventing a disease. These agents, including natural products and stem cells, are essential for health and require a regulatory framework to ensure quality, efficacy, and safety. Legal standards differ globally, with smaller markets using varied accountability methods. Pharmacoepidemiology examines the role of media and ethics, emphasizing the need for compliance with safety laws to improve disease prevention, treatment, and diagnosis (Gupta et al., 2024; Arote, Salade, & Patil, 2023).

The U.S. pharmaceutical sector offers medications and healthcare products, mainly regulated by the American Food and Drug Administration (FDA) under key federal acts. The FDA evaluates new drugs for safety and efficacy before approval and regulates marketing practices. The American Federal Trade Commission (FTC) supervises dietary supplements, while the FDA oversees new foods and beverages. Regulatory committees ensure pharmaceuticals meet standards through laboratory testing, though many drugs fail the test, costing millions. Automated screening aids evaluations, but FDA reviews can take 12 years and over 1.3 billion (Borup & Traulsen, 2016; Zaboli et al., 2016).

India's drug regulation started with the 1940 Drug Act, granting 1,145 licenses after 1947 (Imran et al., 2013). By 2000, Good Manufacturing Practices (GMPs) were mandated. In India, the generic market faced challenges post-1973 due to lack of product patents, which prompted updates to Schedule Y for clinical research.

Collaborative arrangements and national regulatory authorities (NRAs) establish medicinal regulations focused on safety, particularly in developing countries (Keyter et al., 2018). Complementary, alternative, and traditional medicines share regulations with conventional ones, though nontraditional practices vary by country. The NRAs aim to harmonize global governance, but data gaps exist, and World Health Organization (WHO) initiatives for "lifestyle" diseases are insufficient (Wang et al., 2023; Okaiyeto & Oguntibeju, 2021).

1.4.2 Pharmaceutical Security Challenges

Drug shortages disrupt supply chains and expose risks from hostile nations, as seen in the heparin contamination from China (Costantino, 2021). Identifying laboratory vulnerabilities is crucial, and new laws are needed to protect the supply chain. Notably, the WHO stresses a systematic approach to fighting counterfeit pharmaceuticals and maintaining market integrity. Using the same approaches, pharmacists can enhance drug safety and reassure the public about medication origins.

The rise of the illicit drug market threatens legitimate supply chains with counterfeit pharmaceuticals (Mihaila, 2018) leading to significant economic problems. Though regulations exist against fake pharmaceuticals, the WHO estimates that 10% of global medical products are counterfeit (Mackey et al., 2015). Manufacturers suspecting problems must halt sales, take corrective actions, and notify authorities within 72 hours. Essential corrective actions include verifying medicinal products, utilizing detection technologies, and enhancing preventive steps for wholesalers.

Pharmaceuticals and biopharmaceuticals, including controlled substances, impact physiology at low doses and may serve as biosimilars, with unique traits that can cause instability and new agent development (Mantle et al., 2019). Whereas developed nations prioritize safety and quality pre-market entry, out-of-specification agents still pose risks needing careful evaluation. In fact, recent biomanufacturing breaches highlight the need for thorough risk assessments to ensure patient safety and product availability.

Bioanalytical assays encounter challenges with immobilization kinetics, highlighting the necessity to remove poorly soluble fragments. To meet regulatory requirements, the biopharmaceutical sector should implement measures like, for example, screening processes to evaluate the biocompatibility of plastic and rubber components made outside cleanroom conditions, where on-site assessments are lacking (Gastineau, 2020).

Pharmaceutical companies should follow regulatory standards to ensure product quality, safety, and efficacy. However, regulatory noncompliance has increased, posing challenges for leaders and managers who are vital in maintaining compliance throughout the organization (Jagun, 2018).

1.4.3 Chemical Security of Pharmaceutical Agents

Contemporary nations possess chemical arsenals, including German agents from the 1940s to the 1950s. Production of these may continue, with simpler agents replacing complex systems (McManus, 2023; Caves Jr & Seth, 2022). The creation of a chemical agent can be readily accomplished by a pharmaceutical organization if it shifts its production processes to accommodate military testing requirements (Ugwu et al., 2025).

The technique of asphyxiant substances is well known and accessible. Possibly, there could be many compounds within the various categories of those types of pharmaceutical applications (Oakenfull & Sidhu, 2023). These history-based principles provide proper justification for the legal and regulatory inspections imposed upon the pharmaceutical industry. Substantial information about the criminal applications along with legislation directives has been missed to some extent in the respective pharmaceutical professional education. Examples and arguments against chemicals and biological warfare have been presented recently; unfortunately, these cannot be applied universally to support the inspections effectively (Parmar, Narasimha, & Nath, 2024).

Elderly individuals (≥65 years of age) frequently face exclusion from clinical trials for various reasons. Evaluating justifications and ethical concerns is vital in off-label drug use as early drug development often neglects key factors, resulting in disorganized methodologies. Non-investigational drug trials face ethical dilemmas worsened by pre-commercial constraints hindering data sharing. Implementing computerized systems is essential for compliance with quality, safety, and operational efficiency standards. Workflows must prioritize data protection and encourage collaboration among manufacturers, suppliers, and service providers for effective oversight (Westwood et al., 2016).

1.4.4 Risk Management in Pharmaceuticals

Risk management in the pharmaceutical sector involves structured methods to tackle risks related to products and services. ISO 31000 emphasizes integrating risk management into governance, as poor management may distort risk assessments from epidemiological data (Nugroho & Pangeran, 2021).

Assessing risks from medication errors is crucial for improvement, evaluating the likelihood and consequences of adverse events. Error sources can be categorized into ten groups, prioritize failure risks, and analyze metrics such as occurrence frequency, impact severity, and control measures. Cause-and-effect analysis guides mitigation strategies and corrective actions, enhancing pharmaceutical risk management.

Regulatory agencies have established inspection frameworks that focus on risk assessment to evaluate facilities according to risk levels, directing resources toward higher-risk manufacturers. The European Medicines Agency (EMA) assesses Site, Process, and Product Complexity to create a risk profile, prioritizing product classes and facility types rather than individual products. The Biopharmaceutics Classification System (BCS) is a global framework assessing "bioequivalence risk" based on solubility and permeability. It complements the WHO's Essential Medicines List for major oral drugs, aiming to reduce improper drug use and improve access to safe, effective treatments (Xu, Boehm, & Zheng, 2016). Manufacturing deficiencies can cause non-sterility in sterile pharmaceuticals if protocols are ignored. Patients need hazard-free and safe drugs (Boom et al., 2023).

Pharmaceutical companies use strategies to protect products, but tampering and contamination lead to crises. The Fen-Phen case underscores the significant challenges in restoring consumer trust (Schaetz, 2012). Crisis management strategies are vital for reassuring the public about pharmaceutical safety. Detection, transparency, corrective actions, and timely communication are key to fostering trust and consumer protection.

1.4.5 Regulatory Compliance

The pharmaceutical industry is one of the most tightly regulated areas of commerce. All individuals and businesses involved in the manufacture, distribution, and sale of prescription drugs or devices must comply with regulations (Jagun, 2018). Given that failure to follow regulations can result in severe sanctions, commercial firms must build protective walls via compliance systems to prevent being caught in the net. Regulation is often in constant flux, demonstrating the chaos and unpredictability of enforcement.

The Pharmaceuticals and Personal Care Products (PPCPs) must be made according to GMPs to be approved for human use (Covarrubias et al., 2022). Regulatory frameworks for pharmaceutical production differ by region, shaped by FDA, EU, and WHO GMP standards. The WHO GMP guide aligns with EU standards, offering vital guidelines for vaccine manufacturing and sterility tests in facilities that handle gelatin, albumin, and collagen.

Nonadherence to food and drug regulations can lead to severe consequences, such as product seizures, court injunctions, and denied entry for noncompliant goods. Penalties may involve prosecutions, with fines or imprisonment. Besides, noncompliance can damage an organization's reputation.

1.4.6 Technological Innovations in Regulation

Counterfeit pharmaceuticals significantly risk public safety, and current anti-counterfeit measures are inadequate. Blockchain technology is a secure, decentralized ledger in the pharmaceutical supply chain, ensuring immutable transactions that verify medication authenticity and product integrity, and could enhance drug safety in healthcare. Traceability, transparency, and accountability are vital in supply chains for financial integrity and product authenticity. Counterfeit drugs harm public health and cause major losses for companies, while counterfeit chemicals in cosmetics and food also threaten consumer safety, hence the need for robust traceability systems (Mani, Prakash, & Lai, 2022).

Pharmaceutical suppliers must ethically minimize risks related to low-quality medicines. The known negative impacts necessitate further exploration of ethics in these situations, indicating a need for dedicated empirical studies in bioethics (Ravinetto, Pinxten, & Rägo, 2018).

National and regional pharmacovigilance frameworks seek to improve post-market product safety, but patient perceptions of adverse drug reactions are often assessed by healthcare professionals or researchers using third-party data. Patients are vital for drug safety, aiding in understanding adverse effects and personal reactions to medications. Involving them enhances care quality and promotes patient-centered practices. Clear communication about benefits and risks is essential (Smith & Benattia, 2016).

1.4.7 International Collaborative Efforts

There is a substantial and growing international product market that requires collective action on some issues. Accordingly, resources put toward cooperation among countries and organizations can improve regulatory or security systems across those boundaries. The challenge is how to design these international collaborative efforts to be most effective. Standardized pharmaceutical regulations are crucial for improving access to affordable medications, combating counterfeit drugs, and fostering industry innovation (Feeney, Goad, & Flaherty, 2024).

Fundamentally, the international dimension of regulatory systems is aimed at promoting domestic public health and safety by protecting the country's population from adverse effects caused or exacerbated by products that cannot be adequately assured safety, efficacy, effectiveness, or quality over their entire life cycle by domestically run systems. In turn, these demographic, product, and system considerations affect the specific forms taken by international regulations, and also whether those regulations are enforced in practice (Preston, Valdez, & Bond, 2012).

The global consolidation of corporations has led to uniform demands for stability data, animal toxicity details, and specifications for active pharmaceutical ingredients. Translating applications into various languages is essential for compliance with national regulations. While comprehensive guidelines are lacking, most technical data submissions aid export registration (Kumar Badjatya, Bodla, & Musyuni, 2013). The regulatory regions are divided into Asia and other ROW countries regulated by WHO and USFDA. Asia includes ASEAN and Gulf Cooperation Council (GCC) countries.

The HDB drugs must be registered to meet statutory requirements. However, global registration faces challenges from differing manufacturing practices. Many low- and medium-income countries (LMICs) lack proper regulations, while the Asia Pacific shows varied systems with some harmonization through ASEAN and the GCC countries, emphasizing the need for urgent regulatory alignment due to rising healthcare costs and the demand for safe therapies (Alenezi et al., 2023).

Given the global Coronavirus disease (COVID-19) vaccination efforts, it's essential to reassess the research, formulation, and manufacturing of vaccines crucial for managing pandemics. While health agencies have made progress, significant challenges persist in achieving worldwide vaccination coverage (Gyu Kim et al., 2019).

1.5 CHEMICAL AND BIOHAZARD MANAGEMENT

There are numerous complex laboratory procedures exploiting risky chemicals that could result in life-threatening or lethal incidents in the chemical laboratory (D'Angelo, 2023). A chemical production procedure that is sensible for a researcher may become exceptionally complicated and perilous when attempted by an undergraduate, a new employee, or a research assistant (Ananda, Rahmawati, & Khairi, 2023). Chemicals are essential for experiments to be conducted, but can be harmful if not properly managed. Managers' vigilance while overseeing the performance of chemical production experiments, scale-ups, and research processes can effectively prevent chemical accidents.

An awareness of chemical compound hazards, including their intricate nature and potential consequences, would contribute to the avoidance of injuries and the implementation of safety measures. This way, laboratory administrators, workers, and students would understand the critical protective requirements, the development of comprehensive accident prevention initiatives, and the proper response to incidents should they arise (Mahmud, 2022; Isabel, Ellinger, & Griffin, 2024). Similarly, an understanding of the characteristics of the numerous biological agents and their relative safety conditions and boundaries, though complex and fairly confusing, is essential. Such awareness and understanding by safety professionals will help in addressing the small-scale uses and applications of these agents, which are absolutely essential to the practical and efficient management of biohazard projects (Ménard & Trant, 2020; Hassim, Zakaria, & Mahmud, 2024). With this extensive and specialized expertise, biological activities can be seamlessly and successfully integrated into the laboratory setting with the utmost fidelity to precautions and safety protocols (Wang et al., 2024b).

Many laboratories worldwide engage in rigorous biological research, ranging from institutional laboratories at academic and research institutions, to cutting-edge private biotechnology firms, and the inherent hazards posed by these agents demand meticulous handling (Abedsoltan & Shiflett, 2024; Chakraborty et al., 2024). The above teaching, research, and regulatory issues are paramount in ensuring that educational and research institutions prioritize safety and take the necessary steps to create and maintain safer environments. With the proper expertise, the utilization and manipulation of biological agents, which would typically evoke much intimidation and trepidation, can be carried out effectively, efficiently, and most importantly, safely in the laboratory setting (Cordaillat-Simmons, Rouanet, & Pot, 2020).

There is a need to effectively combat the challenge posed by the illicit trade of chemicals across porous international borders. The Chemical Weapons Convention (CWC) regulations and other relevant organizations are recognized as potential regulatory stewards responsible for ensuring the identification, prevention, and response to CBRN (Chemical, Biological, Radiological, and Nuclear) threats in LMICs. These organizations play a pivotal role in fostering collaboration among LMICs, enhancing their capabilities, and promoting the adoption and enforcement of stringent security measures and protocols (Ezeigweneme et al., 2024; Igwama et al., 2024).

The CBRN stewards are actively engaged in capacity-building initiatives aimed at addressing the gaps in both human resources and infrastructural development. From these and LMICs' own collective efforts, the security landscape should be enhanced by establishing robust frameworks for the management, control, and mitigation of CBRN risks (Szklarski, 2024; Mani et al., 2024; Novossiolova, 2024). By so doing, the LMIC's can effectively safeguard its people, valuable resources, and precious environment while simultaneously advancing its long-term prosperity and sustainable growth. The importance of such measures cannot be overstated, as they contribute significantly to ensuring national and regional stability, prosperity, and well-being in the face of mounting global challenges.

In Chapter 8 of this book, we provide a comprehensive and detailed account of the various chemical and hazardous management protocols that have been established by both the CWC and the Biological Weapons Convention (BWC) (Crowley & Dando, 2024; Trapp, 2022). The book highlights the complexities of the protocols and the significant role they play in guaranteeing and safeguarding the safety, well-being, and security of nations globally. The comprehensive framework and provisions outlined by the CWC and the BWC are meant to help nations to uphold international law, promote peace, and combat the threat posed by chemical and biological weapons.

In Kenya, preparations for dealing with chemical hazards are poorly thought out. For instance, ethanol is used instead of paraformaldehyde for extended periods of time, which can lead to potential risks and inadequate containment measures (Domański et al., 2023). The response to chemical incidents is slow, lacking proper coordination and immediate action. In developed countries, there are well-established, comprehensive protocols in place for mass decontamination, transportation of individuals exposed to toxic chemicals, and specialized procedures for industrial pollution prevention. These proactive measures ensure the safety and well-being of the population. However, in Kenya, there is limited education about various chemical disasters and their associated risks, resulting in a significant gap in knowledge and preparedness. Efforts should urgently be made to improve education on toxic chemical hazards, as the chemical industry in Kenya grows (Shem et al., 2021).

By promoting awareness and providing targeted training, individuals and communities can be educated and empowered to mitigate the potential hazards.

1.5.1 Risk Assessment and Management

Daily, many exceptionally committed and industrious individuals engage in the meticulous utilization and skillful manipulation of an incredibly extensive array of diverse chemical compounds (Stein & Luna, 2021; Ripanda et al., 2022; Srinivasulu et al., 2022; Mhlongo, Sibali, & Ndibewu, 2023). These compounds possess a potentially devastating capability to inflict harm, in a myriad of ways, if mishandled or mismanaged. There is, therefore, a need to execute a meticulous, rigorous, and exceedingly thorough "Risk Assessment" of the specific substances they handle (Ardiyanti et al., 2024). The overarching purpose of this comprehensive and discerning risk assessment is traditionally twofold: First, to diligently and accurately determine the viability and feasibility of proceeding with their assigned or designated tasks; second, to judiciously select the most appropriate and fitting personal protective equipment (PPE), commensurate and in accordance with the nature and demands of the given endeavor at hand (Barton-Maclaren et al., 2022; Christiansen et al., 2020).

Within the wider context of inquiry, curiosity, and intellectual scrutiny lies the question of what precisely constitutes, in practical terms and actuality, a risk assessment when executed and enacted within the realm of scientific laboratories and experimental facilities. It is also notable that this process of risk assessment is too complex to be comprehensively encapsulated into readily utilizable management tools. Ergo, there is a need for an unwavering sense of steadfast and vigilant control over any potential, unforeseen, and rogue activities that may potentially arise or manifest on-site (Blacksell et al., 2023; Zhang et al., 2023). This fundamental concept has generated an array of invaluable and indispensable tools and methodologies that will enable even small-scale enterprises to establish an enhanced level of control within their respective risk environments.

Having completed a comprehensive and thorough risk assessment, the next key step is to rank and prioritize the hazards identified within the community, workplace, policy and regulatory framework, or data obtained directly from manufacturers (Zytoon, 2020). This prioritization process is critical, as it enables workers to effectively focus their control measures, which are aptly referred to as "administrative controls," to efficiently manage and mitigate the identified chemical and biological hazards (Che Huei et al., 2020).

Each nation has an extensive array of legislation, regulations, and guidance to effectively control workplace hazards. The United Kingdom provides us with some notable and widely recognized legislations. First is the Control of Substances Hazardous to Health (COSHH) (Shekarloo et al., 2024; Morgan, Stewart, & Bennett, 2020), which meticulously outlines stringent measures to be taken when dealing with dangerous substances. Second is the Manual Handling Operations Regulations (MHOR) that governs and delineates the appropriate techniques for handling loads to prevent injuries and musculoskeletal disorders (Luardo et al., 2024). Third is the crucial Working Time Directive (WTD), which emphasizes the paramount importance of maintaining safe working hours and appropriate rest periods for employees (Sorensen et al., 2021). Fourth is the Control of Noise at Work Regulations (Hansen & Hansen, 2021) that aim to diligently safeguard workers from detrimental noise exposure and its potential adverse effects.

A comprehensive handbook like this one provides invaluable assistance, guidance, and additional support for the management of laboratory and workplace risks (Barling, 2023), hence an outstanding, authoritative, and esteemed resource for employers and employees alike in fostering a truly inclusive, safe, and secure environment. By embracing these principles, diligently applying the provided methodologies, and leveraging the assortment of invaluable tools, organizations can pave the way toward a safer and more secure future for their employees, stakeholders, and the wider community.

1.5.2 Personal Protective Equipment (PPE) and Safety Protocols

Many educational and research institutions as well as industry have begun to appreciate the immense importance of PPE (Bauchner, Fontanarosa, & Livingston, 2020; Key et al., 2020) and the prudence required in the selection and utilization of such PPEs in both chemical and biological laboratories. Unfortunately, there is often a lack of exercise of this prudence, particularly among faculty members, students, and researchers. Suffice it to say, PPE should always be accompanied by judicious, safe, and ethically responsible conduct in all laboratory, research, clinical, and community operations (Cornish et al., 2021; Bauchner et al., 2020). Indeed, the prudent and proper usage of PPE helps in minimizing or guarding against potentially hazardous exposures. While selecting PPE, it is crucial to meticulously examine all available alternatives that may eliminate or significantly reduce the need for such equipment (Clapp et al., 2021).

Innovations in process control and laboratory equipment, as well as substitutions with less hazardous materials; the production and adoption of less hazardous forms of potentially harmful substances; or even alterations in system designs (Teotia et al., 2024), can effectively negate the requirement for PPE. Nevertheless, where the usage of PPE is indispensable, employers and workers alike must demonstrate a comprehensive understanding of what actually constitutes effective PPE (Jafari et al., 2022; McCarthy et al., 2020; Mahmood et al., 2020). Considerations for effective PPE may include levels of protection, associated costs, long-term value, changes in efficacy through prolonged use, proper handling practices, and general as well as specific methods for optimizing the limitations of PPE (Aldrighetti et al., 2021). In a chemical laboratory, the effectiveness of eye and safety showers is significantly increased when accompanied by the immediate availability and proper use of eye and face protection, as well as other indispensable PPE devices (Powell, 2023; Nasrallah et al., 2022).

Despite the skin being particularly vulnerable to chemical exposure, cases of laboratory users suffering from chemical burns on their skin are still very frequent (Niede & Benbi, 2022). It is important to understand that certain chemicals or their air-oxidized derivatives, which are formed upon exposure to air, can cause severe burns. These substances have the ability to attack and damage the skin, resulting in painful injuries and long-term consequences.

Workers will absorb certain chemicals or their derivatives through their skin as rapidly as if they were ingested. This means that even a brief contact with certain solutions can result in severe burns and other health risks (Wu et al., 2022; Bhat, 2024). In the event that they are exposed to hazardous substances, it is critical that workers and other laboratory users are aware of the appropriate actions to take, the potential solutions or solvents that could be employed, and the role of an unobtrusive detection system in identifying and managing risks.

By meticulously adhering to proper PPE and work protocols, workers significantly reduce the likelihood of accidents in their work environment (Win, 2024; Pisu et al., 2024). Therefore, it is crucial for every worker to prioritize safety, diligently follow established procedures and norms, and actively participate in creating a hazard-free working environment.

1.5.3 Emergency Response Procedures

In the event of a *chemical spill*, all personnel should be made aware, and evacuation carried out if the spill is major, or if explicitly defined by the safety data and guidelines (McLeod, 2022; Charlton et al., 2021); and receive comprehensive instruction on how to handle the situation (Ansah et al., 2024). In preparation for any chemical spill emergency, the laboratory or workplace should have the necessary first aid supplies (Wang et al., 2024a) including an eyewash fountain and a deluge shower. Additionally, appropriate materials are needed to effectively neutralize spills of corrosive chemicals. In case of accidental exposure or ingestion, the medical department should be notified promptly and provided with all the essential details of the incident sufficient for proper evaluation and treatment (Igbashio, Idemudia, & Ochoyama, 2022; Abedsoltan & Shiflett, 2024).

Where there is evident or suspected *eye contamination* with acid or chemical powder, affected person should immediately wash and rinse their eyes for at least 15 minutes, while holding their eyelids open, at the nearest eyewash fountain (Dua et al., 2020).

All employees must be familiarized with the location and telephone contact of the nearest *first aid stations* within the institution, adequately equipped to handle and provide initial treatment for minor injuries (Edgerly et al., 2024). Regardless of the size or severity of the incident, the first aid station should always be notified (Wang et al., 2024a).

Knowledge of *First Aid procedures* for minor injuries, including cyanide exposure, is required to respond effectively to biohazard and chemical emergencies. Laboratory users should be trained and empowered on crucial first aid procedures such as relief for airway obstruction, cardiopulmonary resuscitation, treating cuts and wounds, addressing poisoning incidents, managing nosebleeds, handling shock and fainting episodes, and immobilizing fractures (Hu, Chen, & Yu, 2020; Farhat et al., 2024).

In the unfortunate event of an employee sustaining an injury, it is the duty of the supervisor to ensure that the appropriate actions are taken. This includes properly filling out the emergency response form, providing accurate and comprehensive details of the incident. Once the response form has been completed, it should be promptly given to the senior director for further evaluation and documentation. It is also vital to have a copy of the form posted in the designated chemical management database portal to maintain a comprehensive record of all incidents and facilitate efficient monitoring and follow-up actions (ALSG, Gleeson, & Mackway-Jones, 2022).

1.5.4 Cost-Effective Laboratory Operation and Management

Although being successful in attracting and securing funding is an important and necessary goal, being highly effective and efficient in the use of the allocated funding is equally important (O'Dwyer, Filieri, & O'Malley, 2023). The delivery of exceptional and outstanding research outcomes will not only bring about recognition but will also present valuable opportunities to access further resources and expand the boundaries of the ongoing research endeavor. It is universally acknowledged and recognized that the failure to appropriately manage and utilize existing funding can lead to significant repercussions for any institution, particularly within the chemical laboratory and biohazard environments (AlShammari, Alhussain, & Rizk, 2021; Abedsoltan & Shiflett, 2024; Teotia et al., 2024). It is of utmost importance that all research activities should uphold the highest standards of quality (UNDP/World Bank/WHO, 2001) while also being conducted in the most economical manner possible, with safety, health protection, and environmental responsibility taking precedence at all times (Isere & Omorogbe, 2024). To ensure that research management is maximally cost-effective, there exist several prudent strategies that can be employed for optimal outcomes, and they include but are not limited to the following considerations.

High levels of downtime should be minimized, as they are an inefficient use of available resources. Therefore, it is crucial to utilize technology effectively in order to optimize productivity and reduce time spent in the unknown realms of the internet. Time outages typically occur due to malfunctioning equipment, and although a certain level of redundancy is necessary, the primary objective should be to ensure the functionality of all equipment. It is essential to avoid budget wastage and unproductive man-hours caused by having a piece of equipment waiting for a replacement part (Al-Hussain & Al-Ahmed, 2022). By prioritizing swift and efficient equipment repair or replacement, organizations can enhance their overall productivity and make the most of their available resources (Yazdi, 2024; Khosroniya, Hosnavi, & Zahedi, 2024).

1.5.5 Waste and Inventories Management in the Laboratory

A large inventory of hazardous materials is problematic (Nabavi-Pelesaraei et al., 2022; Zhao et al., 2021). It is crucial to effectively monitor, control, and secure the inventory and usage levels of these hazardous materials, which are stored and consumed within the workplace (Vaka, 2024;

Bamakan, Malekinejad, & Ziaeian, 2022). A dedicated and qualified professional could be engaged to meticulously manage the inventory, stock, and its usage. Alternatively, the institution can train and empower its own laboratory users, that is, the skilled scientists and laboratory assistants, to comprehensively understand and grasp the concept of utilizing a specialized inventory or stock control system (Elzagheid, 2023). This approach is important as the trained individuals can take responsibility and be held to account, while at the same time making the laboratory manager's work of stock management less burdensome.

To ensure the continued and safe usage of known hazardous chemicals (e.g., polycyclic aromatic compounds, halogenated aromatic hydrocarbons, and the acids and bases), especially to the aquatic system, it is crucial to implement a comprehensive waste disposal and monitoring program, overseen by the laboratory manager (Anwar, Iqbal, & Saleem, 2020). Work in chemical, biochemical, bioscience, or biotechnology laboratories often leads to the generation or co-use of hazardous, dangerous, or biologically hazardous waste. These wastes mostly consist of solid clinical waste, which may include both nonchemical and nonbiological or medical/laboratory waste.

Proper waste management requires laboratories to adhere to strict protocols and enlist the services of an accredited hazardous waste service company that specializes in the approved and certified disposal of chemical and biochemical waste (Papadopoli et al., 2020; Hantoko et al., 2021).

1.6 EXPOSURE TO ENDOCRINE-DISRUPTING CHEMICALS AND SUSTAINABLE DEVELOPMENT

Endocrine disruptors (EDs), defined by WHO, are harmful substances that alter the endocrine system, affecting health, development, reproduction, and disease rates (Horak, Horn, & Pieters, 2021). The recognition of harmful chemicals on public health has prompted governments and communities to regulate emissions. Chemical pollutants in freshwater systems pose serious public health risks, worsened by improper disposal of PPEs, pharmaceuticals, agricultural chemicals, and pesticides. Urbanization, lack of clean water, and other human economic activities degrade ecosystems, undermining their ecological integrity.

The EDs from various natural and human-made sources harm organisms and offspring by interfering with the endocrine system (Predieri, Alves, & Iughetti, 2022). They include pharmaceuticals, pesticides, and personal care products, which harm hormones and pose serious health risks, especially for developing children. Prenatal and neonatal exposure to EDs like bisphenol A and phthalates disrupts sexual differentiation and delays puberty. Women risk thyroid issues and obesity, with these pollutants linked to neurodevelopmental disorders, emphasizing the need for health risk assessments (Di Pietro, Forcucci, & Chiarelli, 2023).

Endocrine-disrupting chemicals (EDCs) harm growth factors, endocrine systems, and glands such as the thyroid and thymus (Predieri, Alves, & Iughetti, 2022), thereby affecting development, reproduction, and immune functions (Sharma et al., 2022b). Synthetic EDs from petrochemicals harm the environment and accumulate in animals, affecting generations. They disrupt hormonal signaling in embryos, causing defects at low levels, highlighting their cross-species impact (Solecki et al., 2017). Identifying novel EDs requires studying non-vertebrate model systems to evaluate effects. Current knowledge of neurotoxicants on endocrine and immune functions is lacking. Thus, new methods for assessing chemical toxicity on human health and ecosystems are vital.

Exposure to EDCs during critical life stages can disrupt reproductive programming and cause disorders. Multidisciplinary research is needed to connect wildlife findings to humans. Pollution prevention is essential for reproductive health (Street et al., 2018).

In the 1930s, synthetic estrogens raised concerns over ecological and endocrine disruption. By the 1970s, rising testicular germ cell cancer in males from pesticide-polluted areas led to efforts to reduce harmful hormone-active chemicals. Nearly a thousand market chemicals are potential EDCs, contributing to ongoing reproductive health issues.

The EDCs negatively impact hormonal systems in humans and animals, disrupting regulatory mechanisms and altering growth, functions, and health (Predieri, Alves, & Iughetti, 2022). Modifications may cause health issues, especially in developing countries, where vulnerable groups lack awareness of contamination sources. Sustainability reviews are crucial for tackling these health concerns. A systematic scoping review could assess studies on EDCs, summarizing data and identifying key research priorities.

1.6.1 Health Impact of EDCs

The EDCs significantly impact homeostasis, reproductive health, development, and behavior (Predieri, Alves, & Iughetti, 2022), and pose risks to future generations through epigenetic changes from fetal exposure. Since 1996, the U.S. National Institute of Environmental Health Sciences and the Environmental Protection Agency (EPA) have researched these effects. The EDCs interfere with hormones, leading to reproductive problems, neurodevelopmental delays, and childhood obesity. Studies connect organochlorinated pesticides, phthalates, and bisphenols to various health issues.

Wildlife studies and ecotoxicology show EDCs have major impacts on reproductive health, such as hypoplasia, adenomas, and reduced sperm production and egg quality (Fudvoye, Bourguignon, & Parent, 2014) of wildlife, particularly freshwater fish reproduction in Europe. While the United Kingdom has made mitigation efforts, pollution persists. Research on EDC exposure in LMICs is scarce, and early-career researchers often encounter challenges with advanced methods; many undergraduates receive insufficient education on EDCs.

Research shows critical exposure periods for children and longitudinal studies on EDC exposure are vital to assess health outcomes during these sensitive windows. These chemicals impact thyroid receptors and obesity pathways, leading to early puberty and testicular issues. Stakeholder cooperation is essential for addressing environmental changes and remediating polluted areas as EDCs therein harm neurodevelopment. Prenatal exposure to polycyclic aromatic hydrocarbons is associated with lower intelligence quotient (IQ) and behavioral issues. Organophosphate pesticides and high umbilical cord dichloro-diphenyl-trichloroethane (DDT) levels also correlate with reduced IQ and increased hyperactivity (Predieri, Alves, & Iughetti, 2022).

Chronic diseases affect sustainable development, with obesity, diabetes, hypertension, and cancers linked to EDs (Nweke & Sanders, 2024). In many LMICs, weak health regulations obstruct development, with legal decisions favoring media over scientific evidence, leading to a loss of profits to foreign firms. Enhancing legal frameworks and enforcement is key to sustainability.

1.6.2 Environmental Impacts of EDCs

The EDCs are found in food and body products, and can be transmitted from mothers to children, causing widespread exposure. Considering the significance of EDCs on childhood diseases, initiatives should focus on research gaps and child-centric studies. A scoping review is crucial to assess EDCs' environmental impact including the risks from plastics and PFAS during pregnancy.

Access to clean water is crucial for survival and supports social and economic stability, while contamination poses health risks (Horak, Horn, & Pieters, 2021). However, in South Africa, less than 1% of water is safe to drink due to pollution. While municipal water treatment is effective, 3%–10% of contaminants escape through the plants. Anthropogenic chemicals threaten health through water, food, and air. While EDCs disrupt hormones, South Africa acknowledges only pesticides as disruptors. The PPCPs in treated water may affect vertebrate endocrine functions (Metcalfe et al., 2022). Monitoring PPCPs in South African waters highlights their disruptive potential, emphasizing the necessity to address health impact knowledge gaps.

Since EDCs affect development, understanding their role in mammalian reproduction and infertility is essential (Street et al., 2018). In the upper Mississippi River basin, EDCs are known to impede fish growth, energy distribution, and reproduction, impacting ecology (Burden et al., 2021).

Emerging EDCs include alkylparabens, phosphonated surfactants, and various household and agricultural chemicals. Climate change and industrial practices notably complicate their regulation and research. Agricultural EDCs like pesticides, herbicides, and fertilizers create ecological risks through runoff. Generally, regulatory pressures and consumer awareness raise concerns about health and ecosystems. Though uncertainties remain and further research is necessary, public awareness of sustainable practices is lacking.

1.6.3 Regulatory Framework and Policy Responses

Research on EDC regulations in LMICs is limited. Urbanization and climate change threaten Sustainable Development Goal (SDG) 6, while funding issues hinder WASH efforts. African governments, with support from the African Union (AU) Council of Ministers, create policies for women and youth. The AU pushes for chemical safety and stricter agrochemical regulations on EDCs. Recently, Tanzania enacted laws to tackle illegal fishing and tobacco health challenges.

The legal framework for EDCs includes regulations for compliance and cooperation, with varying progress. North Africa began EDC legislation discussions in the late 1990s, while Southern Africa has enhanced awareness and aligned with international standards. The Stockholm Convention is crucial for enforcing restrictions on persistent compounds (Horak, Horn, & Pieters, 2021). Recent publications from Africa indicate greater awareness and engagement.

Research in West Africa studies show EDCs affecting endometriosis and reproductive health in Niger, Nigeria, Sierra Leone, and Ghana. Monitoring shows improvements, but challenges remain despite all the IOC-UNESCO efforts.

International agreements are crucial for addressing EDCs and SDGs 3, 12, 14, and 15. Treaties such as the Stockholm Convention focus on managing EDCs. Although SDG 3 prioritizes health, follow-up actions are insufficient as many agreements have solid frameworks, but unclear provisions hinder their effectiveness (Horak, Horn, & Pieters, 2021).

The EDCs endanger biodiversity and women's health, with 70% of African nations lacking proper healthcare. Limited data hinders understanding, highlighting the need for robust policies and EDC research. A review analyzed Africa's EDC national policies and suggested strategies from limited research. In 2022, nongovernmental organizations (NGOs) in Uganda, Kenya, and Tanzania provided reform insights, and Cameroon and Mozambique emphasized the need for collaboration through an EDC risk assessment workshop (Horak, Horn, & Pieters, 2021).

1.6.4 Public Awareness and Education

Chemicals from pesticides, pharmaceuticals, and plastics accumulate in the environment and human tissues, raising health concerns about EDCs. This research assessed public awareness of EDCs in Delhi NCR, identified stakeholders, and proposed educational campaigns on lesser-known EDCs in advance of new legislation (Boronow et al., 2023). Biomonitoring of EDCs in the United States requires educating participants on findings and promoting health discussions for exposure reduction.

Understanding the origins and applications of engagement strategies is essential for analyzing community involvement in EDC research (Issahaku et al., 2024) and helps to promote effective community engagement strategies for EDC research. Such reviews explore strategies for involving community members in EDC research, focusing on facilitators and barriers.

Subsequent studies are vital for the scoping review. A systematic review explores community engagement strategies in EDC research, emphasizing community-led initiatives (Boronow et al., 2023). The research evaluates studies to establish a baseline for community engagement care. A database search would find articles on community engagement strategies in Education for Development and Change. Reference lists could be reviewed for more studies, with findings shared through publications and presentations.

Educational programs strengthen health agencies and NGOs with tailored workshops. Combining finance and engineering addresses health challenges. Community initiatives boost engagement, while targeted funding aids outreach in low- and middle-income countries. A Massachusetts study showed a program helped women and mother–child pairs understand environmental chemicals better (Boronow et al., 2023). The EDCs affect children's health during organ development; hence, a review would emphasize the necessity of studying their effects from conception to adulthood and enhancing exposure evaluation research (Predieri, Alves, & Iughetti, 2022).

1.6.5 Sustainable Development Goals and EDCs

The United Nations' SDGs tackle interwoven challenges hindering progress. The EDCs pose environmental and health risks, requiring evaluation within the SDGs framework. The 2004 Stockholm Convention highlighted the need to assess chemical impacts as per the 2030 Agenda. However, local efforts can conflict with global goals, complicating progress and threatening multiple SDGs amid complex political dynamics.

Research on EDCs indicates limited understanding of their health impacts in LMICs. Exposure to EDCs is linked to adverse health outcomes that threaten SDG Goal 3, including negative effects on maternal, reproductive, and fetal health, causing neurophysiological disorders, cancer, respiratory issues, and metabolic imprints; and amenorrhea, infertility, low birth weight, and preterm births. These have implications for SDG 3.2 and 3.7, requiring research on developmental effects (Joubert, Mantooth, & McAllister, 2020).

Investigations in a freshwater ecosystem found 44 drugs in the Vaal River, including 16 EDs. The Kraai River basin activities failed to remove PPCPs like metoprolol and venlafaxine (Metcalfe et al., 2022). A review has studied agrochemical prevalence in South African freshwater ecosystems, highlighting health risks and monitoring gaps to enhance detection methods and risk assessment in coastal areas (Horak, Horn, & Pieters, 2021).

The SDG Goal 12 emphasizes sustainable consumption and production. While policy reporting aids integration, many contributions lack depth. Strategies vary by local values and resources, with academics supplying vital knowledge. Evidence-based choices improve disaster management, and collaboration drives change.

1.7 CHEMICALS AND BIOHAZARDS AT THE WORKPLACE

This book provides some insightful examples of workers across diverse industries who frequently encounter chemicals and biohazards within their working environments across the globe, but more so in LMICs. These workers may lack comprehensive knowledge, direct experience, or specialized training in matters related to biology and chemistry. Typically, workers in the production industry receive training that emphasizes efficiency in the creation of end products. Nevertheless, there are distressing situations and concerning statistics that attest to the presence of biohazard and chemical hazard environments within the daily activities of workers in both industries throughout the LMIC's.

There are occupational, hence avoidable risks in the healthcare environment. Occupationally, healthcare workers get exposed to a wide range of cultures, tissues, blood, and other bodily fluids bearing pathogenic agents as part of their extensive patient care and routine diagnostic activities (Smallwood et al., 2022; Blot et al., 2022). There are many cases where healthcare workers participate in activities generating aerosols that unknowingly infect them when inhaled, putting their health and well-being at great risk (Alshehry et al., 2022; Benishek et al., 2023). Furthermore, infected needles, sharps, and hand and cut-splash injuries due to encountering bodily fluids occur far too often, creating hazardous conditions for the practicing healthcare workers (Abalkhail & Alslamah, 2022). Additionally, healthcare workers tirelessly deal with patients who come in carrying and spreading blood and body fluids everywhere, presenting a constant challenge to maintain cleanliness and infection control.

These healthcare settings are unfortunately plagued by highly contagious infections and an array of other communicable diseases that pose an immense threat to both the healthcare providers and the patients they are dedicated to serving (Abalkhail & Alslamah, 2022; Weber et al., 2023). Due to the nature of their work, healthcare workers rely on a combination of non-regulated and reasonably efficient barriers to prevent infection transmission through the skin, mucous membranes, respiratory system, and indirectly through fomites (Sharma, Mishra, & Mudgal, 2020; Cornish et al., 2021). Despite the vigilance and best efforts of workers, healthcare settings present a constant risk for the acquisition of occupational infections that often have serious consequences on the overall well-being and effectiveness of the healthcare workforce.

Case studies have been conducted to evaluate the required cultural change and comprehensive transformational leadership approaches that would yield effective management of personnel in order to drive the use of PPE. Such studies have additionally encompassed the teaching of PPE and addressed various factors such as type, size, and location requirements (Smith, DeJoy, & Dyal, 2020). It is notable that bylaws in connection with government inspectorates play a significant role in carrying out the oversight responsibilities within this context.

1.8 CONCLUSION

In conclusion, all regulatory and higher safety compliance issues should be in place in research and academic laboratories, due to their not always apparent threat from biohazards and/or chemical agents on site. Furthermore, researchers are very interested in the outcome of their experiments, and something so simple as bacteria hand cleaning or face mask usage might hinder that goal. At the building level, a compliance unit should be involved with the laboratory crew and its occupants in order to ensure that all rules are observed, and that research projects and teaching are conducted in a safe and beneficial manner. Also, newer safety laboratory designs are being reviewed, avoiding disasters as seen in recent events. Training needs will increase with different species, strains, or genetic insertions, and eyes and ears will also be required to be present in certain animal experimental protocols. Sharps and perhaps allergic reactions may occur; this is just a par for the course.

As discussed, both historic devastation nuclear techniques and more current chemical prototype fabrication strategies are in place for evaluating toxicity for high-technology and innovative materials at the nanoscale. Together with, or separately, these techniques fulfill preliminary and potentially postmortem risk management strategic solutions. Several derived and innovative computer tools have undergone rigorous evaluations. Their theoretical foundations are sound due to the widespread experimental and literature values of chemical, biological, and physical behavior encoded into their relationships. Reliable computational models and/or correspondence tools for potential high-concentration and low-probability endpoints provide either an acceptable baseline toxicity value, or disappearance behavior parameters which we can translate to a 10% occupational risk-annum of high-qualitative interest dose for identified exposure. The capability can contribute to degree of insurance stabilization, ease of field use control, and the ability to convert potential exposure clouds that surpass toxic dose limits into systems of lesser hazard in the captive assets defined, recognized, and/or insured by generic claim structures of various technologies.

2 Role of Laboratories in Science Curricula Delivery

Apollo Odhiambo Maima

2.1 INTRODUCTION

The teaching and learning of science is crucial in three important aspects of human development: creative thinking, reflective thinking, and critical thinking. Sound knowledge of science enables people to have deeper understanding of nature, environment, and the world at large. One of the ways of developing scientific attitudes (curiosity, open-mindedness, appraisal of results, humility, patience, honesty, skepticism, and accuracy) in children is by having them engaged in practical work (Çilekrenkli & Kaya, 2023). Eze (2006) reports that before students are exposed to practical sessions in the laboratory, even though they have learned a lot of abstract scientific facts, they are almost always blank in terms of possession of basic scientific knowledge and process skills.

Practical work serves to extend the learners' knowledge and often confirms or contradicts their prevailing knowledge. The laboratory experience, through its engaging and interactive nature, allows students to witness first-hand how science emerges from experimental work, consequently fostering a genuine understanding of scientific concepts (Akhtar & Saeed, 2021). Nevertheless, there exist certain limitations that impede learners from fully harnessing the potential of the laboratory for optimal learning outcomes. These limitations encompass a myriad of factors such as excessively crowded student classrooms, inevitably hindering the smooth operation of scientific equipment, as well as concerns regarding insufficient availability of supplies and equipment necessary for conducting comprehensive experiments (Mohzana et al., 2023; Reeves & Crippen, 2021).

The absence of adequate associated support personnel further compounds the challenges faced, potentially limiting the effectiveness of laboratory sessions. Further, time constraints within the timetables of educational institutions may also impose limitations on the extent to which students can fully engage with laboratory activities. It becomes apparent, therefore, that the quality and accessibility of science laboratories within secondary schools play an instrumental role not only in facilitating effective education in the field of science, but also in nurturing a generation of scientifically literate individuals capable of contributing to the advancement of society as a whole (Mohzana et al., 2023; Reeves & Crippen, 2021).

From the foregoing, the need for science laboratories to function effectively is quite compelling. The development of scientific skills in learners, at all levels including secondary schools, can only be achieved if the laboratories are well equipped with a range of basic apparatus and alternative methods to traditional instruction followed by observation and calculation of data. Delivering up-to-date laboratory curricula in chemistry, physics, and the biological sciences is the greatest challenge faced by STEM educators and hence requires innovative thinking (Hamed & Aljanazrah, 2020; Holland & Davies, 2020; Chan et al., 2021; Sarı et al., 2020).

2.2 THEORETICAL FOUNDATIONS OF LABORATORY LEARNING

Laboratory instruction is a central component of science curricula and is generally recognized to provide students with an opportunity to master concepts and processes through manipulative experiences (Supena, Darmuki, & Hariyadi, 2021). A laboratory is not simply a place to experiment, work, and conduct the exercises described in manuals. Rather, it is a place to generate new knowledge by modifying, extending, complementing, and scrutinizing existing theories, predictions, and assumptions, proving results, or coping with design procedures, methodologies, models, discoveries, and unexpected results.

A laboratory can be described as a place where practical scientific exercises (such as experiments, and other activities which help the students in acquiring scientific skills) are conducted by trainers for the benefit of learners, and by researchers for purposes of scientific inquiry. A laboratory can be indoor (well-designed and equipped rooms for science practical exercises seen in schools, colleges, and research institutions) or outdoor (like roadside, workshop, field, or market for conducting scientific studies) wherein participants are involved in a series of experimental, observational and demonstration activities (Hofstein, 2017).

The major reason for maintaining laboratories is the belief that students learn through doing things, just as well as through visual and auditory representations. Learners can manipulate chemical substances, burn, and react materials, produce or treat effects, observe characteristics of matter and its changes, and finally, realize and feel natural subject phenomena in laboratories that would be difficult, if not impossible, to offer in any other way.

If laboratories are not used for students' practical-based curricula, then the methodology becomes traditional. In such scenarios, memorization, application of recipes and procedures, and consequently, learning are confined to lower cognitive levels of tasks, which are generally considered by those involved in the educational process to be the least important aspect of the major role of education (McMurtry, 2024). The reflective aspect of laboratory activities must be accessible, conducting students to self-assessment and making more precise their role as "inquirers," not only of biologists, physicists, chemistry researchers, and lab technicians, but also of any citizen. Research that is experimental in nature and experimental lab findings function to show how scientific knowledge progresses. What students can learn with the scientific materials themselves must be recurrently discussed and considered during curriculum planning and in practical activities.

2.2.1 The Constructivist Learning Theory

Constructivist theory posits that the process of understanding is actively constructed by children themselves. It asserts that the effectiveness of teaching should be gauged based on the demonstrated learning and understanding of students, while also affirming that young learners should have agency and a say in their own understanding process (Tan & Ng, 2021). In essence, this educational model maintains that children construct knowledge and comprehension by drawing upon their pre-existing background knowledge, experiences, and observations.

The cardinal question for science education is "How to create a framework that enables students to enhance their understanding of fundamental scientific concepts, while developing crucial process skills, fostering an inclination toward inquiry, and cultivating a sense of the subject's relevance in their personal lives?" The constructivist perspective offers a compelling answer to this question. It presents science as a way of thinking and emphasizes the importance of delving deep into scientific concepts. Furthermore, the model asserts that for meaningful development of scientific thinking, comprehensive research has consistently highlighted the significance of providing students with support not only in comprehending the facts and concepts of science, but also in locating their learning within the context of genuine, challenging questions, pressing issues, and exemplary phenomena.

A constructivist educator encourages his learners to be active in learning, and plans his lessons by using problem-based, inquiry-based, project-based, case-based, or discovery-based learning techniques (Makgato, 2012). Constructivist learners should be able to gain some new knowledge, experiences or phenomena building on whatever they had earlier learned through physical or social environment. This way, and applying hands-on activities, learners will reconstruct prior understanding to create a new meaningful idea (Demirci, 2010). It has been observed that such constructivist learning environments significantly increase learners' confidence levels, creativity, higher-order thinking skills, and communication skills (Sorathia & Servidio, 2012).

This approach reframes science as a process rather than a mere collection of facts, encouraging students to actively engage in their own learning journey. It empowers learners to *construct* meaning through *creative* hands-on experiences, *collaborative* problem-solving, and *critical thinking* (Supena, Darmuki, & Hariyadi, 2021). In doing so, this model nurtures students' ability to think like scientists, while also equipping them with the necessary skills to investigate and understand the complex world around them. As such, the constructivist approach emerges as the most representative and effective paradigm for achieving these goals in science education.

To achieve the above skills, a scientist needs a learning environment that is conducive to application of constructivism. One of such places is the laboratory where learning occurs and learners are able to implement practical work. Working in a laboratory enhances the learners' understanding of a science subject to augment learned theory during class time (Ekici, 2011). A laboratory is a place where students learn meaningfully from practical experiences in school settings, experience with materials that simulate natural surroundings, communicate and share ideas among peers. These can be done individually or in groups, as indoor or outdoor activities, from teacher-centered to student-centered learning, and using high-tech instrumentation or none.

For best outcomes, school science laboratories should be equipped with the necessary modern equipment, apparatus, aids, and other instructional materials to assist both the teachers and the learners in their teaching and learning process. Besides the physical infrastructure, the laboratories should be adequately staffed with qualified teachers and technical personnel (Emeka et al., 2021). In many instances, the laboratory personnel find themselves having to keep reminding those in management and financial authority to extend their support for the standards to be maintained. Doubtlessly, setting up and maintaining a well-functional laboratory is no mean feat. This seems more so if those in authority do not fully appreciate the role of the laboratories in teaching and learning.

Laboratory exercises have been identified by designers of high school, college, and university curricula as complementary components of content delivery to learners. Inquiry-based laboratory practices have been shown to improve scientific process skills and all creative thinking levels and attitudes toward science experiments of preservice science teachers (Yakar & Baykara, 2014). This guide likewise considers practical activities as immediate and good ways of realizing learning outcomes in institutions of learning. Laboratory work is a crucial component of any science school intending to achieve effective delivery of a science curriculum. It is also useful for scientific research, experiments, demonstrations, testing and analyzing of data.

2.2.2 Experiential Learning Theory

The experiential learning theory (ELT) proposes that learning is the process whereby knowledge is created through the transformation of experience (Agarwal & Bain, 2019; Korucu-Kış, 2021). It suggests that for learning to be effective in instilling a capable mind, one must fully engage in the learning experience and go beyond the boundaries of educational institutions (Bell & Bell, 2020). According to the ELT model, there are four key aspects that encompass the learning process: concurrence and recurring interactive experience, interaction with various individuals and diverse settings, innovative ideas, and meaningful encounters (Morris, 2020). In addition, critical reflection on experience plays a vital role in facilitating the conscious implications of concurrent or newly acquired know-how and skills. This introspection allows individuals to further develop their knowledge and skill competence, thus enhancing their overall learning journey. By embracing a holistic approach to learning, incorporating experiential elements alongside continuous support and monitoring, individuals can maximize their educational endeavors and foster personal growth and success (Anwar & Abdullah, 2021; Fromm et al., 2021).

Most people have a propensity to favor certain modes of learning over others, preferring a particular learning style and selecting representations other than the rest to work with. Perception or our preference from the second dialogue over time shapes our habit of mind and the predominant mode of learning we favor (Shamsuddin & Kaur, 2020). These learning styles are thus influenced by how we approach problems: thinking, feeling, and judging; thinking-feeling; feeling-thinking (Lwande et al., 2021). The first is a preferred learning mode, the mode of engagement that reveals the clearest state of implicit learning. Preferred experience is a mode of critical reflection that maintains one's energy through the completion of a choice cycle and determines whether a learner is likely to act on what has been learned. Preferred processing style is how a learner transforms a concrete experience into a concept and hence has a strong effect on the form of concepts for constructed knowledge (Shamsuddin & Kaur, 2020).

2.3 BENEFITS OF LABORATORY LEARNING

The teaching–learning process involving practical work helps students understand topics better, remember concepts longer, think systematically, acquire better problem identification and solving abilities, and appreciate the complexity of real systems, thereby developing an inquisitive and skeptical approach (Shana & Abulibdeh, 2020). Students also learn the constraints on the proper development of the theories by questioning nature, suitability, and number of parameters (Felder & Brent, 2024). Laboratory experiences can also promote students' interest and capability in science and foster creativity (Agarwal & Bain, 2024). The efficiency of the didactic process can, as such, be enhanced by the performance of laboratory activities that enable the student to develop scientific inquiry, problem-solving, communicativeness, and critical thinking skills. These can stimulate student interest in studying natural sciences by highlighting the building of conceptual relationships evidenced through laboratory procedures (Meulenbroeks et al., 2024).

The hands-on experiences obtained in the laboratory can develop students' ability to recall things seen rather than simply heard, handle objects with ease, and appreciate the physical aspects of complex systems. Exercises undertaken in laboratories can help to demonstrate the applications of theoretical knowledge to real-life situations, and to determine new constants of nature in a classroom environment having low background noise (Chew & Cerbin, 2021). In bench work, questions can be raised and different answers provided, and offered explanations can be properly scrutinized and taken further. Instructions obtained can be repeated if necessary to confirm the conclusion drawn. Current and potential applications of existing knowledge can be emphasized, and students can even be motivated to identify alternative possibilities. The idea is to develop innovative schemes incorporating the feasibility of proposing procedures and strategies (Winkelmann et al., 2023).

It is immensely important that practical work be performed by students in all curricula to be studied, featuring experiments requiring inexpensive materials, and allowing them some freedom to express their acquired knowledge, provide personal opinions, and discuss results.

2.3.1 Enhanced Understanding of Concepts

Laboratories provide unique opportunities for students to design and execute experiments in order to test or validate a set of scientific ideas or principles. The laboratory environment allows students to directly engage with the natural world, encompassing both the physical and life sciences (Stroupe & Carlone, 2022; Hogan & O'Flaherty, 2022). Through experiments, students not only gain a deeper understanding of established theories and principles but also develop the skills necessary to differentiate between unequivocal and uncertain knowledge claims, qualified and definitive conclusions, and fundamental truths versus mere sales pitches or deception (Altmeyer et al., 2020; Lombardi et al., 2021). In doing so, students acquire attitudes that enable them to appreciate the value of scientific methods and problem-solving approaches.

It has been observed that students who actively participate in scientific investigations or experiments are more likely to attain desired learning outcomes, including the acquisition of fundamental scientific knowledge, comprehension of the investigative process, ability to conduct meaningful studies, development of intellectual skills, and the ability to overcome misconceptions or misunderstandings (Harefa et al., 2023; Wahono, Lin, & Chang, 2020; Supena, Darmuki, & Hariyadi, 2021). Through laboratory experiences, students are empowered to enhance their scientific knowledge while also honing their abilities to work safely and efficiently (Al-Zoubi et al., 2023; Amanat, Kanwal, & Akhter, 2024). They learn to recognize and identify chemical reactions; interpret and analyze written research findings; engage in effective written, oral, and graphical communication; utilize the scientific method; and gain an appreciation for the historical significance and limitations of scientific inquiry (Winkelmann et al., 2023). In essence, laboratory experiences serve as a launching pad for fostering a comprehensive understanding of the underlying patterns and regularities that govern the natural world (Mohzana et al., 2023; Gomez-del Rio & Rodríguez, 2022; Seifrid et al., 2022).

2.3.2 Development of Critical Thinking Skills

Various methods such as inquiry-based, guided inquiry, deductive, and virtual techniques have been recommended to encourage the development of critical thinking skills in students (Chu et al., 2021). The inquiry-based method, for example, helps participants understand complex and intricate facts, concepts, or impulses through the process of engagement in experiences. It is also beneficial in making learning participatory, interactive, and student-directed (Chu et al., 2021; Orosz et al., 2023). For most cases, open-ended inquiry works best. There have been many debates about the steps in any inquiry.

The steps in scientific inquiry are really the steps in the scientific method through which detecting a mystery could lead to a participant-led brainstorming session that should generate a whiteboard list of questions and "What ifs…" thereby stimulating natural curiosity (Meulenbroeks et al., 2024). The inquiry cycle can be summarized in three simple words – Inspire, Inquire, and Impact. The cycle stands in this order after mentioning a possibility for breaking out of the cycle; the term Impact refers to putting the new knowledge to use again where it was detected. To my mind, the bicycle analogy is more complete and therefore relates better to how we learn and reason before we have established that we know (Orosz et al., 2023; Lameras et al., 2021; Mohammed & Amponsah, 2021).

Techniques that use query-led learning are also beneficial in promoting critical thinking among students. In other words, students must be able to ask questions as well as seek answers in any learning environment (Ghai et al., 2020). With some students, problem-based learning and issue-based learning have also shown good results and, to some, may be better tools than inquiry-based methods (McCowan et al., 2022).

It is imperative to prime the students about why they are collecting the information, what common knowledge gaps need filling, analyze information to find facts, implications, and unknowns, then carry out the analysis and conclusion to find answers and test the answers in a risk–benefit setting (Mustafa, Hatah, & Makmor-Bakry, 2023; Costello et al., 2023). They should be asked to respond by giving their opinions, displaying their decisions, and, if in groups, learning how to build consensus. For primary and secondary school children, up to five steps can be taken in a given inquiry and learning situation. It is also noteworthy that pedagogy applies some pressure on the environment within which candidates think (Kelley, 2021; Bizami, Tasir, & Kew, 2023).

According to Omiko (2015b), the benefits accrued to learners when taught science using the laboratory can be summarized as follows:

1. *Learning the nature of science and technology* in order to foster the knowledge of human enterprise of science, in the process enhancing aesthetic and intellectual understanding. This also entails the corporality of a myriad of scientific procedures, interconnection between science and technology and among other scientific disciplines.
2. *Acquisition of scientific skills* (manipulative, inquiry, investigative, organizational, and communicative skills) that can help in problem-solving. This is in furtherance of a

basic goal of science education, to help instill skills that can be used in other related life situations.

3. *Appreciation and acquisition of manipulation skills*, through the following:
   a. Indirect observation of objects and materials, like when measuring or weighing substances.
   b. Use of multiple trials, and fiddling with materials and activities without stated theories, learners can make useful findings.
   c. Guiding students in observing given phenomena, based on established theories, mastery of concepts (hypotheses, theoretical models, taxonomic categories), and making predictions of likely occurrences.
4. *Development of scientific attitudes*: generating interests, attitudes, and values for scientific field experience by considering what it entails, including interest, curiosity, objectivity, collaboration, consensus, risk-taking, precision, confidence, perseverance, satisfaction, responsibility, and passion. Field experience offers greatest potential for development of life-long passion for science in students as they get to handle real things. The subsequent interest and passion lead the students to desire to scrutinize and explore more about nature and their surroundings.

There are known aspects of scientific attitudes which can be nurtured in the school science laboratory including curiosity, open-mindedness, objectivity, intellectual honesty, rationality, willingness to suspend judgment, humility, and reverence for life (Omiko, 2007). In the same vein, a well-developed laboratory program can be used to achieve several goals such as intuition development and deeper conceptual understanding. Other goals include application of learned concepts to new situations, basic phenomena experience, skills development for data collection and analysis, learned use of scientific tools and materials, ability to estimate statistical and systematic errors, and better communication and reporting skills.

In order to realize effective learning outcomes from students, proper sequencing and harmonization of theory work with the practical work is vital. The practical classes ought to follow soonest after theoretical coverage of contents in classrooms or lecture theaters. This gives an instant bridging of the content covered theoretically through practical exercises. This therefore demands proper lesson planning which is student-centered and enables learners to understand the expected learning outcomes in every laboratory activity they are involved in.

2.4 CHALLENGES IN IMPLEMENTING LABORATORY LEARNING

Many institutions in Africa face challenges posed by poorly structured curricula, lack of infrastructure to implement practical work, insufficient instructional materials, and exorbitant user charges (Kanwar et al., 2020; Irima, 2023; Moshtari & Safarpour, 2024; McCowan et al., 2022). The pre-service training of learners hardly trains them in the handling and maintenance of equipment but lays great emphasis on the use of equipment, especially the newly acquired tools (Okillan Asianut, 2021; Abedi, 2023; Abedi et al., 2023). This overemphasis on equipment use makes them potential trade technicians.

Peer mentors and established scientists are employed to strengthen the transfer of experiential knowledge and professional ethics. However, only a few institutions have the capacity to perform the diagnostic and practical tests of science equipment, thereby increasing the operating cost of the equipment. The challenge of keeping up with the current knowledge of equipment operation and jargon is daunting as the curriculum is constantly reviewed to remain relevant and up to date; yet institutional support for workshops, seminars, and conferences is inconsistent (Holland & Davies, 2020).

It takes a long time to release funds for laboratory setup; yet it takes building contractors no time to demand payment for jobs done. When educational observers visit science laboratories, they observe formalities that are un-motivational to the teachers and users of the science laboratories (Mohzana et al., 2023; Bergmann et al., 2021). Due to cost-saving measures, laboratory assistants are often replaced with other educational support staff and teaching assistants. The summary of the challenge is that the impact of the planned and implemented activities is not monitored. Staff who are or appear to be performing their duties are credited, while staff who are critical and identifying challenges and providing solutions are demotivated or laid off (Barrasso & Spilios, 2021; Jonnalagadda et al., 2022).

Sometimes the linkage between field activities and the related theoretical explanation is unclear in terms of educational standards and goals. Worse still, institutional strategic plans often lack focus and are inadequate, presenting fragmented activities, placing emphasis on building resources, with low budgetary allocations to conduct research and implement developmental projects (Alexander et al., 2023; Madigan, Kim, & Glandorf, 2024).

2.4.1 Resource Constraints

The integration of laboratories within the science curriculum serves as an invaluable tool for demonstrating and reinforcing scientific knowledge (Idris, Talib, & Razali, 2022; Hogan & O'Flaherty, 2022; Çilekrenkli & Kaya, 2023). Laboratory experiments act as a supplement to textbook descriptions, offering a more comprehensive and in-depth exploration of scientific phenomena. Resource constraints such as finance, time, safety conditions, and availability of materials hinder laboratory work. These constraints pose significant challenges to institutions that strive to provide quality education. Often, though, institutions that have laboratory facilities may lack the personnel trained to handle the diverse experimental needs in a particular subject area. This shortage of skilled personnel can create a barrier to engaging learners in meaningful laboratory experiences (Emeka et al., 2021; Baroudi & Helder, 2019).

Furthermore, some institutions lack the motivation to prioritize laboratory work, and to actively participate in facilitating experimental learning. As a result, learners may have limited contact time with the laboratory during semester programs or find themselves in overcrowded spaces where too many learners undertake experimental work simultaneously. This can greatly diminish the value and effectiveness of experiential learning (Batty & Reilly, 2023; Manyilizu, 2023; Kolil & Achuthan, 2024).

It is crucial to emphasize the importance of safety when conducting experiments in the laboratory. Safety training should be an integral part of any laboratory curriculum, and it should commence with the teacher who is responsible for overseeing the laboratory. A well-trained and safety-conscious teacher can instill a culture of responsible experimentation and ensure that learners are aware of the necessary safety protocols and precautions (Bai et al., 2022; Yu et al., 2023; Emeka et al., 2021).

By addressing these resource constraints and prioritizing laboratory work, educational institutions can offer their students comprehensive and engaging scientific experiences. This will not only enhance their understanding of theoretical concepts but also equip them with valuable practical skills that are essential for future academic and professional pursuits.

2.4.2 Safety Concerns

The implication of laboratory safety concerns test action that is necessary in ensuring the well-being of laboratory users, especially students, teachers, technical staff, and any other laboratory users. Regarding such safety, it is imperative to highlight the findings from an extensive study conducted on 143 science teachers. Astonishingly, safety was mentioned a staggering 83 times (which accounted for some 58%) by these diligent educators (Russell & Martin, 2023; Azanaw et al., 2021). It was also noted that most of these profound discussions originated from biology teachers, further underscoring their awareness and commitment to keeping their students safe (Barron et al., 2021; Nyansa et al., 2024).

Furthermore, these teachers were not just mere proponents of safety in theory, but they also translated their concerns into tangible actions. In particular, they voiced their worries about the physical space within the educational setting, as well as the adequacy, or lack thereof, of the facilities provided. They did a commendable job of successfully addressing the limitations and challenges faced during the instructional process, thereby proving their unwavering commitment to ensuring the utmost safety for their students (DeCoito & Estaiteyeh, 2022; Anto et al., 2023).

Remarkably, the teachers' commitment to safety extended beyond the typical boundaries of science content delivery in the classrooms, even reaching the esteemed junior school teachers. These exceptional educators took an additional step in this collective endeavor by meticulously considering the design of learning environments, examined the classroom setting and furniture arrangement (Anto et al., 2023; Schwarz et al., 2021). This holistic approach allowed the teachers to make valuable connections between the physical learning environment and the effective delivery of educational content. These teachers proactively sought out innovative solutions to potential safety concerns. They keenly explored various laboratory supplies, harnessing their creativity and resourcefulness to devise effective strategies that would nurture a safe learning space for their students (Anto et al., 2023; Schwarz et al., 2021).

By addressing the complexities and intricacies of safety protocols, ensuring comprehensive training, promoting a culture of vigilance and responsibility, fostering collaboration and knowledge sharing, and utilizing innovative technologies and tools, science educators can be empowered to create a safe and dynamic laboratory environment. It is paramount to invest in a robust and sustainable support system, professional development, and resources for science teachers to enable them to nurture future scientists, innovators, and critical thinkers. This way, it will be possible to revolutionize laboratory science education and elevate it to new heights of excellence and safety for the betterment of all (Schwarz et al., 2021).

2.5 BEST PRACTICES FOR LABORATORY INSTRUCTION

The quality of laboratory learning improves if certain rules are faithfully observed. One critical practice to observe is the number of experiments conducted per class. Reportedly, when there are fewer experiments per class, the perceived quality of each laboratory session tends to be higher. Therefore, it is recommended to carefully select and limit the number of experiments to maximize the effectiveness of the learning experience. It is also crucial to observe zero tolerance to laboratory absence; students should attend every laboratory session without fail as absenteeism can hinder the progress of an entire class and disrupt the learning flow. By strictly adhering to this rule, students will get to understand the significance of their presence in the laboratory and be more committed to their studies (Winkelmann et al., 2023; Serrano-Perez et al., 2023; Agustian et al., 2022).

Another key rule is to have an optimum number of students in each laboratory session as student numbers play a significant impact on laboratory learning. This limitation ensures that each student receives adequate attention from the instructor and can actively engage during laboratory experiments. The smaller the student-to-teacher ratio, the better the quality of instruction, supervision, and learning. In cases where excluding students becomes a consideration, it is advisable to explore alternative solutions. It is better to work longer hours to accommodate all students rather than excluding anyone. Every effort should be made to provide equal opportunities for every student to participate in laboratory activities. This inclusivity fosters a sense of fairness and promotes a supportive learning environment (Serrano-Perez et al., 2023; Agustian et al., 2022).

Material completion should not be compromised by students who are absent. It might be tempting to skip certain parts or rush through the material to catch up, but this approach is detrimental to both absent and present students. It is important to ensure that all students have access to the complete laboratory curriculum, even if they were unable to attend a session. By maintaining consistency in the material covered, all students can progress at the same pace (Winkelmann et al., 2023; Agustian et al., 2022).

Supervision of laboratory learning plays a vital role in its success. It is crucial for instructors to actively engage in supervising the laboratory sessions personally. This hands-on approach enables them to assess the progress of each student, provide individualized guidance, and address any questions or concerns effectively. Additionally, it is valuable to train and develop a team of competent trainers who can provide consistent support in the laboratory (Serrano-Perez et al., 2023; Agustian et al., 2022).

Promoting collaboration and healthy competition within student groups is another key aspect of enhancing laboratory learning. Encourage careful group work where students actively collaborate, share ideas, and learn from one another. This teamwork fosters a sense of camaraderie and creates a conducive environment for learning. Additionally, introducing moderate levels of competition can motivate students to strive for excellence and excel in their laboratory skills (Agustian et al., 2022).

To further augment the learning experience, extensive use of demonstrations and development of related video clips are encouraged. Visual aids can enhance understanding and reinforce theoretical concepts (Altmeyer et al., 2020). By utilizing these multimedia resources, students can observe experiments in action, grasp complex procedures more easily, and reinforce their theoretical knowledge with practical applications.

Lastly, it is crucial to adapt the timing and content of laboratory sessions to align with the complexity of the learning objectives. Each laboratory session should be designed to meet specific objectives and learning outcomes. By carefully analyzing the complexity of these objectives, instructors can allocate appropriate time and plan the content accordingly (Marks & Thomas, 2022; Mohzana et al., 2023). This tailored approach ensures that students have sufficient time to comprehend and master the concepts being taught.

In conclusion, the quality of laboratory learning can be significantly improved by rigorously following certain rules. The laboratory experience will be greatly enhanced, leading to better learning outcomes for students, by limiting the number of experiments per class, enforcing zero tolerance for absence, constraining student numbers, working longer to accommodate all students, and not compromising on material completion for absentees. Other rules include supervising laboratory learning personally, encouraging careful group work and moderate competition, utilizing demonstrations and video clips, and adapting the timing and content of sessions to match the complexity of objectives.

To achieve the best experimental learning practices at universities, continuous formative assessment of laboratory practice and constant monitoring and reflection by educators is advised. A variety of learning tasks and objectives distinguish practical laboratory education from other types of learning. Furthermore, what makes laboratory education special are its pedagogical features. The more transparent these pedagogical features are, the better laboratory education is effectively delivered (Kelley, 2021; Garzón et al., 2020). Institutions must design, test, implement, and record innovative practical pedagogical methods and monitor the satisfaction of the teaching community to ensure the practical relevance and learning effectiveness for students (Bizami et al., 2023). An intuitive rubric in laboratory teaching and learning focuses on the aspects of experiments, which contributes to the realization of the set laboratory learning objectives. Such rubric should resonate well with and motivate students to learn and reveal effective teaching and learning tools and resources.

2.6 PRACTICAL SCIENCE SUBJECTS FOR SECONDARY SCHOOLS

The main laboratory-based subjects with practical activities in secondary schools are Biology, Chemistry, and Physics. The other applied and technical sciences that may also require laboratory practical sessions, though of different settings, include Computer Studies and Electrical Technology studies.

The **Biology Curriculum** is designed in a way that requires students to plan and perform simple practical activities and research work. In that case the students need to obtain observation and other experimentation skills, including handling biological materials or specimens. Laboratory work makes it possible for students of biology to develop critical thinking skills to enable them to analyze and interpret results observed during laboratory activity. The concepts and contents of teaching biology ought not to be dogmatic but should adopt a system that enhances holistic understanding of the subject and critical thinking. The practical activities in biology therefore must be based on scientific inquiry method, which involve hands-on and minds-on. Teachers need to go outside their normal laboratory activities and use information and communication technology (ICT) also as a way of improving conceptual delivery to learners.

Chemistry Curriculum requires learners to look for, manipulate, analyze, and evaluate data from varied sources. Chemistry, through observation and measurements, provides students with basic competences of critical thinking. The learners get educated on how to appropriately select and use equipment in order to get good learning outcomes and avoid accidents in the laboratory. The conceptual delivery of the chemistry subject should be tilted toward problem solution and inquiry approaches to maximize learning outcomes. The delivery should involve students' hands-on practical exercises. This learner-centered approach is the way to go, according to chemistry curriculum designers. The delivery approach, just like in the case of biology, should be supported by use of ICT for effective delivery of contents to learners.

The **Physics Curriculum** orients students to gain competencies in the operations of devices based on digital technology and computers, among others. The curriculum requires students to develop skills that enable them to lead a healthy and safe life; and to realize the necessity of experimentation in evaluating physical laws. Just like for biology and chemistry, the teaching of physics concepts is inquiry-based. This calls for student thinking, reasoning, and simple research to arrive at the expected learning outcomes. The use of computer supported approach is very highly recommended to ensure that even the hidden concepts are brought to the realization of the learners.

The role of laboratory work in teaching and learning scientific concepts is of utmost importance and cannot be overemphasized. Consequently, the Government of Kenya in collaboration with that of Japan came up with an in-service training program called Strengthening of Mathematics and Sciences in Secondary Education (SMASSE). Kenyan teachers of science and mathematics have gone through a number of inset training programs under SMASSE schedule whose main objective is to ensure the teaching of these subjects is student-centered (Sikolia et al., 2017). The SMASSE insets train teachers to use Activity, Student, Experiment Improvisation (ASEI)/Plan, Do, See, Improve (PDSI) lesson plans, which are activity-based methods of content delivery (Kiige & Atina, 2016). The program aims to improve student performance in secondary school examinations.

Occasionally, the right materials to be used in carrying out laboratory work may be unavailable or too expensive to acquire. In such cases, the ASEI lesson plan becomes handy in concept delivery. This is because it emphasizes student-centered activities done through experiments, using improvised materials.

2.7 PRACTICAL-BASED COURSES FOR TERTIARY INSTITUTIONS

Many tertiary educational institutions (middle-level colleges and universities) offer science and technical courses leading to diplomas and degrees, respectively. These courses have practical-based curricula that require students to perform certain practical activities in laboratory settings and, when fully delivered to learners using hands-on approach, produce diligent and effective graduates who perform well in assigned employment duties. To produce such competent graduates for the job market, the institutions offering these courses must establish well-resourced laboratories, able to facilitate effective realization of the set learning outcomes. Sciences can be conveniently grouped as physical, life (biological), social, and mathematical & logic sciences. Physical and life sciences are invariably partially taught in the laboratory wherein the learners horn their knowledge, skills, and attitudes for the subsequent practices.

Some of the physical science courses that require laboratory work include the following:

Analytical chemistry whose main objective is to produce analytical chemists needed in various job sectors including the industrial sector.

Industrial chemistry which aims to produce industrial chemists with skills required in research institutes, industries, and other institutions. The course curriculum includes applied and practical chemistry concepts which prepare students for the 21st-century technological challenges.

Control and instrumentation is a rigorous science that prepares learners to have skills necessary to operate, calibrate, and maintain modern equipment used in various sectors. The program offers skills to graduates needed in electronics, energy, environment, and implantation of automated appliances.

Renewable energy and environmental physics, designed to produce graduates with the knowledge and technological skills required in solving energy and environmental challenges.

Geophysics which aims to produce graduates with necessary skills and competencies to work in geo-research institutes, mineral exploration, and oil industry sectors. Just like the other courses mentioned above, laboratory work and practical activities are core in geophysics.

Nuclear physics, a program geared toward producing graduates with knowledge, technological skills, and competencies for solving problems associated with atomic and other dangerous radiation leakages. Such radiation if not handled carefully and safely may be harmful to the biological component of the environment.

Other physical science-based courses including engineering and computer studies and technology are loaded with numerous practical works. In all these courses, laboratory work is designed to complement content delivered theoretically through lectures.

The biological sciences also entail lots of laboratory work offered at various tertiary educational institutions. Some of these include the following:

Fisheries and aquaculture whose curriculum is geared toward producing highly competent graduates in fisheries and aquaculture technologies, and skilled enough to efficiently and effectively manage their produce. This course is skill-intensive, and hence involves lots of practical and laboratory work which require a well-laid-out laboratory for effective content delivery to learners. Graduates of the program learn high-order thinking and manipulative technological skills for improving and managing aquatic biodiversity.

Genomic science programs which focus on creating proficient graduates to work in medicine, forensic, food, and agricultural fields. These graduates develop knowledge and skills which enable them to map genes of diverse organisms, even to manipulate them through DNA sequence analysis, so as to get the desired characteristics that are valuable to humans. Students of these programs are practical-oriented, and hence require a laboratory with the right apparatus, good flow of materials, and other substances.

Microbiology whose curricula involve the study of microorganisms like bacteria, protozoans, fungi, algae, and viruses. Graduates of the various microbiology programs develop technical skills used in health, food, agricultural, ecological, and research sectors. The component courses are designed to include many hours of practical and laboratory work.

Biotechnology which has a curriculum that equips learners with knowledge and skills for microbiology, plant biodiversity, and technological expertise required for solving problems in medicine, agroforestry, agriculture, and food industry (food processing and security), as well as in pharmacognosy and drug research. In order to get effective learning outcomes, the courses apply practical activity approaches.

Environmental science programs that provide students with knowledge, techniques, and skills for acquiring, capturing, processing, analyzing, and interpreting environmental[1] data. The learners acquire knowledge on the value of biodiversity and skills for economic and sustainable use of environmental resources to cater for the ever-increasing human needs and wants.

To meet these specifications, practical activities and field work must be integrated in the curricula design.

Applied biology, a program that enables learners to acquire wide-ranging knowledge of life sciences concepts and attain desirable skills and competencies in laboratory practice. During their training in the program, learners obtain critical thinking skills and techniques that qualify them to work in research, teaching, food industry, agricultural and other biological laboratories. The program curricula provide for substantial time for laboratory and field work to support content delivery.

Biostatistics is a program that offers learners statistical techniques and methodology for data analysis and evaluation for life sciences. While going through the program, students learn how to use computers for data collection, storage, cleaning, and analysis within the life science domain. Hence, these programs need both biology and computer laboratories for effective practical learning.

Other biological sciences include the health sciences (such as medicine, pharmacy, dental surgery, nursing, and medical laboratory sciences, among others). Students undertaking these courses have practical and laboratory work almost on daily basis.

2.8 CASE STUDY 2.1

The illustrative case studies represent 2 of the 15 science learning laboratories that have been meticulously constructed under a comprehensive and groundbreaking science learning initiative in Nigeria, striving to revolutionize the landscape of science education. These remarkable laboratories consist of an exceptional Environmental Science Laboratory in a prestigious first regime university, dedicated to fostering a deep understanding of the intricate connections between humans and the environment, and committed to ensuring their students receive the highest quality education. Additionally, there is an extraordinary instructional improvement unit in a highly esteemed third regime university teacher education, and committed to enhancing the pedagogical practices of aspiring teachers. In examining the broader context of Nigerian tertiary institutions, it becomes evident that these institutions serve as a microcosm of tertiary education in Sub-Saharan Africa as a whole.

The significance of these case studies reverberates beyond their individual scope, shedding light on the immense potential and imperative of the African tertiary landscape in fostering a transformative and innovative 21st-century science instruction. Central to the realization of the case studies and the overarching subthemes lie the dire need to address and effectively overcome the prevalent infrastructure deficit currently plaguing science education (Bugingo et al., 2024; Emeka et al., 2021; Ogunode & Aiyedun, 2020).

The multifaceted nature of implementing science instruction, as encapsulated within the subthemes, necessitates the establishment of well-equipped laboratories. Such laboratories, equipped with state-of-the-art materials and cutting-edge technologies, serve as the cornerstone for experiential and hands-on learning, enabling students to actively engage with scientific concepts and apply their theoretical knowledge in a practical setting (Quainoo et al., 2020). Thus, the provision of these laboratories is crucial in empowering students and educators alike to explore, discover, and innovate in the realm of science. By highlighting the significance of these case studies within the larger context of the African tertiary landscape, the case studies aim to ignite a collective commitment to foster and nurture a thriving culture of science instruction, propelling Africa into a future illuminated by scientific advancements and breakthrough discoveries (Bugingo et al., 2024; Emeka et al., 2021; Ogunode & Aiyedun, 2020).

It is essential that deliberate and concerted efforts are made to allocate substantial resources and funding toward the advancement and expansion of science instruction. By doing so, it not only paves the way for the successful implementation of these case studies but also creates a solid foundation upon which future endeavors in science education can thrive and flourish.

However, there are several challenges and obstacles that arise when striving to provide such exceptional laboratory facilities. The procurement of high-quality equipment, the implementation of robust safety measures, and the maintenance of these laboratories present formidable tasks that require careful consideration and strategic planning. In conclusion, illustrative case studies serve as beacons of inspiration and exemplars of excellence in science education. They reflect the profound impact of investing in infrastructure and resources, not only in elevating the quality of education but also in empowering the next generation of scientists, innovators, and critical thinkers.

NOTE

1 Note: Environment constitutes the biotic and abiotic parts whose interplays lead to a self-sustaining system.

3 Laboratory Safety and Security Management

Apollo Odhiambo Maima

3.1 INTRODUCTION

Laboratory (Lab) safety encompasses a range of systems and practices designed to safeguard both individuals and resources within laboratory environments, and involves preventative strategies to recognize, minimize, or eliminate dangers from various work activities and procedures (Anza et al., 2016). Establishing a culture of safety and security in a lab of any type and size calls for unrelenting commitment to maintenance of high standards at all levels. This should run from top managers of the institution to lab workers. As institutions work to establish safety and security protocols, it is important for them to consider the whole life process of chemicals from planning, procurement, and security to final use and disposal of waste (AlShammari, Alhussain, & Rizk, 2021).

Laboratory personnel in an institution should have well-defined roles and responsibilities that are geared toward establishment and maintenance of safe and secure lab practices. In order for lab workers to show their commitment, good and ethical examples should be set. Responsibility for safety and security of a lab therefore heavily lies with the head of the institution as well as its operating organs. Depending on the type and size of the lab, legal obligations for violations of protocols of operations in terms of safety and security may lead to fines or imprisonment or at worst, the closure of the lab. Violations of government legal mandates are considered as an offence when provision of safe and secure working environment is breached. Any lab management should ensure that an effective safety and security program is followed by members of staff. It should be noted that a well-thought and conceived program may be treated casually by workers if it is not taken seriously by the management.

3.2 LABORATORY SAFETY AND SECURITY

3.2.1 General Laboratory Safety and Security Concerns

The important field of laboratory safety and security includes a set of principles, regulations, and protocols that must be adhered to when engaging in potentially hazardous procedures. By ensuring compliance with these established guidelines, the likelihood of accidents diminishes, financial liabilities are reduced, and a safer, more sustainable workplace is fostered.

Effective management of safety protocols and risk factors is especially crucial in educational laboratories, as students may not fully grasp potential dangers. When handling chemicals, laboratories become hazardous, requiring strict adherence to safety practices (Samaranayake, Nishadya, & Jayasundara, 2022). Many substances, equipment, and materials contained in a lab require specific hazard control measures to ensure that individuals and the community are not exposed to unreasonable levels of risk. Laboratory security, therefore, plays an important role in ensuring that unauthorized persons cannot readily access hazardous equipment, materials, and substances. Security must be maintained within all labs at all times to ensure that the institution's laboratories (labs) are not readily accessible to unauthorized people.

Laboratory safety is crucial in scientific research, academia, and industry, yet many institutions lack strong systems for a secure working environment (Samaranayake, Nishadya, & Jayasundara, 2022).

3.2.2 Laboratory Safety Regulations and Standards

Laboratory safety is guided by laws, standards, protocols, and operational benchmarks that enhance scientific practices. The WHO's *Laboratory Biosafety Manual* provides essential guidelines on biosafety and biosecurity, focusing on principles and containment technologies. A robust biosecurity plan involves risk assessment, access control, personnel evaluations, material management, training, and emergency preparedness, limiting access to biological agents and reducing risks linked to loss, theft, or misuse. This highlights the crucial connection between biosafety and biosecurity for a secure research environment (Baskin & Zelicoff, 2016).

The importance of biological safety lies in its role in safeguarding individuals from potential health threats posed by infectious agents. A biosafety program includes various laboratory protocols and protective measures to prevent contact with biological agents. Practices vary based on specific infectious agents, requiring detailed risk assessment before implementation. The main goal is to create procedures that integrate into lab routines and focus on reducing exposure. These procedures are customized for specific biological materials and their applications, ensuring effective safety measures (Gao et al., 2025).

Laboratory safety regulations play a crucial role in safeguarding both people and property from potential dangers. The Institutional Biosafety Committee (IBC) serves a vital function in reducing risks by implementing Integrated Disaster and Emergency Management (IDEM) practices. Cultivating a strong safety culture emphasizes the importance of prioritizing safety in all operations. To avert catastrophic events similar to the Bhopal disaster, security vulnerabilities are assessed through the Facility Security Level methodology, which ensures adherence to the standards set by the American Disability Act (ADA) (Al Wahshi, 2019; Broughton, 2005).

International and industry frameworks are crucial for advancing global laboratory safety harmonization efforts (AlShammari, Alhussain, & Rizk, 2021). An in-depth examination of these standards is crucial for the clarification of laboratory hazard management systems. Grasping the nuances of these safety standards supports national agencies in formulating regulatory guidance concerning infrastructure and safety protocols.

Esteemed international organizations engage in research and development focused on regulations and safety protocols, whereas national agencies implement globally recognized standards such as ISO 45001. Ongoing assessment is crucial to keep pace with emerging technologies (Samaranayake, Nishadya, & Jayasundara, 2022). Regulatory documents and standards play a vital role in the management of safety laboratories around the world.

3.2.3 Physical and Fire Safety measures

Measures aimed at safeguarding physical safety mitigate risks associated with laboratory equipment, electrical dangers, excessive noise, ultraviolet light exposure, toxic vapors, laser usage, and radiofrequency devices. Personnel are required to adhere to established protocols concerning mechanical tool usage, fire safety procedures, and appropriate conduct within the lab environment.

Fire prevention strategies are crucial for ensuring safety in various environments. It's important to store waste properly by keeping chemicals, paper products, and combustible materials separated from buildings. Taking measures to prevent sparks from electrical devices is essential for minimizing risks. Laboratories need to be equipped with flame-proof stopcocks and storage tanks designed to resist fire. It's also critical to keep matches and lighters away from any chemicals, and to make sure that all containers are clearly labeled, particularly those that hold high-pressure or flammable substances. Additionally, compressed air piping should be covered with protective sheeting, and conducting regular inspections of laboratory equipment is necessary to promote a safe working environment.

3.2.4 Laboratory Security Management

For effectiveness, institutional lab security measures should focus on the following key areas: *People*, *Building*, and *Contents of the Building*.

3.2.4.1 People

Only authorized people will be permitted to enter and/or remain in a lab and/or associated facilities. Any member of staff who has reason to believe that there is an unauthorized person in a lab and/or associated facilities should immediately inform the technologist-in-charge and contact Security Staff. All staff should carry an appropriate identification card and actively monitor the overall security status of the lab and/or associated facilities.

Any person who has been provided with a pin code, card, pass key, lock combination, or otherwise granted access to a lab and/or associated facilities by the institution must not, under any circumstances, share their means of access with a third party.

People who are behaving in a manner that could compromise security or safety within a lab and/or associated facilities should be asked to leave. In the event of the person refusing to leave, Security Staff should be contacted immediately. Access by contractors to a lab and/or associated facilities is conditional upon the contractor following all the security requirements relating to the lab and/or associated facilities. Contractors who provide routine cleaning and maintenance services must comply with the safety requirements in the respective labs they are cleaning. Access should, however, be limited to times when trained lab staff is present.

Only technologists and other trained staff or students are allowed to operate the specialized equipment in the labs. For non-technologists, this usage will be restricted only to the equipment they are trained to handle. Any other person will only use the equipment with assistance and guidance by a technologist.

3.2.4.2 Buildings

Building design and construction is critical in the maintenance of the security of a lab and its contents. Some institutional labs are installed with security alarms, Closed Circuit Television (CCTV), light/movement sensors, smoke detectors, and emergency doors that open automatically whenever the security alarm goes off. Laboratory users in such cases are requested to familiarize themselves with these features in good time. The CCTV should help alert Security Staff of an unauthorized entry or inappropriate use of the facility or that an event requires a particular security response. Emergency response is activated, for example when a fire alarm has been activated.

Staff must familiarize themselves with all alarm systems installed within the lab to ensure that they can respond effectively and without delay. The windows and doors of all labs must be kept locked and secure at all times when the lab is not occupied. These and other forms of building security equipment must be maintained to a serviceable operating standard. Technical staff are required to report any unserviceable or malfunctioning security equipment to the maintenance department for rectification.

3.2.4.3 Contents of Buildings

As discussed earlier, labs contain a range of "at higher risk" substances, materials, and equipment that require different levels of secure storage. Particular area(s) of a lab, for example store rooms, flammable liquid cabinets, cold rooms, dark rooms, equipment rooms, fridges, freezers, etc., have been designed, constructed, and installed to improve security of these substances, materials, and equipment. The storage facilities should therefore be kept locked at all times when not in use and only be accessed by authorized persons.

Staff must also ensure that "at higher risk" substances, materials, and equipment are handled and stored correctly at all times so that the risk of injury to persons and/or theft from a lab is reduced to a minimum. An up-to-date inventory of all "at higher risk" substances, materials, and equipment should be kept, maintained, and regularly audited by staff in control of these items, to ensure they are fully accounted for, safe, secure, and not accessible to unauthorized persons. Technical Audit Reports of High-Risk Substances must be submitted to the lab manager on a Monthly, Semester, and Annual basis, and at any other times on request. Only the minimum quantities of "at higher risk" substances, materials, and equipment should be left in the general work area of the lab and only for the period of time required to achieve the desired learning outcome(s).

The operation of laboratory instruments is crucial for the field of science, as it guarantees adherence to established procedures and the effective management of potential risks (AlShammari, Alhussain, & Rizk, 2021). A thorough knowledge of equipment operation, maintenance, and safety protocol enforcement is crucial for proper laboratory use (Cahyaningrum, Tegar Muktiana Sari, & Iswandari, 2019).

3.2.5 Security of Chemicals

Scientists generally use numerous chemicals in their daily lab work, but these chemicals normally pose some danger to the general public if they are not used for the intended purpose. Some chemicals can be used as precursors in making illegal drugs and even dangerous chemical agents. Other chemicals have hazardous properties that may pose risks to lab staff, especially when they are not adequately knowledgeable about their properties.

General safety and security begin with the individual, and sound security measures for labs should have a system that controls access to authorized personnel. Authorized people should get approval to work in a specific area and have verified authority and access to use chemicals. These people should be granted permission through keys and may also be required to have identification badges. The issuance of access keys and badges should be subject to a well-established process in which lab staff sign for access keys and badges and return them when they are through with the work. Access keys and badges should be designed and made in a way that is readily duplicated.

In the academic labs, security measures can be a big challenge. Although teaching labs have few types and small quantities of chemicals in use, most of them might have a wide range of solvents, some of which may be hazardous. In general, more hazardous chemicals are mostly used in advanced research labs, and as such, greater attention is required to limit accessibility to such facilities. Separation of ordinary labs from advanced research labs makes safety and security much easier.

3.2.6 Physical Security of a Laboratory

In the realm of laboratory operations, the significance of security cannot be overstated, as it plays a crucial role in protecting vital materials and sensitive information. A breach in security can lead to various dangers, ultimately jeopardizing the integrity of laboratory practices. By diligently managing physical access, laboratories are able to shield their staff and valuable assets, thereby thwarting any unauthorized interactions with equipment and materials. To ensure effectiveness, it is essential that security protocols are customized to the unique circumstances of each laboratory, considering thorough risk assessments and specific needs (Brizee et al., 2019).

Physical security is essential in a comprehensive safety framework. With proper training and clear guidelines, security guards play a key role in restricting unauthorized access to chemical storage areas. Secure door locks and electronic locks are crucial for chemical storage sites. Key protocols include locked storage compartments, cabinets, and drawers, along with robust alarm systems. Laboratory security measures should be tailored to specific risks while ensuring they do not disrupt research or safety practices.

The success of a security program is closely tied to the standards set by management. Laboratory personnel are required to be vigilant and attentive to any unfamiliar individuals who may enter the premises, as well as to any behaviors that seem out of the ordinary. It is crucial for employees to lock up laboratories when they are not in use and to properly store any chemicals that are not currently needed, ensuring safety in the environment. Managers have the responsibility to implement a rigorous policy that prohibits the unauthorized use of facilities and materials, making certain that any breaches of this policy are met with significant repercussions.

A comprehensive training program dedicated to lab security management ought to detail the expectations placed upon management while highlighting the importance of a secure working environment. It is crucial for new employees to undergo regular training to thoroughly understand the established security protocols. Additionally, routine inspections should be conducted to confirm adherence to these standards, supplemented by well-defined procedures for responding to any security breaches that may occur.

3.2.7 Access to Laboratories and Associated Facilities

Laboratories must maintain a detailed sign-in and sign-out register for all individuals entering and exiting. This record should include the individual's name, arrival and departure times, accessed areas, purpose of visit, and contact number for future reference.

The institutional labs operate from 8:00 am to 5:00 pm during the weekdays, allowing for evening and weekend sessions upon special request. To gain access outside these standard hours, it is necessary to reach a prior arrangement with the supervising technologist. It is imperative for all individuals entering the facility to comply with the guidelines set forth in the institution's *Laboratory Safety, Operation & Design Manual* consistently.

3.2.7.1 Laboratory Keys

All keys intended for the teaching laboratories are to be securely stored at the security desk. These keys are only available to authorized technologists and specific personnel who have the necessary clearance to sign them up. Entry to critical areas such as chemical store rooms, equipment storage rooms, and other specialized facilities, including cold rooms and gas rooms, must be restricted to the technologist-in-charge.

3.2.7.2 Unattended Work in Progress

A "Notice of Experimental Activity Card" must accompany experimental work in progress that is unattended for any length of time. The card presents essential information pertinent to the current tasks and should encompass the following specifics:

- Characteristics of ongoing tasks
- Essential safety measures and details for emergencies
- Emergency contact numbers
- Vital first aid information for immediate health concerns.

Work in progress cards must not contain shorthand or abbreviations and should be prominently displayed near the experimental area. The person carrying out the task is responsible for ensuring that the card is clearly visible throughout the duration of the work.

3.2.7.3 Overnight Work in Progress

All experimental work involving potentially hazardous material, that is unattended between 6:00 pm and 8:00 am, must be set up in an area that will render it safe if it has any potential spills (e.g., setup of a chemical reaction in a fume cupboard). Security Staff must also be advised in writing that the work in progress will run overnight. The person responsible for setting up the work in progress or work that runs overnight **MUST** ensure that a "Notice of Experimental Activity Card" is completed and displayed as required, and a copy deposited with the Security officers. Notwithstanding this, approval for such works in progress will have to be given in writing by the lab manager.

3.2.8 Housekeeping and Cleaning

Good housekeeping and physical cleaning of the lab is important for maintaining safe working conditions and the safe operation of the cabinets. The following measures will be taken to achieve this objective:

- The lab or room where a cabinet is housed shall be kept clean and tidy, and free of physical hazards that may cause spillages or breakages.
- Waste material and unwanted equipment shall be cleared regularly.
- Cleaning of the lab shall be carried out by trained personnel in accordance with specific, written and approved cleaning procedures.
- Contract cleaning staff should only be allowed access for the wet-cleaning of floors, windows, bench-tops, and any other areas assigned to them by the technologist-in-charge. Cleaning is best done only using neutral detergent and a mop with a detachable head (to allow for sanitization/disinfection).
- Sweeping with brooms and the use of vacuum cleaners should be discouraged, as it produces airborne dust that can increase contamination of work in the lab.
- All cleaning should be done when the lab is not in use, but in the presence of a technologist.

3.2.9 Radiation Safety within Laboratory Environments

The importance of radiation safety cannot be overstated, as it plays a crucial role in safeguarding laboratory personnel who handle radioactive materials or sources of ionizing radiation (Ahmed & Taha, 2016). In this domain, attention is directed toward the oversight of radionuclides, radioisotopes, and X-ray apparatus. The possibility of ultraviolet radiation emitted from germicidal lamps poses a significant hazard, alongside the risk of contamination from widely encountered radioactive materials. Similar to other areas of safety, the framework consists of applicable legislation, regulatory measures, and established standards that are essential for effective execution (AlShammari, Alhussain, & Rizk, 2021).

Radiation safety is acknowledged as a distinct field within the realm of laboratory safety management, necessitating specific expertise and control methods. Nonetheless, it adheres to the essential principles of general safety practices and seamlessly integrates into the comprehensive safety protocols found in typical facilities.

3.2.10 Laboratory Management of Cybersecurity

Cybersecurity plays a crucial role in the realm of laboratory management, serving to protect digital devices that store sensitive information and critical experimental data. The task of safeguarding both intangible data and fragile instruments presents a significant challenge. It is imperative to limit access exclusively to authorized personnel who possess the appropriate passwords, necessitating enhanced monitoring within the laboratories to ensure adherence to these access restrictions (AlShammari, Alhussain, & Rizk, 2021).

Although the presence of civilian security at the entrances of laboratories is robust, there is frequently a notable deficiency in the safeguarding of digital information and technological instruments. This vulnerability puts various research resources and apparatus, including computers, experimental frameworks, and surveillance tools, at risk of potential dangers (Shu, Li, & Gao, 2023). Safeguarding the confidentiality and accuracy of this information is essential for preserving investments, protecting intellectual property, and ensuring continued opportunities for breeders and researchers. There is an immediate need for robust protection measures to enhance the efforts of the scientific community.

3.3 SECURITY VULNERABILITY ASSESSMENT

Carrying out a security vulnerability assessment basically reinforces the establishment of effective safety and security control. This assessment may include whole institution or specific facilities within an institution, and it encompasses a series of comprehensive investigations and an integrated analysis. The purpose of the assessment is to document potential security risks to an institution, determine the magnitude of the risks, and assess the adequacy of security systems that are in place.

The assessment is vital for understanding the security requirements of an institution. A comprehensive evaluation of security vulnerabilities entails a meticulous review of assets, pinpointing potential threats, carrying out physical vulnerability surveys, and executing on-site evaluations. Risk assessments play a crucial role in identifying hazards, which subsequently inform strategies aimed at minimizing risks. Significant threats include harmful agents such as chemical, biological, and radioactive substances. Additionally, insufficient chemical hygiene plans can obstruct effective disposal practices. Laboratories that manage hazardous materials, including flammable and toxic agents, face distinct safety challenges, especially during operational interruptions. Such assessments can be categorized as qualitative, semi-quantitative, or quantitative, depending on the specific context of the laboratory (Fatemi, Dehdashti, & Jannati, 2022).

3.3.1 Evaluation of Assets

The purpose of the evaluation is to recognize and measure the various assets within an organization, including equipment, libraries, and documents, that must be safeguarded against damage, loss, destruction, theft, and the impacts of natural disasters. It is crucial to gather information regarding sources for replacement and alternative resources available within the institution to facilitate uninterrupted continuity of operations.

3.3.1.1 Potential Threat Assessment

This assessment explores various institutional dangers, encompassing acts of terrorism and natural calamities. It is essential for institutions to comply with privacy laws while also keeping a record of possible threats. A thorough evaluation of the magnitude of disasters is vital, considering events such as fires, earthquakes, cyclones, hurricanes, and volcanic eruptions. By measuring the intensity of these threats, one can better understand the seriousness of both intentional attacks and natural disasters.

It is vital to thoroughly assess the durability of a building's outer structure against potential breaches and environmental factors at every location, particularly in areas surrounding utility routes and entry points designed for laboratories. A careful evaluation of air intake positions for proper ventilation and a detailed analysis of storage conditions for hazardous materials are imperative to ensure safety and efficiency.

The analysis of the site needs to encompass a comprehensive vehicle traffic strategy that pinpoints designated delivery zones, parking spots, and the various entrances and exits associated with the laboratory. Recommendations should effectively coordinate the flow of both pedestrians and vehicles, while also implementing necessary security measures. It is essential to delineate the functions of different areas within the building and specify access for authorized individuals in order to evaluate potential vulnerabilities. This assessment will aid in the establishment of threat detection protocols, ultimately safeguarding the laboratory's valuable assets and maintaining its overall integrity.

3.3.1.2 Physical Vulnerability Survey

A vulnerability survey for an institution concerns so many investigations, which includes the following:

- Identification of potential targets and how to gain access to the targets.
- Identification and rating potential threats based on the history of institutional threats. For example, threats that have been acted upon should be given more weight than potential threats that have never occurred. This should be considered in terms of natural factors and malicious threats.
- Identification of all institutional personnel such as permanent and contractual employees, merchants, and visitors who may have conflicts or personal problems with the institution. All these people may be able to have knowledge of the internal physical identity of the facility, hence classified as a potential source of threat.

3.3.1.3 Security Audits of the Lab

Laboratory security audits meticulously evaluate the procedures and records in place to ensure they align with established security plans and regulations. This thorough assessment guarantees the implementation of robust safety measures within laboratory environments.

Progress in laboratory security standards has been made, but implementation remains insufficient. As a result, many standards are not adopted, increasing risks of vulnerabilities, accidents, resource theft, or misconduct (Brizee et al., 2019). The auditing procedure evaluates the efficiency and reliability of security and safety protocols.

Security audits are categorized into two distinct types: internal and external. Internal audits are conducted by representatives within the laboratory, who assess the adherence to internal standards and protocols. Conversely, external audits are carried out by certified inspectors or organizations tasked with regulatory evaluations. The choice between these two auditing methods is influenced by several factors, including the nature of the laboratory; whether it operates in a governmental, academic, or industrial context; the size of the facility; the robustness of its security measures; its safety record; applicable regulatory frameworks; the outcomes of risk assessments; previous incidents; and available budgetary resources.

3.3.1.4 Development of a site security plan

A comprehensive site security plan integrates all the information gained in the surveys, analyses, and investigations. It outlines institutional security measures and an emergency response strategy that provides a physical protection plan to detect and respond swiftly and effectively to neutralize threats of any nature and natural calamities. Methods that are available to the general public provide detailed approaches on how to develop a plan that meets the set standards of access and security control policy.

Institutional labs may consider employing concepts of crime prevention based on environmental design. The location of the lab, properly placed entry points, lighting system, windows, and other physical features can work to prevent criminal activity. They are not expensive and as such are security improvements that do not depend entirely on technology. The basis of site security plan should be security and access control policy.

3.4 BEST PRACTICES FOR LABORATORY SAFETY AND SECURITY

Laboratory safety plays a crucial role in reducing the dangers of hazardous materials to laboratory users and the general public. By putting effective safety and security measures into action, this will safeguard not only the individuals working within the lab but also the surrounding environment from possible threats and harm (AlShammari, Alhussain, & Rizk, 2021). Laboratory safety constitutes an international benchmark that highlights the importance of personal accountability and teamwork. Its primary objective is to cultivate a culture of safety, recognize potential hazards, and mitigate risks effectively (Ma et al., 2021). Adhering to effective safety and security protocols requires that employees follow the established guidelines of organizational procedures and policies.

3.4.1 Management Strategies for Laboratory Safety and Security

Effective strategies for lab safety and security management include various best practices:

a. Ensure compliance with local safety regulations and standards to uphold accountability in safety and security measures.
b. Establish and uphold necessary administrative and engineering safeguards, along with protective resources, to fulfill safety and security responsibilities.
c. Create comprehensive documentation, including standard operating procedures, training logs, and incident tracking systems, that accurately represents laboratory activities.
d. Perform thorough inspections to detect potential hazards.
e. Offer training to staff in order to foster a robust safety culture within the workplace.

3.4.2 Environmental Health and Safety Personnel

A **safety manager** oversees lab protocols while all personnel, including new hires and volunteers, need thorough safety training. Visitors should equally be briefed on key safety principles. The Environmental Health, Safety, and Risk Management (EHRS) department conducts regular inspections to ensure compliance and supports Principal Investigators in maintaining a safe environment. In emergencies, EHRS Response Teams are available to handle hazardous spills and fires.

Safety officers play a vital role in promoting a unified approach to safety management within an organization. They offer expert guidance to all staff, making it essential for laboratories to appoint at least one dedicated safety officer. This individual must have extensive knowledge and accountability, along with the authority to design and implement a strong safety strategy, ensuring compliance across the institution.

Major organizations are mandated by law to create an **environmental health and safety office**, which is directed by experts possessing relevant qualifications in areas such as chemical safety, occupational medicine, toxicology, fire safety, and engineering. This office plays an essential role in developing laboratory policies and promoting safety protocols. Its responsibilities encompass managing hazardous waste, performing audits, ensuring adherence to regulations, investigating accidents, coordinating emergency response efforts, keeping detailed records, and delivering training programs.

3.4.2.1 Students and Laboratory Staff

It is the responsibility of the learners and other lab staff to ensure that they carry out their work in a responsible manner. Even though students and other workers are greatly influenced by authority and guidance by lab managers, their behavior and conduct greatly impact their safety and security in the lab.

3.4.3 Laboratory Safety Culture

Safety culture refers to the collective behaviors and attitudes exhibited by individuals and groups toward the safety priorities within an organization. A robust safety culture plays an essential role in ensuring laboratory safety, reflecting a deep commitment that goes beyond mere adherence to regulations. It facilitates the implementation of safety protocols, improves comprehension of hazards, and helps create secure environments. Solely depending on regulations may limit safety performance to the abilities of individual workers (Edward Steward, 2018).

A laboratory safety culture is characterized by a strong focus on safety during various operations, including ergonomic practices and proper equipment usage. However, cultivating such a culture poses significant challenges. Numerous organizations may claim to have a safety culture, yet this often fails to manifest in actual safe practices, thereby jeopardizing safety initiatives. A dynamic safety culture is propelled by effective leadership that encourages desirable safety behaviors and actively engages staff in managing safety protocols. Employees become aware of the benefits of reducing personal hazards and undergo training in recognizing and controlling potential risks. Conversely, unsafe behaviors may develop due to the existing operational framework, resulting in entrenched habits, as individuals alter their actions in pursuit of enhanced efficiency and productivity. These encompass the following:

a. Standardization of risk exposure levels
b. Drive for operational efficiency eliminating protective measures, resulting in narrow profit margins
c. Lack of awareness of potential risks
d. Failure to follow established protocols and guidelines
e. Inadequate mechanisms for safety-related feedback
f. Rigid adherence to regulations or blind compliance with established rules and procedures.

Redirecting attention can result in lapses in behavior. A robust safety culture within the laboratory cultivates a mindset that prioritizes safety, influencing every task and operation conducted (Samaranayake, Nishadya, & Jayasundara, 2022).

3.4.4 Ergonomics and Ethics in Lab Operations

Recent advancements in technology have significantly enhanced the use of tools and equipment in the workplace, leading to a growing need for ergonomically designed laboratories (AlShammari, Alhussain, & Rizk, 2021). The incorporation of ergonomic principles into laboratory equipment plays a crucial role in reducing the risk of repetitive strain injuries and promoting the overall health of personnel. When equipment is poorly designed, it can result in stress, fatigue, and discomfort in the musculoskeletal system, ultimately impacting productivity. Various industries recognize the significance of these principles and implement them to improve employee health and safety, underscoring their essential role in laboratory settings.

The field of ethics within science and technology outlines a framework of moral principles that prioritize safety and security within laboratory settings. This framework underscores the obligation to safeguard public health, preserve the environment, and uphold integrity. Consequently, the establishment of safety standards becomes crucial for ensuring the well-being of individuals, enhancing community welfare, and protecting intellectual property rights (Samaranayake, Nishadya, & Jayasundara, 2022). They protect valuable resources and the surrounding community by encouraging positive actions and minimizing potential harm. The implementation of stringent scientific regulations is vital, as secure laboratories limit access to only responsible researchers, thereby ensuring both confidentiality and fairness (AlShammari, Alhussain, & Rizk, 2021).

3.4.5 Some Laboratory Safety Case Studies

A laboratory safety case study has explored how theoretical frameworks can be linked to real-world practices, illustrating both the successes and limitations of existing regulations and risk assessments in the effort to prevent injuries, illnesses, and threats. An examination of laboratory incidents revealed that improper storage and handling of materials frequently lead to accidents, besides violations of established protocols. Furthermore, a significant contributor to explosion-related incidents is insufficient expertise and training among personnel. Alarmingly, 9% of those surveyed did not know about appropriate actions to take in the event of a fire or explosion, underscoring the urgent need for comprehensive emergency preparedness. The widespread existence of safety risks is attributed to a deficiency in safety awareness and a failure to cultivate a proactive safety culture. Fostering such a culture within laboratories proves to be problematic, as the principle of academic freedom is often invoked to oppose the enforcement of stringent safety measures (Ma et al., 2021).

Risk management assessments emphasize that each individual bears a responsibility for their own safety as well as the safety of those around them. The process of identifying potential risks can be approached through a variety of methods, including system mapping, brainstorming sessions, analyzing prior incidents, and consulting with industry experts. Tools such as Failure Modes and Effects Analysis (FMEA) are instrumental in evaluating both the likelihood and severity of potential hazards, thereby uncovering areas where safety may be compromised. However, FMEA is not without its drawbacks: it may be influenced by subjective interpretations, can require extensive labor and specialized expertise, and might overlook certain risks due to the biases of those conducting the evaluations. Additionally, educational laboratories play a crucial role in fostering problem-solving skills, enhancing critical thinking abilities, and raising safety awareness, thereby underscoring the importance of implementing rigorous safety protocols. A comprehensive strategy for managing risks and clearly delineating responsibilities is essential for nurturing a robust safety culture (AlShammari, Alhussain, & Rizk, 2021).

3.5 LABORATORY SAFETY AND SECURITY MANAGEMENT PROTOCOL

The most important aspect of establishing an effective safety and security program largely depends on the support and commitment it gets from the top managers of the institution. If leadership of the institution can be able to facilitate the establishment of safety and security plan, and ensure that they are responsible and accountable, then compliance cannot be a problem. The institutional management should ensure the following:

a. *Establishment of a laboratory safety and security oversight committee, with a chemical safety officer*

   The top institutional managers such as those of universities or institutes, and directors of research agencies should establish an oversight committee responsible for lab safety and security.

b. *Communication of a formal policy statement to all members of staff establishing organizational expectations and intentions*

   The policy statement should clearly stipulate how the institution will

   i. prevent human and economic losses that may arise from accidents, severe occupational exposures, and environmental calamities;
   ii. build safety and security considerations into all spheres of operations;
   iii. achieve and maintain strict compliance with existing laws and regulations; and
   iv. continuously improve performance.

c. *Implementation of administrative processes and controls for measuring performance*

   Administrative controls and processes refer to an institution's defined rules and procedures that concern safe and secure practices. They also establish the core mandates and responsibilities of the people involved. Security officers should develop general safety and security rules, lab maintenance procedures, materials and equipment manuals, and other documents to communicate expectations and rules to all lab workers. The protocol should give clear definition of individual responsibilities of learners and other lab personnel, lab managers, institutional managers, emergency service providers, and lab visitors.

   The work of carrying out experiments is quite dynamic and therefore prone to change. Changes in regulations, attrition in students and workers, and evolution of technologies should be considered. Evaluation of the safety and security of lab operations should be part and parcel of everyday operations. Administrative controls and processes should provide logical mechanisms for responding and managing change that may emerge from use of new technologies, procedures, legal obligations, workers, and institutional changes. The managers of an institution should from time to time review formal policy and systems of management on a regular basis to fit the emerging trends.

d *Identification and addressing each hazardous situation*

All stakeholders of an institution should take active roles in ensuring the safety and security of the labs. A status review should be carried out to assess the adequacy, scope, and implementation of safety and security procedures. This is important in the provision of a foundation for having a comprehensive safety and security program as well as setting priorities for the improvement of safety and security. A risk-based assessment should be done to determine the preparedness and adequacy of existing control measures so as to set priorities on the existing and emerging needs. Corrective actions should be incorporated based on available resources and significance.

The role of physical access control in improving safety and security of lab and workers should be addressed specifically. This calls for development of a detailed security policy setting and vulnerability assessment.

e *Establishment of procedures for management and handling of equipment and chemicals*

Management of chemicals and equipment is a critical component of a lab procedure. Safety and security should be integrated into the whole process of chemicals and equipment. This includes procurement, inventory and storage, daily operations and handling, transport, and waste disposal. It should include procedures for screening of chemicals and equipment as part of the normal operations of procurement. An inventory process should be put in place to follow chemical and equipment use until they are completely used and disposed. For example, when equipment and chemicals are procured and received, their identities and quantities are detailed down in the inventory system. The recordkeeping and inventory systems are very important for a lab in several ways:

i. Ensuring the security of equipment and chemicals through accountability of their usage
ii. Providing a resource for consultation on the possible sharing of equipment and chemicals
iii. Providing information that enables lab managers to know exactly when to reorder equipment and chemicals
iv. Providing the exact location of hazards in the labs to enable quick response in case of emergency
v. Enabling determination of future needs of chemicals and equipment
vi. Reducing excess chemical and equipment waste.

3.5.1 Personal Protective Equipment and Engineering Measures

Engineering controls such as gloves, lab hoods, and exhaust ventilation are the primary methods for ensuring that potential hazards are controlled in the chemical lab. In the event these safety measures fail to mitigate hazards, Personal Protective Equipment (PPE) become essential supplementary protection for laboratory personnel (Lestari & Pratama, 2023). It is important for the lab management to ensure that no experiment is carried out without proper hazard control measures.

The PPE includes four categories for safety and health. The first is respiratory protection, which safeguards air passages from harmful substances. The second provides eye and face protection against hazards. Third, hand protection comes from gloves that defend the skin. Lastly, skin protection is vital to shield against harmful materials in risky situations. Respirators can be classified into two main types: air-purifying devices and supplied-air systems. Air-purifying respirators (Figure 3.1) come in both snug-fitting and loose-fitting styles, which include filtering face pieces, elastomeric respirators, and powered air-purifying respirators (PAPRs). On the other hand, supplied-air respirators (Figure 3.2) deliver purified air from external sources or tanks. This category includes snug-fitting air-line pressure-demand systems and self-contained breathing apparatus, as well as loose-fitting alternatives like hoods and helmets (NIOSH, 2019).

(a). Half mask / Dustmask
Assigned Protection Factor (APF) = 10

(b). Half mask (Elastomeric)
AFP = 10

(c). Full facepiece (Elastomeric)
AFP = 50

FIGURE 3.1 Air-purifying respirators (PAPR), which remove contaminants from the air. (*Source of Illustrations* – OSHA 3352-02 (2009).)

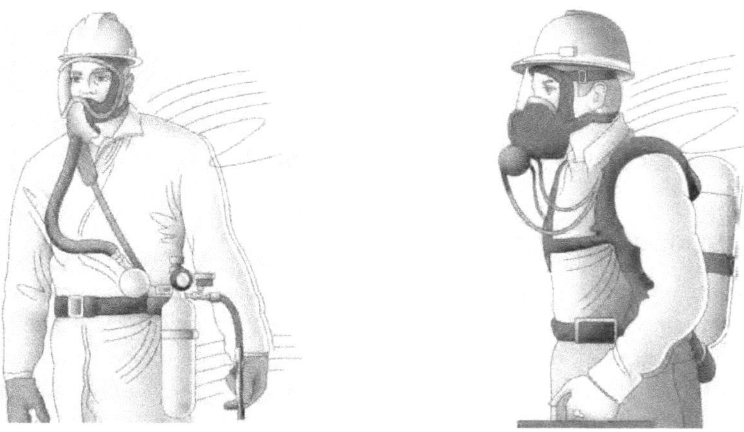

(a). Full Facepiece Supplied-Air Respirator (SAR) with an auxiliary Escape Bottle
Pressure demand mode is APF=1,000

(b). Full Facepiece Self-Contained Breathing Apparatus (SCBA)
Pressure demand mode is APF=10,000

FIGURE 3.2 Supplied-air respirators (SARs), which provide clean air from an uncontaminated source. (*Source of Illustrations* – OSHA 3352-02 (2009).)

3.6 PROTOCOL FOR EMERGENCY RESPONSE

Laboratories and scientific workshops are inherently risky environments, which can cause serious accidents that endanger people, destroy property, and create environmental problems. This would be more catastrophic if these were institutional environments which are prone to dense populations. In such emergencies, timely intervention is crucial to minimize escalation and avert additional damage. Well-defined protocols are in place to manage a variety of incidents, including chemical leaks, power failures, medical emergencies, accidents, and the evacuation of staff (AlShammari, Alhussain, & Rizk, 2021).

Emergency protocols are essential for laboratory safety, and in protecting lives and assets. Swift recognition and resolution of hazards significantly reduce potential damage and financial losses.

For example, in a small chemical spill, organized procedures allow for quick containment, preventing further dissemination of hazardous materials and enhancing safety. These protocols arise from assessed risks identified through evaluations, emphasizing the need to define roles and responsibilities for all personnel involved in emergency responses (Shu, Li, & Gao, 2023).

Laboratories present inherent safety and health hazards, underscoring the importance of rapidly contacting emergency services when significant injuries occur. It is essential for laboratory personnel to receive comprehensive training in emergency protocols related to hazardous materials and equipment. The presence of well-marked emergency exits and a reliable alarm system is crucial to facilitate prompt evacuations. The integration of advanced technologies, such as the Internet of Things and cloud computing, can greatly enhance the efficacy of emergency responses. By employing continuous monitoring within the lab through an array of sensors, data is consistently gathered regarding temperature, humidity, gas concentrations, and the presence of personnel. This collected data is subsequently analyzed using algorithms to establish a systematic security workflow. Further, a fuzzy uncertainty measurement model bolsters real-time surveillance and decision-making processes, improving threat identification and allowing for rapid emergency interventions (Shu, Li, & Gao, 2023).

In the event of a chemical spill, the area should be cleared and a supervisor consulted before starting cleanup. For minor spills, protective gloves and clothing are essential. Detailed procedures are in designated manuals. The chemicals and spill size per hygiene regulations should be assessed before coordinating safety personnel on necessary actions (Anza et al., 2016).

Laboratories should therefore have comprehensive plans to handle and take care of emergencies and implement the plans by timely purchasing and proper maintenance of emergency supplies and equipment such as eye washes, fire extinguishers, sill kits, and safety showers. Emergency response preparedness team should involve government emergency response teams such as fire departments to ensure that they have the right equipment to use in case of an emergency.

3.6.1 Training, Communication, and Mentorship

Establishing training focused on laboratory safety plays a vital role in accident prevention. It is imperative that all staff members undergo education concerning broad safety guidelines as well as the unique risks associated with their work environments. Prior to commencing their respective duties, individuals should successfully complete comprehensive training sessions. Maintaining and continuously enhancing a robust safety program for the trainers is of utmost importance. Additionally, regular refresher courses are advisable to strengthen understanding and adherence to safety practices.

The laboratory staff is required to participate in comprehensive training focused on critical safety procedures. This includes learning the steps to take in the event of an accident, familiarization with the locations of fire extinguishers and alarms, strategies for managing chemical spills, identification of first aid kit placements, and understanding the designated evacuation routes. It is important that no one attempts to handle hazardous chemicals without assistance, and the training should encompass the correct methodologies for handling gas cylinders and valves. Following these guidelines is essential in reducing the potential risks associated with accidents and emergencies.

A comprehensive process for the management of the lab would automatically lead to responsible management of the safety and security aspects of chemicals and equipment. The safety officer should ensure that proper guidelines and processes are established and timely communicated to lab stakeholders. It should, however, be noted that it takes total commitment by management of an institution to design the best safety and security procedures and to establish a protocol that guarantees the safety of lab staff and the general public.

3.6.2 Handling and Management of Chemicals

Chemical management involves the procurement, transportation, manipulation, storage, usage, and disposal of the same. Always consult safety data sheets (SDS) for toxicity, flammability, and hazards before handling. One should order only necessary quantities for experiments and clearly label hazardous containers with chemical names and warnings, using chemical-resistant materials that match SDS guidelines to ensure safety.

Laboratory managers should adopt a thorough approach for authorizing chemical procurement, including online purchases, credit card transactions, and donations. Chief security officers must be informed of all incoming orders, requiring prior approval for new chemicals. They are tasked with supervising the receipt of chemicals and maintaining detailed accountability records from procurement to disposal. Chemicals must be securely stored for accessibility while preventing unauthorized changes. If a centralized log is insufficient, individual laboratories should maintain distinct logs. Proper documentation and compliance are crucial; failure to adhere could compromise safety and security objectives. Each chemical must be clearly labeled to prevent misuse, with routine evaluations by the chief security officer ensuring protocol compliance.

3.6.2.1 Chemical Storage in the Laboratory

Laboratory chemicals should be safely and securely stored based on potential risks and hazards. Accessing chemical storage facility should be only for approved members of staff and authorized by the safety officer and lab manager. Identities of authorized persons should be available at the storage facility. It is also very important to maintain a high level of physical security, for example doors and boxes locks should be thoroughly checked. Laboratory managers should ensure that they regularly remind workers that chemicals should remain secure. They should ensure that those chemicals that have no immediate use are safely locked in the storage.

3.6.2.2 Safe Use of Laboratory Chemicals

There should be established protocols on how to manage the safe and secure use of lab chemicals and other hazardous materials. The critical aspects of safe work with lab chemicals, which should be known to lab workers, provide an important checklist to inform usage of any potential hazardous chemical. This protocol may be made available to each lab staff. The first important step to ensure safety is guaranteed in the lab is to carry out hazard evaluations in order to understand the potential risks being posed by the chemicals in use. Upon evaluation and determination of risks, selection of the best available measures to manage and control them is done. Such measures include development of safe operating procedures and specific safety plan for the specific intended use of every chemical.

It is of great value to ensure that all personnel mandated to handle chemicals are deeply knowledgeable about the potential risks and safe handling procedures. It is also important for the lab manager to ensure that lab workers and learners adhere to established safety and security procedures during any work. It is therefore important that the manager communicates safety procedures to all lab users, and a suitable channel for doing that should be developed. Additional training may be needed for the personnel using chemicals that may be more hazardous.

Incidents or events during handling of chemicals in the lab are unexpected, so a system that calls for and encourages timely reporting of all concerns, problems, or incidents should be established. The system should allow continuous improvement and reviews of the existing systems and processes.

3.6.3 Hazard Control and Management Evaluation

In laboratory experiments, managing potential hazards is vital, primarily through engineering controls such as laboratory hoods, local exhaust systems, and glove boxes for weighing chemicals. For radioactive materials, personal monitoring devices can help prevent hazardous substances from escaping secure areas. Additionally, designing smaller experiments allows managers to oversee hazardous operations more effectively, reducing the chemicals involved and significantly lowering associated risks.

3.6.3.1 Hazard Evaluation

Chemicals are a major cause of lab accidents and about 50% of hazardous chemicals involved have dangerous properties, including flammability, carcinogenicity, toxicity, and explosiveness (Anza et al., 2016). The potential risks posed by use of any lab chemical greatly depend on the quantity in use;

hence, the larger the quantity of a chemical being used, the more the potentiality of risk. Consequently, reduction of the quantity of chemicals in use reduces the seriousness of the danger. The safety officer should evaluate the magnitudes of the hazards posed by various chemicals when stored or in use. The evaluation of hazard should consider the possible routes of exposure such as inhalation, skin contact, eye contact, ingestion, and injection. Most of these possible routes of exposure may be hugely minimized by prohibiting eating and drinking within the facility.

Evaluation of hazard should also take into account health and physical hazards. Health hazards include highly toxic chemicals that may cause chronic toxicity and reproductive toxicants. Laboratory managers should develop specific handling procedures and practices for lab health hazards. Physical hazards encompass explosive or flammable chemicals, high-pressure equipment and systems, hot devices, corrosive materials, and radioactive materials. Chemical reactions that are highly exothermic may lead to explosions if no control measures are in place.

Although the principal methods of safety control are majorly engineering methods, PPE should be used to supplement them. Face and eye protection are important in any chemical lab. Protective gloves are essential if chemicals may come into contact with skin. The selection of gloves depends on the type of chemicals being used since a specified type cannot be used for protection against all potential hazards. Laboratory aprons and coats provide additional protection by preventing exposure of the body to potential chemical hazards. Some labs may also require protective shoes.

Depending on the size and type of the lab, respirators should be the last line of help for controlling a hazard and preventing exposure. They can only be used when other methods of control have been deemed not fit and are also found to be inadequate. Use of respirators should strictly follow guidelines and any personnel using them should be made aware of their shortcomings (NIOSH, 2019). It is also prudent that all personnel receive training on the use of respiratory protection to avoid confusion and poor use that may bring more harm than good.

If the lab manager certifies that the available hazard control measures are not sufficient to prevent hazardous exposure and provide a high level of safety, then no experiment or procedure should be allowed to be done. In such a case, a safe substitute for chemicals can be employed.

3.6.3.2 Reporting and Investigation of Incidents

Establishing a robust framework for investigating and reporting laboratory incidents is critical. This framework should encourage information exchange without consequences, promoting a secure environment for employees to express concerns. Ensuring safety for laboratory personnel is essential. An effective incident management strategy includes detailed reporting and investigations of accidents, hazards, and near-misses. A systematic approach enhances safety protocols and drives performance improvements. Thorough investigations are crucial for identifying root causes and implementing corrective measures. Laboratories must clearly communicate reporting criteria and procedures, ensuring incidents are reported within 24 hours and investigations begin immediately. The analytical process should be methodical, starting with a timeline of events and identifying immediate, contributory, and root causes, as well as latent factors leading to the incident (Cahyaningrum et al., 2019).

The analysis findings enhance laboratory operations by (1) adjusting risk evaluations and protocols, (2) identifying training needs and awareness initiatives, (3) improving equipment, and (4) adopting new technologies and methods.

3.6.3.3 Disposal of Laboratory Chemicals and Wastes

The whole life process of a lab chemical ends with its usage in a lab experiment or disposal after use. The safety officer is mandated to design a program that ensures safety and disposal of chemicals.

Before chemicals are procured, lab workers should ensure that disposal facilities are available. Protocols for handling lab waste from the time of procurement, use, and disposal should be developed. Laboratory managers should train lab workers on safe collection of waste and their packaging

The workers should be guided on waste management and storage. This should include lab storage of waste upon collection and prevention of accidental releases and spillage Waste containers should be kept sealed at all times when not in use.

There should be formulation of procedures for removal of lab chemical waste from the facility for storage in an area solely reserved for that purpose. The formulation should also include chemical waste entrepreneurs. A communication process whereby lab managers inform safety officers of the availability of wastes should be established. Guidance on recordkeeping of chemical waste should also be structured. Disposal records should have disposal date, method of disposal, and quantity disposed. Recordkeeping should be a continuous process and kept for a long period of time until such a time that the management feels satisfied with the specific waste disposal regulations.

3.6.4 Community Engagement and Partnership in Emergency Services

Comparing major disasters like the 2003 Canberra bushfires and the 2013 Chelyabinsk meteor event highlights similarities with lab incidents requiring external aid. Successful responses depend on careful coordination with emergency services, monitoring public and staff access, and fostering a positive safety culture (Rogers, 2013; Sharma & Hossain, 2019). Working with emergency services requires joint exercises to prepare laboratory staff and responders for handling contaminated patients and serious infectious diseases. Effective communication with stakeholders, such as police, fire departments, and public health organizations, is essential. Collaboration includes informing external personnel about access points, restricted zones, decontamination, evacuation, and ensuring laboratory access for post-incident investigations.

Instruction on laboratory safety plays a crucial role in raising public consciousness about the potential hazards associated with scientific activities. By involving the community in various initiatives, meaningful conversations about the advantages and drawbacks of scientific progress are encouraged, thereby enhancing trust between researchers and the general public. This well-informed perspective equips individuals to make more enlightened choices concerning the future trajectory of science and technology (Samaranayake, Nishadya, & Jayasundara, 2022).

3.7 EMERGING TRENDS IN LABORATORY SAFETY AND SECURITY

The detection of new patterns suggests a significant change taking place across various industries that warrants thorough examination. Innovations such as artificial intelligence and automation are fundamentally altering conventional methods. A growing preference among consumers for sustainable and ethical practices is driving organizations to adjust their operations. The increase in global connectivity opens doors for collaboration, emphasizing the necessity for businesses to be flexible and responsive. It is crucial for stakeholders to acknowledge these evolving trends in order to successfully maneuver through the complexities anticipated in the future.

The safety framework enhances laboratory procedures by incorporating advanced sensor technologies. Chemical sensors are employed to identify hazardous vapors in the environment, while flame detectors notify responders via cloud-based alerts. Surveillance cameras are strategically positioned to oversee access points, and remote devices are in place to dispense neutralizing agents when necessary. Additionally, smart lab coats serve as a warning system for staff, indicating instances of unauthorized access. This comprehensive approach not only improves hazard management and security but also aims to elevate overall safety awareness and risk management practices within the laboratory setting (Samaranayake, Nishadya, & Jayasundara, 2022).

A collection of chemical sensors and cameras works together to automate the monitoring of safety within the laboratory, all while conforming to the Common Alerting Protocol standard, which ensures interoperability. The Data eXchange Message (DXM) transport systems facilitate the exchange of information and improve communication in the event of spills. The outputs from the sensors utilize fuzzy logic software, allowing for more nuanced interpretation of data, while diverse technologies are employed to assess environmental parameters, ensuring a dependable monitoring process. Additionally, flash volumetric fire-extinguishing systems, operating with dry powder jets that reach up to 1.5 m, are in place to uphold safety in the lab environment (Ma et al., 2021).

The evolution of laboratory safety systems, enriched by diverse fields of expertise, lays the groundwork for emerging technologies. By integrating optimal methodologies and cutting-edge sensors, facilities enhance their safety measures, security protocols, and operational efficiency. An all-encompassing system manages dangerous substances, monitors environmental pollutants, facilitates fire detection, and regulates access, serving as a benchmark for future safety infrastructures. The fusion of regulatory compliance with a robust safety-oriented culture not only amplifies the protection of personnel but also fosters an environment ripe for innovation (AlShammari, Alhussain, & Rizk, 2021).

Competence in laboratory safety plays a critical role in safeguarding not just the employees but also the surrounding communities and the environment at large. Adherence to established laboratory practices and regulations is of utmost importance. Heightened safety awareness coupled with the ability to recognize hazards significantly contributes to ensuring safe operational procedures. In situations where risks are present, the use of PPE becomes indispensable. Moreover, the meticulous management of bio-hazardous waste is essential in shielding both healthcare professionals and patients from potential harm.

A thorough assessment of safety protocols (encompassing the identification, prediction, evaluation, and control of risks) is fundamental to achieving effective laboratory operations. The potential for serious repercussions from laboratory accidents underscores the necessity for rigorous risk analysis prior to any public investment in research endeavors. Implementing mitigation strategies can lead to reduced expenses, decreased dependence on external services, and an enhanced consciousness regarding safety in the management of biohazard waste.

3.8 LABORATORY CODES OF CONDUCT AND DRESSING

3.8.1 Laboratory Dressing Code

The importance of personal appearance must not be underestimated, as the image of the institution and its student body reflect the profession, in part. The dress code policy varies for classroom and labs, and this has impact, especially on the students that take science courses.

3.8.1.1 Guides for Laboratory Dressing Code

- Students and staff are expected to dress appropriately for the nature of the activity in which they are involved.
- Lab coats must be worn at all times for work in the lab. They should be long sleeved with a buttoned front and a name tag.
- White lab coats are the most preferred to be worn in any science labs. The Institution has provided its preferred designs for lab coats that are available from any lab technologist.
- Students **MUST** put on appropriate gloves for all procedures that may involve direct or accidental contact with blood, infectious materials, or infected animals. After use, gloves should be removed aseptically, and hands must then be washed.
- Masks should be put on while handling harmful gas substances to prevent possible inhalation.
- Flat shoes are required for both men and women; high heels are prohibited while entering any lab.
- Long hair or loose clothes, **MUST** be tied back or confined.
- It is prohibited to wear protective lab clothing outside of the lab, especially in public places, for example in canteens, coffee rooms, libraries, staff rooms, and toilets.
- Hands must be washed before leaving the lab and, again, before eating.

3.8.2 Laboratory Code of Conduct

A lab is a room, building, or area in a building equipped with apparatus, equipment, chemicals, or test animals, and used for research, quality control, performance tests, experiments, measurements, photographic development, or the preparation of drugs or other products in the natural sciences. For example, areas used for clinical, optical, dental, veterinary, electronic, nondestructive, and pharmaceutical testing are all considered to be labs. Working in a lab usually involves working with various chemical, physical, and biological hazards that may pose a health hazard or danger to both students and staff if not handled properly. To ensure a safe and healthy environment in our labs, the following Laboratory Code of Conduct should be adhered to.

3.8.2.1 General Guides

- Before accessing the lab, the lab users **MUST** read the biohazard warning signs so as to be aware of the safety measures that are in place for any case of danger.
- Users should behave in a mature and responsible manner at all times in the lab. All inappropriate behavior is specially prohibited.
- Users **MUST** follow all verbal and written instructions carefully. If one is unsure of any procedure, then assistance should be sought from the lab technologist-in-charge.
- Users **MUST** read through the procedures of each lab activity before starting any lab work so that they are familiar with them.
- Students are required to obtain signatures from the technologist-in-charge on their lab books at the end of each lab session.
- Students should not touch any equipment or chemicals unless specifically instructed or permitted to do so.
- Users **MUST** not eat, drink, apply cosmetics, or chew gum in the lab.
- Students **MUST** perform only those experiments authorized by the lecturer for the session.
- Laboratory users are not allowed to store food or drinks in a lab working area.
- Wearing an iPod, Bluetooth, or any other device that interferes with hearing is prohibited.
- Laboratory users **MUST** know how to locate and use all safety equipment in the lab, including the fume hood, emergency shower, first aid kit, fire blanket, fire extinguisher, and eyewash. They should also be able to locate the nearest exit in case of an emergency.
- Users should not pick up broken glass with their hands. Broken glass should be swept with a broom and disposed of in a container labeled "For glass disposal".
- The "No smoking", "No eating", and "No drinking" signs are displayed clearly inside and outside the lab doors. Therefore, lab users should ensure they have carefully read, understood, and followed those instructions.
- Glassware **MUST** be washed with hot, soapy water and scrubbed with the appropriate type and sized brush, rinsed, dried, and returned to its original location after use.
- Users **MUST** clean up their equipment and work area before leaving the lab.
- Users **MUST** wash hands thoroughly after participating in any lab activities.

3.8.2.2 Guides for Handling Chemicals and Equipment

- Laboratory users **MUST** never enter the science storage rooms or preparation areas unless accompanied by a lab technologist.
- Students **MUST** wear **Institution**-approved eye protection whenever chemicals, heat, or glassware are used by either the lecturer, technologist, or students in the lab.
- Laboratory users should wear appropriate personal apparel at all times in the lab and avoid wearing loose or flammable clothing; long hair should be tied back.
- Users **MUST** check the labels of lab reagents, and double-check to be sure that they are using the correct chemical for the experiment.
- Students **MUST** report any incident (including all spills, breakages, or other releases of hazardous materials) to the instructor or technologist immediately; no matter how

insignificant it may appear. This should include all injuries such as cuts, burns, or other signs of physical harm.

- Students **MUST** never remove chemicals, equipment, or supplies from the lab area.
- Laboratory users should keep their hands away from their faces when working with chemicals.
- Laboratory users **MUST** carefully examine all equipment before each use and report any broken or defective equipment to the technologist immediately.
- Users **MUST** properly dispose of all chemical waste as directed after completion of any practical.
- Laboratory users should be familiar with the hazards of the chemicals being used and know the safety precautions and emergency procedures before undertaking any work.
- Laboratory users **MUST** use volatile and flammable compounds only in a fume hood. Procedures that produce aerosols should also be performed in a hood to prevent inhalation of hazardous material.
- One **MUST** never "smell" a solvent!! It is a requirement to read the label on the solvent bottle to identify its contents.
- Users **MUST** never return used reagents into their original containers. Adding a chemical to its original container can contaminate the entire container. As much as possible, users should take only what they need since chemicals are scarce and expensive resources and should be used economically. If a user has more than needed, then the excess should be disposed of in the correct waste container. Technologists will advise.
- Bottles containing corrosive liquids (e.g., concentrated acids, bromine) should never be carried by their necks. There are baskets specially made for that purpose.
- Hazardous materials should never be poured or disposed of into sinks. Labeled residual bottles should be used and kept in the fume cupboard.

3.8.2.3 Guidelines to Follow While Heating Substances

- Laboratory users **MUST** never reach over an exposed flame or hot plate or leave a flame or hot plate unattended.
- Laboratory users **MUST** never point a test tube or reaction vessel of any type toward another person or themselves.
- Users **MUST** make sure that all Bunsen burners, gas outlets, and water faucets are turned off before leaving the lab.
- To avoid burns, users **MUST** not handle heated glassware or materials directly. Tongs, test-tube holders, or heat-resistant gloves or mitts are available for this purpose.
- There is glassware meant to be used for heating purposes. Laboratory users are **NOT** allowed to use any other type of glassware for heating.
- When heating flasks or beakers over the lab burner, a ring-stand setup with a square of wire gauze should be used.
- Liquids should be heated using a water bath.
- When heating with a lab burner, the test tube should be moved gently over the hottest part of the flame.
- Hot liquids **MUST NOT** be poured into plastic containers.

4 Comprehensive Laboratory Safety Guides

Apollo Odhiambo Maima and Sarah A. Otieno

4.1 INTRODUCTION

Safety is a collective responsibility for which all lab users must fully cooperate. Nonetheless, the ultimate responsibility for safety rests with the individual carrying out a prescribed lab procedure. That person is often the student, in the case of a teaching lab, or a researcher.

Most accidents occur due to carelessness, indifferent attitudes, failure to use common sense, or failure to follow instructions, on the part of the users. Each lab user should be alert to the environment, particularly to what the other users are doing, because a single mistake by one user can be a disaster, and harmful to all the other users. A user must never shy away from pointing out to other users that they are engaging in unsafe practices or operations. Better still, this should be reported to the Lab Instructor or Technologist. Whichever way one looks at it, users will be the worst off in an unsafe lab environment. This bestows upon them the greatest responsibility to ensure their own personal safety.

A range of guidelines exist world over aimed at improving best practices in educational and research laboratories. Regulatory bodies like the National Institutes of Health (NIH) establish safety standards for government-funded institutions. On the other hand, organizations such as the National Research Council, the Organization for Economic Co-operation and Development (OECD), the Council of Scientific Society Presidents (CSSP), and the National Science Teachers Association (NSTA) have produced laboratory safety manuals. However, many existing rules often neglect the essential attitudes needed to maintain a safe work environment.

Chemical hazards can be categorized by how they enter the body: inhalation, skin contact, fire or explosion, ingestion, chemical interaction, and compressed gas hazards. Biological hazards involve handling specimens, with pathogens transmitted via punctures, contact, ingestion, and inhalation. Biosafety levels I–IV manage risks based on the agent's inherent risk level (Bhusnure et al., 2018).

Physical hazards encompass excessive noise, radiation, laser light, extreme temperatures, and pressure variations. Safety glasses are insufficient for radiation; welding helmets, face shields, goggles, and specialized shields offer better protection (AlShammari, Alhussain, & Rizk, 2021). Heating devices like alcohol lamps and Bunsen burners pose fire and chemical hazards from solids, liquids, vapors, or gases. Proper handling and hazard assessments are crucial to mitigate chemical risks due to toxicity, corrosivity, reactivity, and flammability during mixing, dispensing, heating, or storage.

This guide provides the fundamental practices that will significantly decrease chances of accidents and health problems associated with experimental lab work. It also provides, where possible, the ideas and concepts that underlie the practical suggestions. However, the user is expected to become involved and to contribute to the overall solutions.

4.1.1 Biological, Chemical, and Physical Hazards

Chemical laboratory environments contain volatile substances, flammable materials, high-pressure equipment, and other hazardous agents. These portend risks like inhalation of vapors, oral exposure, and dermal contact, as well as dangers from fire and explosions. Adhering to safety protocols is essential to mitigate these hazards (Bhusnure et al., 2018).

Laboratory safety summaries are essential for chemicals in labs, providing specific safety guidance beyond material safety data sheets (MSDSs).

Biological hazards threaten human health via exposure to microorganisms, causing infections or toxic emissions. Laboratory staff manage biological substances from human or animal tissues and cultured microorganisms.

A systematic approach with protocols mitigates biological risks from infectious agents. Good laboratory practices (GLPs) reduce cross-contamination and limit personnel exposure. Procedures are tailored to microorganisms and detailed in safety manuals from organizations like the CDC and NIH, with the BMBL manual as a key reference (Ta, Gosa, & Nathanson, 2019).

Physical hazards can cause injuries like strains, sprains, bruises, cuts, fractures, and eye damage. Commonly, musculoskeletal injuries relate to high-force repetitive tasks and prolonged postures. In educational labs, risks increase from repetitive pipetting and improper tool use, particularly with percussion, heat, electrical systems, or solvents (AlShammari, Alhussain, & Rizk, 2021).

Mitigating biological, chemical, or physical hazards requires structured risk assessments, enhanced reduction strategies, and improved emergency preparedness. Many lab personnel lack proper training in lab safety and often ignore PPE. Regulatory noncompliance frequently arises during emergencies and unsafe practices. Handling flammable substances adds complexity, with significant concerns around economic issues, unawareness, and insufficient stakeholder commitment (Fatemi, Dehdashti, & Jannati, 2022).

4.2 GENERAL LABORATORY SAFETY GUIDES

Maintaining safety is essential in both laboratory settings and surrounding areas. An important aspect of laboratory safety is the integration of knowledge, practical judgment, and careful habits. It is crucial for laboratory personnel to be attuned to the considerable risks present, understand the possible issues that may arise, implement appropriate safety measures, and cultivate a responsible mindset focused on safety practices.

Whereas it is impossible to design lab safety rules that will eliminate all potential hazards and occurrences in the labs, accidents and injuries can be avoided by using good judgment and by following general safety guidelines and applicable prudent practices.

It is with this understanding that these procedures and requirements are established, to improve lab safety, avoid accidental release of hazardous chemicals or biological agents, and to minimize the risk of personnel exposure or environmental contamination. Staff and students who work and/or learn in the labs are required to comply with these guidelines. All lab users must consequently take personal responsibility for ensuring their own safety and the safety of other lab users by:

- Taking the action(s) necessary to minimize, or eliminate, any hazards over which they have control;
- Complying with safety instructions, policies, and procedures while in the labs;
- Completing the appropriate lab safety induction and training to enable them to undertake their work safely;
- Making proper use of all safety devices and PPE;
- Complying with the instructions given by emergency response personnel such as emergency wardens and first aiders;
- Not willfully placing at risk the health and safety of any other person;
- Seeking information or advice where necessary before carrying out new or unfamiliar work including experiments;
- Maintaining high standards of cleanliness while working in the labs;
- All lab users must wear clean, tidy, and fitting lab coats and be in closed, low shoes while in the lab at all times. Those with long hair must clip them at the back;

- No food or drink should be allowed within the precincts of the labs except in the designated areas;
- All lab users must be familiar with emergency and evacuation procedures and the location of, and if appropriately trained on, emergency equipment, reporting all incidents, hazards, and "near miss" incidents to the Technologist in-charge;
- Clearance must be got from the Lab Technologist in-charge of the lab at all times to access and use it;
- No one is allowed to perform unauthorized experiments in the lab.

The general lab safety guides follow next (with a synopsis given as Appendix 1).

4.2.1 Housekeeping in the Laboratory

There is always a definite relationship between safety, performance, and orderliness in the lab. Laboratory users are individually and collectively responsible for maintaining the work area in a clean and orderly fashion. In the minimum, they should aspire to:

i. Keep work areas clean and free from obstructions. Hallways, corridors, aisles, and exit routes must **ALWAYS BE KEPT CLEAR** and free of obstructions. Emergency equipment, electrical panels, and such should not be blocked. Hallways or stairways **MUST NOT** be used for storage.
ii. Clean up after the completion of any operation or at the end of each day.
iii. Clean and dispose of spilled chemicals and clean up waste promptly and properly.
iv. Dispose of unneeded chemicals, empty containers, etc. Unneeded items **MUST NOT** be allowed to accumulate and create clutter.
v. Clean floors regularly. Accumulated dust, chromatography adsorbents, and other powdered chemicals and materials may pose respiratory hazards.
vi. Store equipment and chemicals properly to minimize clutter.

4.2.2 General Safety Measures

Laboratory users and personnel are always expected to:

1. Follow all safety instructions carefully.
2. Get thoroughly acquainted with the location and use of safety facilities such as safety showers, exits, and eyewash fountains.
3. Familiarize themselves with the chemical operations, hazards of the chemicals being used, and the relevant safety precautions and emergency procedures before undertaking any work.
4. Strictly **AVOID** eating, drinking, or smoking while in the lab.
5. Read labels carefully so as to pick only the intended chemical/item.
6. Strictly **AVOID** using any equipment unless they are trained and approved as a user by their supervisors or Technologists.
7. Wear safety glasses or face shields when working with hazardous materials and/or equipment.
8. Wear gloves when using any hazardous or toxic agents.
9. Wear gloves, lab coats, and safety shields or glasses when handling dangerous substances. Shorts and sandals **MUST NOT** be worn in the lab at any time. Coats should be hung in the changing room or placed in a locker.
10. Make sure any long hair or loose clothes are tied back or confined.

11. **Keep their work area clear** of all materials except those needed for their work. Extra books, purses, etc. should be kept away from equipment which require air flow or ventilation to prevent overheating.
12. **Dispose** of used material, if any, properly and in appropriate containers.
13. Report immediately to the Lab Manager or Technologist in-charge any equipment that fails while being used. They MUST never try to fix the problem themselves because they could harm themselves and other users.
14. Turn off all ignition sources and lock the doors if leaving a lab unattended.
15. Never pipette anything by mouth.
16. Clean up their work areas before leaving.
17. Wash hands before leaving the lab and before eating.

4.2.3 Electrical Safety

Laboratory users and personnel **MUST**:

1. Obtain permission before operating any high-voltage equipment.
2. Maintain unobstructed access to all electrical panels.
3. Refer wiring or other electrical modifications to the Lab Manager who will then liaise with the office of the Director in-charge of Lab Operations.
4. Avoid using extension cords whenever possible. If they must use one, they should obtain a heavy-duty one that is electrically grounded, with its own fuse, and install it safely. Extension cords must not go under doors, across aisles, be hung from the ceiling, or plugged into other extension cords.
5. Never modify, attach, or otherwise change any high-voltage equipment.
6. Always make sure all capacitors are discharged (using a grounded cable with an insulating handle) before touching high-voltage leads or the "inside" of any equipment even after it has been turned off. Capacitors can hold charge for many hours after the equipment has been turned off.
7. Use only one hand when adjusting any high-voltage equipment or a laser which is powered with a high voltage supply. The other hand is best placed in a pocket or behind the back. This procedure eliminates the possibility of an accident where high-voltage current flows up one arm, through the chest, and down the other arm.

4.2.4 Laser Safety

Laboratory users and personnel **MUST**:

1. Never, ever look into any Laser Beam, no matter how low power or "eye safe" it may look.
2. Always wear safety goggles if instructed to do so by the Instructor or Technologist. The most common injury when using lasers is an eye injury resulting from scattered laser light reflected off of mountings, sides of mirrors or from the "shiny" surface of an optical table. The best way to avoid these injuries is to always wear goggles.
3. Never lower their heads to the level of the Laser Beam! The laser beam should always be at or below chest level.
4. Always use "beam stops" to intercept laser beams. They **MUST** never allow them to propagate into the lab. No one should ever walk through a laser beam. Some laser beams of only a few watts can burn a hole through a shirt in only a few seconds.
5. Notify the Instructor or Technologist **IMMEDIATELY** on suspecting that they have suffered an eye injury! The ability to recover from an eye injury decreases the longer it takes to initiate treatment.

4.2.5 Laboratory Glassware

The integrity of testing and scientific validity depend on accurate laboratory measurements. Different glassware types offer varying accuracy, with some providing only estimative results. It is crucial to handle lab glassware carefully, using both hands and protective gear like gloves and goggles. When using tubing, employ towels or cotton gloves for a secure grip. If stoppers are stuck, hold the base and twist gently; water can help with removal. Some glassware is for liquid storage, not precise measurement. Maintenance is crucial; don't immerse hot glass in cold water to avoid fractures.

Laboratory glassware presents inherent risks when not handled with caution. The threat of injury from broken glass necessitates meticulous disposal practices. It is imperative for students and other lab users to adhere to established protocols and refrain from transporting glassware without supervision. Adequate labeling of all glassware is critical to avert potentially dangerous chemical interactions. Identifying incompatibilities, such as the degradation of polytetrafluoroethylene in the presence of ozone, can prove challenging. Consistently replacing compromised items is vital for the preservation of safety standards within the laboratory environment. Cracked or worn glassware can harbor bacteria and increase breakage risk, so using damaged glassware is unsafe.

Accidents involving glassware are a leading cause of lab injuries. Laboratory users should therefore always observe the following glassware safety protocols:

1. Carefully follow handling and storage procedures to avoid damaging glassware. Damaged items should be discarded or repaired and reported to the Technologist in-charge or the Lab Manager.
2. Use appropriate hand protection when inserting glass tubing into rubber stoppers or corks, or when placing rubber tubing on glass hose connections. Tubing should be fire polished or rounded and lubricated. Hands should be held close together to limit movement of glass should a fracture occur. Plastic or metal connectors (instead of glass connectors) should be used when possible.
3. Handle vacuum-jacketed glass apparatus with extreme care to prevent implosions. Equipment such as Dewar flasks should be taped or shielded.
4. Use heavy-walled, silica glassware (Pyrex, etc.) for work involving non-ambient pressure or temperature work.
5. Use hand protection when picking up broken glass (and use a brush to sweep small pieces into a dustpan).
6. Seek instruction in the use of glass equipment designed for specialized tasks (e.g., separator funnels containing volatile solvents can develop considerable pressure during use).

4.2.6 Emergency Response Protocol

Emergency protocols in clinical laboratories address fire incidents, chemical spills, and medical crises. Labs dealing with high-risk pathogens must create safe handling protocols for contaminated and uncontaminated specimens. Key measures during fires include securing samples, attempting to extinguish flames when possible, evacuating, and promptly notifying emergency services after evacuation.

Laboratory safety protocols are vital for managing chemical spills, stressing the importance of preparedness and adhering to procedures. Consistent use of PPE is essential. When a spill occurs, notifying designated personnel is crucial. If unsure, the main action is to notify staff, evacuate personnel from the area, and allow only emergency response teams to remain. Small spills can be addressed using proper gear and spill kits. Waste disposal must follow local regulations, and immediate medical assistance is necessary for injuries. If a spill causes skin or eye contact, identifying symptoms underscores the importance of following established safety protocols.

In emergencies, alert nearby individuals, start evacuation if needed, and move to safety. Know the locations of safety equipment like eyewash stations and fire extinguishers. Understand fire extinguisher classes: Class A for cloth/wood, Class B for flammable liquids, and Class C for electrical gear. Familiarity with fire alarms and exits improves response. Safe practices can prevent accidents and help in seeking assistance.

Medical emergencies involve incidents causing injury or requiring medical help. Initial steps include consulting healthcare professionals, notifying authorities, and filling out accident reports. It's important to discuss observations and evaluations from medical personnel thereafter. In brief, procedures for fire, chemical spills, and medical emergencies are vital for protecting individuals and property, and consistent enforcement is crucial.

It is imperative for laboratory personnel to comply with established emergency protocols and to consistently assess the efficacy of protective equipment. The availability of pre-assembled resources is essential for facilitating prompt reactions to emergencies. Continuous enhancements to procedures are made in light of cumulative experience. Furthermore, emergency information is strategically positioned at entrance points and throughout various lab areas to ensure easy accessibility (Iwen et al., 2018).

4.2.7 Use and Maintenance of Personal Protective Equipment

PPE is crucial in laboratories, minimizing exposure to hazardous agents like chemical, biological, physical, and radiological threats. PPE categories include eye protection, face shields, gloves, respiratory gear, and specialized clothing, each targeting specific risks (Raveh et al., 2022). Protective apparel, like laboratory coats, safeguards against contamination of personal clothing and skin. The addition of items such as aprons and sleeve guards depends on specific hazards. Choosing and using all PPE based on the risks is crucial for effective protection.

Laboratory safety guidelines necessitate the use of PPE, including lab coats, safety glasses, and gloves, to safeguard individuals from hazardous substances. It is imperative that gloves are compatible with the specific chemicals being handled, while safety eyewear must comply with the relevant Occupational Safety and Health Administration (OSHA) standards. Additionally, all clothing and gloves should conform to the regulations established by OSHA (Poller et al., 2018).

Grasping the principles of PPE maintenance is crucial for ensuring both its efficacy and user comfort. Individuals can significantly extend the lifespan and protective capabilities of PPE through straightforward maintenance tasks. It is advisable to routinely cleanse goggles, face shields, and masks using soapy water, followed by rinsing and drying to mitigate contamination risks. PPE should be stored in locations that are not frequently trafficked and away from direct sunlight, with regular inspections for any signs of wear or damage being imperative. Adherence to specific maintenance protocols, such as those required for respirators, is necessary. Prompt removal of any compromised PPE is critical, prioritizing timely replacements to maintain safety standards. Engaging with experts for best practice recommendations is also a prudent approach.

4.2.8 Vigilance and Vital Safety Numbers

Laboratory users **MUST** always be alert to, and immediately report, any unsafe conditions or activities to the Manager or Technologist in-charge. Each lab user or worker should remember that injuries can and do occur inside the lab or other work area. It is important that safety be practiced in offices, stairways, corridors, and other places. Safety is largely a matter of common sense, but a constant safety awareness of everyday hazards is vital. Required Vital Safety Numbers are as below:

4.2.9 Institution Emergency Numbers

1. Institution Security Main Gate (Control Room): ________________
2. Institution Health Center: ____

3. Institution Ambulance: ____
4. Institution Transport Office: ________________________________
5. Institution Health Center: ___________________________________

4.2.10 Hotlines Telephone Numbers

1. Private Fire Brigade
   Phone: Emergency Hotline:
2. County/City Fire Brigade:
3. Police Service:
4. Nearest Referral Hospital Emergency Department:

4.3 CHEMICAL LABORATORY SAFETY GUIDES

MSDSs are essential for laboratory safety with chemicals. They provide key information on safe usage, toxicity, reactivity, melting and boiling points, flash points, and reactivity. MSDS documentation helps assess hazards and establish exposure control measures for safe handling. Safety information sheets are vital for developing standard operating procedures (SOPs) for hazardous substances. They help reduce human error in laboratories. The MSDS offers crucial information for proper labeling of hazardous waste in on-site containers. Reviewing the MSDS is essential for occasional chemicals even without an SOP.

Laboratory chemicals can be toxic or corrosive, so safe storage, transportation, handling, and disposal are essential to prevent accidents. Proper labeling of all containers, including secondary ones, is vital. Labels should provide complete chemical names, receipt dates, concentrations, and hazard details clearly. Approved safety containers are necessary for transport; chemicals should stay in original containers unless they are transferred to properly labeled ones. Secondary containment can prevent spills, and labels must be intact, readable, and regularly checked, including assessments of chemical half-lives (Anza et al., 2016).

4.3.1 Chemical Safety

Chemicals can be very hazardous to humans because of their wide scope of possible chemical reactions, fumes, and emissions. Laboratory users are therefore advised to:

1. Treat every chemical as if it were hazardous.
2. Make sure all chemicals are clearly and currently labeled with the substance name, concentration, date, and name of the individual responsible.
3. Never return chemicals to reagent bottles (target withdrawing only the correct amount and share any excess).
4. Comply with fire regulations concerning storage quantities, types of approved containers and cabinets, proper labeling, etc. If uncertain about regulations, the Lab Technologist in-charge can clarify.
5. Use volatile and flammable compounds only in a fume hood. Procedures that produce aerosols should be performed in a hood to prevent inhalation of hazardous material.
6. Never allow a solvent to come into contact with the skin. Always gloves should be used.
7. Never "smell" a solvent!! The label on the solvent bottle should be read to identify its contents.
8. Dispose of waste and broken glassware in proper containers.
9. Clean up spills immediately.
10. Never store food in labs.

4.3.2 Chemical Labeling and Storage

Effective management of laboratory chemicals necessitates precise labeling and storage protocols. A comprehensive understanding of chemical incompatibility enhances overall safety in handling these materials. Each container must clearly display its contents, associated hazards, and pertinent contact information. It is imperative to implement a systematic storage approach based on hazard classification; for instance, flammable substances require storage in spark-resistant cabinets, while corrosive agents must be stored in isolation to mitigate safety risks, as guided below:

1. Chemicals should be stored by hazard class; hence, adequate separation of incompatible chemicals should be maintained.
2. Flammable solvents and strong acids or bases should be stored separately and in appropriate cabinets.
3. Compressed gas cylinders should be secured (by strap, chain, or cylinder stand).
4. All stored chemicals must be in appropriate containers, tightly sealed, properly labeled, and in good condition.
5. Flammable materials **MUST** not be stored in refrigerated equipment, unless the refrigerator is specifically designed for that purpose.

4.3.3 Handling and Disposal of Hazardous Waste

Effective management of hazardous materials in labs is crucial to reduce toxicity and flammability risks. Toxic or flammable substances should be stored in approved cabinets, away from workspaces. Harmful vapors require well-ventilated storage. Materials must be cool, dry, and clearly labeled to avoid misuse, and those prone to decomposition should have expiration dates for safety. Understanding proper protocols for managing and disposing of hazardous substances is crucial for laboratory safety. Safety guidelines cover labeling, storage, and Waste Management Data Sheets. Discarded reagents must adhere to Safety Data Sheet regulations. If allowed, dispose via sink; otherwise, use labeled containers with secure lids to prevent leakage (Emmert & ASM, 2013).

This guide outlines emergency protocols for fires, chemical spills, and medical issues. Workers must understand procedures and supplies for effective responses to hazardous materials as below:

1. Containers used for collection of hazardous waste must be in good condition, must not leak, and must remain closed except when waste is being added. Containers and contents must be compatible. Hazardous waste containers **MUST NOT** be overfilled (at least 10% headspace should be allowed for liquid expansion).
2. Users should triple rinse, deface the labels, and remove lids before disposing of empty containers.
3. Do not combine incompatible chemical waste.
4. Hazardous waste containers must be properly labeled. All components must be identified on the label (e.g., 10% ethanol, 10% xylene, and 80% water) and label as "Hazardous Waste".
5. Hazardous chemicals MUST NOT be disposed of down the drain, intentionally evaporated or disposed in the regular trash.
6. When a container is ready for disposal, a completed waste tag should be attached to the container (specifying location for storage of hazardous chemical waste awaiting disposal).

4.3.4 Response to Chemical Spills

Chemical exposures can cause nausea, vomiting, disorientation, and potentially death. While minor injuries are manageable, critical situations need urgent action. Lab staff should alert colleagues request assistance, and seek medical help promptly. Detailed incident information is vital, and regularly reviewing medical emergency protocols is crucial for all personnel. Acute illnesses from toxic chemical exposure can vary in severity and may need emergency care. Symptoms include nausea, headaches, and loss of consciousness, with some leading to death. Certain substances can show symptoms months or years later. Blood tests and urine analysis are essential for accurate diagnosis (Iwen et al., 2018).

Every reasonable Lab should have direct access to equipment and supplies to clean up the largest container of every type of hazardous chemical material in the work area. Each Institution must make singular effort to ensure that such cleaning is efficient and effective. But this also depends on the swiftness of response from the user(s). In the event of a ***small, contained spill***, the steps below should be followed:

1. If the substance is **dry and/or nonvolatile**, the hoods should be shut off, windows and doors closed, and rooms vacated. The door should then be labeled with appropriate warning. The aerosol should be allowed to settle for about 30 minutes before re-entering room.
2. If the substance is **volatile**, the ventilation should be left on and the room vacated. Then the door should be closed and a warning posted outside the door.
3. The Lab Manager should be notified.
4. Materials necessary for **decontamination** should be assembled and appropriate **protective clothing** (i.e., disposable lab coat, gloves, and splash goggles) donned. If respiratory protection is needed, the Manager or Technologist in-charge should help.
5. For a **liquid carcinogen or other hazardous chemical** spill, the spill should be wiped with absorbent material. All surfaces should be washed down with the decontaminating solution appropriate to the nature of the material.
6. For a **dry chemical or biological spill**, all surfaces should be washed down with an appropriate solvent to neutralize or deactivate and remove the substance.
7. All contaminated materials should be placed in impermeable containers and sealed. Then the protective clothing should be properly removed and bagged and appropriate disposal procedures followed.

4.4 USE AND SAFETY OF LABORATORY EQUIPMENT

4.4.1 Laboratory Safety Equipment

To ensure a safe laboratory and minimize injury risks, understanding safety protocols is essential. This section outlines the proper use of common lab equipment. For more details, consult manufacturers' guidelines and relevant academic sources (AlShammari, Alhussain, & Rizk, 2021). Laboratory equipment presents a range of hazards that necessitate careful evaluation. The risks include injuries resulting from glass breakage, insufficient support structures, pressure devices that may expel hot liquids, and tools that enhance the likelihood of burns. Additionally, ungrounded electrical circuits pose a risk of electric shocks. To ensure a safe working environment, it is important to either disconnect power sources or ensure appropriate grounding. Furthermore, adequate training is essential for the effective management of sophisticated equipment.

The design of bench-top laboratory instruments emphasizes both safety and usability across various states of matter, such as solids, liquids, and gases. This encompasses a range of devices, including flow injection systems and microplate instruments, which are engineered to enhance precision and improve throughput in laboratory settings.

4.4.1.1 Chemical Fume Hoods

Fume hoods are expansive enclosures featuring sliding glass doors or sash designed to safeguard laboratory personnel from exposure to detrimental fumes, gases, and unpleasant odors. They are equipped with an air duct linked to a fan that facilitates the upward movement of air, effectively eliminating hazardous vapors from the workspace.

Laboratory fume hoods effectively reduce exposure to hazardous substances by utilizing appropriate containment strategies and maintaining a specified face velocity for managing vapors and fumes (Emmert & ASM, 2013). Assessing fume hood functionality is essential for maintaining safety in labs (He, Lewis, & Guffey, 2017).

The efficacy of a laboratory fume hood in reducing contaminants relies on three factors: (1) its containment of hazardous substances, (2) air extraction speed, and (3) how contaminants mix with surrounding air. Safety depends on these elements' combined effects on ventilation. Prioritize containment due to its direct link to contaminant release, while release and dilution rates are secondary but crucial for overall safety. It is imperative that any experiments yielding toxic, flammable, or malodorous byproducts are carried out within these hoods. Additionally, deactivation of the fume hood system is restricted to authorized individuals. Prior to operation, it is crucial to verify that the fume hood is functioning optimally.

The sash should be adjusted on the lab side as needed for operations (Emmert & ASM, 2013). The optimal sash height of a fume hood should be kept minimal when not in use. Volatile substances must be securely stored in appropriate containers, avoiding large volumes for extended periods. Various construction designs exist for fume hoods, making it important to understand these distinctions before modifications. If a fume hood does not exhaust efficiently, a charcoal filter may be needed, but this requires frequent replacement. Consulting experts before making changes is advisable, and consistent evaluation of fume hoods' performance is crucial for risk mitigation, highlighting the need for clear operational guidelines for users. Some cautions to note when using the hood include the following:

1. Minimize storage of materials in the hood.
2. A fume hood **MUST NOT** be used for long-term storage of chemicals or equipment.
3. Blockage of vents or disruption of airflow of the hood is **NOT** acceptable.
4. The hood should always be left "**ON**" if volatile or toxic substances are stored in it or if needed to assure adequate general lab ventilation.
5. Equipment and other materials should be placed at least six inches behind the sash to assure proper airflow and to reduce the escape of chemical vapors into the lab.
6. Large equipment should be raised at least 1.5 inches off the work surface to allow air to flow underneath.
7. Loose papers, paper towels, or tissues should not be allowed to be drawn into the exhaust duct.
8. Objects (such as refrigerators or lab coats hanging on the manual controls) that might interfere with airflow **MUST NOT** be placed in front of a fume hood.
9. Any hood or exhaust system must not be installed, removed, relocated, or modified without prior approval.

4.4.1.2 Safety Showers and Eye Wash Stations

Safety showers and eye wash stations are located in each lab. It is advisable to locate and study these facilities carefully beforehand. Laboratory users should know how to operate the safety showers and eye wash equipment *before* an accident happens. The Lab Technologist will demonstrate how the facilities are used during induction.

The treatment for any chemicals which get on the body is to rinse the affected body area for at least 15 minutes under cold running water.

If chemicals are splashed into the face, they should not reach the eyes because users will be wearing safety goggles. If this sort of accident happens, *goggles should be left on* while the affected user is going to the eye wash station. There the face should be washed with the goggles still on until it is reasonable to believe that most of the chemical is gone from the face. Then the goggles can be removed and the face washed again.

If chemicals get into the eyes, the affected user should call out for help. Just in case the user cannot see, someone else should guide him/her to the eye wash station, where the eyes should be washed out thoroughly. The eyes should be blinked continuously and rapidly while washing to aid the flushing action of the water. If chemicals get onto the body, any contaminated clothing should be quickly removed and the body rinsed off in a safety shower.

4.4.1.3 Fire Extinguishers

Fire extinguishers are usually located at the entrance of each lab. Laboratory users are requested to study them carefully. It is notable that they are attached to the wall; hence, it is important to figure out what would have to be done to get them off of the wall. The instructions on the side of the fire extinguisher should be read carefully for familiarization with their use. If there will ever be need to use a fire extinguisher, the following should be remembered: "(A) pull the pin, (B) aim to the side at first, (C) depress the handle, (D) sweep the spray from side to side across the BASE of the fire (where the fire meets the fuel), not just at the flames! When the fire is out, clean up the area"!

Note: *A person should never be sprayed with a fire extinguisher. The chemicals in the fire extinguisher can be harmful.*

4.4.1.4 First Aid Kit

A first aid kit has been provided in each lab, on the shelves. It contains gauze squares, small, adhesive bandages and antibiotic ointment. Each user is expected to check the kit out before using the lab. If any injury occurs which cannot be handled with these supplies, then the injured user can be escorted to the Health Services office to receive treatment from the healthcare professionals there.

Note: *In an institutional lab, the Technologist should demonstrate to students how to use the adhesive bandage and ointment. However, the lab coordinator or instructor can arrange for individual instruction.*

4.4.1.5 Sink

While the sink is used for cleaning glassware and many other tasks, it is also a part of our safety equipment. If any user happens to get chemicals onto his/her hands or forearms, he/she must move quickly to the sink to rinse the chemicals off. The treatment for any chemicals which get on the body is to rinse the affected body area for **at least 15 minutes** under cold running water. When a user is in the lab and notices a **mysterious itch** on the arm which just won't go away, it is always safe to assume that it is a chemical on the skin and should be washed with soap and plenty of water.

4.4.1.6 Broom and Dust Pan

We use a lot of glassware in our labs, especially in the Chemistry Lab. Glassware usually ends up getting broken sometime during the semester or while using. When that happens, it is unsafe to pick up the broken glass with the hands. Instead, a broom and dustpan should be used to collect the broken glass and then disposed of in the specially marked container provided. Cleaning material shall be provided for wiping the bench top, and a big broom for cleaning the floor.

4.4.1.7 Gas Shutoff Valve

Each lab has a master valve which shuts off the gas supply to the entire lab. The Instructor or Lab Technologist should point it out during students' induction. This valve is especially useful in case of a fire in the lab. In case of a fire, *students as well as instructors are authorized to shut off the room's gas supply.*

4.4.1.8 Telephone

There is limited access, emergencies-only telephone in the Chemicals Preparation Room. This room is only accessible by Lab Instructors and staff. In case there is need to call for any help, and there is no staff around, a Lab Instructor (in the neighboring labs), professor, or staff member should help in calling the emergency number.

4.4.2 Laboratory Centrifuges

A laboratory centrifuge is vital in clinical, industrial, and research settings, with some models featuring refrigerated chambers for temperature control. Sample tubes are inserted into a rotor that spins them at high speeds, relying on sedimentation principles. Centripetal acceleration causes denser materials to separate from lighter ones. Below are useful operational guidelines for safe use of the centrifuge:

1. *Selection of suitable centrifuge tubes is critical*:
   Centrifuge tubes must be constructed from heavy wall glass or unplasticized polymers like polypropylene (PP), as materials like acrylonitrile–butadiene–styrene and polyethylene terephthalate degrade in solvents and become brittle at –20°C. The tubes must be made from materials suitable for the sample type and centrifugal forces: polycarbonate tubes can handle up to 12,000×*g*, polymethyl methacrylate up to 2,800×*g*, and PP up to 20,000×*g*. After filling, the tubes should be racked and sealed with leakproof caps to prevent aerosol contamination from leaks or breakage. Overfilling must also be avoided (Bauer & Pickels, 1936).
2. *Achieving equilibrium in the centrifugal tubes and rotor*:
   For optimal rotor performance, tube mass distribution must be symmetrical. Place sample tubes on opposite sides with equal mass tubes directly opposite. Avoid odd numbers of tubes to ensure balance.
3. *Centrifuge cleaning and maintenance*:
   Thoroughly clean the rotor after each use for optimal performance. Use sodium hypochlorite for chemical decontamination and appropriate detergents to remove organic residues (Setioningsih & Setiawan, 2025).
4. *Adhere to the centrifuge manufacturer's operating guidelines*:
   Confirm that the lid and covers are securely closed during operation. Use the correct rotor for the specific tubes and speed. Ensure the centrifuge is stopped before removing samples.
   The operation of a centrifuge could appear to be a safe activity; however, its rotational processes introduce potential hazards. Frequently, these devices are operated for extended durations, which increases the likelihood of accidents. Additionally, inadequate maintenance can lead to elevated air-line pressures, further compounding the risks associated with their use. Monitoring samples is now effectively addressed with an advanced tube holder for real-time observation (Hoang, Moskwa, & Halvorsen, 2018).
   Implementing stringent safety protocols is vital to reduce operational failures. Despite the availability of high-quality equipment, neglecting wear and improper assembly increases accident risks. Heavy bucket handling poses hazards, so using an articulated arm is recommended. Key failure factors include improper speed settings and unbalanced rotors (Bauer & Pickels, 1936).

4.4.3 Fire Prevention in the Laboratory

If a fire alarm sounds, the building must be evacuated!

There should be alarm switches in the lab. It is highly unlikely that there will be need to use them, especially if the labs are all equipped with several smoke detectors. However, lab users should locate them and learn how they work.

Note: *It is an offense to intentionally trigger false alarms!*

In a laboratory, fire poses a significant hazard, along with risks from corrosive chemicals, toxic vapors, intense UV radiation, and cryogenic operations. A "bad fire" poses significant risks to life and emits toxic substances. Clear signage and understanding emergency procedures are vital for effective evacuation from a fire. Avoid fighting fires alone; prioritize prompt evacuation by following responders' instructions (National Research Council, 2011).

It is important that all lab users remain vigilant against possible fire outbreaks in the lab due to acts of man. In that regard, the following rules should be observed when using flammable materials in lab operations:

1. Open flame MUST NEVER be used to heat a flammable liquid or for vacuum distillation.
2. Open flames should be used only when necessary but extinguished as soon as it is no longer actually needed.
3. A flame **MUST NOT** be lit unless all flammable substances have been removed from the immediate area. Containers of flammable materials in the area should remain tightly closed.
4. Other users of the lab must be **NOTIFIED** before lighting a flame.
5. Flammable materials must be stored properly, ensuring they are separated from oxidizers.
6. Only non-sparking electrical equipment should be used when volatile flammable materials are in use or left open.
7. Quantities of flammable solvents in the lab **MUST** be minimized.

4.4.4 Pressurized Systems

No reaction should be carried out in, or heat applied to, a closed system unless it is designed and tested to withstand pressure. Pressurized apparatus should have an appropriate relief device. An inert gas emulator system should be used to avoid bubbling or pressure buildup for reactions that cannot be opened directly to the air.

4.4.5 Unattended Operations

Unattended operations **MUST** be avoided. When it is necessary to carry out lab operations continuously or overnight, it is essential to plan for interruptions in utility services such as electricity, water, and coolant. These operations must be designed to be safe. and plans should be made to minimize hazards in case of failure. Wherever possible, routine inspection of the operation should be arranged and, in all cases, the lab lights left **ON** and an appropriate sign placed on the door.

4.4.6 Working Alone

It is **imprudent practice** to work alone in a lab building. Arrangements should be made between individuals working in separate labs outside of working hours to cross-check one another periodically. Users are, however, **cautioned** against undertaking hazardous procedures when alone in the lab to avoid implosions or other dangers that they may not be able to handle alone. Besides, under unusual conditions, **special rules** may be necessary. The Lab Manager or Technologist in-charge will determine whether the work requires special safety precautions such as having two persons in the same room during a particular operation.

4.4.7 Prior Approval

Prior to engaging in any lab task, students must **at all times** seek approval from the Lab Technologist, with knowledge and concurrence of their instructor. Employees of the institution will also need to obtain prior approval to proceed from the Manager or Technologist whenever:

1. A new lab procedure or test is initiated.
2. It is likely that toxic concentrations could be reached or other hazardous situations could arise.

3. There is a change in a procedure or test, even if it is very similar to prior practices. Change in a procedure or test means:
   a. A 10% or greater increase or decrease in the amount of one or more hazardous chemicals (including catalysts) used.
   b. A substitution or deletion of any of the chemicals in a procedure.
   c. Any significant change in the system or in other conditions under which the procedure is to be conducted.
4. There is a failure of any of the equipment used in the process, especially of safeguards such as fume hoods or clamped apparatus.
5. There have been previous unexpected results or potential or near accidents.
6. Members of the lab staff become ill, suspect that they or others have been exposed, or otherwise suspect a failure of any safeguards.

4.5 BIOHAZARD LABORATORY SAFETY GUIDES

A biohazard is any organism, or material produced by such an organism, that is known or has the potential to cause human or animal disease. Biohazards include: microorganisms such as viruses, fungi, parasites, and bacteria and their toxic metabolites; mammalian blood and body fluids; unfixed and fixed tissues and specimens from humans and nonhuman primates; cell lines and other tissue cultures; certain types of nucleic acids, such as DNA derived from pathogenic organisms, human oncogenes, or transformed cell lines; genetically modified organisms, including plants; and zoonotic agents. Laboratory users may be exposed to these agents through cuts, wounds, or as a result of absorption through the respiratory tract, digestive system, skin, and mucous membranes.

4.5.1 General Safety Measures

The safe handling of microorganisms, hence an adequate degree of protection, in such a lab is top priority and requires containment and risk assessment principles. The fundamentals of containment include microbiological practices, safety equipment, and facility safeguards that protect lab personnel, the environment, and the local community from exposure to infectious microorganisms that are handled and stored in the lab. Risk assessment is the process that enables the appropriate selection of microbiological practices, safety equipment, and facility safeguards that can prevent lab-associated infections.

A specialized research or production lab that deals with infectious agents is called a biosafety lab. The protections required by these types of activities are defined as biosafety levels, ranked from one to four, and are selected based on the level of danger associated with the biohazards used. Most School and Institutional Labs are pegged at Biosafety Levels 1 and 2 (BSL1 and BSL2) which cover work with animals and agents associated with human pathogens or moderately infectious organisms.

Essentially, BSL1 is for biological agents that rarely cause illness, permitting safe lab activities at the bench-top level. BSL1 is for basic education and research, requiring only a sink for hand hygiene. BSL2 is for moderate-risk pathogens that can pose dangers through skin contact, ingestion, or mucous membranes, and is suitable for healthcare and diagnostic services with moderate risk. BSL3 and BSL4 involve high-risk agents that can lead to severe infections, necessitating stricter safety protocols (Gao et al., 2025).

The BSL1 labs control access and stress hand hygiene; BSL2 facilities have stricter rules and limit activities that generate aerosols. Biological safety cabinets address splash risks, and reducing sharp instruments is vital. Immunocompromised individuals should consult doctors before entry. Staff must be trained on safety protocols, including responses to spills and exposures. Updated biosafety references are necessary for compliance (Emmert & ASM, 2013).

BSL3 laboratories are designated for specific activities involving indigenous or exotic pathogens that can cause serious illnesses through inhalation. Access to the laboratory is controlled through secure entry and exit points for personnel, materials, and waste. Airflow is directed from clean to potentially contaminated areas. Infectious materials are handled in biological safety cabinets to prevent exposure. BSL4 laboratories adhere to strict containment standards for dangerous pathogens, ensuring isolation from the environment. Work is conducted in class III safety cabinets or positive pressure suits. These labs also have decontamination showers, secure airlocks, controlled vehicle access, and specialized waste disposal systems.

Decontamination methods are crucial for hazardous waste management. Sealed bio-seals provide airtight containment, and gas discharge makes materials noninfectious. Pathogen facilities utilize layered containment like airlocks and chemical showers. Protective suits require decontamination before leaving high-risk zones, alongside emergency protocols for failures (Ta, Gosa, & Nathanson, 2019).

Disinfection and sterilization are essential for minimizing infection risks in laboratories. Sterilization completely eliminates all microorganisms using thermal or chemical methods. *Geobacillus stearothermophilus* is a biological indicator for sterilization validation. Autoclaves sterilize glassware, surgical instruments, and pre-treat waste. Ethylene oxide sterilizes heat-sensitive plastics by penetrating cells and damaging DNA but poses risks due to toxicity. Microwave energy with hydrogen peroxide sterilizes in around 50 minutes without toxic residues, though at a higher cost. Spaulding's classification aids in choosing between disinfection and sterilization, highlighting method selection for pathogens (Yoo, 2018).

Because of their potential to cause human disease, great care is used to prevent percutaneous injury (needle-sticks, cuts, and other breaches of the skin), ingestion, and mucous membrane exposures. Due to the foregoing, lab users and personnel are always expected to:

1. Read the international biohazard warning sign displayed on the doors of the rooms where microorganisms of Risk Group 2 or higher risk groups are handled so as to be aware of the safety measures that are put in case of any danger.
2. Become familiar with the microorganisms being used, and know the safety precautions and emergency procedures before undertaking any work.
3. Get clearance from the Technologist in-charge to access the animal houses.
4. Wear appropriate gloves for all procedures that may involve direct or accidental contact with blood, infectious materials, or infected animals. After use, gloves should be removed aseptically and hands must then be washed.
5. Wash hands immediately and thoroughly if they become contaminated with microorganisms.
6. Discard razor blades, scalpels, and hypodermic needles ("sharps") into the sharps container.
7. Decontaminate all cultures, stocks, and other regulated wastes by autoclaving before disposal. Liquids (nonorganic) can be decontaminated with bleach, bringing the solution to 10% bleach, and discarded in the sink. No other chemicals can be discarded in the sink.

4.5.2 Guides on Handling Sharp Objects and Broken Glass

Laboratory users and personnel **MUST**:

1. Use dissection kits, scalpels, razor blades, scissors, and microtome knives with great care, and place in a safe position when not in use.
2. **Never** dispose of any sharp object in the regular trash containers.
3. Report all cuts, no matter how minor, to the instructor.
4. Not touch broken glass with bare hands. Put on gloves and use a broom and dust pan to clean up glass. Dispose of all broken glass in the specific container marked for glass. Do not place broken glass in the regular trash.
5. **Never** use fingers to hold a part of the specimen while cutting.

4.5.3 Handling of Biological Specimen and Wastes

4.5.3.1 Guides for Handling Preserved Specimens

Laboratory users and personnel **MUST**:

1. Use gloves (latex) to handle preserved specimens.
2. Notify the instructor if there is a spill of preservative.
3. Clearly label and store specimens in designated containers or cabinets when not in use.
4. Follow the directions of the instructor concerning the proper disposal of preserved specimens after they have been used.

4.5.3.2 Handling Body Fluids

Special precautions should be followed in all labs using any body fluids, such as blood, saliva, and urine, because of the potential to transmit disease-causing organisms.

1. Follow all instructions carefully.
2. Use gloves and goggles in all lab experiments that involve the use of body fluids.
3. All contaminated materials, such as slides, coverslips, toothpicks, lancets, and alcohol swabs, **MUST** be placed in a biohazard bag for proper disposal and should never be reused.

4.5.3.3 Handling of Biological Waste

1. Containers used for collection of biological waste must be in good condition, must not leak, and must remain closed except when waste is being added. Containers and contents must be compatible. Biological waste containers **MUST NOT** be overfilled (At least 10% headspace should be allowed for liquid expansion).
2. Work surfaces **MUST** be decontaminated after any spill of potentially dangerous material and at the end of the working day.
3. All contaminated materials, specimens, and cultures **MUST** be decontaminated before disposal or cleaning for reuse.
4. An autoclave for the sterilization of waste materials is readily accessible in the lab site.
5. Materials for autoclaving should be transported to the autoclave in robust containers without spillage.
6. Contaminated waste should be suitably labeled before removal for incineration.
7. Used lab glassware and other materials awaiting sterilization before recycling should be stored in a safe manner. Pipettes, if placed in disinfectant, should be totally immersed.
8. A high degree of precaution must always be taken with any contaminated sharp items, including needles and syringes, slides, pipettes, capillary tubes, and scalpels.
9. Animal bedding materials must be removed in a manner that minimizes the generation of aerosol and dust.
10. Biological waste containers **MUST** be properly labeled.

4.5.4 Biological Spills

All labs working with biohazards should have a basic biological spill kit ready to use at all times. The basic biological spill kit contains the following and all personnel **MUST** know the location.

- Disinfectant (bleach, bleach 1:10 freshly prepared, or Iodine-based disinfectants)
- Absorbent material (e.g., paper towels)
- Waste containers (e.g., biohazard bags, sharp containers)

- Mechanical tools (e.g., household rubber gloves and forceps to deal with broken glass, dustpan and broom)
- PPE (e.g., gloves, masks, goggles, disposable gowns)

4.5.4.1 General Spill Clean-up Guideline

Laboratory users and personnel are always expected to:

1. Wear gloves and lab coat.
2. Use forceps to pick up broken glass and discard them into sharps container.
3. Cover spilled material with paper towels.
4. Add diluted disinfectant in sufficient quantity to ensure effective microbial inactivation.
5. Allow a 30-minutes contact for disinfection.
6. Wipe spill area with diluted disinfectants.
7. Wash hands with soap and water when finished.
8. Autoclave all towels, gloves, and other materials worn or used to clean up the spill.

4.5.4.2 Spill in a Biological Safety Cabinet

Laboratory users and personnel are always expected to:

1. Leave the cabinet turned on.
2. Wear gloves, spray or wipe cabinet walls, work surfaces, and equipment with disinfectant such as 70% ethanol. If necessary, flood work surfaces, as well as drain pans and catch basins below the work surface, with disinfectant. Allow 30 minutes contact time.
3. Soak up the disinfectant and spill with paper towels, and drain catch basin into a container. Lift front exhaust grille and tray and wipe all surfaces.
4. Ensure that no paper towels or solid debris are blown into area below the grille.
5. Disinfect all items that may have been spattered before removing them from the cabinet.
6. Discard all clean-up materials into biohazard waste container. Wash hands and exposed skin areas with soap and water.
7. Notify Technologist in-charge if the spill overflows into the interior of the cabinet. It may be necessary to do a more extensive decontamination of the cabinet.

4.5.4.3 Spills Inside a Centrifuge

Laboratory users and personnel are always expected to:

1. Shut centrifuge off and do not open the lid for 20 minutes to allow aerosols to settle.
2. Put on the gloves and lab coat.
3. Use a squeeze bottle to apply disinfectant to all contaminated surfaces within the chamber, taking care to minimize splashing.
4. Allow 20-minute contact period and then complete clean-up of the chamber.

4.5.4.4 Spills Inside the Laboratory

Laboratory users and personnel are always expected to:

1. Notify others in the area, to prevent contamination of additional personnel and environment.
2. Avoid inhaling airborne material, while quickly leaving the room. Notify others to leave.
3. Close the door, and post a warning sign.
4. Inform the supervisor.
5. Allow aerosols to settle for at least 30 minutes before re-entering the lab.
6. Have a complete biological spill kit ready to go before you start the clean-up.
7. Put on protective clothing gown, aerosol mask, gloves, and shoe covers.
8. Initiate clean-up with disinfectant.

4.5.4.5 Spills Outside the Laboratory

Laboratory users and personnel are always expected to:

1. Ensure that viable organisms should only leave the lab in a well-sealed primary (inner) **and** secondary (outer) container with a closable top.
2. The exterior of the secondary container should be wiped down with disinfectant prior to leaving the lab so that it can be transported without wearing gloves.
3. Carry paper towels and if a spill occurs, use the towels to cover the spill but do not attempt a clean-up without appropriate disinfectant and PPE.
4. Notify people in the immediate area and collect clean-up material and proceed with clean-up.

4.5.4.6 Personal Exposure Takes Priority over Clean-up

If you are exposed, immediately remove contaminated clothing and other protective equipment and wash affected areas with soap and water. If medical follow-up is warranted, it should be sought immediately.

4.5.4.7 Extra Vigilance and Immunization

Laboratory users will be informed by the Technologist in-charge whenever there is need for extra vigilance including, for example, any vaccinations. Immunocompromised, immunosuppressed, and other persons with increased risk for infection may be denied admittance at the discretion of the lab director.

4.5.5 Safety Equipment and Their Usage

A biosafety cabinet (BSC) is an enclosed, ventilated lab workspace for safely working with materials contaminated with (or potentially contaminated with) pathogens requiring a defined biosafety level. Several different types of BSC exist, differentiated by the degree of biocontainment required. The primary purpose of a BSC is to serve as a means to protect the lab worker, the sample, and the surrounding environment from pathogens. It allows all exhaust air to be *HEPA*-filtered, as it exits the BSC, removing harmful bacteria and viruses. Anytime when dealing with infectious agent, use the BSC.

4.6 POSITIVE SAFETY CULTURE IN THE LAB

Achieving safety for laboratory personnel necessitates a systematic method focused on the identification and mitigation of potential hazards. The engagement of staff in safety initiatives cultivates an environment centered on risk reduction and accountability. When empowered to report potential dangers, workers significantly contribute to the improvement of the workplace safety culture and its long-term viability (Steward, 2018). In this section, we outline strategies to foster a safety-oriented culture.

An effective safety program is contingent upon fostering a culture that seamlessly incorporates safety into laboratory practices, backed by institutional reinforcement. This culture emerges from shared values alongside a methodical framework for identifying and addressing potential hazards. It emphasizes the significance of organizational frameworks, collective attitudes, and personal accountability, thereby prioritizing safety as the central element of all work-related activities. Work culture, organizational culture, and safety culture set norms in an organization. Improving safety culture requires addressing unsafe behaviors to lower injury incidents. A management-led approach is vital for shaping the lab environment. Daily practices represent the organization's core values. The Australian Institute of Criminology highlights that managers and employees must recognize their safety responsibilities, prioritize safety, collaborate on objectives, and effectively communicate safety protocols (Steward, 2018).

Laboratories require collaboration and engagement for effective safety programs. The WHO *Laboratory Biosafety Manual* (2020) highlights the importance of integrating biosafety into personnel practices. Individual attitudes shape safety and should align with safety goals. Assessing safety attitudes and knowledge is crucial for identifying improvements, while regular audits ensure personnel understand and follow SOPs.

Laboratory safety necessitates not only compliance with established procedures but also the cultivation of a safety-oriented culture and robust interpersonal relationships among staff. The commitment of management, coupled with adequate training on safety protocols, is crucial. It is imperative for laboratories to promote prompt reporting of incidents and to develop an environment conducive to learning without fear of blame. Establishing a non-reprisal policy is vital to safeguard individuals who disclose unsafe conditions from any forms of retribution (Emmert & ASM, 2013; Samaranayake, Nishadya, & Jayasundara, 2022).

Accidents or injuries in the laboratory must be reported to a supervisor or safety officer without delay, followed by completion of an official report following institutional guidelines. Hazardous conditions should also be reported immediately. Timely reporting allows for corrective actions, preventing accidents and improving laboratory safety for everyone.

4.7 TRAINING AND CONTINUOUS LABORATORY SAFETY EDUCATION

Effective laboratory safety requires thorough training and continuous education about hazards and control measures. Employers must offer initial training and regular refresher courses to maintain safety standards and minimize risks.

In the USA, the OSHA mandates that employers offer training and document key laboratory safety areas. These encompass hazard communication, chemical hygiene, blood-borne pathogen handling, biosafety, biological safety cabinets, evacuation, electrical safety, compressed gases, equipment operations, ergonomic training, and medical surveillance. Documentation must be retained for a minimum of three years.

A key aspect of a secure laboratory environment is tailored safety training for diverse user needs. This may include formal sessions, practical demonstrations, and review of safety documents. Lab staff knowledge depends on the task risks, and differing training needs may lead to issues (Kocak, 2019). The training should align with specific laboratory hazards to equip personnel with skills to identify, assess, and manage risks as general principles are inadequate. Graduate students and postdoctoral researchers require extensive training before research begins, as training for experienced researchers often lacks depth. Early programs focus on hazard identification, risk mitigation, key information sources, and essential emergency protocols (Emmert & ASM, 2013).

An effective laboratory safety program requires strict adherence to regulatory frameworks, consent agreements, and institutional guidelines. Key components include Compliance and Consent Agreements, Personal Safety Protocols, Chemical Safety Procedures, Hazard Communication Standards, and Emergency Policies. Implementing Medical Surveillance Programs and an Access Control Program is also crucial for personnel health and workplace safety, along with a System for Reporting Unsafe Conditions and Emergencies (Stitt-Fisher, 2015).

Despite technological advancements, the principles of GLP set by OECD remain relevant today. For example, research institutions are still required to keep laboratory notebooks for experiments. The transition to computerized systems has led to increased data volumes, making electronic notebooks essential for storage and analysis. In biological research, recording sample details such as location, name, type, and acquisition date is vital. Notebooks need to be user-friendly and integrate physical and digital components, project documentation, and protocols while ensuring interconnected online provenance trails to document methodologies and support analyses and inquiries (Ney, Kloss, & Schreiber, 2011).

4.8 CONCLUSION

An evaluation of safety protocols at Wolaita Sodo University showed a widespread unfamiliarity with proper practices to protect experimental procedures (Anza et al., 2016). Many individuals ignore basic safety procedures, such as labeling and protective gear. In Sri Lankan chemical labs, 83% consider their environments hazardous, highlighting poor safety protocols (Samaranayake, Nishadya, & Jayasundara, 2022). Many acknowledged safety regulations' importance, with 32% favoring stricter rules while being concerned about productivity impacts. The study stressed tailored training, PPE, inspections, and incident reporting, along with the need for hazard identification and compliance in biomedical labs (AlShammari, Alhussain, & Rizk, 2021). Techniques like system mapping and failure modes and effects analysis are important for risk assessment in processes, but their subjectivity and dependence on expertise present challenges. In academic labs, clear safety protocols are essential due to students' varying levels of hazard awareness, necessitating customized safety guidelines for different educational settings, particularly in science.

Part 2

Chemical and Biohazard Security

5 Evaluation of Chemical Security Status

Apollo Odhiambo Maima

5.1 INTRODUCTION

Evaluating chemical security standards is essential for organizations to perform self-assessments. This framework supports training within the global community and allows for further evaluations. The rise in chemical terrorism risk underscores the necessity for specialized expertise, enhanced knowledge, and dependable technologies to tackle these threats effectively (Szklarski, 2023). Assault strategies can erode public confidence in government and threaten economic stability. For example, insurgents in Iraq used chlorine gas to instill fear in US military areas, revealing societal security vulnerabilities (Blakey et al., 2013). Domestic terrorists use materials like gasoline to threaten crowded areas, endangering lives while avoiding law enforcement. As tactics change, both private and public sectors must urgently address these threats. Research should prioritize hazardous substances from industrial or natural disasters, which can pose greater risks than intentional attacks. An incident in India highlights the dangers of accidental spills, underlining the need for continuous vigilance and proactive public health measures.

Laboratory security stops unauthorized access to hazardous materials that threaten public safety. Evaluating protocols is key for enhancing safety (Housni et al., 2023). An assessment of chemical security needs a thorough method to evaluate attack scenarios, threats, and consequences, emphasizing public safety and emergency preparedness.

5.2 OVERVIEW OF CHEMICAL SECURITY

Chemical Security Emergencies include accidental releases and terrorism-related acts, necessitating strict safety regulations to safeguard people and infrastructure (Khanizade et al., 2025; Soltanzadeh et al., 2024). Emergencies, especially transportation accidents involving hazardous materials, can arise suddenly, causing significant damage and public alarm. Common substances include chlorine and ammonia, complicating identification and release. Continued research and training are vital for improving responder readiness and minimizing public health effects (Blakey et al., 2013).

Determining if an event is localized or expansive relies on location traits and agents. The Chemical Hazard and Risk Framework Assessment (CHoRFA) framework shows that chemical emergencies may cross borders, affecting various countries. Isolated activities, such as tank sludge disposal, can cause public health crises, highlighting complexities in emergency classification and the global impacts of industrial practices (Grobelak et al., 2024). This situation highlights the challenges in predicting post-adverse events and emphasizes the need for reactive strategies to minimize damage and prevent operational disruptions. Continuous rehabilitation is crucial for addressing health and environmental risks. A systematic review of emergency preparedness for disaster management, mainly related to chemical, biological, radiological, and nuclear incidents showed that inadequate documentation hinders effective risk management (Farhat et al., 2024).

In recent decades, advancements in chemical process safety and facility security have emerged. The 1970s–1990s witnessed catastrophic events like the Sandoz explosion, the Seveso dioxin leak, Bhopal disaster, Texaco refinery explosion in the Netherlands, and the Buncefield fire in the United

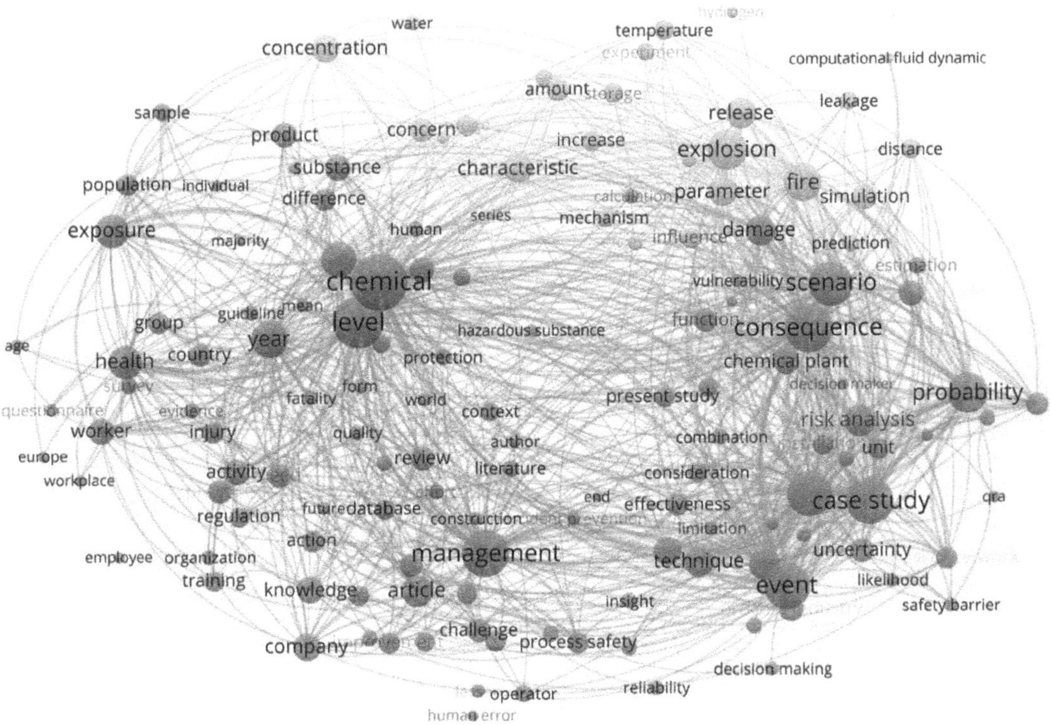

FIGURE 5.1 Network visualization map with keywords of chemical accidents (Liang, Yan, & Zhao, 2024).

Kingdom (Reniers & Khakzad, 2017). These incidents exposed vulnerabilities in the chemical industry, prompting a re-evaluation of safety protocols.

In recent years, China's chemical manufacturing industry has grown significantly, positioning the nation as a leader in global production. However, this growth has resulted in major chemical accidents, causing substantial casualties and damage. Addressing these incidents demands a comprehensive approach to safety, risk assessments, and regulations.

Figure 5.1 represents an in-depth analysis of a co-occurrence network map that was conducted to extract relevant terminology from scientific publications in China via the Web of Science database. The map visualization was created using VOSviewer software, showcasing key topics and themes in the chemical accidents research field. The assessment produced three distinct keyword clusters, each marked by a unique color. *Cluster 1 (Red)* includes keywords related to metrics of chemical accidents—examining incident frequency, trends, geographical patterns, and casualties, as well as regulatory frameworks for chemical management and safety protocols. *Cluster 2 (Green)* contains keywords relevant to case study methodologies, focusing on accident analysis and risk assessment practices, while *Cluster 3 (Blue)* highlights keywords representing various classifications of chemical incidents, reflecting their diversity (Liang, Yan, & Zhao, 2024).

5.3 HISTORICAL CONTEXT

The journey to enhance chemical safety regulations has been shaped by key incidents, notably the Seveso Directive after the 1976 dioxin leak in Italy, which underscored the need for rigorous oversight. Similarly, the Bhopal disaster (the catastrophic methyl isocyanate release in Bhopal, India) in 1984, with an estimated 2,500–6,000 fatalities and 500,000 affected due to toxic exposure, prompted a global response. This tragedy led to the establishment of comprehensive process safety management (PSM) standards to prevent future disasters and safeguard communities near chemical facilities (Lee et al., 2021).

In response to international developments, the UN Security Council Resolution 1540, enacted in April 2004, established requirements for all 191 member states to regulate chemical, biological, and nuclear weapons, preventing unauthorized transfers and ensuring responsible use of these substances. The resolution aims to prevent terrorist organizations from obtaining hazardous weapons by imposing strict limitations on their procurement of essential materials for their manufacture. It mandates nations to secure and limit access to facilities involved in the production or storage of these crucial materials. Countries failing to report their measures to the Council must provide a valid reason for noncompliance, promoting accountability and transparency among nations (Holmes & von Winterfeldt, 2025).

The UN resolutions were vital in encouraging numerous voluntary professional organizations at global, regional, and national levels to enhance government adherence to guidelines. Support was provided by the United Nations and groups such as the Organization for the Prohibition of Chemical Weapons, WHO, and NATO, with significant contributions from the EU, G-8, and various NGOs promoting these initiatives globally (de Vries, 2023). The authors launched their project against weapon proliferation by engaging key organizations and figures. Their understanding of arms dissemination and commitment helped identify regional gaps. Insights from government and industry were enriched through roundtable discussions with intelligence, academia, and media participants, providing essential perspectives. This collaboration laid a strong foundation for developing and publishing articles (Edwards et al., 2022).

In the subsequent years, chemical weapon risks in East Asia rose significantly. Publications highlighted maps of precursor stockpiles, stressing their potential danger if misused. Nations importing these precursors were noted as weaponizing toxic gases. An intelligence report compiled on chemical facilities and security vulnerabilities (Rozsypal et al., 2025) noted that several East Asian countries were hesitant to collaborate, leading to their absence from security initiatives. A comprehensive evaluation for protecting chemical facilities was developed using UN guidelines, focusing on their features, operations, and hazardous materials. This report, endorsed by the Trinidad & Tobago's National Energy Corporation (NEC), incorporated insights from engineers and security experts, aiding project advancement and enhancing safety measures for aging facilities (Trinidad & Tobago, 2018).

In the United States, the Chemical Security Improvement Act (CSIA) amended and significantly enhanced the Toxic Substances Control Act (TSCA). It mandates safety assessments for existing and new chemicals, thus expanding the Environmental Protection Agency's (EPA) regulatory powers. The act also emphasizes transparency and better public access to information on chemical safety (Guenther, 2022).

Finally, the Chemical Facility Anti-Terrorism Standards (CFATS) initiative improves chemical facility security by fostering collaboration between the Department of Homeland Security (DHS) and operators to identify vulnerabilities. Without CFATS, many facilities risk increased chemical terrorism due to insufficient oversight. A robust risk assessment framework with facility operators, law enforcement, and emergency teams is essential for effective threat management. This cooperation results in customized security plans, necessitating shared safety responsibilities among facility management, stakeholders, regulatory bodies, and communities, focusing on proactive risk mitigation (Gough, 2021).

5.4 REGULATORY FRAMEWORK

The United States set Chemical Security Design Standards under the Chemical Weapons Convention (CWC) to enhance safety, though facilities handling hazardous chemicals under 225,000 pounds can be exempted from the regulations. A specialized tool is used to evaluate security, while high-risk situations are assessed on-site for data collection. Recent updates stress cost-effective strategies for smaller entities against evolving threats (Caves Jr & Carus, 2022). Recent governmental transitions have highlighted significant gaps in overseeing high-risk chemical installations handling hazardous materials, an often-overlooked aspect of national security. Many facilities operate under lax regulations, endangering public health, especially near agrochemical and water treatment sites. This lack of proactive strategies increases risks nationally. Past incidents demonstrate that safety measures alone are insufficient. Improved prioritization and resource allocation could reduce terrorist activities, shifting threats from national crises to local incidents while maintaining risks.

To effectively convert significant chemical substances into practical scientific knowledge, screening mechanisms have been established to evaluate potential threats and associated facilities. This screen has been based on an analysis of available facility data and an assessment of associated threats. The many screening protocols and containment strategies resemble standard business practices and can identify areas needing oversight, serving as benchmarks for collective risk assessments during heightened threat levels. For specific international facilities or security assets, direct communication would help in the processing of export documentation for third-party evaluations, while ensuring critical information retention is minimal. Protecting chemical facilities and their corresponding security measures is vital for maintaining safety during chemical assessment, manufacturing, knowledge management, storage, and transportation (Crowley, Shang, & Dando, 2018; Van Ham, van der Meer, & Ellahi, 2017).

5.4.1 International Regulations

The International Health Regulations (IHR) provide a binding and comprehensive framework for global health security among 194 member countries. Their design emphasizes the enhancement of nations' capabilities to prevent, detect, and respond effectively to health threats. However, compliance data from 2009 revealed that only approximately 50% of these countries were aligned with the IHR, indicating a significant compliance gap. This situation calls attention to the pressing need for increased support and resources, particularly aimed at developing nations. Additionally, a comprehensive report has highlighted the necessity of establishing a strong legal framework to ensure effective global health security. This report further elucidates the complementary and interdependent nature of the IHR in relation to other legal instruments that play a crucial role in international health security (Patel & George, 2021; Elhakim, Ghazy, & Samhouri, 2024).

UN Security Council Resolution 1540 compels all member states to introduce and enforce thorough measures aimed at preventing non-state actors, especially terrorist groups, from gaining access to these highly dangerous substances. The Resolution acted as a crucial impetus for the establishment of the Global Initiative to Combat Nuclear Terrorism, thereby enhancing global collaboration in tackling the risks associated with nuclear terrorism. Furthermore, it facilitated the creation of the Counter-Terrorism Committee and its Executive Directorate, entities dedicated to improving international security and promoting cooperation among nations in their efforts against terrorism (Holmes & von Winterfeldt, 2025).

The Biological Weapons Convention (BWC) represents a pivotal and essential multilateral treaty that explicitly forbids the development, manufacture, and stockpiling of biological and toxin weapons. Signed initially in 1972, this significant accord has attracted the involvement of 183 states parties as of June 2019, showcasing a profound level of interest and engagement from the global community. This high level of participation reflects a considerable and unwavering international commitment to mitigating the spread of these dangerous substances, which pose serious and multifaceted risks to both global security and public health systems.

The primary objective of the BWC is to prevent the misuse of harmful biological agents in ways that could jeopardize human welfare and overall well-being, underscoring the treaty's critical importance in safeguarding humanity against potential biological threats (de Vries, 2023). The BWC represents the inaugural multilateral agreement aimed at prohibiting the development, production, and accumulation of biological weaponry. States that are signatories to the convention bear the responsibility to provide assistance in the face of biological threats, encourage global collaboration in the areas of disease prevention and pathogen management, and engage in ongoing dialogue following an attack. Furthermore, member nations must ensure the maintenance of robust national biosafety systems and actively take part in verification processes designed to monitor adherence to the stipulations laid out by the BWC (Crowley & Dando, 2024).

The CWC represents a meticulously crafted legal structure that categorically forbids the creation, production, possession, storage, retention, or distribution of chemical weapons across all modalities. It also firmly disallows their application in manners that could potentially endanger or affect human life and safety. This treaty, inaugurated in 1993 and operational since 1997, enforces adherence through a systematic series of review conferences that convene every five years. The most recent of these conferences, conducted in 2018, allowed participating nations to evaluate their compliance with the predefined regulations. During this event, representatives analyzed the existing situation, assessed levels of adherence, and engaged in discussions regarding prospective measures aimed at bolstering international safety and security concerning chemical weapons. This continual dialogue is vital for averting the potential abuse of these hazardous materials (Krutzsch & Trapp, 2023).

5.4.2 National Regulations

The Chemical Security Index serves as an essential tool for evaluating and thoroughly tracking the overall security condition of various chemical substances. It facilitates the meticulous identification of potential security weaknesses or vulnerabilities that may exist within the system, which can then be effectively rectified through relevant regulatory actions or comprehensive management strategies. These strategies may encompass the critical formulation or significant enhancement of national laws and regulations specifically related to chemical security to ensure safer handling of hazardous materials (Soltanzadeh et al., 2024).

Across numerous countries globally, there exists a significant and troubling lack of cohesive legislation specifically addressing the crucial issue of chemical security. Although some nations may opt to establish legal frameworks that address specific components or dimensions of chemical security, such regulations frequently suffer from a lack of coherence, being dispersed across various statutes and legal texts. This prevalent fragmentation can greatly hinder their overall functionality and efficacy, largely due to the vague and indistinct definition of roles and responsibilities assigned to the relevant authorities and organizations responsible for managing these essential domains (Bruinen de Bruin et al., 2022). Consequently, the presence of suitable regulations at the national level, along with government accountability concerning chemical security, may be equally significant as possessing a national action plan and a comprehensive framework for chemical security.

Regulations are vital for executing national chemical security action plans, coordinating initiatives, and establishing accountability frameworks. Effective regulations define authorities' roles, ensuring they can implement measures to develop strong national chemical security capabilities. Additionally, regulations must guarantee resource allocation for these actions. Precise chemical security regulations improve clarity and accountability, ultimately strengthening structures for managing chemical risks and enhancing national safety standards for hazardous substances (Fouzi et al., 2024). This will help clarify the various uncertainties that are related to government duties in this specific context and will also facilitate the conversion of broader national goals into quantifiable or enforceable objectives that can be implemented effectively at the operational or agency tiers.

Establishing the roles of appointed authorities in chemical security is crucial within the regulatory framework. Many countries have multiple agencies at various levels, each with numerous responsibilities related to chemical security. These include registering and categorizing chemicals, managing sensitive information, overseeing chemical inventories, developing emergency response strategies, managing emergencies, monitoring, and enforcing legislation, all vital for safety and compliance (Fouzi et al., 2024). To establish a robust national chemical security framework, it is imperative to facilitate coordination among the relevant authorities. Clearly delineated roles and responsibilities associated with chemical security must be outlined in national regulations or pertinent official documents. The absence of such clear assignments could result in the underutilization of existing chemical security resources owing to insufficient coordination and ambiguous guidelines. This lack of clarity may ultimately lead to severe consequences, including loss of life and property due to chemical incidents.

5.5 RISK ASSESSMENT METHODOLOGIES

The thorough assessment of security within chemical facilities that manage hazardous materials can be well effected through a comprehensive quantitative approach, which emphasizes the utilization of an automated and streamlined risk assessment procedure (Soltanzadeh et al., 2024). This method has been carefully incorporated into a robust risk assessment instrument, specifically created in alignment with the existing relevant regulations and safety standards. This instrument has been effectively utilized in a variety of international and national chemical security initiatives, showcasing its significant capability to assist in the prioritization of key projects focused on safeguarding some of the most at-risk facilities, especially in developing nations. By leveraging data and systematic evaluation, this innovative instrument aims to greatly enhance security measures within the chemical sector (Landucci et al., 2022).

The target users of this access risk assessment methodology and tool are technical decision-makers at the facility level, who may not have a detailed technical education. The tool should be simple to use, requiring only about 15–20 minutes for risk assessment, helping them to conduct a first critical review of the chemical security status of a facility. Sources of hazard and threat scenarios were provided as hints to help users fill in the tool's forms. After completing the mandatory fields, the users could process the assessment and immediately receive the following five outputs: a two-dimensional graphical presentation of results focused on risk levels and risk mitigation recommendations; a list of action items, including a brief description, current status, and priority ranking; an interactive matrix indicating vulnerability levels for each scenario; a two-dimensional diagram focused on residual risk; and a concise risk assessment report, which can be printed as a formal document (Mahdavi et al., 2022).

With regard to access, it allows direct self-help chemical security assessments by users. Decisions can be based purely on the results received from the tool, and it can be employed worldwide with no restrictions on access and use. There are also some limitations of the tool. Confidentiality of the processed data cannot be ensured or protected. Public posting of the outputs linked to a specific facility online could endanger security. In addition, potential terrorists could use this tool too, because it raises awareness and targets facilities with already lower levels of chemical security.

5.5.1 Qualitative Risk Assessment

Next-Generation Risk Assessment (NGRA) methodologies, which are grounded in New Approach Methodologies (NAMs), are currently being employed in industry for decision-making purposes in specific areas. The cosmetics sector, for example, shows many cases of NGRA related to systemic toxicity and skin sensitization. The utilization of NAMs in safety assessment frequently relies on in vitro methods. Therefore, the integration of kinetic data and quantitative in vitro–in vivo extrapolation (QIVIVE) is essential for merging NAMs into decision-making structures. QIVIVE plays a vital role in establishing connections between the responses observed in cellular systems and the biological responses in entire organisms (Hewitt et al., 2022).

A new tool has been developed to assess qualitatively the public health risk associated with accidental or deliberate release of large amounts of chemicals into the atmosphere. It is relevant for municipalities and other bodies interested in quickly and easily comparing chemicals in terms of risk independently from release scenarios, types of events, and choice of consequence assessment tool. Built as a spreadsheet, the tool estimates risk based on a limited number of available input or external factors. Long but readily available lists as well as a reference risk rankings document are provided. The tool also gives insight into the uncertainty associated with the listed chemicals

(Stucki et al., 2022). Although it is implicated for use by technical experts, it has been designed to allow easier use by public health officials and risk managers who have an interest in pesticides, cleaning products, large storage tanks, and other chemical installations or industries.

The tool is intended to list a relatively small number of chemicals with the potential to adversely affect public health due to an accidental or deliberate release into the atmosphere. The exposure considered is inhalation exposure. This exposure could arise from several kinds of accidents or terrorist acts ranging from the explosion of a tank, fire, suicide bomb, downed aircraft, or other means causing a release of large quantities of a chemical over a site large enough that a cloud could disperse to a populated area. While intoxication depends on release scenarios and effluents conditions, affected populations, topography, weather, distance, and time, long-term risk comparisons require simplifying assumptions about some of these factors which would lead to arguments over subjective factors applying morbidity or mortality weights (Stucki et al., 2022).

5.5.2 Quantitative Risk Assessment

Risk assessment is a process in which the risk associated with relevant threats and vulnerable scenarios is evaluated. The basic components of the process are as follows: (1) Definition of Risk Assessment Scope, (2) Identification of Hazards and Characterization of Chemical Agents, (3) Identification of Vulnerable Scenarios, and (4) Evaluation of Risk Related to the Threats and Vulnerable Scenarios. Risk assessment is executed within the scope of evaluation of risk regarding toxicological effects on human health, habitat, and ecosystem, regarding the effect on citizen's economy, and regarding terrorist threats such as chemical suicide incidents. The above data comprised both qualitative and quantitative assessments to evaluate the risk. An important task of the assessment is to prioritize the analysis subject or data focused on assessment and evaluation (Zodrow et al., 2022; Okoro et al., 2023).

Utilizing an analysis of published literature, risk assessment devices were developed with AcuSim employed as a computer tool to evaluate the risk regarding the relevant threat events and vulnerable scenarios. Computer code modeling atmospheric dispersion releases are employed except for those expert systems used in analyzing gathering on specific data inputs of the facility to solve the risk related to a terrorist attack. Prioritizing Public Health Risk Assessment is accomplished with a series tool that consists of recommended chemical apologies from considering both health and public safety perspectives and are used in testing and providing further data inputs. Logical specifications of the chemicals are set to determine at which endpoints the chemicals and/or release scenarios are subsequently assessed in detail (Blakey et al., 2013).

5.6 CHEMICAL THREAT ANALYSIS

Chemical security identifies threats by distinguishing terrorism risks from military actions, crucial for managing chemical hazards (Zhu et al., 2020). Chemical security threats range from speculation on toxic substances to attacks on critical infrastructure like rail systems. State actors may engage in espionage, while non-state groups often strike at economic assets to provoke violence. Despite differing methodologies, outcomes can be similar. Assessing chemical security necessitates evaluating risks and security integrity, considering political and military implications, usually neglecting specific examples and actionable insights vital for understanding chemical security dynamics. The evaluation analyzed questionnaire data to identify risks linked to available and synthesized chemicals, detailing their associated threats as below:

1. *Easily obtainable industrial chemicals*: 2,4,5-trichloro-1,3,6-triazine; propylene oxide; 2-chloropropan-1-ol; bifenthrin; refined oil; pyridine; diacetone alcohol; heterocyclic amines; chloroacetyl chloride; propanol; 1-chloro-2-methyl-2-butanol; 4-(carbamoyl-2,4-dihydro-1H-quinolin-8-ylsulfonyl)-2-phenyl-3(2H)-isoindolinone; peginterferon; natural gas.

2. *Commercial chemicals easily available upon request*: 2-methyl-2,4-pentanediol; acetic anhydride; methacrylate-imida-2-siloxan; sodium hydroxide; dichloromethane; acetic acid; ortho-phthalaldehyde; diisopropylamine; fluorophosphoric acid; acrolein; 4-amino-butyric acid.
3. *Chemicals for commercial use but under strict regulation*: chloroacetone; chlorine; ciphro; NH_3; potassium cyanide.
4. *Acquiring hazardous precursors and infrastructure needs substantial investment*: phosphorus trichloride; phosphorus pentachloride; lithium phosphide; tetraethyl lead; 3-phenylisoxazole-5-carboxylic acid; carbon disulfide; cis-3-hexen-1-ol; vigabatrin.

An analytical structure was developed to assess public health threats from atmospheric chemical releases by examining hazards. It classifies substances by hazard and ranks them by risk through event analysis, hazard evaluation, exposure assessment, and risk calculation. The scenario tree explores outcomes during a release, influenced by physicochemical properties. A characterization matrix enhances hazard evaluation, and combined with exposure assessment, it enables thorough risk calculation. Its effectiveness was confirmed in a study of 139 hazardous substances with varied toxicological characteristics.

5.6.1 Emerging Threats

The rise in chemical terrorism poses significant security challenges involving toxic industrial chemicals (TICs), chemical warfare agents, and chemical products. Attack strategies focus on target identification, while safety frameworks emphasize source identification; hence, disposal strategies must adapt to target classification. The global terrorism threat is evident through initiatives like the US "Goosport" and violent incidents in France. The chemical industry has faced 35,000 noncompliance cases over the past decade, with 1,137 linked to terrorism. Serious incidents occur 8.4 times more in chemical clusters than standalone sites, revealing higher fatalities and injuries. Key challenges include inconsistencies in safety culture, design flaws, and security gaps (Zhu et al., 2020). To achieve destruction effectively, a thorough analysis of target information is essential. This involves examining factors like the facility's geographical location, structural organization, personnel numbers, traffic patterns, and commute durations.

Terrorists choose targets based on factors such as chemical toxicity, with higher toxicity posing greater risks to aquatic systems. Target processing facilities evaluate population density, transport patterns, and geography. Target desirability is inversely linked to chemical hazard classifications. Attack success rates depend on environmental conditions, chemical availability, tactics, countermeasures, and socioeconomic contexts. Atmospheric conditions influence chemical dispersion and casualty estimates. Law enforcement actions can disrupt attackers' plans. Analysts propose a multi-stage risk assessment to foresee potential attack scenarios. Grasping terrorist ideologies, organizational traits, and historical precedents is essential for predicting targets and methods.

Emerging Paradigms in Chemical Threat Mitigation examines research on countermeasures against chemical threats (CounterACT), emphasizing medical countermeasures (MCMs) for under-represented chemical agents. It addresses various TICs and toxic industrial materials (TIMs) that are vital for industries but pose significant risks of misuse, possibly leading to catastrophic events (Casillas, Tewari-Singh, & Gray, 2021). Chemical warfare's significance is growing due to easy access to TICs and TIMs, and heightened terrorist risks from lax regulations. Industrial accidents exacerbate these threats. The lack of FDA-approved MCMs for terrorism-related chemicals highlights the urgent need for development and deeper understanding of these chemicals' mechanisms.

Heavy metals in the environment, from natural occurrences and anthropogenic activities, harm ecosystems in soil, plants, water, and air. Their toxic nature raises concerns due to bioaccumulation in the food chain, affecting various organisms. With increasing interest in herbal medicine, heavy metal contamination threatens herbal material integrity, extract efficacy, and product safety, affecting market viability. Effective monitoring and management of heavy metal levels in herbal plants for pharmaceuticals and food are essential. There is a need to probe ecological pollutants, cultivation methods, and production practices that affect herbal quality and traditional medicine, proposing management strategies for sustainability and safety (Chen et al., 2021a).

5.6.2 Natural Threats

In chemical production facilities, safety is essential, demanding a culture of vigilance and accountability. The industry faces numerous incidents yearly, with over three significant ones indicating serious risks, community effects, and even loss of lives. In 2015, there were 323 fatalities and 532 injuries, resulting in evacuations and critical water shortages. These tragedies highlight the need for strict regulatory compliance and heightened stakeholder awareness. Effective risk management depends on strong communication and collaboration among industry leaders, communities, facility workers, and regulatory bodies to ensure the safety of personnel and the surrounding community (Ricci, Moreno, & Cozzani, 2021). Wang, Fan, and Niu (2022) have written a paper that explores the application of the Human Factors Analysis and Classification System (HFACS) to understand the pathways leading to chemical accidents.

Many facilities worldwide deal with large volumes of hazardous substances, which are used, produced, and/or stored in facilities that are often located near towns and cities. Damage to a facility due to a Natech event can potentially result in significant impact on the surrounding community from exposure to hazardous materials (HazMats) (Varshney, Singh, & Yadav, 2021). It is therefore necessary to conduct safety and risk assessments of chemical facilities. In addition to large-scale industrial processes, trans-storage, mixing/separating/heat-exchange processes of HazMats, or continued risk assessments of such facilities are necessary.

Large-volume storage and processing of highly hazardous materials is a global phenomenon as it is used for most industrial chemicals. Serious incidents due to failure of containment structures in such industries have resulted in significant radiological consequences. Similar investigations of unintentional releases of material used in processes other than nuclear would contribute to a broader understanding of human factor issues associated with these incidents in general (Li et al., 2022a). Risk assessments for chemical facilities are widely discussed where processes are modeled quantitatively and meteorological consequences affecting nearby land areas are calculated. Large incidents where a rupture and subsequent release of a container lead to a cloud explosion and large-scale dispersion of HazMat should be modeled diversely, and the modeling results would be useful for preparedness of such events.

Besides dust explosions, pool evaporation of flammables due to dam failure of storage tanks should also be modeled on a similar scale. Another major group of Natechs that have led to responses are the hazardous material incidents resulting from transport activities. Safety measures have been implemented to minimize hazardous material spillage in transport due to natural hazards. The transport of concentrated hazardous chemicals and the regulation model of transport safety are in the context of risks to land areas (Popov et al., 2023). Predicting the consequences to urban infrastructure currently requires development.

5.6.3 Human-Made Threats

Human-made threats (man-made threats) entail all illegal acts committed deliberately, negligently or through error by individual(s) or group(s) of individuals resulting in immediate damage, destruction, and/or loss of target assets, and hence have deleterious effects on human health, the environment, and organization of societies (Farahi-Ashtiani et al., 2023). Actions that are not intended by the actor(s) to cause damage, destruction, or loss are usually not considered as threats, for example bad weather or accidents. Threats of chemical origin will be limited to chemical agents including chemical weapon compounds/toxins and related delivery systems. The development and/or use of weapons of mass destruction (WMDs) and their means of delivery by terrorist organizations would cause catastrophic damage. Thus, as a preventive action, there is a need for risk assessments of the danger, based on as rational information as possible (Zhu et al., 2020).

In developing nations, agrochemicals are sold in open markets, creating serious chemical security risks due to poor protective measures. A survey assessed concerns about chemical security during the marketing and handling of these products. It targeted supply and demand factors, yielding valuable insights from participants that included vendors and consumers from three semi-urban markets, ensuring validity and reliability. By synthesizing views from suppliers and end-users, a comprehensive understanding of the agricultural chemical landscape has emerged. This research utilized a multifaceted methodology, revealing key deficiencies needing attention. The insights gathered aim to guide strategies and policies to enhance safety protocols for agrochemicals, promoting dialogue for a sustainable agricultural framework (Anaduaka et al., 2023).

Major findings were linked to poor chemical handling facilities, no regard for hazard labels, mixing of money and chemicals, exposure of displayed chemicals, visibility of children, carrying of unwrapped containers, and related issues. These results were further analyzed to meet international discussion and decision point requirements and then integrated for further national discussions and decisions on chemical security in agricultural open market places (Scott & Pocock, 2021; Anaduaka et al., 2023; Atinkut Asmare, Freyer, & Bingen, 2022).

Among the four types of threats, it should be stressed that the Human-Made Threats can be considered closest to terrorism scenarios. Even though there are no publicly available single-purposed chemical weapons in West countries, terrorist organizations might still have the capability to design and fabricate such agents as long as there are accessible precursors, knowledge about CW production, and an appropriate way to weaponize agent dispersal (Ubani, Madu, & Umeifekwem, 2024). About disbanding terrorist organizations, unfortunately, such acts are, most importantly, motivated by political, ideological, or religious reasons of significant grievances. Thus, the ultimate factor determining the degree of Human-Made Threats is the motives and the capabilities of terrorist organizations. These two key factors can determine if CW or its delivery systems are sought and acquired by terrorist organizations, if any attacks are to be performed, and if such attacks, once performed, would be successful (Farahi-Ashtiani et al., 2023).

5.6.4 Chemical Security in Africa: A Circular Economy Perspective

The management of chemical safety and security in Africa aligns with Circular Economy principles, shaped by historical chemical weapons experiences. Whereas Morocco has been compliant with the CWC since 1994, many regions lack sufficient legal frameworks for hazardous materials. Though progress has been made in eliminating chemical weapons, ongoing threats from some nations and non-state actors pose risks of fatalities and environmental harm. Compliance with safety regulations is vital due to the export of chemical-grade products. The 2018 CWC Africa Capacity-Building Forum introduced a "Chemical Security Framework" with sixteen "Progress Indicators" to enhance chemical security, particularly in informal sectors, emphasizing awareness and green development initiatives (Bertrand et al., 2016).

Regulation of pesticides in sub-Saharan Africa, particularly West Africa, faces significant challenges. Countries like Senegal, Mali, Côte d'Ivoire, Nigeria, and Ghana are establishing frameworks like Nigeria's National Agency for Food and Drug Administration and Control (NAFDAC) to enhance productivity and minimize risks. Enforcement is hindered by limited resources (Amusat et al., 2023). Sustainable agriculture necessitates following environmental guidelines and integrating eco-friendly practices with traditional methods. Educating farmers on pesticide risks and promoting Good Agricultural Practices are crucial. Collaborations among governments, NGOs, and stakeholders are vital for responsible pesticide usage and sustainable farming (Jepson et al., 2014). Inadequate monitoring of pesticide regulation raises concerns, while weak oversight and untrained customs inspectors result in unchecked imports. This reliance on user knowledge risks pesticide safety, emphasizing the need for a robust regulatory framework to safeguard public health and the environment.

The oversight of 192 pesticides by the Central Sanitary Department (CSD) reveals a critical resource gap. Only four are WHO class 1a or b, restricted to Russian research. However, unregistered pesticides and misuse persist, with at least three unauthorized products identified. Their low cost and regulatory lapses raise concerns, necessitating stronger enforcement. The ongoing use of older chemicals frustrates analysts, as agencies respond mainly to poisoning incidents or fatalities.

An investigation was conducted into the use of 266 chemicals within three specified enterprises in Kenya over a thirty-day timeframe. Various aspects of chemical security, including skills and pertinent information, were evaluated and reported upon efficacy regression (alpha), logistic regression (beta), and a chemical score assessment. The scrutiny of regression alphas with a 0.99 efficacy score is essential. There's insufficient dialogue on scoring plant pesticides and the sharing of hazardous materials by insurance companies. Analysis for nine highly toxic substances termed "utmost imhotep" showed 25 harmful chemicals from 266 data points. Risk categorization recognized two alpha classes: flammable powders and inhalable toxicants, emphasizing the need for effective management (Lekei, Ngowi, & London, 2014).

The 25 hazardous chemicals were categorized into distinct groups: Pesticides and Pesticide Plants, encompassing insecticides (5), herbicides (5), and fungicides (5). Additionally, the analysis identified another category comprising Acute Toxicants, which are characterized by their potential to induce anaphylactic reactions or produce toxic fumes upon combustion (5), as well as a final category consisting of Flammable Powder agents (5).

Understanding pesticide toxicity involves knowing that organophosphates and carbamates block acetylcholinesterase, causing acetylcholine buildup. This hampers neuromuscular functions and risks health. A systematic chemical management approach stresses awareness, education, and safety protocols. Clear labeling and communication improve safety and encourage user responsibility, making informed choices critical for safeguarding public health and the environment (Lekei, Ngowi, & London, 2014).

Kenya adheres to UN agreements on chemical management for responsible usage and disposal, ensuring safety and aligning with global standards. These treaties include the Basel Convention on the Control of Transboundary Movements of Hazardous Wastes and Their Disposal, and the Rotterdam Convention on the Prior Informed Consent Procedure for Certain Hazardous Chemicals and Pesticides in International Trade. Others include the Stockholm Convention on Persistent Organic Pollutants, and the Montreal Protocol on regulation of the production and use of ozone depleting chemicals (Shem et al., 2021; Muigua, 2022). The protocols promote compliance and awareness, emphasizing shared responsibility among government, industry, and communities.

The above require party states to (1) mine and control the production, import/export, and disposal of all chemicals listed in the treaties; and (2) develop and disseminate Training Manuals on the implementation and compliance to the protocols. In Kenya and Nigeria, there exist no compassionate policies or legal frameworks that cater for the vulnerable like the youth, women, and disabled. Such policies and frameworks would seek to evaluate the chemical security situation in the agricultural farms, informal settlements, and the light industry. If offered classes on chemical hazards, the youth and women can adapt safety and protective measures to suit the working environment. Consequently, surveys should be conducted to actively seek these groups' views regarding current training efforts and to identify topics of concern, which may be integrated into safety programs (Saina, Odimu, & Otara, 2017; Ekwempu & Anderson, 2019).

To enhance chemical safety and security, raising public awareness of chemical hazards is vital, as this can empower communities and individuals to address critical issues. However, the lack of training resources for vulnerable populations significantly limits progress. Currently, chemical management training predominantly occurs in specific industries of developed nations. The Kenyan Action Plan promotes inclusive chemical management but often overlooks important safety factors.

For instance, fishermen face pesticide contamination threats but are frequently excluded from compliance discussions and remain uninformed about safe application distances. Ignoring local educational resources and community dialogue hampers vital information exchange and the integration of traditional methods. The exclusion of indigenous knowledge shifts the conversation toward expert opinions, often influenced by political agendas, further sidelining community involvement in tackling these urgent challenges.

5.6.5 Analysis of Chemical Threats in Manufacturing Facilities

In chemical manufacturing facilities, a primary safety issue arises from the potential hazards associated with explosive synthesis areas. To systematically analyze this risk, a consequence tree has been constructed to pinpoint critical events and evaluate their repercussions. Relevant data have been gathered to analyze the probability of these occurrences and incorporated into a comprehensive risk assessment model. Additionally, a methodology has been established to identify noteworthy risks, alongside a risk-grading framework designed to enhance safety management practices and proactively mitigate possible dangers (Sperry, Azuma, & Stone, 2021; Iaiani et al., 2021; Collett et al., 2021).

Chemical safety involves recognizing various threats, which may not lead to major explosions or gas releases. There are eleven significant threats categorized into five risk levels. A new framework aims to assess the likelihood and consequences of consequence-based chemical threats. This method prioritizes safety rankings, highlights the need for precise input analysis, and utilizes simple techniques for probability estimation and consequence scaling. Integrating this approach with system modeling enhances practicality and adaptability across different safety scenarios (Chen et al., 2024a).

In assessing chemical storage tank failures or explosions, even trees can be used to illustrate complex relationships. Bayesian methodologies offer advantages over decision trees, and provide quantitative analyses, forming consequence models based on varying severity inputs. Such methodologies clarify interdependencies among variables, boosting predictive accuracy vital for decision-making in complex situations (Zhu et al., 2020). These models enhance comprehension of outcomes and improve strategic planning, aiding effective risk management.

An improved threat classification model like the Bayesian methodologies enhances analytics with generic graphs, and custom event tree software that calculates probable frequencies, facilitating a cohesive model. Notably, effective risk assessments depend on thorough data collection and management, and high accuracy in data analysis with defined protocols clarifying stakeholder roles. Recognizing hazards and prioritizing practices heighten safety, while dedicated training enhances personnel competence. Regular evaluations of the plan increase management flexibility, allowing proactive assessments to mitigate hazards and foster a safety-focused organization (Segarra, Bensi, & Modarres, 2021).

In event tree construction, a flexible framework is crucial for predicting implications for post-construction. Effective methods to mitigate adverse outcomes include employee training, engineering adjustments, and proactive measures. This underscores the cascading risks from interconnected sub-events influenced by material properties. Addressing risks requires stakeholder collaboration and regular assessments to promote improvement. Advanced modeling enhances prediction accuracy, while feedback loops enable adaptation. A proactive risk management strategy fosters resilience, allowing organizations to navigate uncertainties confidently.

5.7 VULNERABILITY ASSESSMENT

The vulnerability assessment procedure evaluates a facility to identify and prioritize chemical sources and their security measures. A preliminary assessment takes half a day by the Chemical Manager or trained personnel. Specialized teams, after three to four days of Organization for Economic

Co-operation and Development (OECD)-aligned training, conduct thorough evaluations with at least two team members and required materials (Blakey et al., 2013). Guidelines for a comprehensive vulnerability assessment framework for chemical sources are being developed for training. Regulatory bodies will use these methodologies to evaluate the regional chemical industry. The document outlines the assessment process, includes an OECD-compliant chemical risk assessment template, and covers data collection while identifying high-risk chemicals and areas for further investigation.

A chemical security risk assessment (CSRA) aims to identify and prioritize vulnerabilities for effective risk management, adhering to principles like transparency, adaptability, and timeliness (Yuan, Yang, & Reniers, 2024). Countries have varying chemical security laws but share concerns about vulnerabilities. The CSRA framework identifies potential unwanted events linked to hazardous substances. Scenario trees generate WEs, yet the complexities of diverse substances and locations call for qualitative methods and expert insights. Investigating WEs thoroughly entails examining both internal and external causal factors.

Assessing chemical security system's vulnerabilities is crucial for identifying information to develop countermeasures and enhance security. It should address key issues: (1) current state of chemical security, (2) types of potential chemical threats that could impact this system, (3) specific vulnerabilities that exist within the system regarding these threats, and (4) vulnerabilities that pose the greatest risk, and the tactical measures that can mitigate them. Each inquiry comprises several more specific questions aimed at exploring the subcomponents of the primary question. Alternatively, a partially more quantitative series of questions on the detection of possible chemical attacks against a chemical security system can be developed. An initial assessment of vulnerabilities within a chemical system may be performed utilizing these inquiries, either relying on the expertise of professionals or through the administration of a straightforward questionnaire (Huang et al., 2022a).

Responses to these issues generate a deeper understanding, helping to establish a stronger security posture. It is noteworthy that a thorough approach vulnerability assessment ought to address all the issues raised in either of the two methods offered. Stakeholders must assess and counter emerging threats using adaptive strategies. This proactive approach enhances chemical security and fosters personnel awareness, as vigilance is a shared duty of operators and regulators, emphasizing collaboration to safeguard against threats (Rahim et al., 2024; Emkani et al., 2024).

Detailed system diagrams are vital for grasping a system's characteristics and weaknesses, as they improve stakeholder understanding and inspire creative solutions. In chemical facility security, attacks fall into exit-mode and non-exit-mode categories. Exit-mode attacks endanger environmental safety, while non-exit-mode attacks threaten operational integrity and can lead to disasters. Each type has risk assessment indicators to facilitate proactive measures.

The concept of impact evaluates consequences from attacks and sets criteria for assessing chemical security threats. An analytical approach considers factors like attack scale, duration, and secondary effects, which may lead to cascading repercussions. A model focusing on specific regions identifies chemical properties to find immediate and subsequent targets. Damage evaluation outlines exposure scenarios while factoring in significant ancillary effects. Damage functions should reflect attack specifics for accurate assessments. Evaluating potential attackers linked to the identified targets is crucial for risk assessments. Attack trees assess capabilities and guide evaluations. Effective self-assessments require collaboration among stakeholders, while peer assessors enhance awareness through evaluations. Small-scale assessments can lessen participant burdens, and ongoing training improves expertise.

The primary security of hazardous chemicals involves physical security, cybersecurity, and personnel security. Effective measures enhance overall security, while neglecting external factors or improper implementation can diminish security investments. Compliance with legal and societal norms, alongside cost-effectiveness in relation to risks, is vital. Robust chemical security demands specialized knowledge and training in each category, revealing the complexities and integration challenges. Inadequate or discordant oversight can lead to fragmented security initiatives, underscoring the necessity for increased advocacy and understanding of chemical security.

5.7.1 Physical Vulnerability Assessments

Physical security involves off-site assessments and threat evaluations to inform on-site protocols and collaborate with law enforcement, especially in oil, gas, and petrochemical sectors. Access control uses guardhouses, barriers, and ID protocols. Detection systems require intrusion and fatality detection, with fencing at least 3 m high and 2 m deep. Regular maintenance and secondary fencing are crucial to minimize vulnerabilities. Compliance with electric signaling standards is key, with perimeter alarms utilizing varied detection methods, including dual-layer motion detectors.

Assessing vulnerabilities in physical security infrastructures is necessary for robust chemical security. The effectiveness of chemical security relies on adequate physical protection measures. Current standards for evaluating physical safeguards in extreme chemical situations often have deficiencies, overlooking plausible threats (Rahim et al., 2024; Amin et al., 2022). The investigation into targeted assault methodologies for specific physical security systems is underexplored. Consequently, a systematic framework has been created to evaluate physical vulnerabilities. In this framework, authentic attack scenarios are generated using a scenario generation algorithm based on game theory principles, analyzed through the attack tree methodology for vulnerability assessment (Yuan, Reniers, & Yang, 2022).

To assess vulnerabilities in physical protection systems, detailed attack trees must be created to outline potential scenarios, defensive measures, and system attributes. This quantifies vulnerabilities and employs analytical techniques to identify weak points. Methodological advances enable multitiered configurations, lowering computational costs. A case study illustrates the practical efficacy of these methodologies. To manage threats, developing a Strategic Defense Framework, Risk Management Program (RMP), or Proactive Defense Strategy is essential. Response protocols for unauthorized access and breaches should involve local security agencies and be guided by Security Planning Committees. The Security Management Team (SMT) ensures public safety, while critical infrastructures require robust communication and coordination with law enforcement. Effective security and risk management necessitate ongoing resources like surveillance equipment, access controls, alarms, and situational awareness tools (Wang et al., 2021b).

Evaluating a test facility security protocol necessitates a thorough examination of its security architecture, pinpointing system weaknesses that may be exploited through various attack vectors. This analysis delves into the security implications, aiming to bolster comprehension of potential attack scenarios while fortifying chemical security measures. The proposed framework delineates a comprehensive approach to vulnerability assessments, applicable to a multitude of international physical security infrastructures.

In alignment with regulatory standards, a Strategic Threat Mitigation Plan is essential for addressing potential security breaches and safeguarding sensitive materials. Each distinct threat scenario should be meticulously assessed, with tailored response strategies developed in collaboration with security experts and influenced by industry's best practices. The SMT plays a pivotal role in ensuring organizational integrity and operational continuity. Integral support mechanisms encompass robust communication channels across multilevel facilities, and collaboration with law enforcement and emergency services. The dynamic nature of security needs necessitates preparedness for diverse threats and incidents. Such preparedness may involve an array of resources, including surveillance systems, protective gear, emergency communication devices, and clear signage (Tripathi et al., 2021).

5.7.2 Cyber Vulnerability Assessments

Modern industrial reliance on computerized systems highlights the critical role of industrial control systems (ICSs), such as Supervisory Control and Data Acquisition (SCADA) and Programmable Logic Controller (PLC) technologies. Protecting these systems from cyber threats is vital, as exploitation can disrupt infrastructure. A robust framework has been developed for assessing cyber risk in ICS architectures, using an enhanced common vulnerability scoring system (CVSS) for accurate evaluations. Creating a Cybersecurity Risk Management Plan or Incident Response Strategy is essential for handling cyber incidents affecting ICSs. A tailored response protocol for breaches, developed with cybersecurity experts and best practices, is necessary. A comprehensive framework must include robust communication networks and collaboration with law enforcement and cybersecurity professionals, along with continuous resources like intrusion detection systems, firewalls, incident response tools, and personnel training (El-Hajj & Mirza, 2024; Wang et al., 2021a).

Tripathi et al. (2021) investigated model-based security verification of Cyber-Physical Systems (CPS), focusing on a Nuclear Power Plant case study. They used Petri nets to model and assess security aspects, providing a comprehensive evaluation of vulnerabilities and defenses in complex environments through qualitative and quantitative analysis. They highlight the crucial importance of security analysis in CPS, especially in safety-critical areas like nuclear power facilities. Security breaches in such critical infrastructure can cause major disruptions, with possible casualties.

5.7.3 Cybersecurity Vulnerability Assessments in Industrial Control Systems

Navigating cybersecurity challenges in critical infrastructure, particularly in ICS, necessitates frameworks like the CVSS to identify and assess risks. A detailed methodology is vital for evaluating cyber vulnerabilities in a continuous stirred tank reactor (CSTR) system. The refined CVSS model integrates scenario-based assessments associated with attack vectors. This analysis strengthens risk assessment strategies, providing tailored templates for SCADA and PLC systems, leveraging insights from previous assessments and IT operational technology experience. A summary of detected vulnerabilities and engineering changes explores cross-domain attack patterns in CSTR systems, with service process scripts for practical evaluations and a CSTR case study illustrating the framework's significance for industrial assets, highlighting the urgency for proactive cybersecurity measures.

The cyber vulnerability evaluation of chemical production equipment mainly includes three aspects: the identification of chemical devices and the corresponding equipment network topology, the collection of cybersecurity vulnerabilities of devices and security configuration defects, and the vulnerability penetration route generation based on the assets and vulnerabilities. The vulnerabilities are described by the CVSS, which is a standardized severity scoring system. CVSS is a coding system with several dimensions, including attack vector, attack complexity, privileges required, user interaction, scope, confidentiality impact, integrity impact, availability impact, and the base score (El-Hajj & Mirza, 2024; Wang et al., 2021a).

The first phase in vulnerability assessment is cataloging assets and creating a protective framework for their topology. Identifying assets reveals vulnerabilities, such as those from default credentials. The next phase involves using vulnerability intelligence and databases, leveraging asset implementation protocols for automated retrieval, leading to a comprehensive identification report. Understanding each vulnerability through specific exploitation conditions is crucial, utilizing parameters from the CVSS. Cataloged vulnerabilities are analyzed by a threat intelligence service. Severity scores, reflecting common security issues tied to specific equipment, are calculated from exposure metrics. These scores inform a risk assessment model, combined with dynamic scores, resulting in a definitive risk score for each device. The final phase develops control measures for identified exploits and designs test cases to emulate potential attacks, ensuring strong defenses against threats (Nikolaou et al., 2023).

5.7.4 Enhancing Personnel Security in Chemical Facilities

The safeguarding of personnel security is essential for protecting chemical facilities from unauthorized access. It requires thorough assessment of personnel, timely identification and reporting of unusual behaviors, and the establishment of a solid security framework. Evaluating personnel security involves detailed risk analysis aligned with job functions and regulatory requirements, focusing on high and moderate risk roles to quickly identify and address vulnerabilities. Following these assessments, corrective actions must adhere to strict documentation and recruitment policies for Medium Hazard positions, including the regular updating of background checks for those moving from high-risk categories to maintain vigilance.

Clear criteria for supporting documentation and careful screening methods reflecting best practices are essential. A robust risk management approach must differentiate security measures for low- to medium-risk personnel from those designed for high-risk individuals, requiring careful scrutiny. The screening strategy should include deterrents against unauthorized activities and proactive evaluative techniques to catch potential issues early.

A comprehensive Security Plan must include strategic measures, extensive risk assessments, systematic screening protocols, and established mechanisms for reporting incidents, alongside regular post-hire screenings to maintain compliance. Furthermore, the security framework should promote operational transparency through regular meetings focusing on monitoring concerning behaviors and the thorough documentation of records over time. Clear Risk Assessment and Reporting procedures must identify behaviors that require reporting, ensuring the information gathered is utilized to bolster the organization's security stance, thus creating a strong defense against potential threats and enhancing the safety of chemical facilities through a proactive approach to personnel security.

5.8 DOCUMENTATION AND TRAINING IN OPERATIONAL SAFETY

Preparedness in chemical facility operations is critical for preventing and mitigating incidents. Understanding chemical hazards, safety protocols, and equipment functionality is essential. Alarm systems, communication protocols, spill response strategies, and emergency procedures must be integrated into operations. Training staff increases preparedness, equipping them to manage crises effectively. Continuous education fosters a safety culture, enhancing facility resilience against unforeseen events. Preparedness transcends procedures; it involves all stakeholders uniting in commitment to safety and excellence (Blakey et al., 2013). To foster a secure operational environment, meticulous records of on-site conditions, inspections, incidents, and maintenance activities must be diligently maintained. Regular training sessions, ideally on a biannual basis, should be structured around effective documentation practices. Establishing minimum participation thresholds for pertinent authorities is vital, alongside the consistent reinforcement of safety protocols.

The success of incident response frameworks is tied to their execution and unifying them into an operational structure enhances efficacy. A thorough framework identifies vulnerabilities, sets goals, and defines tactical approaches, clarifying roles within the organization. Scenario-based training is vital for improving response skills. Preparedness involves establishing communication with authorities, assessing strategies among responders, and engaging in collaborative drills. Involving evaluators early ensures targeted interventions and data integrity while creating a centralized repository for performance metrics strengthens the response infrastructure and promotes oversight, fostering a culture of vigilance and readiness for staff to manage incidents effectively.

5.8.1 Accident Mitigation Strategies in Chemical Handling

In chemical handling, accidents can occur during storage, transportation, or use, each with distinct indicators needing careful examination. Mitigating incidents starts with minimizing false alarms, accurately interpreting olfactory cues, and engaging personnel skilled in hazard recognition. When emergency responders arrive, strong communication with on-site and off-site specialists is crucial. This collaboration aids in evaluating chemical hazards, implementing safety protocols, assessing transportation risks, and collecting data for future analysis. The initial phase demands efficient information collection and communication, as complex techniques require more time.

The assessment process involves ongoing evaluations: on-site inspections for visible signs, thorough intelligence reviews, operational guidance provision, safety evaluations of these recommendations, outcome estimations, and adjustments to emergency response plans. It's essential that the information provided remains robust and accessible to all involved stakeholders. Additionally, a well-structured communication plan should engage stakeholders and outline follow-up actions to enhance situational awareness. This comprehensive approach is vital in creating an environment that prioritizes safety and sustains preparedness (Hua et al., 2021).

5.8.2 Navigating Social Challenges to Build Resilience

Navigating social challenges necessitates collaboration, community engagement, risk assessments, and emergency preparedness, which build resilience. Engaging diverse stakeholders enhances initiatives with effective solutions. Risk evaluations inform decisions, enabling communities to foresee problems and create mitigation strategies, while preparedness empowers them to manage unexpected situations through cooperation.

To transcend mere incident-focused evaluations, one has to engage in thorough risk assessments that traverse the multifaceted landscape of economic, environmental, and public safety considerations. Such in-depth evaluations can uncover potential risk trade-offs and facilitate informed decisions regarding the substitution of hazardous substances. Established protocols for vulnerability assessments and contingency strategies ought to be recognized and applied. The crafting of Environmental Emergency Plans (E2Ps) calls for a synergistic collaboration between community stakeholders and industry allies. It becomes paramount to delve into recognized frameworks for developing community-oriented E2Ps. Moreover, the advancement of more integrative hazardous chemical release prevention initiatives is essential, accompanied by robust methodologies for analyzing their relative costs and benefits.

In the realm of emergency preparedness, fostering partnerships among various community entities is vital. These collaborations not only enhance the resilience of local infrastructures but also fortify public awareness and involvement. Engaging diverse stakeholders in the development of emergency protocols ensures that the voices of all affected parties are heard, promoting a comprehensive understanding of risks. Educational campaigns and workshops can serve to elevate community knowledge regarding hazardous materials, their potential impacts, and effective response strategies. This proactive approach cultivates an informed populace, equipped to contribute to safety initiatives and emergency responses. Ultimately, the synergy of community engagement and collaborative frameworks establishes a more secure environment for all, facilitating a culture of safety that extends beyond regulatory compliance (Soltanzadeh et al., 2024).

5.9 TRAINING AND STRATEGIES FOR OPERATIONAL EXCELLENCE

An effective training regimen begins with engaging site operators in development. A thorough user community assessment follows, to gather constructive feedback on training efficacy. The site must provide comprehensive pre-training, including detailed security protocol documentation and timely operational intelligence updates. Implementing Standard Operating Procedures (SOPs) is crucial. A full-day evaluation should cover all operations and lead to training sessions for staff, documenting pre-training deficiencies for analysis. Ensuring sufficient overlap between senior and junior initiatives promotes collaboration. Uniform training sessions should be designed to target specific knowledge gaps, and systematic observation reports will be created to refine documentation and improve operational practices continuously.

To achieve operational excellence, facilities need a strong risk management process to identify hazards and comply with site criteria. Rigorous product testing is essential after a review. Operators must undergo critical training evaluations, adopting updated SOPs based on recent threat assessments. Third-party audits and retraining protocols examine security plans and personnel skills, ensuring effectiveness without personal blame for breaches. Measurable outcomes, prompt feedback, adaptability, and proactive engagement strategies are vital for enhancing operational integrity and continuous improvement (Hawkins, 2010).

5.9.1 Enhancing Chemical Security in Low- and Middle-Income Countries

The endeavor to fortify chemical security within low- and middle-income countries (LMICs) has gained formidable traction, yet the pathway to success is fraught with multifaceted challenges. Central to this initiative is the imperative to nurture political will in these regions, thereby ensuring that chemical security occupies a prominent position on national policy agendas. This chapter meticulously outlines a comprehensive framework designed to assess the chemical security landscape across LMIC operations, enriched by preliminary findings from four illustrative case studies. It encapsulates a series of activities undertaken from 2020 to 2022, providing insights into the strategic methodologies employed. Furthermore, it articulates a coherent approach for bolstering LMICs in their quest to markedly enhance their chemical security protocols and elevate their overall operational integrity. Engaging stakeholders at all levels is pivotal, as a unified commitment can cultivate an environment conducive to sustainable improvement in chemical safety and security practices (Abbate et al., 2025; Ball et al., 2022).

The LMICs are confronted with an imperative to bolster their chemical security measures, yet they frequently find themselves devoid of the essential knowledge and strategic frameworks (SFs) necessary for effective enhancement. Despite a robust enthusiasm from donor nations and international organizations to extend their support, a universally recognized methodology to pinpoint the most impactful avenues for intervention remains absent. In a concerted effort to address this gap, the Chemical Security Program has unveiled the "Chemical Security Status Tool," a comprehensive instrument featuring 81 question–answer pairs designed to appraise the chemical security landscape within LMICs. This tool initiates its investigation by scrutinizing the chemical security conditions present in establishments engaged in global health initiatives.

The insights garnered from assessments are pivotal in identifying critical areas that necessitate more thorough evaluations and in steering focused interventions. The report delineates four specific instances of this approach in action: a fragility assessment undertaken in universities across Kenya, Tanzania, Uganda, and Zimbabwe; the application of mass spectrometry techniques within university laboratories in these nations; an appraisal of quality control practices in private laboratories situated in Tanzania and Uganda; and a thorough examination of blood bank operations in Uganda and Tanzania (Rutebemberwa et al., 2020).

The outcomes of these exploratory initiatives reveal a concerning reality: the overall chemical security status within LMIC operations is deficient, with chemical security often relegated to a position of low priority. This predicament is further compounded by insufficient regulatory frameworks and weak organizational infrastructures. Moreover, economic constraints and the lack of a definitive strategic pathway for advancement pose significant challenges. Consequently, it is imperative to offer these nations clear guidance on methodologies for elevating their chemical security protocols and to cultivate a robust intervention framework tailored to support LMICs in this vital domain.

5.10 TECHNOLOGICAL INNOVATIONS IN CHEMICAL SECURITY

Chemical security professionals are continuously encountering evolving and increasingly complex threats that reveal various gaps in response capabilities. The existing literature often oscillates between providing detailed analyses and presenting simplified risk assessments. This approach frequently overlooks the moderate threats that can highlight critical vulnerabilities within systems. In the Netherlands, inspectors prioritize procedural risk management while often overlooking organizational dynamics and human factors, compromising operational safety. Improving resilience in the sector requires combining procedural rigor with insight into human behavior, promoting a safety culture that incorporates technical and human elements essential for preventing incidents (Soltanzadeh et al., 2024; Szklarski, 2023).

5.10.1 Emerging Technologies

Improving safety management in the Dutch chemical industry entails observing best practices through international trips and researching quality enhancement methods. These actions bolster safety protocols. Compliance with legal site selection and operational procedures, alongside national regulations, influences safety outcomes. Facility attributes and data from recognized databases are vital. Moreover, input from local communities, corporate stakeholders, and emergency response teams is crucial for establishing an effective safety management framework for a secure operational environment (Reniers & Khakzad, 2017).

In risk management, software enhances decision-making by allowing quantitative risk assessments. Some tools cater specifically to the chemical industry, while others assess incidents across different sectors. However, a major limitation is their focus on areas near chemical facilities, usually functioning effectively only within a few kilometers. Publicly available tools often rely on limited datasets and broad assumptions. Recent developments highlight the need for improved safety organization and risk governance in larger facilities. Engaging stakeholders is essential; without a strong risk assessment framework, compliance may suffer. Fostering a safety-driven culture requires participation from technical experts and collaboration among users, regulators, and suppliers (Calabrese et al., 2024).

Advancements in Interactive Risk Assessment Models have evolved from focusing on water quality to addressing chemical hazards. Using Monte Carlo simulations, these models improve risk analysis by evaluating uncertain variables' probability distributions (Jiang, Nan, & Yang, 2013). A three-year initiative across 25 European locations enhanced our understanding of chemical risk pathways, highlighting the need for security, prevention, and mitigation in assessments. This issue demands strategies to counter terrorist threats, leading to priority financial support for managing explosive chemical risks.

5.10.2 Assessment of Water Quality through River Graphs

In the evaluation of water quality, the river graph plays an essential and crucial role in effectively displaying the interactions among various components. This graph creates a dynamic and comprehensive model designed to clarify the intricate chemical pathways, which include water sources, diverse chemical constituents, and the transportation systems they navigate. Furthermore, it delves into how various mechanisms effectively dilute chemical concentrations, which is pivotal in assessing exposure risks posed to surrounding communities. Public health officials actively assess these risks, diligently addressing potential threats that may arise from terrorism and warfare scenarios. Moreover, an advanced and interactive model for chemical risk assessment produces detailed visuals in a digital format, emphasizing the importance of continuous monitoring and proactive measures needed to safeguard both water quality and public health in the face of potential contaminants (Blakey et al., 2013).

5.10.3 Artificial Intelligence and Molecular Development Landscape

The rise of advanced industrial artificial intelligence (AI) tools significantly lowers knowledge barriers for Toxic Compound Design, surpassing traditional methods. This shift marks a new era of science-driven hazards. The development of AI and Machine Learning (ML) presents challenges regarding technology proliferation, necessitating strategies to limit their reach. These innovations threaten nonproliferation efforts related to small molecules, novel protein construction, and genetic material. Advanced AI and ML tools pose diverse proliferation risks, including the unintended spread of information about toxic compounds, which a skilled engineer could have controlled. Moreover, industrial AI technology facilitates hazardous molecule assessments like benign compounds, creating further concerns (Hunter, 2024). Harnessing AI and ML in quality control necessitates understanding their principles.

Data-driven decisions enhance inspections, while visual cues emphasize critical factors. Clear guidelines are vital in complex settings. Basic alerts fall short; training staff on AI/ML develops a robust quality assurance culture. Foundational programs are essential, as mere vigilance is insufficient. A SF and IPs are critical for incorporating AI/ML technologies per industry standards. Addressing data integrity and algorithmic transparency requires expert collaboration and updated benchmarks. The Quality Management Team ensures effectiveness and alignment. Implementing AI/ML needs strong data networks, ongoing support, and tools like validation, analytical software, training, and documentation.

5.11 CASE STUDIES

Assessing chemical security conditions is crucial for analysis insights. Various multi-criteria assessment methods, like the fuzzy analytic hierarchy process, are employed for detailed evaluation. An illustrative example highlights the methodology, clarifying complex dynamics. These techniques emphasize systematic evaluation and interconnections among key factors essential for effective chemical security management, including the cases of terrorist attacks that could cause accidental releases of hazardous materials at industrial facilities. Analyzing these aspects helps stakeholders gain insights, leading to informed decisions for enhanced safety and regulatory compliance (Santella & Steinberg, 2011).

In specialty chemical production, a meticulous approach is essential. Subcontracting for innovative chemical reaction technologies requires commitment to synthesizing organic chemicals without animal or vegetable substrates. Precise creation of specialty chemicals is crucial for specific applications, emphasizing sterility, high purity, and cleanliness for optimal performance. Comprehensive materials safety data sheets are updated to support safety and regulatory adherence, ensuring compliance with standards, while classification of chemicals according to Chinese regulations reinforces safety in specialty chemical manufacturing (Chen et al., 2024a).

A thorough examination of protocols is essential for chemical safety and management. This includes meticulous documentation of equipment safety, measurement and analytical controls, and emergency response strategies for hazardous material releases. Establishing lockout/tagout procedures and deploying gas detection and fire alarm systems, along with executing emergency drills with trained personnel, are vital for a robust safety framework. Routine inspections and certifications of fire suppression systems and monitoring devices, coupled with phasing out methylene chloride for safer alternatives, are pivotal. Ultimately, Company A reflects commendable chemical safety but has notable gaps and industry challenges to address. The comprehensive review offers an in-depth analysis of safety parameters and a user-friendly graphical interface to enhance data manipulation (Furxhi et al., 2023).

5.11.1 Case Study 1: Chemical Facility Incident

Following the best practices of a major industrial specialty chemical manufacturer, an incident occurred at a manufacturing facility. A chemical reaction involving an intermediate compound went out of control in a reactor and when the pressures exceeded the safety limits, the safety vent on a pressure-dependent vessel opened. The vent went on to release fumes of an intermediate compound, which was categorized as both toxic gas and vapor by human health criteria and as a reactive material. The original safety vent was designed with a vertical 8-inch pipe properly directed skyward. However, its position was later modified to connect directly with a common manifold which services multiple vessels, instead of directing the vent outward away from flammable storage field tanks. Ingress water during the subsequent violent reaction expanded rapidly. The heterogeneous mixture became a white cloud of toxic vapor, expanding laterally within minutes toward housing estates and the workplace on site, with no effective on-site response in place to contain the event (Coker & Kayode, 2001).

Data on the incremental aspects of the incident were reconstructed from recordings made by a factory supervisor in the operators' room. No operator was directly present near the site of the incident due to the toxic hazard, but they were monitoring the automatic safety controllers. The recording contained events that led to the secondary explosion in the reactor module, which involved the original scenario before the cloud was detectable half a mile away on the weather radar.

Animation of incremental reconciled observations helped describe how one large offset cloud expanded rapidly toward residences, which were around 3,250 feet away—far more than its predicted travel time of 2 minutes based on weather data. A hidden cause-and-effect analysis of why the initial toxic gas cloud, intended as a quick response, failed was conducted. Signs of abnormal behavior that occurred before the failure mode were explained. Investigation into the constraints that led to hidden defects among the participants found that subjective know-how, inadequate justification, and a lack of foresighted analysis all contributed to the weaknesses underlying the failures.

In 2020, sodium dichloroisocyanurate dihydrate (NaDCC dihydrate), known as CDB-56®, exploded, causing a fire and releasing toxic chlorine gas. The explosion severely damaged the facility and scattered debris up to 800 m away. This incident occurred during a dehydration process to convert the compound into anhydrous sodium dichloroisocyanurate. A decomposition reaction unexpectedly raised the internal pressure in the dryer beyond safe limits, leading to the explosion. Fragments from the dryer struck a methanol pipe, igniting it (Chemical Safety and Hazard Investigation Board, 2023).

5.11.2 Case Study 2: Regulatory Compliance Failure

Any chemical company that fails to comply with regulatory controls can be said to have breached regulatory compliance, an act that could cause chemical contamination in surrounding neighborhoods. Problems regarding detection and diagnostic measures for regulatory compliance are discussed by focusing on the study of a Korean chemical company located downtown in an urbanized neighborhood.

The hydrofluoric acid accident in Gumi, Korea, in 2012 was a catalyst for the Chemicals Control Act in 2015, leading to the Chemicals Accident Prevention System (CAPS). This legislation categorized all facilities managing hazardous substances as high-risk, raising concerns about its broad scope. It included smaller entities like laboratories, prompting debates on safety and operational feasibility. Discussions highlighted the need for refined regulatory approaches and the challenges in risk assessment, aiming to protect public health while supporting scientific research.

An investigation assessed the 2021 revision of CAPS on medium-sized plating and paint manufacturing operations. The study compared the former Korean Occupational Chemical Accident (OCA) Prevention System with the new Preparation of Off-Site Consequence Analyses (POCA) framework, which includes off-site consequence evaluations. POCA incorporates aspects of OCA and the US RMP, differentiating risk assessment by business size. Results showed that the POCA system offers a more rational approach for medium-sized organizations (Lee et al., 2021).

The Korean CAPS, established under the Occupational Safety and Health Act (OSHA) and the Chemical Control Act, aims to reduce chemical-related incidents in occupational and community settings, and the environment at large. Key CAPS elements include (1) *PSM*—initiated in 1996 to prevent accidents via hazard identification and safety protocols; (2) *Chemical Control Act*—revised in 2021 to categorize businesses by risk profiles (high, medium, or low) based on hazardous chemicals used; (3) *Chemical Accidents Response Information System (CARIS)*—tracks hazardous materials during emergencies, providing real-time data for effective responses; (4) *Risk Analysis and Management (RAM)*—uses a unique Korean approach for assessing potential hazards across business sites, identifying improvement opportunities; (5) *Reporting and Management Plans (RAMP)*—businesses handling hazardous chemicals must create comprehensive prevention plans that consolidate reporting obligations; (6) *Grading System*—PSM also employs a grading framework to encourage compliance among organizations.

The CAPS evolution is categorized into three phases: (1) *Prior to 1996*—focused mainly on worker safety, (2) *1996–2021*—introduction and refinement of PSM and the Chemical Control Act expanded regulations, and (3) *2021 Revision*—updates included mandatory off-site consequence assessments and exemptions for smaller businesses from some regulations. Overall, the KAPS integrates preventive measures, emergency response, and ongoing improvements to manage chemical risks effectively.

According to statutory records, a chemical company had passed compliance evaluations and safety inspections for all water, air, and toxic substances-related laws but later discovered a leaking pond full of unpermitted chemicals. Upon evaluation of the company, it was determined that toxicity and pervasiveness were among the impairing factors due to the presence of rudimentary regional contamination assessments.

Reliance on official records consequently results in a type of stupidity: blind faith in a system or a dataset that could be corrupted (Lee et al., 2021). Another problem is enforcement discretion error and insufficient consideration of regional conditions during the evaluation process. Validation and review were lacking, such that legal compliance records alone could not excuse the chemical company from accountability. It should be noted that laws exist to protect human life and the environment, not vice versa. Failure of a substance to meet the formalistic definition of a hazardous chemical should not allow punishment-free exploitation of the substance to corrupt communities. Another problem is attempting to comply with laws in a country where state-backed discrimination persists.

Verifying the missing link of a problem is accompanied by indirect detection and broadening of a management scope, and effective use of the identification-based scheme results in proactivity. Furthermore, there is a need for the active use of analysis tools involving spatial analysis along with the collection of hydrochemical data, such that existing enforcement strategies can be effectively combined with updated detection measures to reinforce monitoring of the compliance status of chemicals in a regional context.

In a nutshell, understanding the needs of smaller businesses is crucial in addressing regulatory challenges. Exempting very small organizations, such as labs and hospitals, from strict regulations can create a framework for mid-sized enterprises. Research shows that the POCA system effectively manages chemical risks and improves safety and security culture. A strong risk management framework should be adaptable for various sizes, supporting medium-sized businesses without overwhelming them. Engaging stakeholders in strategy development enhances safety outcomes while aligning with business needs. The goal is to establish a safety infrastructure that helps organizations manage chemical risks and comply with safety standards.

5.11.3 Strategies for Chemical Safety and Environmental Stewardship

The Trinidad and Tobago Chemicals Industry and the Environmental Management Authority initiated a self-assessment program focusing on managing chemical substances. This program analyzed objectives, demographics, frameworks, methodologies, and procedures essential for risk identification and mitigation in chemical usage. The initiative involved six self-assessments, bolstered by discussions with government, industry, and NGOs. The primary goal was to cultivate a robust self-assessment culture within the chemical industry. Strategic initiatives have since been established to promote partnerships and self-assessment techniques, ensuring involvement and progress (Trinidad & Tobago, 2018).

The self-assessment project, a collaboration between the chemicals sector and the Environmental Management Authority in Trinidad and Tobago, sought to raise awareness on industry vulnerability issues. This analysis outlined the chemical landscape of Trinidad and Tobago, the structure and implementation of the initiative, and provides insights from its application.

The chemical and environmental sectors in Trinidad and Tobago include a wide variety of industries, such as petrochemicals, agriculture, and multiple manufacturing operations. Before the launch of this initiative, a considerable lack of awareness existed concerning environmental impacts. This gap was systematically addressed through a series of comprehensive workshops and targeted meetings within specific sectors, focusing on essential topics like waste management, air quality, and chemical usage. The initial phase of the self-assessment initiative featured a targeted workshop designed to fit the unique needs of Trinidad and Tobago. This workshop aimed to gather insights from various regional stakeholders (Trinidad & Tobago, 2018).

Additionally, outreach strategies were implemented to engage potential donors, with the goal of increasing recognition and support beyond local areas. A well-crafted press release, along with strategic communications to stakeholders, outlined the program's objectives, expected outcomes, and complexity. In April 2000, a thorough proposal was developed and presented to key members of the Chemicals Sector and the Environmental Management Authority for evaluation, feedback, and funding identification. This evaluation led to a new project proposal aimed at Canadian consultants for expertise, culminating in the appointment of a Trinidadian facilitator from the EMA's staff in July 2000, marking significant project progress.

The initiative was carefully organized around two key elements: developing an environmental policy for the domestic chemicals sector to catalyze transformation, and implementing an inclusive workshop approach that involves diverse stakeholders, emphasizing the complexities of chemicals management.

Understanding the multifaceted nature of environmental impacts is paramount, necessitating a collaborative approach that fosters dialogue and collective action. Comprehensive guidelines should be outlined to facilitate effective communication among all participants, ensuring that insights gleaned from varied perspectives are harnessed to inform best practices. Additionally, ongoing education and capacity-building efforts are essential to empower stakeholders, promoting awareness of the environmental ramifications associated with chemical usage. This proactive engagement not only cultivates a shared responsibility but also enhances the overall efficacy of policy implementation in the realm of chemical management (Trinidad & Tobago, 2018; Santella & Steinberg, 2011).

5.12 CHALLENGES IN CHEMICAL SECURITY IMPLEMENTATION

To improve chemical security and safety, initiatives must assess each facility's unique conditions, especially regarding chemical WMDs disposal risks. Identifying vulnerabilities and creating customized strategies are crucial. Security programs should be facility-specific, as standardized methods often miss out on key details. Tailored programs, while complex and resource-heavy, are necessary for effective risk management in processing environments (Zhu et al., 2020). Traditional chemical risk assessment primarily focuses on safety assessment, and hence cannot be directly used to address chemical security assessment, as the mindset is different.

Toxic substances don't always indicate abuse potential, yet conversations on chemical security highlight misuse risks. It's essential to identify facilities that could be vulnerable to WMDs, relying on public resources like data about nations with chemical agents, official reports, and intelligence assessments. Comprehensive evaluations require analyzing the geopolitical landscape with both qualitative and quantitative data. Collaborating with regulatory agencies improves risk assessments and promotes proactive prevention. Continuous monitoring is vital due to evolving threats, and engaging stakeholders enhances communication and promotes shared responsibility for preventing misuse of hazardous materials.

5.12.1 Resource Constraints

To ensure the security of hazardous substances, various initiatives must be enacted, including the development of robust policies, careful procurement methods, and the creation of fortified storage solutions alongside oversight of laboratory settings. Studies indicate a strong correlation between chemical security protocols and GDP per capita, supporting existing theories. However, quick security assessments often overlook the complex international landscape governing these practices. A comprehensive examination is essential, as chemical security involves complex systems crucial for enhancing global safety standards related to the management and use of chemical agents.

Exploring economic indicators reveals crucial insights into global trade dynamics. Similar states of chemical security efforts are often explained by similar GDP per capita levels across countries. For instance, South Korea, Japan, and France, with GDP per capita of between 25,000 and 40,000 dollars in 2010, have high national chemical security investment efforts and with relatively low numbers of laboratories or agents. In their vicinity are exploration countries of Russia, North Korea, or India, where the reliance on chemical capabilities and investments in relevant security efforts is rapid. The reverse observation is true for low GDP per capita emerging economies, such as Afghanistan or Bangladesh (Najafabadi et al., 2022).

5.12.2 Policy Gaps

To enhance chemical security monitoring in the United States, addressing key policy deficiencies identified is essential. These deficiencies relate to data integrity, regulatory frameworks, and research funding. Establishing solid data collection mechanisms is critical for accuracy and timely dissemination. Reforms should promote a cohesive governing structure to encourage compliance. Moreover, increased research funding is vital for exploring new methodologies to strengthen chemical security. Collaboration among policymakers, industry leaders, and academic institutions is necessary to bolster national chemical security and create resilience against potential threats.

Establishing a comprehensive chemical inventory addresses the urgent need for a detailed database of chemicals in circulation. This inventory catalogs crucial information on substances available in the market, their applications, and quantities in local and regional facilities. Many organizations, especially educational institutions, often overlook the hazardous chemicals in their premises and surrounding areas. A well-organized chemical inventory would streamline data collection for vulnerability assessments, enhance outreach accuracy, support community risk evaluations, and reduce regulatory overlap for less critical substances. Additionally, it would enable the integration of data from state and federal databases, leading to a geographic information system that provides public access to these essential records.

For most nations, especially the LMICs, there is notable deficiency in the public's access to crucial information regarding chemical safety and security. Therefore, Enhancing Public Access to Chemical Information is crucial. Such an initiative signifies a vital leap toward establishing a more integrated framework for the surveillance and dissemination of data pertinent to chemical security. Currently, many progressive states engage in the exchange of information that remains limited in scope. A related policy gap exists concerning the enhanced regulation and availability of mobile chemical data, which often traverses various ownerships in a manner that is transactional rather than consistently reliable. Within this framework, it is prudent to consider the establishment of an interagency entity dedicated to the oversight of the portability and distribution of chemical information nationwide.

5.13 BEST PRACTICES FOR CHEMICAL SECURITY

Many countries have passed laws regulating toxic substances. Enforcing these regulations is vital for preventing misuse and ensuring proper storage and logging to trace chemical exposure in incidents. However, the effectiveness of these frameworks is limited. They can deter improper use, but creating a complete system for secure storage and documentation to fully eliminate misuse proves impractical in an open society (Blakey et al., 2013).

The need for absolute security in chemical facilities makes conducting susceptibility and vulnerability assessments essential. These evaluations identify critical deficiencies in current security frameworks and form the basis for developing effective mitigation strategies. A systematic vulnerability assessment identifies security gaps by examining procedures against worst-case breach scenarios. This process outlines necessary short-term, medium-term, and long-term measures for vulnerabilities. It is crucial for the assessment team to grasp security policies set by management and operational practices. Key areas include the types of identification credentials, verification processes, visitor logs, audit frequencies, and consequences of discrepancies. If individuals show frequent inconsistencies, protocols need establishment. Analyzing security measures is vital; even "redundant" procedures may lack oversight and require attention.

5.14 STAKEHOLDER ENGAGEMENT AND PARTNERSHIPS

Effective assessments, like audits, require understanding of the implementation level of variables to determine status and drive future progress. Organizations develop chemical security audits and assessment tools with significant outcomes. Clear methodologies and protocols are vital to improve these tools, while knowledgeable personnel are essential for a robust organizational framework for evaluators (Soltanzadeh et al., 2024). Chemical security entities often lack protective measures and tools, highlighting the need for effective resource use in assessments. Evaluating protocols and engaging experts are vital. Tailored assessments can drive improvements, and fostering participation is key. Building stakeholder capacity for collaboration with local experts and partnering with resourceful organizations is essential for enhancing protocols. Collaboration among stakeholders (law enforcement, emergency responders, and the chemical industry) is crucial for effective educational campaigns. Government policies should establish a framework to raise awareness about chemical security issues. Enhanced effectiveness comes from broad stakeholder involvement complemented by annual public reports highlighting successes and contributions from partners.

5.14.1 Public Engagement and Educational Framework in Chemical Security

The promotion of public engagement and educational programs within a national chemical security framework is a critical element for success in enhancing safety protocols. It is imperative for nations to establish comprehensive initiatives aimed at improving awareness across a broad spectrum of demographics, including the general populace, chemical manufacturers, distributors, educators, and young people. These programs play a vital role in effectively communicating the nature of chemical security threats, the associated risks, and the requisite protective measures. An effective public engagement approach should specifically target different demographic groups while avoiding the provocation of unnecessary alarm.

This engagement fosters community–industry partnerships to address concerns. Participation builds trust, essential for a secure environment, while diverse perspectives enhance safety measures in chemical management (Blakey et al., 2013). Communities get concerned about risks from local facilities, highlighting hazards and resource challenges. Public health authorities need to evaluate the chemicals produced, used, stored, disposed of, or transported to understand their security. This involves documenting chemical distribution and volume, while considering demographics and sensitive sites like schools and hospitals.

Effective communication is crucial for sharing findings and engaging stakeholders, using meetings, newsletters, and digital platforms to foster trust. Public awareness campaigns should educate residents on risks and safety practices. Collaboration among local authorities, industry, and community members is vital for the management of hazardous material risks. Therefore, collaborations between governmental bodies and industry should concentrate on creating and delivering educational materials focused on chemical security. While traditional training approaches are significant, there is a pressing need for innovative outreach methods, especially tailored for smaller entities in developing countries. Engagement from civil society in these initiatives is crucial, as it can facilitate partnerships with government entities and participation in specialized training programs aimed at officials.

Collaboration between public and private sectors is essential for responding to key events, with media playing a critical role. Engagement from both sectors and media integration in simulations help understand the impacts on public perception. Identifying target audiences is vital in planning, while gamification offers insights on safety and health. Ongoing evaluations must assess data and media coverage, comparing traditional and innovative practices for sustainable solutions.

5.14.2 Government and Industry Cooperation

A unified framework combining government, scientific, and industrial efforts is vital for managing chemical security and incidents. Strategic initiatives should aim to create sustainable partnerships for better information exchange and customized training programs. Thorough damage assessments are essential for response optimization, while strong communication with local authorities enhances public awareness and community self-defense. Developing early warning systems and effective alert protocols is crucial. Identifying the most effective collaborative structure, whether formal coalitions or broader groups, will effectively tackle chemical security challenges.

Enhancing educational initiatives and knowledge sharing is vital for chemical management. Establishing a national institution to evaluate educational frameworks is crucial to ensure alignment with regional strategies and a secure future. The Chemical Safety Initiative highlights the importance of strong government oversight in the chemical sector, calling for strict regulations to protect public health and environmental integrity. Regulatory bodies must support private entities until they achieve operational maturity and can adopt self-regulation. This approach should be viewed as a strategic investment that advances security in the chemical industry while easing government oversight. Nations should develop cohesive regulatory frameworks in line with international standards. Additionally, governments need to enhance safety protocol supervision in rapidly evolving sectors to ensure innovations do not compromise safety.

5.14.3 Advancements in Global Chemical Security

Enhancing global chemical security requires international cooperation in several areas: performing national risk assessments, creating robust regulations, developing strategies for chemical threats, and offering extensive training. These actions promote vigilance and readiness, enabling stakeholders to effectively manage risks. Strong communication among regulators, security experts, and facility operators is vital for coordination. A proactive stance with ongoing assessments and flexible strategies is key to confronting emerging threats. Chemical security is a collective duty that nurtures a culture of safety to protect communities and ecosystems from hazardous chemicals (Alavandar, 2025).

Navigating the Complexities of International Collaboration in international relations reveals that nation interactions impact strategic effectiveness. Information exchange fosters security awareness, crucial for safeguarding cross-border chemical substances. A unified strategy, integrating diverse perspectives and expert insights, is vital for strengthening chemical security and global vigilance. Examining the complexities and intricacies of diplomacy reveals its fundamental reliance on building and maintaining trust. The establishment of shared objectives facilitates collaborative efforts, and a deep respect for cultural differences significantly contributes to the efficiency of communication. A thorough risk assessment addresses challenges collectively. Emphasizing shared resilience and proactive engagement enables nations to form alliances that tackle immediate concerns and encourage sustainable cooperation (Fabia, Fabia, & Piotrowski, 2024).

To tackle existing challenges, it is crucial to engage stakeholders across sectors, ensuring public and private entities contribute to the collaborative effort. Integrating advanced technologies and new methodologies can strengthen security protocols in line with international standards. The progress in global chemical security depends on a unified strategy focused on raising awareness, fostering cooperation, and implementing measures to navigate the complexities of a connected world (Blakey et al., 2013). Enhancing chemical security through international cooperation requires global expert involvement and effective strategies, including security practices and training programs to strengthen measures. These initiatives support national authorities and promote capacity building. Regional organizations foster collaboration, enabling partnerships through workshops. Challenges highlight the need for alliances with national authorities to tackle urgent issues.

Organizations confront challenges in safeguarding chemical substances due to threats such as corruption, theft, espionage, and insider risks, which differ by region and are often ignored in wider discussions. Acknowledging local entities that provide training in chemical security is crucial for fostering collaboration and knowledge sharing. Partnerships with disarmament-focused institutions are essential to enhance current organizations and ensure ongoing training through cooperative rather than new initiatives.

5.14.4 Innovative Approaches to Chemical Security in Education

To innovate educational programs, integrating chemical security into higher education and professional development is essential. Defining chemical security in curricula promotes a focused approach. Approaches for conducting vulnerability assessments and managing crises can lead to the development of compelling content, while frameworks for risk analysis provide essential understandings. Cognitive exploration in new fields can expand offerings, while outreach in public institutions can raise awareness. The supportive structure of public education, backed by regulations, strengthens these efforts. Analyzing existing educational programs allows for a thorough evaluation of their ability to incorporate emerging themes. It is essential to create innovative resources for educators to effectively engage the public in discussions surrounding chemical security issues. By tackling these challenges, the educational sector enhances its preparedness to navigate the complexities associated with chemical security.

The landscape of pre-college educational programs is vast and complex, making a singular analysis challenging. However, we can categorize these initiatives into three main groups: (1) programs organized by governmental bodies, (2) those led by nontraditional organizations, and (3) specialized programs focusing on technical or scientific skills rather than standard curricula. This classification reveals the varying roles of stakeholders—workers, agencies, and organizations—in creating new educational opportunities, with involvement ranging from minimal to extensive. This framework is relevant for entities yet to start projects related to chemical security, indicating a clear potential for innovative program development in this important area.

5.14.5 Gaps in Chemical Defense Strategies

The primary goal of chemical security is to prevent harmful incidents from happening. However, it is essential to acknowledge that no plan can fully prevent catastrophic events, especially those threatening human life. If a breach occurs at a High-Risk Facility (HRF) meant to prevent terrorist attacks or sabotage with hazardous materials, effective measures must be in place to mitigate the impact of such violations. Infiltrations into HRFs require a clear response protocol to handle the aftermath. Most terrorist plans aim to avoid detection, and if successful, the next critical phase may lead to severe outcomes, highlighting the importance of timing to prevent irreversible releases of hazardous materials.

In chemical security, a gap exists in understanding how to collaborate and resource for mitigating the impacts of chemical weapon attacks. While focus often lies on immediate threats, any chemical incident can disrupt industries and critical infrastructure. Such disasters could surpass previous terrorist events and raise concerns for first responders at high-risk sites. Addressing these threats requires strategic planning and collaboration among agencies, with a response framework that involves local government, emergency services, and industry leaders to establish protocols and long-term resilience strategies. Enhanced training and regular drills for first responders are crucial, as is strong interagency communication for better information sharing. Identifying vulnerabilities and developing robust contingency plans are vital for community safety against chemical threats (Blakey et al., 2013).

Mitigating emerging threats in chemical incident response addresses the challenges of chemical incidents, despite existing safety measures. The initial phase involves assessing detection needs and safety protocols for first responders against deliberate hazardous attacks. Key steps include identifying contaminants, utilizing detection tools, ensuring protective gear, and implementing decontamination methods. This highlights the need for precautions to prepare responders for complex situations effectively.

5.15 FUTURE TRENDS IN CHEMICAL SECURITY

The landscape of chemical security is disabled by practical barriers such as limited resources for regulatory compliance and information sharing, and commitment to adhering to security regulations. Other barriers include communication and logistical considerations associated with transportation security, limited organizational buy-in for health and safety issues, challenges associated with implementing protective strategies in user facilities and distribution networks, trade secret concerns, and the influence of legal environments (Reniers & Khakzad, 2017). However, the future of chemical security is enabled by several trends in and impacting the private sector. These trends include heightened awareness of terrorist threats, rapid growth in commercial information technologies, increased emphasis on information security in chemical and utility sectors, advancements in biotechnology, and greater involvement of economic and industrial sectors in government security discussions.

The prediction of costs and availability of component parts is essential for advancing modern technologies. A situation arises where these components can be either overly expensive or too accessible, which sharply contrasts with the tightly regulated and secretive aspects of nuclear weaponry. In this environment, the capability of terrorist organizations to develop chemical weapons is greatly reduced. Additionally, the availability of education in less technical societies adds another layer of difficulty, as it becomes harder for these groups to properly analyze the complex and detailed information required for the effective design and production of a weapon. This complexity also affects the development of future technologies, which necessitate even deeper technical understanding and the added responsibility of maintaining weapons.

Importantly, terrorist groups typically need at least two uninterrupted years to gather the necessary resources and knowledge to create a small-scale reprocessing and storage facility. Consequently, it is crucial for relevant parties to sustain strict design limitations. Research suggests that discouraging research activities entirely is often more economical and effective than merely tackling problems related to the availability of supply chains and quantities. By examining and exposing the inherent weaknesses in various proposed strategies while underscoring the significance of theoretical research, a more effective method for reducing the substantial risks linked to the proliferation and misuse of chemical weapons can be achieved.

5.16 CONCLUSION AND RECOMMENDATIONS

Frequent incidents in the chemical industry highlight the need for improved safety protocols Governments and organizations aim to implement effective practices, such as the Seveso Directive in Europe, mandating robust safety management systems. Despite these efforts, challenges remain, with over three major accidents annually resulting in fatalities, injuries, evacuations, and infrastructure damage. Continuous improvement in chemical safety is essential for industry leaders and regulators globally.

Concerns about chemical facility security have surged due to threats such as espionage, sabotage, cyberattacks, and terrorism. Experts warn of severe consequences from potential attacks, raising doubts about the sector's security. After the September 11 attacks in United States, nations invested billions in critical security infrastructure, and improved safety and security legislation. Yet, assessments show that physical security at chemical facilities remains critically weak. Geopolitical tensions, especially following Russia's 2022 invasion of Ukraine, have further complicated regional safety challenges.

5.16.1 Critical Components of Chemical Security Frameworks

There have been concerted efforts to create a community of practice focused on chemical security, uniting individuals from diverse national and institutional backgrounds to enhance the effectiveness of security measures. Their initial goal was to optimize resources by addressing recurrent mistakes seen in multiple countries. Although many networks recognize the establishment of such communities as significant, practical execution has often been hindered by disappointing outcomes from previous assessments and initiatives. This has led to a lack of confidence in the area, despite some stakeholders believing in the potential for meaningful progress in chemical security.

Advancements in chemical security are evolving amid increasing global complexity. New developments are set to boost collaboration with international stakeholders and enhance proactive measures. More participants will likely engage in documenting effective practices and fostering robust knowledge management resources. Current contributors are expected to promote communities of practice, emphasizing multilateral collaborations that enrich collective responses to emerging challenges in chemical security.

5.16.2 Proposed Areas for Future Research

Before re-evaluating chemical security status, researching current evaluation methodologies is essential. A chemical facility inventory can identify potential concerns, but its effectiveness depends on the manager's knowledge of the chemicals, the inventory methodology, and the chemical list used. Vulnerabilities may also go beyond hazardous pollutants, highlighting the need for better guidance on information collection and potential gaps in the inventory.

In chemical containment facilities, vulnerability assessments are crucial, especially for transfer, storage, and transport of liquid chemicals. Identifying potential weaknesses is vital for enhancing security measures. However, other chemical security challenges also need detailed examination and may require varied evaluative techniques. These considerations highlight that current guidelines may not encompass all relevant issues. A comprehensive approach is essential to address every vulnerability adequately, thereby ensuring a robust system for chemical safety and security.

A promising future investigation area involves utilizing historical incident records to enhance vulnerability assessments without relying on proprietary intelligence. Such data can be sourced from engineering risk assessments or legal cases, focusing on the probability and impact of chemical emissions. Anonymized groups handling similar chemicals could share experiences, improving risk management processes. Integrating past insights with current methodologies offers a valuable opportunity to bolster safety protocols in chemical management.

5.16.3 Policy Recommendations

To ensure effective chemical safety management, a systematic methodology is essential for developing politically viable and operationally sound strategies. This involves strict regulations, regular security inspections, comprehensive audits, and meticulous reporting. Engaging stakeholders from industry, regulators, the public and customs authorities is vital to limit access to dangerous chemicals. Incorporating risk management principles into governmental decisions and fostering innovative discussions between public institutions and the chemical industry are crucial. Identifying unaddressed hazards and implementing proactive measures within a broad risk reduction framework is fundamental. As many hazards arise from industrial activities, refining safety and inspection protocols in production facilities is necessary.

Systematic examination of hazardous incident reports and investigations into root causes is critical for improving safety measures. Although establishing chemical safety authorities is recognized as essential, urgency in this matter is lacking. Embedding safety into societal norms and shifting from reactive to proactive prevention strategies is vital for creating a resilient safety management system with clear accountability. Furthermore, advancing emergency preparedness against potential terrorist threats regarding chemicals requires international dialogue, with effective communication between industry stakeholders and security agencies, in addition to stringent oversight of chemical transportation and hazardous materials management.

The introductory section of this report has undergone a thorough reassessment in light of a detailed examination of a hypothetical chemical security incident. This examination enables TRENDS to evaluate the present state of chemical security more accurately. The insights gained from this analysis are incorporated within this section, which also outlines potential strategies for the enhancement of chemical security protocols. The scenario analysis concentrated on discerning relevant signals and signs, culminating in the detection team successfully identifying all significant indicators, classifying them as observations. However, a critical point of concern is that the detection team failed to enact any follow-up measures post-identification. This omission raises significant questions regarding the implementation efficacy of the detection protocol, suggesting it may either lack completeness or be insufficient in its execution.

Consequently, the evaluation of the signal and sign detection process has been classified as Moderate, indicating performance that does not meet the anticipated standards on the Evaluation Scale. This finding emphasizes the urgent necessity for chemical facilities to develop more explicit protocols to respond appropriately following the detection of signals and signs, thereby improving their preparedness for analogous incidents. Additionally, it may prove advantageous to establish a preliminary system for the classification or assessment of observations before proceeding with actions during detection scenarios. Furthermore, an in-depth exploration of the secondary scenario could be enhanced by incorporating additional data and more nuanced narrative contexts. Implementing this evaluation within a workshop framework could also facilitate increased stakeholder engagement in the process. On a broader scale, it is advisable for chemical facilities in Ethiopia to adopt comprehensive chemical security measures, utilizing these scenario evaluations as a fundamental strategy for capacity enhancement.

6 Management of Biological Specimen and Biohazard Risks

Apollo Odhiambo Maima

6.1 UNDERSTANDING BIO-SPECIMENS AND THE BIOHAZARD RISKS

Clinical work encompasses the careful management of biological specimens, which can originate from humans, animals, or environmental sources. Biological specimens, either for laboratory use or for medical diagnostics, come in a variety of forms, including blood, urine, feces, tissues, sputum, and other bodily fluids, as well as swabs and more. These specimens, regardless of infection status, present significant biohazard risks in terms of potential infections. Mishandling or accidental exposure to these specimens can lead to unexpected infections among the handler, healthcare units, and the wider community, particularly if the specimen provider harbors infectious agents.

Clinical specimens are commonly present in the laboratory, with clinicians routinely interacting with human samples; often, these specimens are addressed within the laboratory environment. Nonetheless, the biosafety and biohazard risks associated with the manipulation and processing of biological specimens, whether in clinical or research settings, can occasionally be overlooked by clinical personnel. The outcomes generated in the laboratory play a crucial role in diagnoses, predicting patient outcomes, and tailoring personalized treatment plans (Cornish et al., 2021; Destura, Estrella, & Bello, 2025).

When immersed in the laboratory atmosphere, it is vital to maintain an awareness of the scientific context, which includes an understanding of the risks tied to biological specimens and the necessary biohazard safety protocols (Vourtsis et al., 2025). Biohazardous agents encompass a range of harmful organisms, such as bacteria, rickettsia, viruses, fungi, parasites, prions, and transmissible spongiform encephalopathy agents, all of which pose a threat of infection or disease in vulnerable hosts. Furthermore, there exists a potential risk to both the laboratory personnel and the surrounding environment. Certain groups, including immunocompromised individuals, pregnant women, the elderly, and those with preexisting health conditions, face a heightened risk of infection.

To ensure safety, hazardous agents should be managed within a specialized containment laboratory, designed with the protection of both workers and the environment in mind. The consequences of exposure to biohazardous agents can vary, ranging from infections and toxic exposure to injuries from sharp objects, allergic reactions, and skin irritations (Igbashio, Idemudia, & Ochoyama, 2022). Current biomedical laboratory settings are equipped with various pathogenic and biohazardous entities. Behavioral aspects often exacerbate the risk associated with biohazardous agents, as seen with preventable illnesses like hepatitis B virus and *Streptococcus pneumoniae*.

Biohazards can also emerge from unexpected samples or materials, exemplified by a cerebrospinal fluid sample exhibiting unusual characteristics upon arrival in the microbiology department. In such cases, direct handling is inadvisable; it is better to consult with a supervisor to initiate appropriate handling procedures (Paul, 2021). It is crucial for all staff members to be well versed in and comprehend the biohazard risks associated with biological materials. Cultivating awareness of biohazards and adhering to safety measures are vital within the workplace environment.

6.2 DEFINITION AND TYPES OF BIOLOGICAL SPECIMENS

The concept of a biological specimen should correspond to very specific criteria. Most definitions range from a simple morphological definition, which refers to the physical appearance, shape, and structure of the specimen, to a more precise definition that could integrate the operational management system and the collected data. This comprehensive definition takes into account not only the physical attributes of the specimen but also the information gathered during the collection process (Shang et al., 2023; Rank et al., 2020).

However, it is important to note that the concept of the biological specimen is not static and can vary depending on the qualification and expertise of the collecting agency. Different organizations may have slightly different interpretations of what constitutes a biological specimen. While some may adhere to a stricter definition, considering only individual organisms as specimens, others may include organs, tissues, or even smaller components such as cells, molecules, or genetic material.

In a broader sense, a more general concept of a biological specimen would be that it represents an individual, organ, tissue, or any other biological entity that is indicative of a larger group of specimens that have been or will be studied. Each specimen serves as a representative sample, providing valuable information about the characteristics, behavior, or traits of the group as a whole. These specimens are carefully selected to ensure they accurately reflect the diversity and characteristics of the larger population under investigation.

Biological specimens have become increasingly essential in the fields of biotechnology and biomedical research. In fact, biological specimens are routinely employed in contemporary life sciences. This growing reliance has led to a pressing need for effective management of both the specimens and the associated biohazard risks during their utilization. The preservation process often involves freezing the specimens or mounting them on slides, which are then safeguarded under varying temperatures and conditions. This meticulous preparation is crucial for ensuring the long-term viability of the specimens, allowing for future investigations. Traditionally, such practices have been a cornerstone of bio-banking and clinical pathology laboratories. By diligently overseeing both the specimens and the potential hazards they may present, researchers and laboratory managers can enhance the overall efficacy of clinical investigations and biological research while simultaneously ensuring the safety of workflows and personnel (Kawamoto & Kawamoto, 2021).

A study exploring the environmental safety considerations related to biohazard risk management in research laboratories revealed that a significant 70% of biohazard signage in the settings examined was not directly linked to the storage of specimens. The primary challenges in managing studies involving biological specimens often stemmed from the deterioration or loss of numerous specimens. Furthermore, in instances where specimen storage lacked regulation, many of the warning signs were specifically installed to alert personnel to the biohazard risks associated with genetically modified microorganisms (GMMs) (Ta, Gosa, & Nathanson, 2019). The discussion centered on the various biohazard risks linked to the specimens and their storage environments, aiming to create a more suitable backdrop for conducting biological specimen investigations. Additionally, considerable attention was devoted to improving the management of biohazard specimen labels alongside their related materials. To enhance safety by reducing misunderstandings or accidental exposure to biohazardous specimens, the study also presented multiple instances of biohazard-related incidents that have transpired in research laboratories.

6.2.1 Human Specimens

Upon receiving the requisite approval, a select group of trained staff members, specifically from the Laboratory and Patient Care teams, will be tasked with the collection of specimens. These individuals will have undergone specific training and drills to ensure proficiency in the proper use of personal protective equipment (PPE) and the established collection procedures. Before transporting specimens beyond the confines of the patient isolation area and through any public spaces, they (specimens) should be securely enclosed in triple packaging. This packaging protocol adheres to established safety standards and comprises three layers: a primary receptacle, along with secondary and outer packaging. The primary receptacle, essentially the specimen container obtained directly from the patient, must be designed to be both sealable and leak-proof to prevent any contamination. Furthermore, a pre-printed information card label should be placed within a biohazard bag on the external surface of the secondary packaging for clear identification.

> In this procedure, Partner 1, who serves as a non-patient caregiver, will provide assistance to Partner 2, the designated patient caregiver, by exclusively collecting the specimen(s) from them. Partner 1 will don the same type of PPE as Partner 2 to maintain consistency in safety measures. This individual will also be responsible for holding the large biohazard bags designated for specimen containment. Meanwhile, Partner 2 will carry out the task of collecting the specimen(s), placing them into smaller biohazard bags whilst protected by the appropriate PPE. To further ensure safety during cleanup, Partner 2 will wear an additional gown layered over their powered air-purifying respirator (PAPR). Additionally, Partner 2 will be in charge of holding the specimen collection containers and will meticulously document the collected specimens using the bag receptacle card, which is conveniently positioned on the specimen tray for easy reference (Iwen et al., 2018). Partner 2 will not open the door to the patient room at any time, unless in a dire emergency.

When Partner 2 is prepared to gather the necessary specimens, it is essential that he/she effectively communicates with Partner 1 regarding the specific types and volumes of specimens required. Once Partner 1 receives the large biohazard bags, the door may be securely closed to ensure safety. The specimens are to be transported to the patient's room within a covered bio-specimen tray, maintaining their integrity throughout the process. Partner 2 will then proceed to collect the specimens at the patient's bedside, while Partner 1 remains positioned at the doorway, ready to assist when the collection is finalized. Following the collection of all specimens, it is crucial that every human specimen is properly contained by designated patient care staff members who have undergone thorough training and have been fit tested for the use of N95 respirators.

For the safe handoff of the contaminated specimen containers, it is imperative that personnel trained in Company Emergency Response Team protocol, equipped with appropriate PPE, oversee the biohazard containment procedures (Tun & Sim, 2021). After the patient has been appropriately placed in airborne isolation and this status has been verified, the Environmental Service team will be contacted to conduct further cleaning of surfaces. To access the room for clinical visits or similar purposes, an employee must have received requisite training and have been fit-tested for face masks, and should don N95HSR respirators prior to entering the space. Furthermore, these individuals are advised to perform a walkthrough to familiarize themselves with local control measures and should consider having a hand respirator available in case the PAPR experiences a malfunction.

6.2.2 Animal Specimens

When handling clinical specimens, it is essential to adhere to the protocols established for Biological Safety Level 2 (BSL-2). This is particularly important for the handling of clinical specimens that may harbor microorganisms capable of causing disease in humans. The BSL-2 guidelines promote sound microbiological practices aimed at minimizing the risk of ingesting infectious materials or the same coming into contact with mucous membranes. Furthermore, these guidelines emphasize the necessity of avoiding direct interaction with infectious substances and their associated waste. To ensure safety, appropriate facilities for waste disposal must be readily accessible, and it is crucial that laboratory personnel receive comprehensive training to navigate these practices effectively (Mossmann, Nicklas, & Hedrich, 2008).

On activity completion, the Biological Safety Cabinet (BSC) requires thorough decontamination to mitigate any potential contamination risks. It is crucial to categorize laboratory waste into infectious and noninfectious types. Waste that has not been exposed to hazardous materials can be safely disposed of after undergoing autoclaving and appropriate disinfection treatment. For materials that have come into contact with contaminants, a 10% hypochlorite solution should be used for decontamination (De, 2017).

Minimization of needle use is highly advisable, with potentially contaminated needle waste collected in designated, secure containers to promote safety. Items categorized as sharps originating from BSL-2 environments must be decontaminated by autoclaving and shredding, to reduce accidental inoculation. Alternatively, the sharps should be incinerated using an open-barrel incinerator, ensuring that measures are in place to prevent any exposure to harmful aerosols.

Biological Safety Level 3 (BSL-3) is a rigorous framework for investigating endemic and exotic biological agents that pose significant health risks, especially those capable of aerosol transmission. This classification includes viruses like the yellow fever virus, human immunodeficiency virus (HIV), and herpes simplex virus type B, all of which have serious public health implications. Bacterial pathogens like *Mycobacterium tuberculosis*, *Yersinia pestis*, and *Brucella melitensis* are also acknowledged as major risks. This underscores the need for strict containment and safety protocols to ensure personnel are aware of the hazards linked with these pathogens (Garg et al., 2025).

Research on hazardous species requires strict adherence to Biological Safety Level 3 (BSL-3) protocols, which enhance BSL-2 standards to minimize risks. Essential components include biosafety training, protective gear, and ongoing monitoring to ensure public health and scientific integrity are maintained (Wan et al., 2021). The BSL-3 framework enhances ventilation and safely extracts airborne pathogens. Following procedures in a BSC is essential to prevent material release. Annual evaluations of hood functionality confirm effectiveness, ensuring compliance with safety standards. Regular checks promote safety culture and prepare personnel to manage biological hazards and risks.

6.2.2.1 Biohazard Class 3: Application in Veterinary Pathology Laboratories

In biohazard management, handling animals at Biosafety Levels 3 and 4 demands careful methods, especially for infected or immunocompromised specimens. Humane euthanasia aids dissection, typically conducted in Biosafety Level 3 (BSL-3) environments, with some at Level 2 (BSL-2). Adhering to BSL-3 protocols is crucial for personnel safety, scientific integrity, and ethical obligations (Joseph, 2021). In outbreak situations, deploying a crisis management team is vital. Animal management operations must follow BSL-4 safety standards, especially in anesthesia and critical procedures, while laboratory staff should understand BSL-2 protocols for handling sharp instruments and microbiological waste. Diagnosing wild animal tissue now requires collaboration among technical staff, researchers, and regulators for safety and compliance. Knowledge and proactive teamwork improve laboratory responses to emerging threats (Franson et al., 1999).

In urban ecosystems facing species decline, identifying recent biological specimens, such as plant materials and ingested substances, is vital. A thorough analysis of internal structures (nerves, organs, gastrointestinal tracts, musculature, and adipose tissue) is essential, often requiring computed tomography imaging for accuracy. Accumulated toxins in adipose tissues necessitate nervous tissue sampling to detect metallic contaminants and organic compounds like pesticides. For more populous species, the spleen and liver are key examination areas, while other sites, including the ear and limbs, also require attention. Preserving lesion integrity during tissue extraction poses a challenge for veterinary pathologists, sometimes necessitating complete organ retrieval for accurate diagnosis. In such a situation, it is recommended that formaldehyde should be used to preserve the brain and lungs (Meuten et al., 2021; Hawthorn et al., 2023).

In toxicological research, preserving tissue samples is vital. Nonspecific effects can influence the nervous systems, necessitating the maintenance of the brain and spinal cord with preservation methods like formaldehyde. Organs like the liver and kidneys need separate storage for histological analysis, while preserving a single ocular specimen is recommended despite challenges. This careful preservation ensures reliable future data (Blobner et al., 2023).

6.2.3 Microbial Specimens

Oversight of specimens from individuals under investigation or confirmed agents requires strict adherence to BSL-3 protocols for efficient processing during public health emergencies. Laboratories may request waivers for specimens with documented alternative exposure scenarios, such as *Bacillus anthracis* linked to animals. Proper direction of all waivered specimens is vital, requiring compliance and accountability from laboratory staff. Training enhances awareness, teamwork, and clear communication that is suitable for emergency roles and management of specimens from persons under investigation (PUIs) and confirmed cases (Cornish et al., 2021).

In microbiology laboratories, handling infectious agents requires strict standardization protocols to protect personnel and address public concerns. Most clinical labs deal with noninfectious specimens, yet best practices exist for hazardous agents. Training and adherence to these protocols are vital for cultivating a safety culture. Following guidelines enhances efficiency and supports public health against biological threats, while vigilance and comprehension of these practices are essential for risk mitigation and ensuring a secure lab environment (Vounba et al., 2022).

A clinical laboratory evaluation initiative can adopt a decentralized structure to meet guidelines. A state-level risk assessment can analyze laboratory adherence and effectiveness, focusing on varied specimens, excluding special agents. Evaluations are critical for performance appraisal, identifying strengths and areas for improvement. Qualitative and quantitative metrics enhance accountability and reveal important findings. Insights gained information about strategic decisions and bolster regulatory compliance (Burch et al., 2022).

6.3 UNDERSTANDING KEY BIOLOGICAL BIOHAZARD RISKS

The administration of biohazards in labs presents challenges affecting human and animal welfare due to pathogenic agents like bacteria and viruses, requiring strict safety protocols. Understanding these agents is crucial for management; hence, training programs could enhance laboratory personnel's readiness. An organized framework is essential for biohazard management. Collaboration among staff, researchers, institutions, and regulatory agencies is similarly key to maintaining safety standards in biological risk management (Kumar et al., 2023; Sharma et al., 2023). Biological agents are classified into five severity levels, each with specific safety protocols. Evaluating transmission pathways, like skin punctures, is crucial for effective interventions. Proper training reduces disease transmission risks, while stringent lab safety standards depend on personnel readiness, fostering a robust safety culture. Education and training are therefore key to successful management (Ta, Gosa, & Nathanson, 2019).

A strong proficiency test guidance framework is vital for managing biological specimens. It must tackle transportation challenges and fixative use while complying with best practices. Adhering to waste disposal rules and engaging stakeholders promote sustainability (Mendonça, 2024).

Such an initiative would aim to create a strong environmental risk assessment framework for a biosafety lab, including emergency management and security protocols. Preparedness enhances safety and operational integrity, and enforces strict personnel standards for potential challenges (Kothinti, 2024).

Environmental risks can be primarily classified into two: direct risks and indirect risks (Zerizghi et al., 2022). Addressing environmental risks involves understanding pollutants, harmful materials, equipment failures, and infectious agents. Identifying these is key for assessment. Comprehensive protocols ensure safety, and ongoing monitoring with collaboration enhances accountability for proactive risk mitigation (Singh et al., 2024b). During assessment, it is crucial to identify indirect risks such as a ship carrying infectious materials that might face unexpected weather, risking grounding and cargo release. Recognizing these risks highlights the need for proactive planning to prevent serious ecological and health issues (Yazdani & Haghani, 2023). Urban water collisions pose safety and environmental threats, and managing them requires analytics, training, resources, and better communication for public awareness and collaboration.

6.3.1 Definition of Biohazards

These are biological organisms and their components that pose a threat to the health of humans, crops, forests, livestock, and aquatic species are classified as biohazards (Ta, Gosa, & Nathanson, 2019). Protocols for biohazard materials are crucial for pathogen elimination. Training staff promotes compliance, while researchers work with suppliers and regulators to uphold safety standards. The Emergency Response Plan (ERP) outlines spill and exposure procedures, with support from health authorities. Communication with law enforcement is vital for the Emergency Management Team (EMT) during health crises, ensuring access to protective gear (Choudhury et al., 2025).

Managing biohazard agents in biorepositories requires strict decontamination protocols like ethanol disinfection and safe disposal of medical waste. Risk mitigation includes limited access and an ERP involving local responders and Local Emergency Planning Committee. The EMT ensures safety during crises with preparedness measures like sorbents, PPE, and accurate labeling (Chipangura, Abdussamad, & Lewis, 2024). The presence of biological hazards linked to bio-banks is considerable, warranting systematic oversight to safeguard both personnel and the environment. It is imperative to comply with applicable governmental guidelines and ethical norms upheld by society. Given the diverse range of biological substances managed by bio-banks, the task of classifying the related risks emerges as a multifaceted dilemma (Akyüz et al., 2021; Annaratone et al., 2021).

Laboratory practices must prioritize the protection of personnel and the environment by minimizing exposure to biohazard agents. This involves conducting a thorough risk evaluation, focusing on the hazardous properties of each agent, particularly leading pathogens. Important factors include the ability to infect susceptible individuals, virulence indicating the severity of illness, and the existence of effective treatments and preventive strategies. Other considerations include transmission routes, required infectious doses, environmental stability, host diversity, endemicity, and safety measures for potential risks. Notably, the classifications of risk groups differ from the biosafety levels (BSL) used in labs (Cornish et al., 2021).

6.3.2 Categories of Biohazards

In clinical settings, managing biological specimens and biohazard risks is crucial. This chapter defines biohazard terms, categorizes hazards, and outlines preventive measures and PPE to minimize lab risks. It addresses safe packaging, careful handling of biohazardous specimens, precautions against blood-borne pathogens, waste disposal, and exposure protocols. Effective management necessitates training and protocol adherence, enabling staff to identify biohazards. Collaborating with safety experts enhances vigilance, while an ERP details actions for spills and exposures. Coordination with health authorities ensures effective communication and access to supplies, and regular drills are critical for preparedness against biohazard threats (Wahab et al., 2021). Biohazardous agents include pathogenic microorganisms, viruses, cell lines, tissues, and bodily fluids, which potentially endanger the health of staff, volunteers, and patients, and pose risks to the environment (Choudhury et al., 2025; Obayori & Salam, 2023).

Besides the four risk levels, an additional fifth group has been created for agents that pose a significant threat to both humans and animals. An analytical approach to biohazard risk evaluation is therefore essential in medical projects with hazardous biological agents. The assessment should begin with inquiries, particularly in new labs, noting the key factors that influence the risks being infection probability and transmission routes (Gemler et al., 2022).

Pathogenic biohazards can be classified into risk groups and their defining characteristics outlined. The most common human pathogens fall under Risk Group 3. Furthermore, biohazardous agents are classified into "Category A", "Category B", and "Exempt" organisms, which must be accurately identified and labeled during packaging for transport.

Contact with Category A pathogens can lead to serious, life-threatening illnesses or lasting disability in otherwise healthy humans (e.g., *Bacillus anthracis*, *Brucella abortus*, *Clostridium botulinum*, and Ebola virus [EV]) or animals (e.g., *B. abortus*, Rinderpest virus, and Lumpy skin virus). Exposure to Category B infectious substances generally does not cause lasting disability or life-threatening illnesses in healthy individuals or animals, suggesting moderate to low associated risks. Category C includes bacteria, viruses, or toxins that threaten the health of immunocompromised individuals exclusively. Exempt samples consist of standard tests for noninfectious conditions, such as cancer biopsies, with a low chance of being infectious. They are not classified as hazardous, lack a UN-ID, and pose minimal risk (Osei & Zubair, 2025).

The transfer of pathogens from animals to humans can cause zoonotic diseases, with humans as vectors for transmission back to animals, a process referred to as "spillback". Figure 6.1 shows the discourse on spillback phenomena, highlighting two key adverse effects. First, it increases disease prevalence in animal populations from human pathogens, causing morbidity or mortality (a–c) which threatens conservation efforts. Additionally, pathogens may show heightened virulence in distantly related hosts, posing significant threats to endangered species. The second primary issue is the risk of "secondary spillover" from animal reservoirs to humans, a major public health threat. This concern, highlighted in severe acute respiratory syndrome *coronavirus 2* (SARS-CoV-2) discussions, is shown in Pathway 2 (d–e). When a capable host species is infected, it can circulate enzootically, raising the likelihood of transmission to humans (Fagre et al., 2022).

The sequence above oversimplifies numerous potential transmission pathways within complex metapopulations, as dynamics can operate bidirectionally among hosts, either as singular species or community reservoirs.

6.3.3 Emerging Infectious Diseases: An Exploration of the Growing Threats

New infectious diseases challenge public health and emergency response, necessitating strategies that must align with health guidelines and ensure cooperation among organizations. Key steps in the process include developing response capabilities, evaluating mass casualty possibilities, and managing fatalities. Healthcare professionals need to handle deceased individuals carefully, collaborating closely with the coroner's office for patient surges and medical facility security. Fair resource distribution is essential for respectful body treatment and ethics. Clear public communication on outbreaks, preventive measures, and healthcare services is crucial to mitigate illness and death. Whereas pandemic preparedness has focused on the highly pathogenic H5N1 avian influenza strain since 2003, ongoing animal outbreaks pose continued risks to human health (Clements & Casani, 2016).

Formulating strategic frameworks to address emerging infections in avian and swine populations is critical, along with proactive measures to prevent zoonotic disease transmission. The emergence of new viruses may require community adaptations during recovery. Some emerging infectious diseases (EIDs) might establish endemicity, necessitating their inclusion in public health initiatives. To reduce resurgence risks, interventions must implement new practices and protocols. Public health has historically faced EIDs amid global human migration and trade, which increase pathogen exposure risks. Rapid transportation has accelerated infection spread, creating new challenges. While predicting future EIDs is complex, preparedness can leverage lessons from past outbreaks to mitigate impacts (Ryan, 2016).

Current pandemics did not arise abruptly but evolved gradually over time. Evidently, more cohesive interventions could have helped in mitigating this situation. Future pandemics are certain to occur, underscoring the necessity for different sectors to formulate localized strategies aimed at achieving a coordinated response, given that the repercussions will vary among different communities.

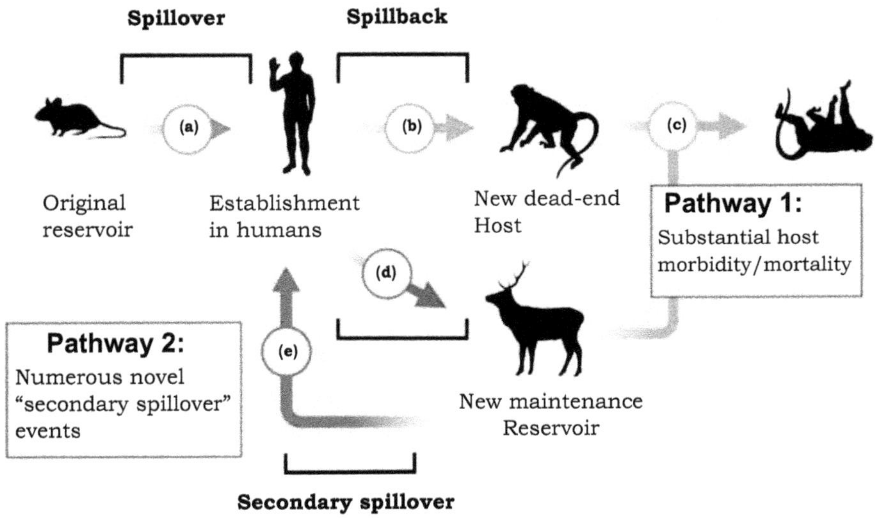

FIGURE 6.1 Analysis of human-to-wildlife pathogen transmission challenges – two notable scenarios (Courtesy of Fagre et al., 2022).

6.3.4 Implications of Biohazard Risks

Biological specimens are essential in biotechnology and bioengineering, including fresh, frozen, and preserved tissues, cells, blood, and various plant and animal samples. As their relevance and sensitivity grow, so do the storage volumes, increasing potential biohazard risks. Effective management strategies for storing these diverse specimens are crucial for advancing biotechnology and biomedical research (Gao et al., 2024).

Numerous programs promote biological specimen donation, yet medical diagnostics and blood processing facilities struggle to manage patient specimens with those containing viral and bacterial pathogens that need strict storage at 4°C and –20°C. Pathogens like *Escherichia coli* O157:H7, *Y. pestis*, *Neisseria meningitidis*, *B. anthracis*, *Staphylococcus aureus*, and *Ricinus communis* are notable examples, each having an *American Type Culture Collection* (ATCC) identification number (Krishan, 2021). The procurement of animal and plant specimens requires scientific permits for their importation, preservation, and distribution. It is crucial to eliminate pathogenic organisms after use. Many lab incidents occur due to staff not following aseptic protocols or proper disposal of contaminated materials. Thus, maintaining strict biosafety protocols is essential during processing, storage, and transfer of such specimens (Shishodia et al., 2023).

In genetics and bioengineering, transformation, transfection, electroporation, and transduction change genetic material in host cells. Sterilizing biohazardous materials before disposal is crucial, often using ultraviolet irradiation or hydrogen peroxide. Laboratory wastewater should be collected in designated containers and disposed of according to protocols.

6.4 REGULATORY FRAMEWORK OVERVIEW FOR BIOHAZARDS

Those involved in producing biological materials have a significant responsibility to ensure their safe management during storage and handling. This duty is shared by recipients, custodians, and researchers, all of whom must emphasize safety measures. If package damage is suspected, especially with visible leakage, recipients must quickly notify the shipping entity. Regulations on hazardous materials apply to everyone involved in shipping, and it's essential to understand that hazardous biological materials include not just biohazardous substances but a wider range of etiologic agents (Health Organization, 2024; Guo & Luo, 2022).

Manifestation of infectious disease involves various pathogens and cultural materials. Hazards are classified through guidelines, identifying biohazardous materials as Category A, including infectious substances. Organizations must register with the Public Safety Office, align with Institutional Biosafety Committee standards, and adhere to storage protocols. Stringent regulations govern the transport of biological materials, controlling hazardous material flow (Shapira et al., 2025).

6.4.1 National Regulations

In biological specimen management, regulations exist at national and state or county levels guiding management, cataloging, and disposal. National laws provide a foundational framework that influences state or county policies, promoting consistency in regulatory guidelines. The national government also implements diverse regulations across various sectors, indicating some frameworks may be exclusively national (Stitt-Fisher, 2015).

In the USA, regulations for managing biological pathogens align closely with federal standards. A key feature is a specialized platform that assists laboratories in navigating the complexities of biological specimens. When encountering biohazards in research or clinical settings, labs must complete a Biohazardous Waste Disposal Agreement Register (Parida et al., 2022). In waste management, those responsible for disposing of "waste" should follow strict protocols, noting that certain items require sanitization with approved cleaning agents that are verified, as typical disinfectants like Clorox wipes may not meet the requisite standards (Le et al., 2022).

National agreements mandate checklists for compliance, requiring lab supervisors to sign off biannually. Compliance entails documenting training on biological hazards, keeping a personnel register, and appointing a biohazard manager; hence, unregistered labs must halt biological activities. An audit found significant protocol deficiencies, notably with needle capping, which has a 30-day compliance grace period. The continued violations are clear indications of challenges in protocol adherence (Le et al., 2022).

6.4.2 International Guidelines

In June 1976, the U.S. National Institutes of Health (NIH) significantly advanced biotechnology regulations by issuing the Guidelines for Research Involving Recombinant Nucleic Acid Molecules. This began with an Interim Guideline, followed by finalized Guidelines published on July 5, 1976, introducing containment strategies for recombinant DNA (r-DNA) methodologies. By November 26, 1978, the guidelines were revised to include three levels of containment: (1) physical, (2) biological, and (3) experimental protocols with specific requirements. Further updates occurred in 1979, 1982, and 1986, leading to the establishment of the code 42 CFR Part 72 in December 1986. A framework for Large-Scale Recombinant DNA Containment was introduced in 1987, concluding with revisions on June 12, 2008, and the adoption of new "Guidelines for Research Involving Recombinant and Synthetic Nucleic Acids" on June 19, 2009 (NIH, 2024).

The *Laboratory Biosafety Manual* was first published in 1983 to provide internationally accepted guidelines for managing pathogenic microorganisms. It outlined essential biosafety principles and became a key reference in the field (Sun, Wu, & Fan, 2021). The Manual was revised in 1993 and 2004, the latter incorporating biosecurity elements. The comprehension and implementation of these fundamental biosafety practices are, however, significantly inconsistent in resource-constrained laboratories. Further, developing countries face notable biosafety challenges due to insufficient documentation of their National Biosafety Policies and Programs (Altıntaş & Koluman, 2024).

Countries must adhere to international biosafety standards and integrate risk management strategies into domestic policies for the protection of health and life, and to secure trade in biotechnology products (Salonga, 2025). The Sanitary and Phytosanitary Agreement highlights member nations' commitment to ensure their sanitary measures are reasonable and compatible with those of their trading partners, thus avoiding trade barriers. It also recognizes the importance of enhancing global biosecurity through strong biosafety practices (Safdar et al., 2023).

6.5 RISK ASSESSMENT PROCEDURES

The International Organization for Standardization (ISO) 3100:2018 standard articulates risk as "the effect of uncertainty on objectives", emphasizing the intricate relationship between unpredictability and goal achievement (Redinger, 2024). ISO 15189:2012 highlights the need for labs to assess how their procedures and shortcomings affect examination results, especially regarding patient safety (Pereira, 2023; Alaamri et al., 2024).

Laboratories must reassess protocols to mitigate risks, emphasizing the value of risk management, which includes analyzing activities to protect patients and staff (Majlesi, Khaloei, & Rabouri, 2015). A proactive approach involves risk identification, severity assessment, prioritization, strategy development, and continuous monitoring, all while respecting quality assurance standards. According to ISO 15189:2012, the first step is identifying potential laboratory risks, then quantifying them by likelihood and consequences without control measures (Alaamri et al., 2024). A prioritization framework is established based on risk assessment, with strategies implemented to reduce or eliminate risks.

To minimize risks effectively, assessing actions through an established indicator framework after implementing risk management measures is crucial. In a clinical laboratory, attention should focus on three core processes: preexamination, examination, and post-analytical stages. A systematic methodology for risk identification involves mapping these processes to uncover potential risks (Majlesi, Khaloei, & Rabouri, 2015). However, conducting risk assessments often requires personnel skilled in quality management, creating challenges for smaller laboratories without trained staff. Additionally, the ISO 15189:2012 standard requires that risk assessments also include laboratory suppliers and external services (Alaamri et al., 2024).

To maintain safety in laboratories, it's essential to evaluate supplier risks every two years. A user-friendly assessment tool should be created to help all staff, even those without expertise, identify task-related risks. This requires the research, development, and validation of a tool to detect biological, chemical, and biohazard risks in clinical settings. Despite some hazardous materials being seen as low risk, their potential threats are significant and require careful examination. Mishandling chemical reagents can have serious, immediate repercussions, highlighting the need for strict oversight in the validation process (Jayamani et al., 2022).

6.5.1 Identifying Hazards

Initiating biological risk management requires a thorough assessment of potential exposure to sprays, splashes, or aerosols in laboratory activities. This evaluation relies on the expertise of trained professionals operating within predefined assumptions. Conducting a biological risk assessment is essential for laboratories to identify and mitigate hazards, following specific protocols. Additionally, using a certified BSC is recommended for effective containment when handling patient specimens (Iwen et al., 2018).

In a regulated laboratory, reducing biological risks with patient specimens is crucial. Mishandling specimen containers can cause biological agents to escape, particularly if fluids like blood become aerosolized. To mitigate this, tightly sealed containers should be used, favoring syringes over open containers. Additionally, a BSC is vital for minimizing airborne pathogen exposure during specimen examination, as patient specimens can harbor harmful biological agents, posing significant risks to laboratory staff (Ta, Gosa, & Nathanson, 2019).

High-risk samples may unintentionally reach the laboratory, posing biohazard risks. Most laboratory personnel lack knowledge of these implications. Moreover, patients' medical histories are expected to remain confidential. The USA's Health Insurance Portability and Accountability Act

(HIPAA) regulations limit sharing information about patients' high-risk behaviors with laboratory staff (Wagner et al., 2022; Marron, 2024). Implementing essential safety measures during specimen processing poses challenges for laboratories. Identifying biohazard "red flags" is crucial, as these may indicate high-risk agents. Personnel opening specimen packaging must scrutinize the requisition and ensure all relevant information is meticulously documented in the laboratory report.

6.5.2 Evaluating Risks

In numerous professional environments, the complete eradication of all risks is typically unfeasible due to the intrinsic characteristics of activities that carry risk. Nonetheless, methodical risk assessments can considerably mitigate identified dangers. A comprehensive high-risk assessment necessitates a thorough comprehension of both processes and hazards, which is generally acquired through consultations with managers, supervisors, and staff, complemented by technical knowledge. Following the identification of hazards, it is critical to assess and document the associated risks, as this step is vital for the implementation of effective controls. This anticipatory strategy guarantees that high-risk activities are systematically managed, safety protocols are put in place, and initiatives aimed at risk reduction are actively pursued (AlShammari, Alhussain, & Rizk, 2021).

A risk assessment systematically examines potential threats by evaluating both their probability of occurrence and the degree of harm they may inflict. It assesses the severity of potential damage in conjunction with the mechanisms through which this harm might manifest. Typically, risk calculators employ a multiplication of pertinent metrics associated with these factors to derive a risk priority number. It is essential to distinguish between the likelihood of an event occurring and the magnitude of the possible consequences (Pascarella et al., 2021).

A thorough analysis of risks necessitates the identification of control measures aimed at their effective mitigation. It is vital to document the processes of hazard identification and risk assessments, as this provides a foundation for informed discussions and facilitates the evaluation of potential hazards. Although completely eradicating risks may not be feasible, striving for their reduction remains a key objective. Identification of significant hazards, requisite control measures, and associated timelines for implementation should be achieved through collaborative discussions or consultations.

6.5.3 Analysis of Risk Mitigation Strategies and Optimal Practices

Upon arrival at the clinical microbiology laboratory, a new biological specimen is presented within a specialized bag, which is marked by a unique submission label. This submission is comprehensively examined to confirm that all requisite components are present, each aspect being systematically documented in the laboratory's computerized system. If necessary criteria aren't met, the submission is marked incomplete and not processed. Once the specimen is officially logged, part of the original sample is placed in a separate container for analysis. The original specimen's container is sealed and returned to the lab bag. Biological residues are cleaned; if the workstation isn't available, cleaning wipes can air-dry for later use (Ta, Gosa, & Nathanson, 2019).

A laboratory chair designated for biohazard activities necessitates the presence of a sizable biohazard bag and should be positioned in proximity to the primary laboratory bench to facilitate convenient access. It is imperative to prohibit any unauthorized relocation of these chairs, as they must never obstruct access to fire extinguishers. To maintain hygiene standards, a thorough disinfection must be performed monthly, or immediate cleaning should occur following any visible contamination, utilizing a new disinfectant wipe specifically treated with HB101, which must then be allowed to air-dry. The workspace should remain unobstructed by oversized boxes, bags, or personal belongings to enhance operational efficiency. To minimize the risk of contaminating personal items, unauthorized access to this area is strongly discouraged. Furthermore, meticulous handwashing is crucial after interaction with biohazardous materials prior to engaging with the L-computer (Iwen et al., 2018).

6.6 PROPER HANDLING AND DISPOSAL OF HAZARDOUS MATERIALS

Workers have examined the biological hazards and associated risks involved in the management and disposal of specimens derived from humans. It is imperative for researchers to adhere to hazardous materials (hazmat) regulations established at the federal level in the USA. This regulatory framework encompasses several critical processes including the appropriate storage, transportation, treatment, recycling, and disposal of such materials. In addition, timely disinfection is essential in the event of spills that involve biohazardous substances (Ta, Gosa, & Nathanson, 2019).

The responsibility of the Principal Investigator encompasses the critical task of implementing adequate safeguards against the hazards posed by biohazardous agents. This involves providing comprehensive training to all laboratory personnel, including the Principal Investigator themselves, on the protocols for biohazard containment. Additionally, it is vital to ensure that all biohazardous materials are stored and disposed of in a manner that adheres to safety regulations (AlShammari, Alhussain, & Rizk, 2021).

The Principal Investigator and staff responsible for managing biohazardous materials are required to participate in annual training focused on hazardous waste management. This training is crucial for reducing potential health hazards and safeguarding public health noting also that the effective processing of waste is paramount in removing any infectious agents. Decontamination techniques, such as autoclaving and incineration, play a critical role in this process. It is mandatory for biohazardous waste to be managed within a Biosafety Level 3 (BSL-3) laboratory environment. Materials that can be autoclaved, including culture media, needles, and scalpels, must be placed in designated biohazard bags and kept in secondary containment to prevent exposure. Furthermore, liquid biomedical waste necessitates treatment with an equivalent amount of undiluted bleach for a minimum period of 30 minutes to ensure thorough decontamination (Callihan et al., 2021).

6.6.1 Guidelines for Handling and Care to Ensure Safety

The careful management of biological specimens throughout their collection, transportation, and processing is of utmost importance. This diligent approach not only ensures the accuracy of subsequent testing and analysis but also protects the health and safety of all individuals who may interact with the specimens (Asrani et al., 2021). The following guidelines provide a detailed framework designed to ensure the secure handling and upkeep of biological specimens in both clinical and laboratory settings.

6.6.1.1 Clinical Protocols and Procedures

From the onset, it is crucial to communicate to all members of the healthcare team as well as the patient that a biological specimen is being collected in the room. *Second*, all necessary supplies should be removed beforehand and the availability of at least two pairs of gloves for the procedure confirmed. *Third*, the specimen should be collected and placed on a sanitized surface equipped with an absorbent area to manage any potential spills. *Fourth*, following the collection, both gloves should be meticulously removed by rolling them inward to contain any possible contaminants, after which they should be discarded in a designated trash can. *Fifth*, all waste resulting from the specimen collection procedure should be properly disposed of. *Sixth*, a comprehensive hand sanitization should be performed using soap and water or a disinfecting wipe. *Seventh*, before leaving the room, one must verify that all specimen labels are accurately and completely filled. *Lastly*, the collected specimens should be transported to the hospital laboratory in a lab pouch, ensuring that it includes the requisite requisition form.

6.6.1.2 Laboratory Processing

First, the specimen collection process occurs in the patient's room, requiring the collection container to be placed on an absorbent pad to prevent spills. *Second*, the use of a second pair of examination gloves is mandatory. An individual in the isolation area should be designated to oversee specimen management and transport, ensuring protocol adherence. *Third*, prior to the collection process, it is essential to remove all packaging materials from the specimen container, which includes labels and caps. *Fourth*, careful labeling practices must be employed consistently, alongside verification of patient identification, as any inaccuracies may result in the need for redundant collections. *Fifth*, in a well-sanitized area of the room, the external surfaces of the specimen container should be disinfected, paying particular attention to the lid and neck. The container should be immersed in the cleaning solution for at least 15 seconds and, while wearing protective gloves, rotated from base to top for thorough exposure. *Sixth*, two cleaning passes are performed beneath the lid of the container, ensuring to include the bands in the process.

6.6.2 Specimen Collection

The meticulous collection and evaluation of patient test samples are conducted systematically across a variety of healthcare environments, including outpatient clinics and hospitals. Test samples may originate from diverse facilities and private practices. Upon collection, these samples are transported either to centralized laboratories or processed in smaller units tailored to particular regions. There is a growing trend in developing countries to utilize economical, one-way cell/gram digital cameras for capturing high-resolution images of blood tests, which facilitates the electronic transmission of results by rural health professionals. Nevertheless, this approach remains relatively uncommon in the public health systems of rural Africa, where it primarily serves to enhance the use of more costly glass slides. Biological specimens are conveyed in Unplasticized Polyvinyl Chloride (UPVC) pipes by vehicles following specific transport standards to specialized laboratories for analysis (Andersen, 2018).

Samples are frequently dispatched in inadequate containers, placed in poorly maintained cages, or abandoned carelessly within vehicles. These practices result in operational challenges within laboratories, impacting the handling of commonly utilized packages and culminating in incidents such as spills, compromised blood tubes, and impaired equipment (Sylverken et al., 2022).

In sub-Saharan Africa, incidents of HIV seroconversion due to lab errors emphasize the need for careful specimen handling and strict safety protocols to protect patients and staff. Adhering to guidelines is essential to safeguard all individuals from infectious threats. Hospital administrators must establish comprehensive procedures and allocate resources effectively for managing biological materials.

Protecting both patients and visitors is crucial, and lab personnel must thoroughly understand and enforce protocols during sample processing. Key practices include proper collection, transportation, analysis, and disposal of specimens to prevent infection spread. It's vital to keep patients informed for reassurance during procedures. Strict personal hygiene, including handwashing before and after sample collection, is essential. Proper use of gloves and gowns is mandatory, with masks required in high-risk situations. Incidents involving organic material spills necessitate immediate disinfection and careful sampling instrument handling. Following sampling protocols and maintaining aseptic techniques are paramount for ensuring safety in the lab environment.

6.6.3 Storage of Hazardous Materials

In bio-banking, the preservation of specimens using formaldehyde introduces considerable biohazard risks, notably associated with lymphoma and carcinogenic implications. There is a potential for formalin containers to crack, or for the pressure relief valve to malfunction due to excessive steam pressure. To mitigate these hazards, it is imperative that containers are securely stored and adequately distanced from safeguard personnel from possible incidents. The magnitude of the biohazard risk is correlated with the required bio-containment level; therefore, bio-banks that handle biohazardous materials must adhere to a biosafety classification of level 2.2 or higher, necessitating stringent safety and compliance protocols (Ta, Gosa, & Nathanson, 2019).

Specimens that are highly sensitive and potentially contaminated are inadequately preserved by liquid preservation methods. In contrast, lyophilization is favored because it upholds RNA integrity when specific conditions are met. Furthermore, it is necessary to maintain a minimum distance of one meter between flammable materials and electrical equipment to mitigate ignition risks associated with high-voltage devices. The apparatus utilized in bio-banking must adhere to rigorous control standards, as it relies on electrical energy to achieve precise regulations of temperature and humidity. Moreover, using short-power-supply cords for devices further diminishes the potential fire hazards that may arise from overcurrents in the electrical system (Aarattuthodi, 2025; Thobela, Maseme, & Duma, 2022).

A critical safety criterion mandates a minimum distance of one meter from flammable substances during operations, regardless of electrical cord presence. When disconnecting devices, they should be pulled from the plug to prevent injury and equipment drops. Gently detaching electrical devices ensures safety. Improper handling of cassette bottles can cause contamination from biohazardous aerosols generated by spills. The sampling line's configuration may expel residual fluid forcefully, creating aerosols. To minimize risk, contact between the cassette bottle and the sampling line should be avoided during storage, reducing aerosolized contaminants from contaminated hands. Proper hygiene protocols, including handwashing and glove usage, are essential when handling cassette bottles. After using bottles containing biological specimens, thorough cleaning is required for safety.

Bio-banks have biohazard disposal containers, but improper positioning can lead to contamination since misalignment can cause spills, increasing exposure to biohazardous agents. Access to appropriate Biohazard Level 2 instruments is critical; their absence can lead to poor handling in a Biohazard Level 3 environment. Those exposed to spills should be treated with sterile saline to prevent contamination from microorganisms. Caution is necessary with detergents or antiseptics to avoid skin damage, followed by washing affected areas with soap and water. Waste tissue from bio-banks may carry bacteria like those in specimens, requiring careful disposal according to biohazard warnings. Biohazardous waste management from bio-banks must be prompt due to its infectious waste status, demanding careful oversight, especially during surgical operations.

6.6.4 Transportation Regulations

The guidelines established provide a structured approach for the transport of biological materials and infectious agents via commercial carriers, prioritizing compliance with regulatory requirements. When selecting air transport as a mode of delivery, it is imperative to adhere strictly to applicable regulations. Additionally, all personnel involved are required to engage in annual training sessions and to re-certify every two years. For those managing biological materials, it is crucial to possess comprehensive understanding of the current regulations imposed by relevant governing authorities (Arora, 2004).

Infectious substances are defined as materials that are either known or anticipated to harbor pathogens capable of causing illness in humans or animals. It is important to note that shipments containing infectious substances that do not include microorganisms or toxins are exempt from certain regulations. However, while biological specimens that are neither confirmed nor suspected of containing infectious agents are typically not deemed hazardous, they may still carry a potential risk for disease transmission, thereby classifying them as Biological Substance Category B (Health Organization, 2024).

This classification includes samples utilized for diagnostic purposes, those derived from known causative agents, and specimens obtained from individuals under observation. It is noteworthy that biological materials like urine, saliva, and blood from healthy donors need not encounter any regulations. However, specific regulations apply to materials sourced from animals. Researchers must adhere to established national and international guidelines on the transport of animal-derived materials. To maintain safety and ensure the integrity of specimens, items should be securely packed within leak-proof containers designed to reduce the likelihood of spillage during transport. Additionally, the volume of liquid specimens should not exceed 30 mL, and if a preservative like formalin is included, its concentration must remain below 10% of the overall volume. The external packaging should not reveal any indication of biological contents. For inquiries regarding manifest requirements or transportation protocols, it is advisable to consult the biosafety officer (Ragavan et al., 2023).

6.7 EMERGENCY RESPONSE TO BIOHAZARD INCIDENTS

Public health laboratories are crucial in addressing biological threats, playing a key role in managing infectious diseases and enhancing biodefense strategies (Olver, Bohn, & Adeli, 2023). These laboratories vary in structure, typically being small with limited Biosafety Level 2 (BSL-2) capabilities. They face challenges like inadequate funding and must engage various stakeholders, including hospitals, epidemiologists, first responders, and law enforcement, necessitating a comprehensive operational strategy. A nationwide training initiative for front-line personnel, including lab staff and first responders, is urgently needed. In contrast, advanced biodefense laboratories operate at higher biosafety levels, often requiring BSL-3 facilities. Establishing BSL-4 labs nationally could enhance responses to biological threats. Experiences from France and Italy indicate that resource-poor nations often depend on external assistance during biological crises (Gao et al., 2024).

Seven strategic actions must be implemented to enhance laboratory risk preparedness and response. First, laboratories should maintain a sufficient inventory of critical reagents and consumables for quick access during crises. Second, trained rapid response teams, including epidemiologists, microbiologists, and modelers, are essential for effective emergency management. Third, a strong communication framework is vital for timely information exchange among healthcare providers, public health officials, and laboratory personnel. Fourth, laboratories must be capable of quickly scaling their operations in response to threats and hence require implementation strategies for increased demand. Fifth, developing standard operating procedures to identify (secure, manage, test, and refer) biological specimen is a key requirement. Sixth, collaboration with law enforcement helps uphold chain-of-custody protocols. Finally, a comprehensive strategy to address bioterrorism concerns is necessary, backed by extensive surveillance and healthcare support (Knezevic et al., 2022).

Improving the capacity of laboratories to effectively identify Category A biological agents yields considerable advantages. Collaboration with security organizations and academic institutions possessing pertinent knowledge is critical. The utilization of rapid screening methodologies, including PCR testing and sophisticated proteomics, is indispensable. Laboratories must create informatics systems that facilitate the swift characterization of microbial agents, given their prevalent use as bioweapons. In summary, enhancing screening capabilities is vital for a timely and effective response (Iwen et al., 2018).

6.7.1 Incident Reporting

The Occupational Safety and Health Administration (OSHA) stipulates that employers must report particular injuries or incidents within a 24-hour timeframe. An incident is characterized by substantial occupational exposure or injuries that necessitate medical intervention resulting from work-related activities. Employees who are affected are obligated to inform their immediate supervisor within one hour of identifying the problem, and certainly prior to the conclusion of their shift. Healthcare personnel are tasked with promptly filling out the appropriate incident reporting documentation. Furthermore, staff and supervisors are required to communicate the event to the Office of Chief Medical Examiner (OCME) by the next business day or within the designated 24-hour period post-incident. Failure to report to a supervisor is treated as insubordination, which may result in disciplinary measures in accordance with the Department's Discipline policy, along with the possibility of further investigation into the matter (Tejada-Pérez et al., 2022).

In a six-month trial study, designated morning periods were allocated for healthcare workers to fill out Accident Reporting Forms following their exposure to blood, bodily fluids, or sharps within a hospital environment. Personnel from the healthcare facility received training aimed at ensuring the accurate completion of these forms. As the trial approached its conclusion, a select group of employees conducted visits to multiple hospitals with the objective of standardizing the reporting forms utilized by exposed healthcare workers.

Some three months later, an assessment of patient vaccination records pertaining to anti-hepatitis B Surface (anti-HBs) antibodies was conducted, specifically targeting individuals who had not received vaccinations. The monitoring efforts placed a particular emphasis on periods of potential viral transmission, resulting in the implementation of stringent precautionary measures. An analysis of the reliability of data derived from the trial on the reporting forms was conducted in conjunction with an examination of the documentation maintained by healthcare workers. The task of completing Accident Reporting Forms was integrated into newly established procedural standards. Furthermore, to bolster oversight, Clinical Trials announcements mandated that reporting forms be included in treatment protocols and stipulated the necessity of telephonic consultations between hospital physicians and patients for the execution of these forms.

This investigation aimed to evaluate the fidelity, precision, and promptness in adhering to the recommended post-exposure prophylaxis (PEP) protocols, with the intention of reducing the likelihood of infection by the HIV and other blood-borne pathogens. Within this framework, only 24% of the 85 retirement hospital physicians eligible for the study pursued post-exposure consultations related to PEP. Furthermore, among 3,104 healthcare professionals who encountered percutaneous exposures, only 34 (1.9%) completed the requisite documentation within the designated six-week follow-up period. The necessity of obtaining written consent from the source patient for HIV serological testing led to a notable decline in adherence, with the implementation rate plummeting to 7.1% (amounting to 50 cases) due to instances in which consent was refused (Tejada-Pérez et al., 2022).

6.7.2 Containment Strategies

Biological specimens comprise an extensive array of materials that necessitate meticulous classification, containment strategies, and compliance with federal regulations. The NIH Guidelines provide a framework for managing these specimens through specific containment protocols related to Recombinant DNA Research, highlighting the explicit division of responsibilities among federal entities that supervise laboratory investigations. Adhering to NIH Guidelines is essential for maintaining safety within the realm of Recombinant DNA Research. Failure to comply, particularly in the case of agents that may threaten public health, can lead to significant repercussions (Gillum, 2024).

The potential consequences of noncompliance can incorporate the halting of research initiatives, the requirement for safety assessments, and possible punitive measures against persons or organizations identified as negligent. Therefore, it is critical for all researchers involved in, especially NIH-funded projects that utilize biological specimens to rigorously follow the prescribed guidelines and regulations. The dangers associated with infectious agents, even in the presence of effective vaccines, warrant careful consideration and should not be overlooked.

Investigations involving biological specimens should be conducted within specialized laboratories adhering to rigorous containment standards and staffed by qualified personnel. These laboratories must be equipped with suitable safety gear and protective clothing, and they should prominently feature biological hazard signage to safeguard staff. Moreover, any experiments utilizing r-DNA within a "host-vector system" must take place in a Biosafety Level 4 (BSL-4) containment environment (Martins & Ribeiro, 2025).

When gene products derived from recombinant vectors present potential threats to human health, it is crucial to implement Biosafety Level 4 (BSL-4) containment protocols. There are guidelines provided that specify the standards for GMMs, including transgenic organisms such as animals and plants, thereby offering valuable guidance for researchers engaged in these areas. Nations with advanced scientific frameworks have developed regulatory systems that delineate biohazard containment levels ranging from level 1 (BSL-1) to level 5 (BSL-5). Nonetheless, the financial burden associated with the construction and upkeep of such laboratories can restrict their establishment in regions with limited resources, especially in low- and medium-income countries.

6.8 ENHANCING TRAINING AND CAPACITY BUILDING

The acknowledgment of the critical importance of biological specimens in disease diagnosis, research, education, and investigations has prompted the establishment of an ethical framework governing their collection and preservation. This framework imposes requirements for waste disposal oversight and stipulates that donor consent is necessary for the use of surplus samples. Nevertheless, the practical implementation of these ethical guidelines presents various challenges. In addressing these issues, a project should employ a self-assessment checklist intended for personnel responsible for managing specimens. This approach encourages dialogue with senior management and clarifies the operational environment.

A considerable number of respondents provide additional insights, emphasizing practices that both affect health outcomes and mitigate risks to the community. Future reports should be planned to evaluate strategies aimed at reducing public health vulnerabilities. The information gathered would function as a valuable tool for laboratory managers and service providers, enhancing their engagement with health authorities and relevant stakeholders. Furthermore, these activities would undergo scrutiny, with adherence to ethical standards potentially impacting funding allocations (White et al., 2021). Enhancing the distribution of biosafety and biosecurity knowledge may be more effective with a focus on practical training exercises instead of just theoretical instruction (Rutebemberwa et al., 2020).

Biosafety and biosecurity education is systematically incorporated into training initiatives, encompassing hands-on activities that pertain to biologically relevant situations. A variety of teaching methodologies are utilized to bolster dialogue and showcase practical demonstrations, augmented by supplementary materials. The Participant Hot Zone Training immerses participants in the practical use of pedagogical strategies within authentic contexts, facilitating the interpretation and evaluation of intricate technical information. Furthermore, research on sleep-based training modules aim to enhance readiness for novel infectious outbreaks. It is essential to fortify local biosafety and biosecurity frameworks, as effective responses that influence both human and animal health require a harmonized strategy involving local authorities at the subnational tier (Safdar et al., 2023).

The orchestration of consultations and capacity-building efforts among subnational stakeholders in the realms of biosafety and biosecurity is executed with strategic intent to bolster professional expertise. This systematic alignment is vital in fostering long-term resilience within laboratories by incorporating biosecurity considerations and promoting collaborative ventures. Research shows that the adoption of well-articulated biosecurity protocols significantly enhances overall developmental maturity. Furthermore, training initiatives emphasize elevating public consciousness regarding the Nagoya Protocol, thereby underscoring the community's pivotal function in the stewardship of traditional knowledge. Ultimately, these concerted efforts strive for a sustainable influence within a unified biosecurity framework, equipping partners with a variety of training options, customized educational experiences, and up-to-date resources to facilitate informed decision-making (Tang et al., 2024).

6.8.1 Staff Training Programs

Community Hospital (CH) was the inaugural comprehensive medical facility worldwide to effectively handle a case of Crimean-Congo hemorrhagic fever. This achievement highlighted the hospital's clinical competencies and has since generated heightened interest among its personnel regarding high-risk infectious diseases. In light of this development, the hospital collaborated with the education department to implement a joint training program. The curriculum underwent a systematic revision informed by educational assessments, which now includes real-time presentations from specialists affiliated with pathogen organizations. As a result, the training modules have undergone transformation into a standardized format, ensuring consistency across various subject areas (Risi et al., 2010).

Aerosol Generating Procedures have been integrated into healthcare training programs, reflecting an evolving approach to infection control. A notable online training initiative has emerged, featuring an advanced patient simulator designed specifically for containment scenarios. Efforts are underway to establish protocols for safe workplace exposure for patients and healthcare personnel. Since June 2024, CH has documented four incidents classified as Category A exposures. The CH teams, leveraging their experience with the Crimean-Congo hemorrhagic case, collaborated with top institutions, including a UK containment facility, the Porton Down hospital, the Centers for Disease Control and Prevention (CDC) Special Pathogens Branch, and the U. S. Army Medical Research Institute of Infectious Diseases (USAMRIID) BSL-4 laboratory. Following an incident in September 2024, clinical staff from CH conducted visits to the BSL-3 labs at both the CDC and the University of Virginia, and interacted with the Viral Hemorrhagic Fever unit at the Royal Free Hospital in London (Muzammil et al., 2024).

6.8.2 Public Awareness Campaigns

Clinical laboratories play a critical role in the realms of public and personal healthcare, significantly improving the accuracy of patient diagnoses and the oversight of treatment efficacy. Professionals engaged in the analysis of biological samples for microbiological research frequently encounter a wider array of possible pathogens than their counterparts directly interacting with patients. Consequently, the meticulous handling of specimen containers is imperative. It is essential for all involved stakeholders to collaborate and communicate efficiently to mitigate biohazard risks. This imperative is notably acute in state public health laboratories, which have observed an increase in specimens being received improperly.

The application of biosafety protocols in clinical laboratories exhibits significant inconsistency. The potential dangers posed by biological agents are exacerbated by factors such as bacterial sporulation, the presence of mycobacteria, and the spread of fungi. To mitigate these risks, it is essential to implement protective measures, such as the use of negative air pressure environments for the safe handling of hazardous pathogens like *B. anthracis* and *M. tuberculosis.* It is imperative for clinics to forward specimens to specialized reference laboratories, as state public health laboratories often lack the requisite infrastructure for secure processing. Adhering to the requirements of Standard *Bacillus thuringiensis* (BT) Safety Level 2 necessitates the use of non-permeable surfaces, mechanical pipetting techniques, and appropriate biohazard containment bags (Munson et al., 2018). After removal, containers should be decontaminated with ninety-percent isopropanol and powder from transport carriers. Bags are usually sterilized and disposed of in a steam autoclave.

6.8.3 Building Capacity for Success

To commence the beta testing and evaluation stage, the *Sequencing the Food Supply Chain* initiative underscores the necessity for developing a comprehensive "working plan" as well as an action plan aligned with *Rogers's Diffusion of Innovation Framework* (Zhao et al., 2024a). The strategy requires tailored adjustments to be in congruence with the organization's inherent strengths and priorities. Employing Rogers's framework, it is essential to pinpoint two novel strategies aimed at improving operational efficiency by integrating new knowledge and solutions pertinent to particular domains of expertise. Having developed the proposed actions, it is time to complete and submit the Survey Questionnaire. Thereafter, data from the survey should be analyzed for modern specimen management techniques, and to ensure regulatory compliance, integrity, and traceability. This examination aims to identify biohazard threats and guide improved safety measures (Marwaha et al., 2022).

The laboratory procedures outlined in this document are derived from established construction methodologies that are commonly recognized in the field. These practices have been modified to align with the specific safety, biological, and quality management standards necessary for operation within a BSL-3 facility (Dickmann, Sheeley, & Lightfoot, 2015).

6.9 ETHICAL CONSIDERATIONS IN SPECIMEN MANAGEMENT

Ethically managing bio-specimens involves significant challenges that can hinder access to invaluable samples for research. Key issues include intellectual property rights, safeguarding sensitive information, and addressing noncancerous conditions that may need treatment. Overcoming these obstacles is crucial for advancing research while maintaining ethical standards and protecting privacy. This commitment will foster groundbreaking discoveries and enhance healthcare outcomes (Erasmus, 2012). Numerous initial donors may agree to contribute under broad conditions; however, they could later encounter ethical dilemmas regarding specific enhancements. Concerns related to dignity frequently emerge in relation to biological specimens obtained from deceased persons. One potential approach to address these concerns is to restrict access to these specimens and prohibit the dissemination of results derived from research utilizing them (Toews, 2024; Nicolás, 2023).

International frameworks aimed at resolving ethical dilemmas in biological and biomedical research underscore the critical need for the effective management of bio-specimens. Nonetheless, these guidelines may inadvertently limit the accessibility of bio-specimens for cancer research purposes. They delineate multiple directives pertaining to the responsible handling and governance of bio-specimens, highlighting the necessity of voluntary involvement from donors within a structured regulatory framework. Moreover, it is imperative to uphold a consent management system to guarantee that the utilization of bio-specimens aligns with the intentions of the donors (Rush, Byrne, & Watson, 2024; Tourapi et al., 2024).

6.9.1 Informed Consent

Historically, diagnostics and research have relied on biological specimens and the demographic data of the individuals from whom they were sourced. However, recent modifications to the Biomedical Research Framework have facilitated the utilization of leftover materials from clinical testing for research purposes. The successful application of these specimens is contingent upon their quality and processing, underscoring the importance of employing meticulous preservation methods. Adhering to established laboratory protocols is crucial for ensuring the integrity of the specimens and reducing the likelihood of losing essential samples (Borovečki et al., 2018).

Before an individual opts to employ residual clinical specimens for research purposes, laboratory testing, or any additional processing, adherence to a defined protocol is essential. It is imperative to meticulously follow these protocols while possessing a comprehensive understanding of the regulations related to biohazard risks associated with biological samples. These specimens may present various hazards, including biological, chemical, and physical risks. Consequently, individuals engaging with these biological materials are required to comply strictly with established procedural guidelines and to exercise appropriate caution throughout the handling process (Begum et al., 2024; Amaresh, Ganachari, & Devarinti, 2024).

The key components of pertinent information involve assessing possible risks alongside the execution of strategies aimed at accident prevention and the establishment of protective protocols. It is essential to authenticate adherence to these procedures to comprehend the associated risks, particularly during the procurement of substantial quantities of high-quality samples mandated by research institutions or funding bodies. Oversight is critical in interventional research to uphold safety and compliance standards. The Principal Investigator is responsible for ensuring that all research conditions are satisfied; neglecting this duty could lead to a "conditional" ruling from the ethics committee, which might necessitate further instructions from the sponsoring organization. The process of acquiring informed consent demands meticulous attention to detail.

6.9.2 Privacy and Confidentiality

The construction of a sophisticated database is crucial for the accurate reconstruction of the operational timeline within a Modern Forensic DNA Laboratory. Since the year 1994, forensic laboratories have developed multiparametric databases that encompass vital information regarding biological samples utilized for DNA profiling and quantitative polymerase chain reaction assessments. Initially, the primary focus of data inquiries was on statistical analyses. However, this focus has evolved in response to new security imperatives and the management of Category III biological specimens that present biohazard risks, such as blood, semen, and hair. This database serves as a pivotal tool in forensic science, prompting significant discussions surrounding biosecurity measures and privacy protection. Therefore, comprehensive strategies designed to address biohazard risks while safeguarding data privacy are scrutinized in detail (Zhao et al., 2024b).

The evaluation of risks posed to individuals by data privacy infringements necessitates an examination of the possibilities for illicit exposure or exploitation of DNA profiles, in conjunction with associated data that may connect a profile to a particular event. Privacy serves as both a legal and ethical benchmark, with legislative structures empowering individuals to control the accessibility of their personal information. This legal dimension intersects with the overarching principle of confidentiality, which encompasses the safeguarding of personal data throughout its collection, storage, transmission, utilization, and analysis while implementing measures to thwart unauthorized entry and safeguard data integrity (Kang & Hwang, 2023; Williamson & Prybutok, 2024).

6.10 TECHNOLOGICAL ADVANCES IN SPECIMEN MANAGEMENT

Recent progress in blood collection methodologies has resulted in the enhancement of certain techniques, particularly the vacuum blood collection tube. This sophisticated design features an inner chamber that sustains a vacuum, encased within an outer chamber that is appropriately treated with anticoagulants when required (Bossi et al., 2024; Cortijo-Alfonso et al., 2025).

Despite significant developments in bio-banking, the influence of dual-container systems on the stability of blood components has not been adequately investigated. The application of a capping method for vacuum blood collection tubes effectively minimizes the incorporation of ambient air, decreasing mechanical stress and enhancing the processing of specimens. This advancement holds promise for improving the acquisition of clinical blood samples intended for diagnostic purposes. Additionally, careful management of collected tissue samples is critical to ensure the integrity and relevance of the resulting data. Research concerning pre-analytical errors underscores the necessity of handling biological specimens within a 30-minute window post-collection. Typically, samples are obtained in clinical environments and are subsequently transported to diagnostic laboratories for further analysis (Lerch et al., 2019).

The conditions under which these samples are stored, particularly temperature, play a vital role in preserving cellular integrity and the chemical properties of the samples. Extended exposure to ischemic conditions can lead to a deterioration in the quality of the samples. Therefore, the use of cold transport devices could establish standardized procedures, thereby minimizing inconsistencies arising from processing protocols (Lerch et al., 2019).

6.10.1 Bio-banking Technologies

Bio-banks play a pivotal role in the gathering, processing, preservation, and distribution of human biological samples. These entities function within dedicated facilities located in either private or academic medical institutions, adhering to stringent quality standards essential for diagnostic purposes, tailored healthcare, and biomedical research initiatives. The availability of high-quality bio-specimens and associated data is critical for achieving a comprehensive understanding of patient biology and the mechanisms of disease (Dagher, 2022).

Bio-specimens play a critical role in the advancement of personalized medicine and translational research, necessitating meticulous collection and processing methods to uphold quality standards. Each specimen must be accompanied by comprehensive annotations that include particular details regarding the patient and the specimen itself. However, there exists a significant disparity between the increasing demand for well-annotated bio-specimens in research and their actual availability, which remains constrained. The pharmaceutical sector, however, requires a variety of biological samples that adhere to stringent criteria to fulfill its research and development (R&D) objectives (Qusairy & Rada, 2023; Rush, Byrne, & Watson, 2024). In the last two decades, demand for biological resources has surged, recognizing their vital role in drug discovery. However, the regulatory framework for bio-specimens is still basic. Despite this, industry stakeholders are pushing for clinical collaborations that ensure stringent quality control for these bio-specimens.

Human biorepositories are collections of biological specimens and associated health data derived from participants in cohort studies, designed to facilitate advancements in biomedical research concerning diseases, including cancer and infectious diseases. A study examined the significance of biological specimens and the necessity of their appropriate storage methods across different categories, such as whole blood, red blood cells, plasma, serum, DNA, and RNA, within the context of bio-banks. Biological specimens are crucial for understanding germline mutations, epigenetic alterations, and the interactions between genes and proteins associated with specific diseases. Furthermore, the preservation of the integrity of bio-bank specimens is critical to ensure the accuracy of biomarker identification (Mohamadkhani & Poustchi, 2015).

6.10.2 Data Management Systems

The integration of biological specimens with human subject data is essential for scientific research but has major risks, requiring implementation of regulations to mitigate. An intramural research program utilized community resources for efficient specimen handling. Established foundational policies have been uniformly implemented across various institutions, with no introduction of new experimental projects or data collection occurring at remote locations. All newly submitted biohazard records must be routed to this institution following the submission of a new protocol or environmental oversight request. The process of accessing these documents reflects a dedication to enhancing laboratory record-management systems in compliance with regulatory standards. These documents encompass essential policies for the handling of hazardous biological materials within clinical laboratories (Alkhatib & Gaede, 2024).

The materials utilized for such projects comprise data obtained from health and safety organizations, recognized standard operating protocols, educational curricula, empirical research findings, and methodological frameworks. Curated collections of these resources are available for review, particularly focusing on strategies aimed at the mitigation of bioterrorism.

6.11 CASE STUDIES OF BIOHAZARD MANAGEMENT

The existing medical and public health protocols concerning the transportation of biological specimens frequently demonstrate a lack of sufficient measures for managing biohazardous materials, particularly regarding the liabilities and responsibilities associated with these exchanges.

A comparative analysis reveals that these protocols are notably less robust than those established for other biohazardous transfers, such as those involving blood-borne pathogens. To mitigate these shortcomings, it is imperative for public health authorities to collaborate in addressing biohazard risks and improving specimen handling practices, emphasizing the need to protect proprietary information while actively involving research partners in biosafety initiatives.

An examination of eleven case studies reveals various challenges and gaps in management practices, along with proposed strategies for enhancement. Since the 1970s, the implementation of institutional biosafety programs has significantly lessened the risks associated with pathogen research, bolstered by national and international regulations that oversee the movement of human specimens during disease outbreaks. Nevertheless, the current governance frameworks inadequately tackle the biohazard risks inherent in the exchange of biological materials, exposing critical gaps in health and safety planning as well as in the delineation of partner responsibilities. Consequently, thorough assessments of biohazard risk management may yield valuable insights for addressing these governance deficiencies effectively (Weng et al., 2024; Mendonça, 2024). These insights aid public health strategists and organizations managing emerging pathogens and ensuring biosafety accountability.

6.11.1 Successful Interventions

The study above underscored the critical necessity for enhanced adherence to cleaning and disinfection protocols specifically for high-touch surfaces in outpatient environments to reduce contamination risks. There is a clear requirement for substantial improvements in maintenance practices within these settings.

An evaluation conducted in the Outpatient Unit analyzed surface cleanliness before and after the implementation of a targeted intervention program. Post-intervention, cleaning protocols were systematically reviewed over a three-month period, involving weekly evaluations through fluorescence markers, visual inspections, and ATP bioluminescence measurements. Additionally, an educational initiative aimed at staff from outsourced cleaning services was introduced. Average fluorescence assessment scores exceeded 3, signaling inadequate performance and highlighting the necessity for corrective actions. The application of fluorescence markers effectively illustrated the intervention's efficacy, offering constructive feedback to housekeeping staff and enhancing awareness regarding the risks of healthcare-associated infections among administrative personnel (Furlan et al., 2019).

Current assisted reproductive technology (ART) centers expose both patients and laboratory personnel to significant infection risks. This research outlined strategies aimed at minimizing infection rates for ART practitioners and patients. It analyzed recent incidents within a particular ART facility and utilized data from a national database of ART practices. Strict adherence to established best practices is crucial for maintaining laboratory safety and safeguarding the health of all individuals involved. However, the application of these protocols often proves difficult, particularly in commercial settings characterized by high patient turnover and insufficient infrastructure (Huyser, 2014).

6.11.2 Lessons Learned from Failures

In the examination of failures, it is essential to identify critical mistakes associated with major risks. Although the 2019 containment failure involving the smallpox virus in three U.S. laboratories may not rank as one of the more notorious biosafety incidents, it is important to contextualize it within the larger framework of public health milestones. The 1976 eradication of Variola major represents a significant achievement in public health. This success contrasts with various incidents that tend to obscure basic childhood experiences, such as engaging in play with Legos (Lieberman et al., 2022; Bell et al., 2024).

Virtually a decade later, apprehensions regarding safety in biological research intensified, particularly following the establishment of the Blood-borne Pathogen standard in 1991. This heightened concern prompted an increase in academic publications suggesting that safety-related anxieties could be overstated, contending that established "good microbiological practices" were incompatible with what were perceived as "advanced" modern methodologies. In this context, the company LabTAG conducted an assessment of routine bacteriology laboratories in relation to the biosafety standards outlined by the Department of Homeland Security (DHS).

The Blood-borne Pathogen standard played a pivotal role in shaping the Laboratory Biosafety Guidelines (LBG), leading to a reassessment following the development of a novel data-collection tool. This new instrument was designed to rectify eleven identified deficiencies within the original guidelines. The approach involved conducting a risk-assessment survey that included site visits to 140 clinical laboratories, all engaged in specimen processing, across Wisconsin. These laboratories varied significantly in size and scope, from large hospital bacteriology departments to smaller individual practices (Perkins & Bakanidze, 2024; Bayrak, 2025).

The assessment systematically examined laboratory protocols pertaining to six biohazardous materials classified as high-risk. Critical components of the evaluation comprised aspects such as the proper packaging of materials, the procedures for unloading them, methods of disinfection, strategies for waste management, techniques for preparing tissue homogenates, and approaches to plate streaking. A total of fifty criteria were established, grounded in LBG and Blood-borne Pathogen standards, with an emphasis on ensuring the safe handling of hazardous materials (Munson et al., 2018).

The 1991 LBG introduced significant biosafety protocols that received recognition from various authoritative bodies, including the OSHA, the American Society for Microbiology (ASM), the CDC, and the NIH. However, it is notable that fewer than 6% of laboratories adhered to the eight critical "prudent" practices outlined in these guidelines. These practices encompassed the use of protective clothing, which includes suits, gowns, gloves, and footwear, as well as additional items like finger cots, as specified in the guidelines issued by ASM, CDC, and NIH.

6.12 FUTURE TRENDS IN BIOLOGICAL SPECIMEN MANAGEMENT

The examination of clinical specimens within laboratory environments is crucial for hospitals catering to a varied patient demographic, encompassing both inpatients and surgical individuals. Critical stakeholders such as facility managers, nursing directors, and public health authorities rely heavily on laboratories for precise diagnostic outcomes. Notably, safety considerations are frequently overlooked until adverse events transpire, highlighting a pattern in the approach to laboratory safety that tends to prioritize reactive measures rather than adopting a proactive stance (Iwen et al., 2018).

Biological specimens present potential issues regarding Select Agents and Toxins. These agents correspond to particular biological agents and toxins identified within the regulatory framework. This guidance focuses on the protocols for treating individuals diagnosed with Ebola, as well as the proper management of remains belonging to those categorized as PUIs for EV disease in healthcare settings across the US (Chavez et al., 2023; Levine & Marshall, 2024). This provides guidelines to prepare healthcare practitioners and public health officials for managing specimens that may contain the EV and other risk group-4 (RG-4) pathogens. The survey was conducted to assess the knowledge and practices of community infectious disease physicians regarding handling suspected RG-4 contaminated specimens.

6.12.1 Emerging Trends

The examination of biological specimens within laboratory settings plays a critical role in the healthcare infrastructure, prevalent in both urban and rural hospitals. Laboratory administrators are responsible for delivering results that are not only timely but also precise, all the while maintaining a secure environment for testing. Numerous laboratories adopt standardized operating procedures that facilitate the safe management of clinical specimens. It is crucial to recognize that every specimen has the potential to contain harmful pathogens, which underscores the importance of utilizing appropriate laboratory equipment to effectively handle these materials (Iwen et al., 2018). Newly identified high-risk pathogens cause significant health challenges in humans, with few strategies for prevention, treatment, or containment.

In November 2018, a rise in severe acute flaccid myelitis cases among U.S. children, averaging 4 years old, was noted (Hayes et al., 2023). This case report underscores the pressing concerns articulated by experts in the aftermath of the 2014 Ebola outbreak, drawing attention to the persistent inadequacies in the handling of hazardous biological materials. Although there has been an uptick in focus on the management of laboratory specimens in both research and policy contexts as a result of the Ebola crisis, the essential domains of laboratory biosafety and biosecurity continue to be inadequately prioritized within the scientific community. Concurrently, CareDx and the laboratory refute the claims of patent infringements (Moedritzer, 2023). It is crucial to recognize that they operate as rivals within the field of blood specimen analysis.

6.12.2 Innovative Practices

In 2018, considerable apprehension arose regarding the preparedness of U.S. healthcare professionals to effectively monitor, evaluate, and address instances of exposure to the EV (Mason et al., 2021). In October 2014, documents were released to enhance readiness within the healthcare sector of the U.S. system. These documents delineated specific protocols concerning the use of PPE, as well as strategies for the management and disposal of infectious waste and the processing of clinical specimens (Khatri et al., 2023).

The guidance delineates the procedures associated with the collection, transportation, and submission of specimens. Primarily designed for the EV, these guidelines function as overarching frameworks applicable to the management of several high-risk pathogens. Additionally, detailed protocols are formulated for other EIDs, with a specific focus on highly pathogenic strains of influenza and the coronavirus responsible for Middle East Respiratory Syndrome (Aguanno et al., 2022).

Significant academic medical institutions are capable of processing specimens according to established protocols. However, it is imperative to develop a comprehensive strategy that ensures healthcare providers lacking sufficient resources or expertise in safe handling can access comparable services (Iwen et al., 2018). As the demand for healthcare services increases, there is a corresponding rise in the potential for healthcare professionals to be exposed to infectious diseases, which can subsequently facilitate their transmission within the wider community. Support staff are particularly vulnerable, especially when processing clinical samples from patients infected with highly contagious pathogens. Prior to engaging with any specimens that may pose health risks, it is imperative for healthcare administrators, in conjunction with laboratory safety officers, to conduct a comprehensive review of institutional guidelines regarding biological specimens. This evaluative process guarantees that safety protocols are both sufficient and rigorously adhered to.

6.13 CONCLUSION

The containment of bio-specimens and biohazards is a major problem for biomedical research world over. The current situation is such that without a long-term strategy in place, the problems will continue to grow in the future. It is important that every country supports biomedical research to enable better science and health for all people. Therefore, there is a need for the establishment of international and regional collaborative strategies among experts in R&D to ensure reliable prevention and security measures for people and the natural environment in the containment and transportation of delicate and potentially risky specimens. It is crucial for researchers, healthcare personnel, and other stakeholders, to be fully cognizant of the potential biohazards in order to effectively protect themselves, their families, their immediate surroundings, and to bear personal responsibility for the welfare of their nations and mankind.

In order to mitigate biohazard risks and adhere to safety guidelines, vocational schools are required to implement comprehensive safety training programs that clearly articulate essential safety protocols. It is imperative that the security and safety regulations are not only documented but also prominently displayed in each laboratory setting, ensuring that instructors actively oversee compliance with these rules by all laboratory users. Additionally, it is crucial that the meanings of warning labels and hazard signs used in educational biomedical laboratories are thoroughly communicated to all. Schools that offer laboratory programs should designate qualified occupational health and safety specialists to oversee these efforts. Furthermore, the implementation of regular inspections is necessary to effectively identify and manage sources of potential hazardous materials.

Adherence to continuous professional development is key, with regular refresher courses, workshops, or seminars focusing on the applicable regulations and management of biohazardous materials. Furthermore, students and other laboratory users should receive instruction in first aid practices. Given that accidents often occur toward the end of lab sessions, it may be beneficial to schedule laboratories more than once per day. To avoid overcrowding, it is recommended that a limited number of students be assigned to each laboratory. Additionally, by designating an assistant in every laboratory, it is possible to distribute the responsibilities more evenly and minimize disruptive behaviors.

Finally, it is imperative that comprehensive awareness campaigns are conducted to educate the masses about the inherent risks posed by bio-specimens and biohazards. Extensive research is warranted in the domain of telehealth diagnostics, therapeutics, and public health, paving the way for cost-effective, accessible, and user-friendly preventive and prophylactic options, with a special focus on marginalized populations and underserved nations.

7 Exposure to Endocrine-Disrupting Chemicals and Sustainable Development in Africa

Patrick O. Onyango and Collins O. Asweto

7.1 INTRODUCTION

Increasing prevalence of endocrine-disrupting chemicals (EDCs) in the environment poses a significant challenge to public health (UNEP/WHO, 2012). The World Health Organization defines EDCs as exogenous substances that alter the function of the endocrine system, resulting in adverse health effects in individuals or populations population (Kortenkamp et al., 2011; UNEP/WHO, 2012). A large body of evidence shows that EDCs negatively impact reproductive and neural function, suppress immune response and alter the timing or onset of life history milestones and other negative health outcomes (Di Renzo et al., 2015). For instance, exposure to EDCs have been linked to preterm births, low birth weight, cancers and infertility (Di Renzo et al., 2015; Fudvoye, Bourguignon, & Parent, 2014; Kortenkamp et al., 2011; UNEP/WHO, 2012). The EDCs cause such adverse health outcomes by interfering with key components of the endocrine system in the synthesis and secretion of hormones (UNEP, 2017).

The health and economic burden of environmental toxicity including those arising from EDCs is substantial and has been shown to be disproportionately higher in poor nations. For example, it was projected in 2013 that the cost of pesticide poisoning in 37 African countries would be US$ 66 billion for the period 2005–2020 (UNEP, 2013). Beyond direct burden of exposure to EDCs on which most estimates are based, it is important to note that negative health outcomes due to exposure to EDCs may not only affect exposed individuals but also the individual's dependents and future offspring (UNEP/WHO, 2012). This is particularly the case when exposure occurs during critical life history stages such as pregnancy or infancy (Di Renzo et al., 2015). Consequently, pregnant and lactating mothers, women planning to have children, and infants represent vulnerable populations (UNEP, 2017). Furthermore, adverse effects of EDCs on wild flora and fauna may have ecosystem-wide implications.

EDCs occur in a broad range of products and environmental matrices. They are found in industrial products, pharmaceuticals, personal care products and cosmetics, agrochemicals such as fertilizers and pesticides, foods and food packaging materials (UNEP, 2017) from where they ultimately find their way into wastewater treatment plants (WWTPs), surface water, groundwater, aquatic systems, soil and air. The implication of the high diversity of products containing EDCs as well as their wide distribution in the environment is that efforts to reduce their public health burden must be multi-sectoral so as to target the production, distribution, use, disposal and removal of such products from the environment.

In Africa, there are several reasons that may exacerbate the adverse effects of exposure to EDCs. First, a large proportion of the African population does not have access to basic sanitation services. The implication is that a significant proportion of the population is exposed to untreated or semi-treated wastewater and carry a disproportionately heavy burden of preventable sanitation-related diseases. Second, the continent has a rapidly increasing urban population, growing at a rate of 2.4% compared to the global average of 1.5% (UNDESA, 2019), which overstretches the capacity of existing WWTPs.

Coupled with the fact that removal of EDCs from wastewater requires advanced and costly wastewater treatment technologies, which is a major constraint in many African municipalities, there is a high risk of exposure to untreated or semi-treated wastewater including those that contain EDCs. Lastly, Africa is second only to Australia with respect to water scarcity (UNEP, 2017) with the implication that effluents from respective wastewater generating sectors drain into few water systems potentially resulting into increased concentration of pollutants.

Given its vulnerability to environmental chemicals, strategies and goals of any development blueprints in Africa must thus have provisions to mitigate against the burden of EDCs in the continent. For example, efforts to achieving Sustainable Development Goals (SDG) 6—ensure access to water and sanitation for all—must appreciate the complexity of the burden of EDCs to public health. Therefore, understanding the knowledge base on EDCs is critical at the research, policy and practice levels. In this chapter, we conduct a scoping review so as to document the status of knowledge on EDCs in Africa. We discuss our findings in the context of the SDGs and the Africa Union (AU) Agenda 2063.

7.2 METHODOLOGY

The Preferred Reporting Items for Systematic Reviews and Meta-Analyses extension for Scoping Reviews (PRISMA-ScR) Checklist on methods and reporting of results (Tricco et al., 2018) was followed so as to answer the question, "What is the focus of research on EDCs in Africa?" A search strategy using the search phrase ("endocrine disrupting" OR "endocrine-disrupting") AND Africa* (the character * allows the search to return terms including such as Africa and African) was used to extract relevant literature. Literature search was conducted in July 2022 and was not restricted by year of publication because a deliberate attempt was made to collect as much data as possible on the focus of research on EDCs in the continent. Literature was retrieved from PubMed and Web of Science databases.

7.2.1 Eligibility

Empirical studies primarily focusing on EDCs as pollutants or contaminants, studies where full texts were available or where targeted information (aim of study, place of study, and findings) were available in the abstract, and studies conducted in Africa were eligible for the scoping review. Citations for all publications that were retrieved were then exported into Microsoft Excel for further processing. In the first stage of screening, reviews and duplicate publications were removed. Both authors screened the publications for inclusion in this review. A total of 454 publications were retrieved (322 from Web of Science and 132 from PubMed). The review process is summarized in the PRISMA diagram (Figure 7.1). Although we developed a protocol for the review, it was not registered.

7.2.2 Data Extraction and Analysis

We summarized publications on the basis of the year of publication, African country in which work was conducted, the goal of the study and research design. In addition, eligible studies were classified, firstly, on the basis of whether they were about methodological approach for measuring EDCs, monitoring of EDCs in environmental matrices, health risk (in both plants and animals, including humans). Secondly, we classified publications on the basis of recommendations that the authors made.

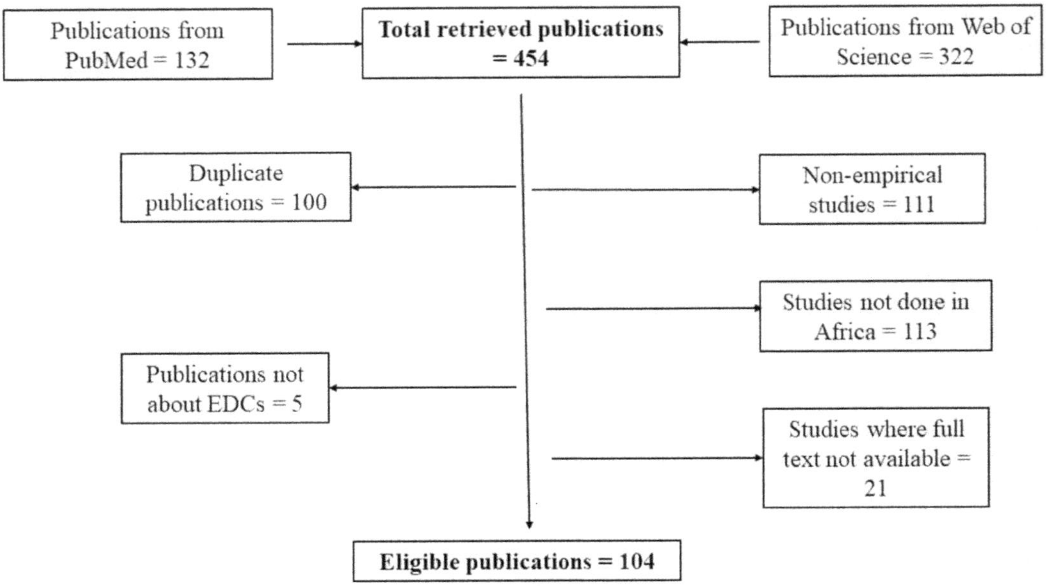

FIGURE 7.1 PRISMA flowchart of study selection process.

7.3 RESULTS AND DISCUSSION

In total, we reviewed 104 eligible publications on empirical studies on EDCs across Africa. A majority (77.9%) of the eligible publications were conducted in South Africa (n=81) followed by Nigeria (n=10); Namibia and Tunisia each had two publications, whereas Algeria, Egypt, Ethiopia, Ghana, Lesotho, Senegal, Tanzania, Uganda and Zimbabwe each had one publication.

The yearly frequency of eligible publications increased, from 3 in 2003 to a maximum of 13 in 2019 (Figure 7.2). The modest increase in yearly frequency of publications on EDCs may reflect both an appreciation of the threat of EDCs to both human and wildlife health. However, the distribution of the studies appears to be heterogeneous across the continent. The increase in yearly frequency of publications on EDCs reported in this chapter mirrors that reported by UNEP for global trends in frequency publications (UNEP, 2017). In terms of EDCs, the studies covered a broad range of chemicals including phenolic compounds, phthalates, parabens, pesticides, fungicides, pharmaceuticals and heavy metals.

Nearly half (46%) of the publications were cross-sectional, longitudinal (12%) and experimental (42%) in design. In part, the low number of long-term studies could be attributed to the continued burden of traditional causes of morbidity and mortality in the continent and therefore the continued investment in their research. Alternatively, it could be attributed to the high cost of quantifying EDCs from environmental or biological samples.

7.3.1 Areas of Focus of Studies on Endocrine-Disrupting Chemicals in Africa

In terms of the four thematic areas of research that we focused on, 42% of the studies were on health risks, methodology (19%), monitoring (29%) and remediation (10%). Next, we present and discuss findings in each thematic area.

7.3.1.1 Methodological Approaches for Quantifying Endocrine-Disrupting Chemicals

Twenty publications representing 19% of eligible publications were about methodological approaches for detecting or quantifying levels of EDCs in environmental matrices. Slightly over half of the 20 publications were published later than 2015.

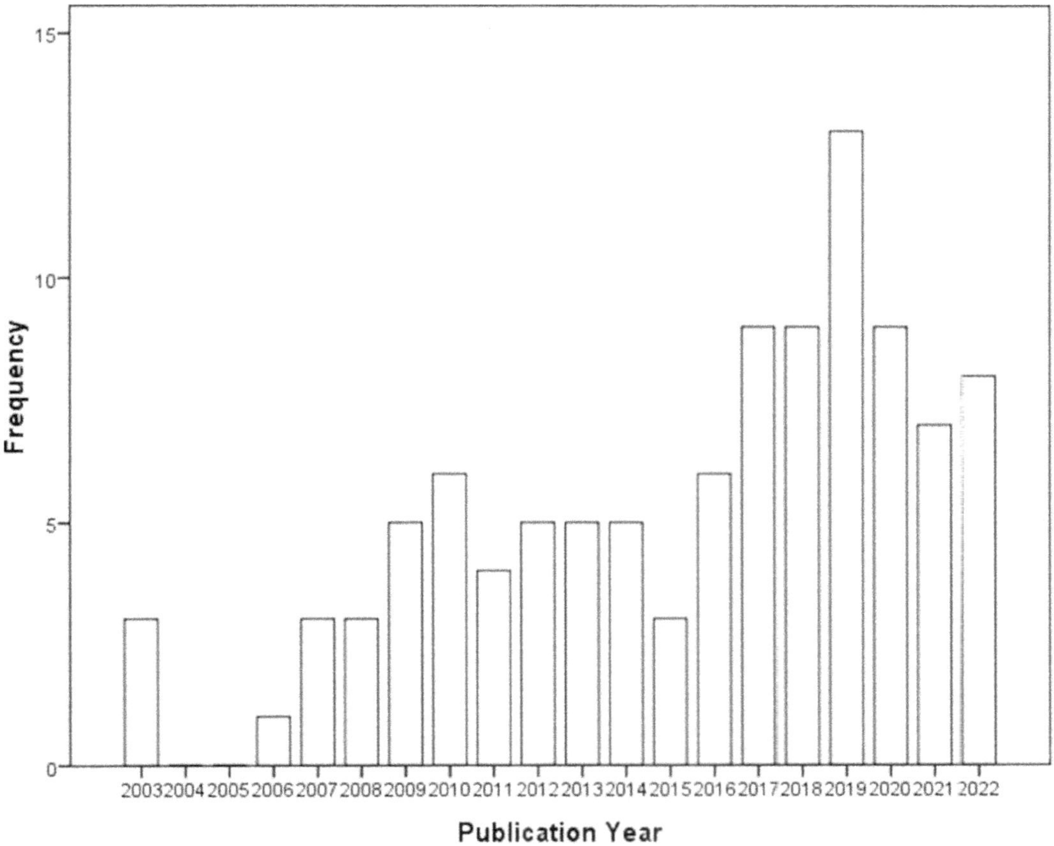

FIGURE 7.2 Frequency of publications on endocrine-disrupting chemicals in Africa through to July 2022.

The apparent continued interest and research on innovative ways of detecting and measuring EDCs may reflect the high cost, both technically and financially, of measuring EDCs in samples of environmental matrices. Consequently, the reviewed publications appear to address the issue of cost and portability (Wooding, Rohwer, & Naude, 2017a), and broad-range detectors that can quantify different EDCs simultaneously (Spath et al., 2021) as discussed in Table 7.1. In terms of cost and portability, for instance Wooding, Rohwer, & Naude (2017a, 2017b) developed a relatively cheap (US$ 1) sampler that was capable of detecting a broad range of micropollutants including pharmaceuticals, personal care products, pesticides and alkylphenol. There is also interest in the use of biological tools to degrade pollutants, bioremediation, from wastewater. For example, Qhanya et al. (2017) used bacteria to biodegrade nonyl-phenols from wastewater; Kasonga et al. (2022) employed a consortium of fungi to biodegrade pharmaceutical products from wastewater.

7.3.1.2 Monitoring of Endocrine-Disrupting Chemicals in Environmental Matrices

The publications that were reviewed investigated concentrations of EDCs in organisms such as brown mussels (Castro et al., 2022), eland (Bornman et al., 2010), crocodile (Humphries et al., 2021), fish (Ssebugere et al., 2014), in animal feeds (Masgoret et al., 2009) and human food (Oloruntoba et al., 2019). Other EDCs were found in fertilizer, surface, ground and packaged water (Aneck-Hahn, Bornman, & de Jager, 2008; Archer, Wolfaardt, & van Wyk, 2017; Awofolu & Fatoki, 2003; Bolujoko et al., 2022; Faul, Julies, & Pool, 2014; Mahomed et al., 2008; Ohoro et al., 2021; Tongu et al., 2021), sediment (Tongu et al., 2021), wastewater (Archer et al., 2020; Etteieb et al., 2015; Farounbi & Ngqwala, 2020; Kasambala, Rwiza, & Mdegela, 2019; Salaudeen et al., 2018) and in settled dust in interiors of homes and institutions (Boudehane et al., 2016).

TABLE 7.1
Methodological Approaches Used to Detect or Measure Endocrine-Disrupting Chemicals in Africa

Publication	Country of Study	Study Design	Goal of Study	Main Finding
Akharame et al. (2022)	South Africa	Experimental	Development of a novel polymeric nanocomposite (PNC) based on hydrothermally synthesized akaganeite *in-situ* supported on polyamide for the heterogeneous catalytic ozonation of the organic pollutant 4-chlorophenol in aqueous media	The synthesized PNC eliminated Fe leaching with stability extending to six cycles
Archer et al. (2017); Kibambe et al. (2020)	South Africa	Cross-sectional	Isolation and enrichment of six bacterial strains from six different soil samples collected from the Mpumalanga Province	Isolated bacteria degraded nonylphenol
Braathen et al. (2009)	South Africa	Experimental	Vitellogenin (VTG) induction in *Clarias gariepinus* assessment in order to develop a method for monitoring estrogenic pollution in freshwater systems	VTG is induced in male *C. gariepinus* after exposure to both estradiol and 17α-ethynylestradiol
Chokwe et al. (2012)	South Africa	Experimental	Investigation of a gas chromatography-mass spectrometry method for the simultaneous analysis of two types of EDCs	Good recoveries for targeted compounds
Esterhuyse et al. (2009)	South Africa	Experimental	Investigation of utility of VTG as indicator of estrogen exposure in *Oreochromis mossambicus*	VTG is an indicator of estrogen exposure in *O. mossambicus*
Fatoki and Awofolu (2003)	South Africa	Experimental	Evaluation of different extraction methods for the determination of fifteen organochlorine pesticides in water and sediments	Liquid–liguid extraction with *n*-hexane and dichloromethane was accurate and reliable for detection of organochlorine pesticides from environmental waters
George et al. (2019)	Lesotho	Cross-sectional	Comparison of Soxhlet and reflux solvent extraction techniques for the screening and characterization of soil for EDCs	Extraction solvent and extraction procedure were critical in screening real dumpsite samples for EDCs
Hlekelele et al. (2019)	South Africa	Experimental	Synthesis and characterization of materials consisting of polypropylene, polyaniline and ferrous oxide for the activation of hydrogen peroxide for Fenton-like reactions toward removal of bisphenol A (BPA)	Synthesized material demonstrated a removal efficiency of 57% for BPA
Hlekelele et al. (2018)	South Africa	Experimental	Use of coal fly ash as a catalyst for the synthesis of nitrogen-doped carbon nanotubes and carbon nanofibers for photo-degradation of BPA	Removal of BPA was demonstrated using a known hazardous material
Kasonga et al. (2022)	South Africa	Experimental	Evaluation of biodegradation of carbamazepine, diclofenac sodium and ibuprofen through fungal granules developed in a sequencing batch reactor	High removal efficiencies reported

(Continued)

TABLE 7.1 (*Continued*)
Methodological Approaches Used to Detect or Measure Endocrine-Disrupting Chemicals in Africa

Publication	Country of Study	Study Design	Goal of Study	Main Finding
Martin et al. (2016)	South Africa	Experimental	Determination of four groups of EDCs in human hair	Detection of BPA and parabens but poor or inability to detect perfluoroalkyl compounds and hexabromocyclododecane
Mnguni et al. (2018)	South Africa	Experimental	Development of methodology for detecting trace levels of estrogens in raw and treated portable water samples	Estrogens were detected in raw but not treated portable water samples
Ndangili et al. (2011)	South Africa	Experimental	Presentation of a 3-mercaptopropionic acid-capped zinc selenide quantum dot/cytochrome P450 3A4 enzyme electrochemical biotransducer for 17β-estradiol	The biotransducer was capable of detecting 17β-estradiol
Olowu et al. (2010)	South Africa	Experimental	Development of a simple and highly sensitive electrochemical DNA aptasensor with high affinity for 17β-estradiol	The aptasensor was able to distinguish 17β-estradiol from structurally similar EDCs
Regasa and Nyokong (2022)	Ethiopia	Experimental	Preparation of electrochemical sensor based on the molecularly imprinted poly(*p*-aminophenol) supported by silver nanoparticles capped with 2-mercaptobenzoxazole for the selective and sensitive detection of 17β-estradiol	The sensor showed high recovery of 17β-estradiol and reusability
Sidwaba et al. (2019)	South Africa	Experimental	Development of an electrochemical manganese peroxidase biosensor fabricated on a glassy carbon electrode modified with a nanocomposite comprising conducting polyaniline, polymethyl methacrylate and titanium dioxide nanoparticles	The biosensor electrochemically degraded and oxidized BPA
Spath et al. (2021)	South Africa	Experimental	Development of a novel method to detect and measure 71 micropollutants using online solid-phase extraction liquid chromatography coupled to tandem mass spectrometry	Targeted micropollutants including pharmaceuticals were detected
Stopforth et al. (2007)	South Africa	Experimental	Development of a sensitive and solvent-free method for the measurement of estrone and 17β-estradiol in human urine samples	Increased levels of the endogenous estrogens were detected after administering a standard dose of Premarin to a female volunteer
Wooding et al. (2017a)	South Africa	Experimental	Development of a portable sampler for EDCs	A wide range of EDCs detected from surface water
Wooding et al. (2017b)	South Africa	Experimental	Development of a low-cost, disposable, in-house-made sorptive extraction sampler was compared to stir-bar sorptive extraction using a commercial Twister sorptive sampler	Performance of the low-cost sampler comparable to stir-bar sorptive extraction for determination of micropollutants

Although the reviewed publications analyzed samples from a broad range of sources, there are glaring gaps with regard to known major sources or pathways of exposure to EDCs. For example, the range of food items covered was quite narrow. Similarly, there also appears to be less focus on cosmetics and personal care products and agrochemicals as potential sources of EDCs. A disproportionately higher number of studies on monitoring investigated level of EDCs in water and wastewater. Interest in water and wastewater is in part guided by the argument that they contain EDCs arising from virtually all sources and therefore should be the focus of research on environmental exposure to EDCs. Although there is merit to that argument, focus on water and wastewater obscures identity of exposure sources, which would imperil targeted interventions aimed at reducing exposure to EDCs as well as remediation efforts.

However, it must be appreciated that the diversity of EDCs, and products that contain these chemicals, as well as the diversity of their general health endpoints may account for the broad range of sources of samples for analysis of EDCs. In addition, the shotgun approach depicted by the reviewed publications may be attributed to the widespread distribution of EDCs or their metabolites in virtually all environmental matrices and continued release from industrial and pharmaceutical, and consumer products.

7.3.1.3 Health Risk Associated with Exposure to Endocrine-Disrupting Chemicals

Research reported by publications that focused on health risks associated with exposure to EDCs were conducted on animals (64% of publications on health risk), food (7%), food package (2%), humans (11%), water (9%), *in vitro* studies using cell lines (5%), and plants (2%). In terms of the aspect of health that was investigated, almost half of the publications (45%) focused on effects of exposure to EDCs on reproduction. Table 7.2 summarizes the focus of studies in Africa on health risk associated with exposure to EDCs.

Studies that investigated reproductive endpoints associated with exposure to EDCs focused on sperm quality (Marchand, Pieterse, & Barnhoorn, 2008, 2010; Wagenaar, Botha, & Barnhoorn, 2012), mating behavior (Hoffmann & Kloas, 2010) androgenic and estrogenic responses or activity (Genthe et al., 2013; Horn, Pieters, & Bohn, 2020; Makene & Pool, 2019; Patrick et al., 2020; Truter et al., 2016, 2017; van Wyk, Pool, & Leslie, 2003), urogenital distance (Kilian et al., 2007; Patrick et al., 2020), steroidogenesis (Bello et al., 2014; Otitoju et al., 2011), intersex (Barnhoorn et al., 2010; Bengu et al., 2017; Brink, van Vuren, & Bornman, 2012; Pieterse et al., 2010; van Wyk, Pool, & Leslie, 2003), gonado-somatic index (Brink, van Vuren, & Bornman, 2012; Kruger et al., 2013), histology of the testis (Asem-Hiablie et al., 2013; Bengu et al., 2017; Bornman et al., 2010; Das Neves, Barnhoorn, & Wagenaar, 2018; Kruger et al., 2013; Marchand, Pieterse, & Barnhoorn, 2010; Patrick et al., 2016; Pieterse et al., 2010; Wagenaar, Botha, & Barnhoorn, 2012), timing of puberty (Nkomo et al., 2018) and risk of microbial infection (Genthe et al., 2013).

For the most part, the above studies demonstrated that exposure to EDCs adversely affect reproductive physiology. However, the adverse reproductive effects cannot be generalized. For instance, exposure to DDT was not found to imperil reproduction in catfish, *Clarias gariepinus* (Brink, van Vuren, & Bornman, 2012).

Other studies investigated association between exposure to EDCs and age at maturation (Babalola & van Wyk, 2021) and growth (Kim et al., 2022), and residues of DDE in fat tissue of crocodiles (Humphries et al., 2021). The DDE is a common component of insecticide used for residual indoor spraying where malaria burden is high. Others have studied the impact of exposure to EDCs including lower maturation for age associated with lead exposure during development in humans (Nkomo et al., 2018); effects on brain (Johnson et al., 2021), gills (Ibrahim & Harabawy, 2014), heart (Johnson et al., 2021), liver (Asem-Hiablie et al., 2013; Bengu et al., 2017; Ibrahim & Harabawy, 2014; Kilian et al., 2007; Marchand et al., 2009; van Wyk, Pool, & Leslie, 2003), kidney (Ibrahim & Harabawy, 2014), thyroid (Babalola & van Wyk, 2021; Truter et al., 2016), brain gene expression (Thompson et al., 2018) and broader ecological risk (Ojemaye & Petrik, 2019).

Although food is a critical pathway to exposure to EDCs (UNEP, 2017), only three studies of the 44 studies that focused on health risk quantified levels of EDCs in foods. Furthermore, the

TABLE 7.2
Focus of Studies in Africa on Health Risk Associated with Exposure to Endocrine-Disrupting Chemicals

Source of Samples	Study Design	Aspect of Health or Targeted Organ	References
Animal	Cross-sectional	Reproduction, fat tissue, liver, gill, kidney, nonspecific	Barnhoorn et al. (2010); Bengu et al. (2017); Bornman et al. (2010); Brink et al. (2012); Humphries et al. (2021); Kruger et al. (2013); Marchand et al. (2008, 2009, 2010); McHugh et al. (2011); Ojemaye and Petrik (2019); Thompson et al. (2018); Wagenaar et al. (2012)
	Experimental	Reproduction, liver, thyroid activity, kidney, muscles, heart, brain, cholesterol	Asem-Hiablie et al. (2013); Babalola and van Wyk (2021); Bello et al. (2014); Das Neves, Barnhoorn, and Wagenaar (2018); Hoffmann and Kloas (2010); Ibrahim and Harabawy (2014); Johnson et al. (2021); Kilian et al. (2007); Otitoju et al. (2011); Patrick et al. (2016, 2020); Truter et al. (2016, 2017); van Wyk, Pool, & Leslie (2003)
	Longitudinal	Reproduction	Pieterse et al. (2010)
Cell line *in vitro* assay	Experimental	Androgenic and estrogenic activity	Horn et al. (2020); Makene and Pool (2019)
Food	Cross-sectional	Wheat, fruits, vegetables, packaged and non-packaged fish and bread	Dalvie and London (2009); Mutengwe et al. (2016); Tian et al. (2022)
Food package	Cross-sectional	Nonspecific	de Jager et al. (2019)
Human	Cohort	Growth	Kim et al. (2022); Nkomo et al. (2018)
	Cross-sectional	Wristband, reproduction	Dixon et al. (2019); Aneck-Hahn et al. (2007)
	Longitudinal	Risk of microbial infection, reproduction	Genthe et al. (2013)
Plant	Experimental	Morphology of plasma membrane in lettuce	de Bruin et al. (2017)
Water	Cross-sectional	Nonspecific	Inam et al. (2019); Onyekwere et al. (2019)
	Experimental	Estrogenic activity, nonspecific	Aneck-Hahn et al. (2018); Olujimi et al. (2017)

Note: Nonspecific is used in reference to studies that did not focus on a specific aspect of health but are included here because they referred to ecological risk reported as risk quotient, the standard for assessing health risk associated with exposure to environmental pollutants.

range of food items that were considered was very narrow—bread, fish, fruits and vegetables—thus highlighting the need for expanding the range of foods to be investigated for EDCs. Perhaps more worrying is that only one study (de Bruin et al., 2017) on health risk associated with exposure to EDCs was carried out on plants. The minimal research on health risk that exposure to EDCs poses to plants is a nontrivial concern given the key of plants being primary producers such that adverse effects on plants are likely to cascade through subsequent trophic levels. Similarly missing were studies on the effect of EDCs on microorganisms that support biogeochemical processes such as the carbon and nitrogen cycles upon which life depends.

Overall, reviewed studies on health risk associated with exposure to EDCs reveal three areas for future research in the continent. First, there are few published studies on effects of EDCs throughout the life course of individuals or on intergenerational effects. Second, gaps exist on studies on occupational exposure to EDCs. This is particularly a concern given the absence or weak regulatory framework on the production, distribution and use, and disposal of products that contain EDCs in most African countries. Additionally, occupational exposure to EDCs should be a major focus of research given increasing rate of industrial chemical production in developing countries (Signe, 2018). Third, a majority of the studies appear to focus on reproduction, yet exposure to EDCs has been shown to be associated with other pathologies including cancer as well as the onset and attainment of key life history milestones such as age at puberty.

7.3.1.4 Remediation of Endocrine-Disrupting Chemicals from the Environment

Studies that investigated removal of EDCs from wastewater are summarized in Table 7.3. Removal of EDCs from the environment and particularly from what can be considered to be the sink of EDCs, water, is a critical component of any efforts to reduce exposure to these toxic chemicals. In our review of the literature, researchers raised concerns on the inability of WWTPs to remove EDCs from wastewater (Archer et al., 2017; Kibambe et al., 2020). For example, Archer et al. (2017) found that WWTPs lacked the capacity to remove emerging environmental contaminants such as those contained in personal care products. However, Kibambe et al. (2020) demonstrated that using a combination of wastewater treatment approaches has higher removal efficiencies compared to single-technology approaches.

In total, all these studies point to the complexity of removing EDCs such as micropollutants or pharmaceuticals from wastewater. In part, this complexity derives from the diverse range of EDCs and their chemical and structural nature as well as their quantity in environmental matrices. Yet, in most African municipalities the major approaches to wastewater treatment include constructed wetlands, aerated lagoons and wastewater stabilization ponds, which have been shown to lack capacity to remove inorganic pollutants. However, some studies reported high removal efficiencies for targeted EDCs such as estrogens and testosterone (Aziz & Ojumu, 2020), BPA (Kruid, Fogel, & Limson, 2018) and pharmaceuticals with estrogenic activity (Kasonga et al., 2019) through reverse osmosis and nano-filtration, catalytic oxidation and biodegradation, respectively.

The above findings are consistent with the prevailing body of literature on the substantially low removal capacities by traditional technologies or approaches compared to advanced technologies. The challenge, however, is that advanced wastewater treatment technologies have very high capital and operational costs that are beyond the budgets of most municipal wastewater management agencies across Africa.

7.4 RECOMMENDATIONS OF STUDIES ON ENDOCRINE-DISRUPTING CHEMICALS IN AFRICA

Recommendations of studies on EDCs in Africa can be summarized into categories of studies that recommended the use of biomarkers for exposure to EDCs, educational intervention, further research, need for innovative ways for waste management, need for regulation, safe interventions and utility of methodology from the respective studies.

TABLE 7.3
Studies in Africa on Removal of Endocrine-Disrupting Chemicals from the Environment

Authors	Place of Study	Study Design	Study Goal	Main Findings
Archer et al. (2017)	South Africa	Cross-sectional	Investigation of removal of emerging contaminants in wastewater	Inability of wastewater treatment works to remove emerging environmental contaminants
Arukwe, Eggen, and Moder (2012)	Nigeria	Cross-sectional	Analysis of sediments and run-off water samples from a solid waste dumping site in Owerri, Nigeria for organic load and compared these with data from representative world cities	High concentration of substances typically released from man-made products such as plastic, textile, household and consumer products
Aziz and Ojumu (2020)	South Africa	Experimental	Investigation of reverse osmosis pilot plant for the removal of selected natural steroid hormones 17β-estradiol and testosterone as a potential indirect water recycling application	Reverse osmosis and nano-filtration exhibited excellent removal rates (>95%) for estrogen and testosterone
Faul, Julies, and Pool (2013)	Namibia	Cross-sectional	Removal of estrogen, testosterone, cytotoxin and cholinesterase inhibitor during reclamation of sewage to drinking water	Sewage treatment removed estrogen, testosterone and acetylcholinesterase inhibition
Kasonga et al. (2019)	South Africa	Cross-sectional	Investigation of the removal of estrogenic activity of pharmaceuticals and their intermediates in a sequencing batch reactor	Use of the fungal consortium can remove the estrogenic activity of pharmaceutical metabolites
Kibambe et al. (2020)	South Africa	Cross-sectional	Assessment of the removal efficiency for the selected estrogens and reduction of estrogenic activity in wastewater samples from wastewater treatment plants	A combination of different technologies demonstrated highest pollutant removal efficiency
Kruid, Fogel, and Limson (2018)	South Africa	Experimental	Investigation of oxidative catalytic activity for BPA removal	Iron and manganese phthalocyanine showed ability to remove BPA from wastewater
Manickum and John (2014)	South Africa	Longitudinal	Investigation of the occurrence, fate and environmental risk assessment of endocrine-disrupting compounds at the wastewater treatment works	Baseline levels of steroid hormones in raw sewage, effluent and surface water at the Darvill Waste Water Works
Salaudeen, Okoh, and Okoh (2019)	South Africa	Cross-sectional	Investigation of the performance of five different wastewater treatment plants in some rural areas	Low capacity of WWTPs to remove targeted EDCs
Surujlal-Naicker and Bux (2013)	South Africa	Experimental	Investigation of the use of radio-immunoassays (RIAs) to determine the fate of estradiol in a laboratory batch test and the three natural estrogens in wastewater treatment plants	RIAs can be employed as a rapid technique for natural estrogen EDC detection in wastewater samples

Biomarkers of pollution and contamination that were recommended include the use of specific glands or organs such as the nuptial pad in *Xenopus laevis* (van Wyk, Pool, & Leslie, 2003) or organic molecules or messengers such as vitellogenin (Braathen et al., 2009; Esterhuyse et al., 2009) or mating behavior such as the calling behavior in male frogs (Hoffmann & Kloas, 2010). Closely related to studies recommending the use of respective biomarkers of exposure to EDCs are recommendations on the utility of methodologies for measuring or removing EDCs from the environment. In their study, for instance, Wooding et al. (2017b) demonstrated the versatility of their low-cost portable sampler in detecting and quantifying a broad range of EDCs in wastewater.

As mentioned earlier, the diversity of products containing EDCs and the ubiquitous distribution of these chemicals together with the challenges of removing them from the environment all signal the need for multi-sectoral involvement of key stakeholders. Consequently, the one study (Arukwe, Eggen, & Moder, 2012), on the need for education and therefore behavior change, provides an invaluable direction for mitigating release of EDCs into the environment.

It must be noted, however, that success from educational interventions and further research to improve our knowledge on EDCs must be anchored on evidence-based regulatory frameworks governing the production, distribution, use, disposal and remediation of EDCs. Studies that recommended the need for regulations pointed to the employment of a precautionary approach (Johnson et al., 2021; Kibambe et al., 2020), establishment of strict regulatory frameworks (Onyekwere et al., 2019) and the need for decision-makers to prioritize reducing the release of EDCs containing products into the environment (Ssebugere et al., 2014).

7.5 EXPOSURE TO ENDOCRINE-DISRUPTING CHEMICALS AND SUSTAINABLE DEVELOPMENT IN AFRICA

A review of the African Union Agenda 2063 indicates that sustainable development in Africa is anchored on the continent's demographic (a relatively youth population) and natural resource base including vast arable lands in several countries. In this section, we discuss, in light of the reviewed literature in this chapter, the potential demographic and environmental implication of exposure to EDCs in the context of both the Africa Union Agenda 2063 and SDGs.

7.5.1 Exposure to Endocrine-Disrupting Chemicals and Africa's Labor Force

In Agenda 2063, the AU recognizes, as outlined in Aspiration 1—a prosperous Africa based on inclusive growth and sustainable development, that investment in the manufacturing industries is a sure way to stir economic growth in Africa. In fact, there is evidence that manufacturing unlocks a country's economic potential (UNIDO, 2020). For example, with an investment of US$ 1.3 billion, Ethiopia established several industrial parks that have created about 50,000 jobs in sectors deemed to be environmentally friendly (Arthur, 2019).

There are projections of an impending relocation of manufacturing industries and labor force from China to Africa (Signe, 2018). In order to capitalize on the shift in manufacturing to developing countries including those in Africa, the AU places the continent's large youthful labor force at the core of Africa's industrial renaissance. The demographic base is expected to provide labor to drive the manufacturing and industrial production sectors. However, the reviewed studies on health risks associated with exposure to EDCs may constrain the available labor force in the continent unless deliberate efforts are made to fashion industrial development in Africa in a way that recognizes the adverse health and environmental efforts that characterized industrialization in Europe.

A seemingly obvious area of attention with regards to safeguarding the labor force is occupation exposure. Although none of the reviewed publications focused on occupational exposure to products containing known or potential EDCs, this remains an important point of exposure to EDCs. It is expected that the quest for industrial advancement as espoused in Agenda 2063 will be associated with an increased pool of individuals exposed to EDCs and who will themselves and their descendants carry a high burden of the pathologies that arise from exposure to EDCs. The burden of exposure to EDCs will manifest in increased healthcare costs arising from increased morbidity among exposed individuals.

The challenge is that large segments of the population in Africa are still grappling with traditional causes of morbidity and mortality. For example, the AU recognizes that diseases such as malaria and HIV/AIDS are a significant threat to Agenda 2063. For its part, malaria associated mortality in the continent remains high even in the face of huge investments to control and eradicate it. Ironically, one of the interventions against malaria, the use of indoor residual spraying using insecticides such as DDT, a known EDC, has been linked to adverse health effects. The irony, however, is that malaria remains a significant epidemiologic condition in Africa and so the use of insecticides, including those containing DDT and its metabolites, remains key tools for reducing the burden of this tropical disease.

Overall, although gaps in knowledge persist on population-level effects of exposure to EDCs, in part because of the absence of long-term studies on EDCs, increased incidences of congenital reproductive conditions such as epispadia and hypospadias (Di Renzo et al., 2015), altered pace of maturation and increased incidence of cancer are likely to negatively imperil the available labor force upon which most national and international development agenda are anchored.

7.5.2 Exposure to Endocrine-Disrupting Chemicals and Africa's Natural Resource Base

Studies of EDCs in Africa confirm the global concern on the ubiquitous presence of EDCs in virtually all environmental matrices. However, it is the presence of EDCs in water that presents perhaps the most serious threat to both human and wildlife health. A large pool of studies confirmed the presence of EDCs in aquatic ecosystems where these chemicals have been demonstrated to cause effects on biodiversity. Negative effects of EDCs on biodiversity are likely to have ecosystem-wide consequences including declines in fisheries abundance and diversity resulting in further decline in capture fisheries.

Although aquaculture has gained traction in recent decades, capture fisheries continues to contribute the largest segment of fish and fisheries-derived proteins to meet an ever-increasing demand of animal protein both in the continent and globally. Threats to capture fisheries, and to the ecosystem more broadly, will have serious implications on food security and on livelihoods of communities primarily dependent on capture fisheries. Threatened livelihoods are likely to drive activities such as slash and burn, and deforestation as more forested areas will be cleared to provide land for alternative agricultural production systems. In addition, farmers are likely to respond by increasing the level of intensification of their production systems through increased use of EDC-containing agrochemicals such as pesticides, fertilizers and additives that enhance growth.

7.6 CONCLUSION

The publications reviewed in this chapter highlight the broad focus of research on EDCs in Africa and identify existing gaps at the research, policy and practice levels. Broadly, data are missing from four areas. First, occupational exposure studies are largely missing yet policy and regulatory framework for the production, distribution and use of products containing EDCs are generally weak in the continent. Second, equally lacking are long-term studies such as those on the link between DDT exposure and the development of breast cancer in daughters, on monitoring of the effects of exposure on the development of health or disease.

Third, the burden of the insidious exposure to EDCs contained in personal care products and cosmetics especially in light of increasing use of skin whitening products remains largely unexplored. Fourth, there are gaps in research in the transboundary nature of the threat from exposure to EDCs. These gaps have serious implications on the labor market, health and food security in ways that have potential to derail sustainable development in Africa. Targeted measures are urgently needed to guide the production, distribution, use, disposal and removal of EDCs from the environment both within nations and internationally.

8 Regulatory Classification of Hazardous Agents

Stanley N. Ndwigah

8.1 INTRODUCTION

Chemical security refers to all measures designed to safeguard chemical substances from theft, misuse or accidental release, which can lead to environmental damage, chemical terrorism and health hazards. Modern society relies heavily on chemical substances for manufacture of medicines, agricultural products and for industrial purposes. Chemical security protects public health and the environment, and is of great concern for governments all over the world. Biohazard security on the other hand refers to all measures and policies intentionally designed to respond to the risks posed by biological hazards ranging from microbes (bacteria, viruses and fungi), and toxins or other biological agents that are a threat to human, animal or environmental safety. Biohazard security is vital as humanity faces threats like pandemics and bioterrorism.

Requirements for the classification and consequently the regulation of hazardous agents exist not just based on the intrinsic hazardous properties of a given substance but often as a prerequisite for the assignment to a specific risk class of packaging, labeling and documentation for transport (Ostad-Ali-Askari, 2022; Kassotis et al., 2020). It is not uncommon that hazards represented by chemicals and biological agents are also addressed by other legislation aiming at the protection of workers, the general public or the environment. Consequently, goods affected are subject to a range of regulatory labeling and documentation requirements. Especially, in the area of transport law, the classification of hazardous agents is often directly derived through reference requirements to other legislation, but often a classification-based hazard identification prevails (Stucki et al., 2022).

Hazard classification at defined criteria helps to protect public health, workers and the environment through support of chemical and biological safety, and hence, it is included in various regulations (Wohlleben, Mehling, & Landsiedel, 2023; Adamopoulos & Syrou, 2022; Wahab et al., 2021). Furthermore, the data reporting variance described for these regulations indicates a potential requirement for hazard classification information at specific data requirements. Although the classifications concerning release to the environment via industrial effluent discharges as well as to the end consumer via the product itself are variable, the protection goal is the same in both cases (Saravanakumar et al., 2022). Both sectoral legislations regulate in terms of environmental protection requirements according to hazard and/or risk-sensitive criteria, and the classification of the relevant regulatory fields is carried out via a valid method.

This chapter explores the significance of the security, and legal and regulatory classification of hazardous chemical and biological agents. This chapter provides examples of substances, which have been scheduled by various regulatory authorities, such as chemicals, poisons, biohazards and regulated laboratory chemicals.

8.2 INTERNATIONAL CLASSIFICATION FRAMEWORKS

The legal and regulatory framework for the classification of hazardous agents is established by international frameworks and national regulation initiatives. The need and importance of having harmonized approaches across regions and countries are underlined in the definition of "good regulation", promoting safety at work and consumer protection, and approaching agent classification to write or review safety data sheets (SDSs) and finally to ensure the free movement of goods (Aidonojie et al., 2024; Marx-Stoelting et al., 2023).

Several international frameworks have been developed, or are in the process of being developed, around the world. The Globally Harmonized System of Classification and Labeling of Chemicals (GHS), the Hazard Communication Standard (HCS), and the European Union's (EU) Classification, Labeling and Packaging (CLP) Regulation are currently the most influential at the international level (Hoffman & Dicks, 2023).

GHS is intended to ensure a high level of protection of human health and the environment. It establishes classification criteria and standardized hazard communication elements, including labels and SDSs. The criteria define health, physical and environmental hazards. It addresses the classification of chemicals for their hazards and proposes harmonized hazard communication elements, including labels and material SDSs, guidelines on the transport of dangerous goods and numerous guidance and technical publications (Ta et al., 2022; Giovanni, Marques, & Günther, 2021). The European Commission (EC) has partially adopted the GHS through Regulation No. 1272/2008, the so-called CLP Regulation, which is directly applicable in every member state of the EC. The CLP Regulation adapts the GHS to the technical and legal situation in the EC. It introduces the GHS criteria for physical, environmental and health hazard classification, and bridging principles exist to adapt the GHS to the EC and EU (Yazid, Ta, & Mokhtar, 2020; Berggren & Worth, 2023).

While those international frameworks are setting the theoretical dangerous goods classification, the practical hazardous constituent classification is required for healthcare workers in order to define medical surveillance procedures in the workplace. Harmonization of agent classification under national programs with the global minimum standards of classification, as outlined in the international frameworks, should be the basic strategy. It may also be necessary to align such programs with the national regulatory structure of agents, in harmony with the CLP approach (Alquier, 2021).

A number of national initiatives have been reported, though they are not activity areas. Fortunately, current or future area-specific large-scale collection or classifications and related harmonization toward the international frameworks will help to document knowledge in this section. In Europe, the Cooperation Centre for Scientific Expertise (CCSE) provides support for the classification and labeling of hazardous substances and mixtures in accordance with the CLP Regulation (Kättström et al., 2024; Hungerbühler et al., 2021; Kirillova, Ahmad, & Wilden, 2024).

In 2002, the imminent introduction of new chemical classification and labeling systems designed to replace more than 14 independent national and global systems was described. This international approach to hazardous classification was perceived to offer the opportunity for substantial progress in effective communication and management of hazardous agents. Since the development of this document, a number of trend changes are now being observed in the international scene (Rashidi et al., 2023).

8.2.1 United Nations' Globally Harmonized System of Classification and Labeling of Chemicals (GHS)

In the late 1980s, based on the realization that globalized trade meant that substances and mixtures deemed safe in one jurisdiction might well be marketed elsewhere, the United Nations took steps to harmonize the treatment of chemicals and adopted the first version of the GHS (Jonai et al., 2021). The key objective of the GHS is the enhancement of human and environmental safety as well as facilitating international or domestic trade, and the protection of proprietary interests of regulated industry.

The GHS lists 11 hazard classes of substances and mixtures, where substances are chemicals in pure form or products containing known and relatively high levels of a single chemical. Hazard classes include explosive materials, gases, flammable liquids, flammable solids, self-reactive chemicals, reactive substances with water, self-heating or water-reactive materials, organic peroxides and mixtures thereof. Also, hazard classes address oxidizing gases, oxidizing liquids or solids, and mixtures thereof. Finally, substances or mixtures that can cause harm directly to people in the short term are classified as harmful chemicals, and can cause skin or eye irritation, or are sufficiently caustic to cause skin corrosion.

The GHS criteria are supported by a number of communication tools, including packaging symbols, pictograms, signal words and risk management statements (Jagadeeswaran & Sriram, 2022). The aforementioned tools will maximize a label or other document's effectiveness as a warning device, but under the current GHS protocols, different hazard class and category criteria may generate differences in labeling, thus undermining one of GHS's objectives. Whereas the GHS has been designed to help ensure the flow of harmonized hazard and classification data through product value chains to support appropriate risk assessments and management, such data hindrances have slowed the implementation process in some jurisdictions.

Furthermore, classification differences are seen as a nontariff trade barrier, as they represent discounts on a potentially dangerous product's compliance costs. For the foregoing reasons, countries are making efforts to converge their respective legal systems with the GHS. Finally, the hazard-communication components of the GHS are structured so as not to discriminate against particular packaging or labeling technologies or materials. In this manner, the GHS is designed to facilitate rapid changes to hazard-communication requirements as science and technology evolve. The GHS system is continuously updated based on contemporary dictates (Zhong et al., 2020).

8.2.2 European Union's CLP Regulation

The EU has a set of legal tools to regulate risks to humans, animals and the environment from exposure to hazardous substances such as chemicals, toxins and pathogens (Arnesdotter et al., 2021; Blok & Balk, 2020; Pistollato et al., 2021). Many directives are in force for the regulation of hazardous agents that are, on the whole, specific, such as single agents associated with an identified sector of industrial activity or a particular working process (van Dijk et al., 2021; Authority, 2024; Hale et al., 2022). Regulatory authorities have established limit/level values, such as maximum workplace concentration levels for these noxious agents. The EU limits currently apply to approximately 250 such individual chemical substances (Pistollato et al., 2021).

There is considerable variation among EU member states in the legal and administrative measures taken to classify and regulate hazardous agents. Legislation may be general or special and apply to specific agents or industries, or to populations at risk (World Health Organization, 2020a; Puri, Gandhi, & Kumar, 2023; Stucki et al., 2022; Parra-Arroyo et al., 2022). A categorical distinction may not always be made, and national practices may be governed as much by administrative convenience as by logical or scientific principles. A lack of meaningful definition of "hazardous agent" or "hazard", based on quantum and duration of exposure, has resulted in wide differences in approach and inaccessibility to workers' entitlements to compensation for disease or other benefits. Where legislation exists, issues of interpretation, administrative competence and issues of consulting with workers and their organizations on what regulations are needed have led to accidents.

In 2008, the EU established the regulations for CLP of substances. The CLP currently provides for several types of derogations from classification and labeling provisions, and for their justification under specified conditions, which considerably reduces the administrative burden. The regulations contained harmonized criteria for physical, human health and environmental hazards classification, as well as requirements for packaging and labeling of hazardous chemicals (Groh et al., 2021; Ostad-Ali-Askari 2022; Hou et al., 2021). The European regulations were complemented in 2009 by the publication of a corresponding regulation for the classification of mixtures, including the implementation of a specific calculation method for health hazards, generally called Bridging Principles. Additionally, transition periods were provided to ensure the smooth implementation of these new rules.

Through CLP Regulation, the EU aimed at safer handling of chemicals, human health and protection of the environment. These regulatory provisions align with the GHS, to a certain extent contain additional EU-wide provisions, and all substances and further mixtures are bound to be labeled according to these rules. Sector legislation on specific products, such as detergents, cosmetics, pharmaceuticals or plant protection products, may contain additional requirements (Berggren & Worth, 2023; Kirillova, Ahmad, & Wilden, 2024). The supplier of a substance or mixture on the European market is obliged to classify, label and package in accordance with the EU regulations. The consumer should know the dangers of a chemical without opening the outer packaging. This entails package labeling for all classified products, and clear understanding of the hazards inherent to a weighted communication cycle (Vinceti et al., 2021).

Several levels of classification, packed with derogations, are established, and the acceptable classification should be substantiated based on the available data determining the chemical's empirical and theoretical properties (Berggren & Worth, 2023). A categorization of more severe effects to less with a following assignment to hazard phrases (or H-phrases) will guide the severity of the recommendations on the package. The classification list tables are risk-based, relying on the knowledge of routes of entry and exposure which can be considered representative for human exposures (Vinceti et al., 2021; Kirillova et al., 2024). It is important to match health hazards with the packaging based on the amounts present.

The principles outlined above result in an accurate system guiding the obligations for labeling, which might differ in some details from the requirements in the HCS. The **H-phrases** (hazard statements), **P-phrases** (precautionary statements), **R-phrases** (risk phrases) and **S-phrases** (safety phrases) are standardized phrases used in chemical safety labeling to communicate hazards and provide guidance on safe handling. The standardized phrases are accompanied by numbers. Furthermore, the R- and S-phrases also occur in combination (combined R- and S-phrases).[1]

The R-phrases present in the earlier legislation will have been changed to have one for each hazard category if the classification enters the system. The regulation also foresees adaptation to scientific and technical progress by the implementation of already agreed sufficient international data (Wohlleben, Mehling, & Landsiedel, 2023; Wiesner, Araya, & Lovsin Barle, 2022).

8.2.3 Canadian Workplace Hazardous Materials Information System

The Workplace Hazardous Materials Information System (WHMIS) is Canada's national HCS for workplace chemicals. Under WHMIS, Health Canada administers the Hazardous Products Act (HPA) and associated regulations, which set labeling and material SDSs requirements for suppliers. The WHMIS is a comprehensive framework for providing information on the hazardous products in Canadian workplaces and ensuring all workers have access to this information (Tenkate et al., 2021; Ho & Tenkate, 2024; Liao & Helen, 2021; Rachkowski, 2024).

The key elements of WHMIS include the following:

- *Hazard classification:* Products are classified based on their hazardous properties.
- *Hazard communication:* Containers are labeled with cautionary information.
- *SDSs:* These provide more detailed information about a product's hazardous properties, physical and chemical characteristics and necessary handling precautions.
- *Worker education and training:* Workers are educated on the information on labels and SDSs, so they can apply it on the job.

The federal HPA and related regulations focus on requirements for hazard communication using labels, SDSs and worker education and training. The federal government adopts the requirements for suppliers and employers of workers at the federal level and works with the 10 provinces and three territories in Canada to establish the same requirements in their jurisdictions. Approved training materials for WHMIS were updated to reflect WHMIS in 2015 to include the GHS and regulator-added criteria. In 2019, further updates to reflect the incorporation of regulatory amendments to WHMIS were applied (Liao & Helen, 2021; Rachkowski, 2024).

Regulators, suppliers and employers all have legal obligations to ensure a system of compliance exists in their workplaces and that the content of a WHMIS program meets regulatory requirements.

In Canadian federal, provincial and territorial laws, there is a harmonized approach for WHMIS that all jurisdictions conform to. In the implementation of laws for WHMIS, focusing on workplace policies, hazard assessment, education/training audits and employer/worker consultations should occur. The WHMIS Association recognized that the success of the WHMIS program is directly proportional to the educational efforts associated with the care, handling and use of hazardous materials. This goal is supported through the addition of new tools available to assist in education/outreach endeavors.

The WHMIS was deliberately developed as a national hazard communication system to ensure that workplace and chemical safety are concomitant (Liao & Helen, 2021). It is believed that the common-sense approach to hazard identification and the introduction of the rational use of precautionary measures will ultimately lead to the reduction of injury, disease and disability. The full, active participation of all workplace parties is critical to the effectiveness of WHMIS. It is necessary, however, to recognize that WHMIS is designated as a federal responsibility.

Provincial and territorial interests in the development, maintenance and delivery of WHMIS-related programs have been protected and fostered. Further, WHMIS does not prevent the right of any jurisdiction to establish other measures of protection for employee right-to-know that differ from those included in WHMIS. Although the layouts of the Before Reconciliation Risk Phrases and the WHMIS Risk Phrases are the same, there is no direct relationship between the words used in each.

8.2.4 Japanese Industrial Safety and Health Law

In Japan, regulation of hazards or risks that can be generated in the workplace is extremely important for labor safety and health and is done based on the Industrial Safety and Health Law (Jonai et al., 2021). The law covers machinery and equipment, noise and dust, safety and health equipment, control standards for chemical work, and specifications for tanks for hazardous materials to ensure labor safety and health. This law also has a chapter entitled "Handling of Injurious Materials", which mentions the classification, identification, containers, handling, storage and record keeping of "Injurious Materials" (Patel et al., 2022; de Bienassis, Slawomirski, & Klazinga, 2021). The term "hazardous materials" had to be changed to "Injurious Materials" based on an agreement, as the law is based on the belief that work should be carried out safely. Safe and improved workplace regulations can be spread universally around the world. It is necessary to govern the workplace in accordance with international standards.

The law classifies hazardous chemicals with some reliance on the international unified classification criterion. A new version of GHS has just been adopted, and effective regulations have been developed to enhance the safety of workers (Rashidi et al., 2023; Jonai et al., 2021). Government agencies and sectors that have jurisdiction over the law are given the power to impose criminal penalties on companies or individuals who do not comply with the law or violate safety regulations during an accident inspection. In Japan, all workplaces must appoint an occupational health physician and an occupational health nurse. Workers are required to undergo a biennial health examination, and resultant health statistics are analyzed to calculate the implied level of harm, though the number of factories and companies in Japan is enormous. Guidance is provided since the percentages of exposure may fall within the acceptable range, in compliance with the law (Igarashi et al., 2022; Sato, Jakobsson, & Midlöv, 2024; Ikegami et al., 2023).

8.3 REGULATION OF BIOHAZARD AGENTS

Humanity has lived surrounded by pathogens of significant health importance. The significance of biohazard regulation lies in its capacity to prevent the release of harmful biological agents into the environment or human populations. These agents cause disease, environmental damage and deaths. As the world has become a global village, biological agents can spread quickly across borders causing pandemics like COVID-19, making international collaboration essential for effective biohazard security. Biohazard regulation is also vital in preventing bioterrorism. The threat of bioterrorism has prompted governments to establish stringent regulations and practices (Murch, 2022). The 2001 anthrax attacks in the United States, in which spores of *Bacillus anthracis* were sent through the mail, for example, led to increased awareness and policy changes (Jernigan et al., 2002).

8.3.1 Classification of Biohazard Risks

Biohazards pose a threat to human, animal or plant health. These agents can be divided into several categories (Ta, Gosa, & Nathanson, 2019):

- Pathogens which include viruses, bacteria, fungi and other microorganisms that can cause diseases. Examples include the Ebola virus, the H1N1 influenza virus and *Mycobacterium tuberculosis*. These agents can cause outbreaks and pandemics, as seen with the SARS-CoV2 virus that led to the COVID-19 global pandemic.
- Biological agents which produce toxins that can be harmful to living organisms, e.g., *Clostridium botulinum* that produces botulinum toxin, one of the most potent biological toxins known to man. Other toxins include Ricin, a highly toxic protein found in the seeds of the castor plant bean that can cause severe illness or death. Aflatoxins are a group of poisonous, carcinogenic and mutagenic toxins produced by certain fungi, mainly *Aspergillus flavus* and *Aspergillus parasiticus*, which contaminate cereal crops like peanuts, maize and tree nuts. These toxins can be and have been weaponized in bioterrorism attacks.
- Advances in biotechnology have enabled development of genetically modified organisms (GMOs) and other gain-of-function microorganisms for various purposes. There is great concern over the potential misuse of these as biological weapons; for example, there have been allegations that SARS-CoV2 COVID-19 might have been created in a laboratory (Dyer, 2024).
- Finally, there are zoonotic diseases that can be transmitted from animals to humans. Examples include rabies, brucellosis, SARS and avian influenza. Zoonotic diseases can cause severe public health emergencies, as evidenced by past outbreaks of Ebola and swine flu (Rahman et al., 2020).

8.3.2 Challenges in Biohazard Security

Challenges that make it difficult to effectively manage biological risks may include the following:

- Early identification of a biohazardous agent, which is crucial in preventing its spread. However, many agents remain undetected and unidentified, complicating early warning systems (Institute of Medicine, 2002; Murch, 2022). The difficulty of quickly identifying pathogens brings a big concern especially in the cases of bioterrorism or early disease outbreak as has been seen with COVID-19 and Ebola.
- Globalization has made travel and trade become global, making the spread of biohazardous agents more rapid than ever. A disease outbreak in one part of the world can quickly spread to other regions, making it harder to contain and control (Madhav et al., 2017).
- The rise of synthetic biology and biotechnology makes it possible for individuals or groups to engineer biological agents with the potential for harm (Trump et al., 2021). Thus, accessibility of biotechnological tools poses a significant challenge to biohazard security.

8.3.3 Measures for Biohazard Security

Various measures can be implemented across different sectors to enhance security. These include the following:

- Surveillance and monitoring to ensure early detection and prevention of the spread of harmful biological agents. Surveillance systems include the World Health Organization's (WHO) Global Outbreak Alert and Response Network (GOARN) (Christensen et al., 2021) and the Centers for Disease Control and Prevention (CDC) (Thacker, Dannenberg, & Hamilton, 2001; Thacker, Sencer, & Jaffe, 2017).
- Establishment of biosafety regulations to ensure that biological research and biotechnology practices are conducted safely. The WHO International Health Regulations (IHR) (WHO, 2016) and the Biological Weapons Convention (BWC) (Walker, 2003) are examples of international frameworks designed to prevent the misuse of biological agents.
- Countries have developed response plans to prepare for biohazard events. For instance, Kenya has established The National Biosafety Authority (NBA)[2], and United States the Biomedical Advanced Research and Development Authority (BARDA)[3] to support the development of medical countermeasures against biological threats.
- Laboratories handling biological agents are subject to strict security protocols. The Biological Safety Level (BSL) system classifies laboratories based on the level of containment required to handle different types of biological agents (Ta, Gosa, & Nathanson, 2019).
- Educating the public about the risks of biological hazards and how to respond in the event of an outbreak is also crucial. Public health campaigns like vaccination programs and hygiene education will reduce the impact of biological threats.

8.3.4 Biohazard-Dependent Classification of Laboratories

The BSL system classifies laboratories based on the level of risk associated with the biological agents they work with, ranging from BSL-1 (lowest risk) to BSL-4 (highest risk), defining safety practices and equipment needed (Bayot & King, 2025; Ta, Gosa, & Nathanson, 2019).

- *BSL-1:* Used for agents that pose minimal risk to personnel and the environment, requiring basic safety practices like handwashing and standard lab coats.
- *BSL-2:* Designed for agents that pose a moderate risk of infection, requiring additional safety measures like gloves, eye protection and potentially biosafety cabinets.
- *BSL-3:* Used for agents that can be transmitted through the air and cause serious or potentially lethal diseases, requiring specialized ventilation systems, sealed windows and researchers to perform manipulations in a gas-tight enclosure.
- *BSL-4:* The highest level, used for highly dangerous and exotic microbes for which there is no vaccine or therapy available, requiring full-body, air-supplied suits and shower exits, and isolated zones within a larger building.

In summary, **Regulation of Biohazards** is an essential aspect of modern public health and safety given the increasing threat posed by infectious diseases, pandemics and bioterrorism. While significant progress has been made in terms of detection and response, challenges such as globalization, technological advancements, regulation and international cooperation continue to complicate efforts to prevent and mitigate the risks of biological threats. By strengthening regulation, surveillance systems, improving laboratory security and fostering global cooperation, it is possible to enhance biohazard security and protect against the potentially devastating effects of biological agents.

8.4 REGULATION OF CHEMICAL HAZARDS

Chemicals are used literally in all industries including pharmaceutical, food & agriculture, manufacturing, mining and even energy. These chemicals range from active pharmaceutical ingredients, preservatives, electrochemicals, pesticides, poisons, explosives, laboratory chemicals and their precursors. Many are toxic, corrosive or reactive. For these reasons, their misuse or improper handling can lead to contamination of foods & drugs, severe accidents or intentional harm like terrorism. Industrial accidents, such as chemical spills or explosions, can result in immediate casualties and long-term environmental damage, as evidenced by disasters like the Bhopal gas tragedy in 1984, which killed thousands of people in India (Broughton, 2005).

Chemical weaponry involves use of toxic chemicals as weapons as was seen during World War I (Williams, Armstrong, & Sizemore, 2025) which demonstrate the devastating potential of chemicals to be used for modern terrorism. Many times, terrorists have used explosives and their precursors to improvise explosive devices as was seen in Iraq war (National Academies of Sciences, Engineering, and Medicine, 2018). Other chemicals if not controlled strictly can be used to poison water sources leading to catastrophic effects (Dehkordi et al., 2024). Chemical weapons are prohibited under the international treaty referred to as Chemical Weapons Convention (CWC), which is aimed at eliminating the use and development of chemical weapons (Kelle, 2021; Ahmed & Al Diab Al Azzawi, 2024). However, the possibility of rogue states or terrorists using chemicals as weapons remains a big threat, underscoring the need for enhanced chemical security. Thus, chemical security is a matter of environmental and public health preservation.

8.4.1 Chemical Threats and Risks

- Compounds like ammonia, chlorine and hydrogen sulfide used in many industries are toxic or hazardous (Gabel & Rubin, 2019) and pose significant risks if improperly handled or released into the atmosphere. The transport and storage of such chemicals require strict security protocols to prevent accidents or sabotage.
- Deliberate use of chemicals by terrorists to harm civilians or disrupt society is a growing concern. Terrorist groups seek to acquire hazardous chemicals to create mass casualties, similar to the 1995 Tokyo subway Sarin attack (Okumura et al., 1996).
- Chemicals like pesticides, solvents and heavy metals can contaminate the environment if they leak or spill during manufacturing, transport or disposal processes. This leads to contamination of water, soil and air leading to long-term ecological damage posing health risks to populations. Managing chemical waste and safe disposal practices are critical aspects of chemical security (Dehkordi et al., 2024).

8.4.2 Challenges in Chemical Security

- It is difficult to effectively manage and mitigate risks because of the vast chemical inventory and sheer volume used in industries, making it difficult to track and secure every chemical substance.
- Many chemical plants and storage facilities in developing countries lack sufficient security infrastructure because of cost, and lack stringent regulatory oversight (Hock et al., 2022). Inadequate fencing, monitoring and personnel training can leave facilities vulnerable to theft or sabotage.
- Chemical plants, like other critical infrastructure, are vulnerable to the impact of natural hazards which can trigger Natech (natural and technological) accidents and the release of hazardous substances, fires and explosions (Krausmann, Girgin, & Necci, 2019). Such accidents can result in significant health effects, *environmental pollution* and economic losses.
- While international treaties like CWC aim to regulate the use and development of chemical weapons, the implementation of these agreements varies across countries and continents. Some nations may not have the resources or distrust the international order leading to decreased political will to enforce security measures creating gaps in global security (Amadi, 2020).

- Advances in chemical technology have created dual-use chemicals, i.e., both industrial and military applications, making it harder to monitor and control potentially dangerous substances. Some chemicals used in legitimate industries can also be repurposed for harmful purposes, adding complexity to chemical security efforts. Best example is the Oklahoma City bombing on April 19, 1995, in which a homemade explosive device (ED) made using ammonium nitrate fertilizer and fuel oil exploded heavily damaging a Federal Building leaving 168 people killed and more than 500 injured (Mallonee et al., 1996).

8.4.3 Measures to Enhance Chemical Security

Measures require a combination of regulations, technology and international collaboration, such as the following:

- The CWC was enacted in 1997 as an international treaty to eliminate chemical weapons and prevent their use. It required countries to declare and destroy their chemical weapons stockpiles and provides a framework for the peaceful use of chemicals. The Organization for the Prohibition of Chemical Weapons (OPCW) oversees the implementation of the CWC (Kelle, 2021).
- Chemical facilities must implement physical security infrastructure to prevent unauthorized access, theft and sabotage. This includes fencing with surveillance cameras, personnel screening and regular security audits. Operators on such facilities must be trained in emergency response procedures and the facility must have contingency plans in place in case of chemical leaks or spills.
- The security of chemical shipments is crucial in ensuring proper security. Use of GPS tracking, secure packaging and proper labeling can help prevent accidents and unauthorized access during transportation. Transporters should comply with international safety standards, such as the International Maritime Dangerous Goods (IMDG) Code and the European Agreement concerning the International Carriage of Dangerous Goods by Road.
- Making of international laws and regulations to control the production, storage and transport of hazardous chemicals, e.g., the EU's Seveso Directive which aims to prevent major industrial accidents involving hazardous chemicals (Vierendeels, Reniers, & Ale, 2011). International cooperation is essential to combat chemical threats, particularly in the case of chemical weapons and terrorism. The United Nations, the OPCW and other international organizations facilitate cooperation between governments, security agencies and industries to strengthen global chemical security.

In summary, **Regulation of Chemical Hazards** is a critical aspect of modern safety and public health, given the widespread use of hazardous chemicals in various educational, research and industrial setups. While significant progress has been made in securing chemical substances, regulatory challenges remain especially due to the vast inventory of chemicals, the potential for terrorism and the dual-use nature of many chemicals. By strengthening regulatory frameworks, enhancing security at chemical facilities and fostering international collaboration, the global community can better address the risks associated with chemical threats and ensure a safer future for all.

8.5 REGULATION OF POISONS, CHEMICALS AND REAGENTS IN KENYA

These substances are non-medicinal poisons, chemicals and reagents whose possession and access should be controlled because of the potential for misuse that can lead to physical hazards, toxicity or terrorism. They are categorized into five parts in their order of decreasing risk according to Government of Kenya, Ministry of Health, Pharmacy and Poisons Board (2023), Guidelines for scheduling of Health Products and Technologies, Version 1, 2022 and United Kingdom Government, Supplying explosives precursors and poisons (Gavrila et al., 2024).

i. Regulated explosive precursors
ii. Regulated poisons
iii. Reportable explosive precursors
iv. Reportable/common use poisons
v. Laboratory chemicals.

These products are not intended for use in human beings through any route of administration. Their level of risk is dependent on many factors including concentration of the chemicals in a product, complexity of a product and amount of the product. The risk of causing harm by explosive precursor chemicals and poisons can be reduced through the use of appropriate packaging with simple warnings and safety directions on the label.

Regulated poisons and regulated precursors bear the warning that "Acquisition, possession or use by the general public is restricted". Additional labeling includes the terms "CAUTION" for low-risk chemicals, "POISON" for medium-risk chemicals, "SPECIAL PRECAUTIONS" for high-risk chemicals and appropriate safety symbols for hazardous products.

8.5.1 Regulated Explosive Precursors

A regulated explosive precursor refers to a chemical substance which can be made into an explosive with relative ease by mixing or blending with other substances, or by simple chemical processing. The majority of these chemicals are used for legitimate purposes but could potentially be misused for the illicit manufacture of homemade or improvised explosives. Relatively small amounts of these chemicals are required to manufacture explosives as shown in Table 8.1.

8.5.2 Regulated Poisons

These substances refer to chemical weapons that can be used to cause harm or terrorism. They can cause severe harm or death to human beings even in low amounts if ingested, breathed in, injected into the body or absorbed through the skin. Table 8.2 provides examples.

TABLE 8.1
Regulated Explosives Precursors and Concentration Thresholds

- Hydrogen peroxide, over 12% w/w
- Nitromethane, over 30% w/w
- Nitric acid, over 3% w/w
- Potassium chlorate, over 40% w/w
- Potassium perchlorate, over 40%w/w
- Sodium chlorate, over 40% w/w
- Sodium perchlorate, over 40% w/w
- Sulfuric acid, over 15% w/w

TABLE 8.2
Regulated Non-medicinal Poisons and Concentration Thresholds

- Aluminum phosphide
- Arsenic and its compounds (other than calcium arsenites, copper acetoarsenite, copper arsenates, copper arsenites, lead arsenates)
- Barium salts (other than barium sulfate, barium carbonate and barium silicofluoride)
- Bromomethane
- Chloropicrin
- Fluoroacetic acid and its salts. Fluoroacetamide
- Hydrogen cyanide; metal cyanides, other than ferrocyanides and ferricyanides
- Lead acetates
- Magnesium phosphide
- Mercury, compounds of, the following:
  - Nitrates of mercury
  - Mercuric cyanide oxides
  - Mercuric thiocyanate
  - Ammonium mercuric chlorides
  - Potassium mercuric iodides
- Organic compounds of mercury which contain a methyl group directly linked to the mercury atom
- Oxalic acid, 10% w/w
- Phenols
- Phenols (phenol; phenolic isomers of the following: cresols, xylenols, monoethylphenols) except in substances containing less than 60% weight in weight of phenols; compounds of phenols with metal, except in substances containing less than the equivalent of 60% weight in weight of phenols
- Phosphorus yellow
- Strychnine and its salts
- Thallium and its salts

8.5.3 Reportable Explosive Precursors

The substances listed in Table 8.3 may be used for illicit manufacture of explosives. The ease of illicit manufacture of explosives from these chemicals and level of risk is less than that of regulated explosive precursors. In addition to professional use, the products are used in common consumer products within a vast supply chain. Suspicious transactions, significant disappearances and thefts of these poisons shall be reported to regulatory authorities.

TABLE 8.3
Reportable Explosive Precursors

- Ammonium nitrate
- Acetone
- Hexamine
- Potassium nitrate
- Sodium nitrate
- Calcium nitrate
- Calcium ammonium nitrate
- Aluminum powders
- Magnesium powders
- Magnesium nitrate hexahydrate

8.5.4 Reportable Poisons

Reportable poisons in Table 8.4 are commonly used poisons and can be supplied to purchasers if the purchase is not suspicious. Suspicious transactions, significant disappearances and thefts of these poisons shall be reported to authorities.

8.5.5 Laboratory Chemicals

Table 8.5 includes all non-medicinal chemicals, poisons and reagents used commonly in laboratories in teaching and research institutions. They should be supplied only to licensed institutions such as training institutions, quality control laboratories, research institutions and manufacturers within their approved scope of operations.

TABLE 8.4
Reportable Poisons

- Aldicarb
- Alpha-chloralose
- Ammonia 10% w/w
- Arsenic, compounds of, the following:
  - Calcium arsenites
  - Copper acetoarsenites
  - Copper arsenates
  - Copper arsenites
  - Lead arsenates
- Barium, salts of, the following:
  - Barium carbonate
  - Barium silicofluoride
- Carbofuran
- Cycloheximide
- Dinitrocresols, their compounds with a metal or a base
- Dinoseb, its compounds with a metal or a base
- Dinoterb
- Drazoxolon, its salts
- Endosulfan
- Endothal, its salts
- Endrin
- Fentin and its compounds
- Formaldehyde, 5% w/w
- Formic acid, 25% w/w
- Hydrochloric acid 10% w/w
- Hydrofluoric acid and
  - Alkali metal bifluorides
  - Ammonium bifluoride
  - Alkali metal fluorides
  - Ammonium fluoride
  - Sodium silicofluoride
- Mercuric chloride
- Mercuric iodide; organic compounds of mercury except compounds that contain a methyl (CH_3) group directly linked to the mercury atom
- Metallic oxalates
- Methomyl
- Nicotine, its salts, its quaternary compounds
- Nitrobenzene, 0.1% w/w
- Oxamyl
- Paraquat, salts of
- Phenols
- Phosphoric acid
- Phosphorus compounds:
  - Azinphos-methyl
  - Chlorfenvinphos
  - Demephion
  - Demeton-S-methyl
  - Demeton-S-methyl sulfone, dialifos
  - Dichlorvos
  - Dioxathion
  - Disulfoton
  - Fonofos
  - Mecarbam
  - Mephosfolan
  - Methidathion
  - Mevinphos
  - Omethoate
  - Oxydemeton-methyl
  - Parathion
  - Phenkapton
  - Phorate
  - Phosphamidon
  - Pirimiphos-ethyl
  - Quinalphos
  - Thiometon
  - Thionazin
  - Triazophos
  - Vamidothion
- Potassium hydroxide, 17%
- Sodium hydroxide, 12%
- Sodium nitrite
- Thiofanox
- Zinc phosphide

TABLE 8.5
Other Laboratory Chemicals

- Acetic acid
- Acetic anhydride
- Acetone
- Ammonium hydroxide
- Benzene
- Calcium carbonate
- Calcium chloride – granular form
- Copper sulfate – pentahydrate
- Corn starch – powder form
- Decon 90
- Dextrose or monohydrate
- Dichloromethane
- Ethanol
- Ethyl acetate
- Ferric chloride
- Hexane
- Hydrochloric acid
- Hydrofluoric acid
- Hydrogen peroxide
- Iodine
- Isopropanol
- Iron chloride solution
- Isopropyl alcohol
- Magnesium
- Methanol
- Nitric acid
- Naphrax
- Phenolphthalein
- Potassium chloride
- Potassium hydroxide
- Potassium iodide
- Sodium acetate or trihydrate
- Sodium bicarbonate
- Sodium borate or decahydrate
- Sodium carbonate or anhydrous sodium chloride
- Sodium hydroxide
- Sulfuric acid
- Water – HPLC grade
- Zinc

8.6 EMERGING ISSUES AND FUTURE TRENDS

The legal and regulatory classification of hazardous agents has been continuously evolving science, adapting to new scientific knowledge and technology. This trend can be expected to continue as legislators also strive for greater public participation, public involvement in decision-making and systematic improvement of the activities established in the future. One of the emerging issues of the last years is the growing complexity of legal frameworks and protocols regulating health and environmental risks associated with the production, use and waste of chemical and biological agents.

A trend becoming more prevalent is the use of international guidance to facilitate routine implementation of the requirements of various domestic regulators (Grillo et al., 2021; Cordaillat-Simmons, Rouanet, & Pot, 2020). Fundamental to this is the definition of criteria and threshold values for classification that are established or evolved with a continuum of international expertise. Increasingly, consumer awareness, media stories and activist interests have become engendered by the pressing issues of climate change, polychlorinated hydrocarbons, e-waste, GMOs and the offshoring of dirty industries. This is increasingly a topic of regulatory action, partly because it is topical and significant to the general public, but also because governments have concluded that it is good regulation (Cordaillat-Simmons, Rouanet, & Pot, 2020). This latter trend, while arguably of no significant scientific impact, seems a trend worth noting.

Looking into the future, further and continued scientific advances will provide risk assessors and regulators with enhanced tools for evaluating safe uses of chemicals and biological agents in the environment. In particular, a greater understanding of underlying toxic, biological and mechanistic issues that relate to toxicity for more robust alternative/substitute assessments using soft systems can be expected. Following current trends, there is need for protocols to evolve in tandem with the scientific and technical knowledge base.

Currently, challenges are presented by the classification of nanomaterials with, particularly, far-reaching developments of general applicability as critically examined criteria are established in light of further experimentation. In the longer term, evolutionary trends seen in the area of assessing chemicals that can accumulate in the human body may also occur with regard to chemicals that are bioactive but may not bio-accumulate. Finally, the shift in appraisal from societal risks implies a change in natural resource management and use. It should therefore be an important emerging issue in the same vein as measures to ensure health and environmental safety that also contribute to a more sustainable economy.

8.7 CONCLUSION

Robust regulatory frameworks for ensuring that laboratory users, industry workers and communities in general remain safe are essential. In an ever more interconnected global economy, the need for international cooperation becomes even more apparent. This is partly because individual countries do not have all the resources and expertise required to set up and operate effective regulatory mechanisms and partly because countries are no longer totally self-sufficient in chemicals. It is imperative that systems for the classification of chemicals according to health and environmental hazards should be harmonized to the greatest extent possible at both national and international levels. Hence, an Intergovernmental Forum on Chemical Safety was established to further progress this issue.

First, the community requires robust systems for classifying hazardous agents so that business can be regulated. Second, the systems must be operational if they are to be acceptable; it is a major problem when "everything" is classified as an R48, "danger of serious damage to health". Nonetheless, it is noteworthy that each country's legal system requires a bespoke approach to incorporate the international criteria so as to attribute liabilities for those found not to have applied the principles to current practice whenever an injury occurs.

Public pressure no doubt forces authorities in these countries to ratify international conventions, and this will no doubt force production practices to improve. Compound toxicity testing and introduction of data from genomics into the risk assessment also force the industry's hands in adopting methods for assessing the toxic effects of chemicals on humans. Finally, if due to terrorism threats more chemicals are to be classified as toxic, we must pray that funding for effective regulation, monitoring and control is secured to prevent the spread of more hazardous chemicals throughout the globe.

NOTES

1 https://www.msds-europe.com/r-phrases-s-phrases/
2 https://www.linkedin.com/company/national-biosafety-authority/?originalSubdomain=ke
3 https://aspr.hhs.gov/AboutASPR/ProgramOffices/BARDA/Pages/default.aspx

9 Chemical Security in the Pharmaceutical Industry

A Focus on Kenya and Somalia

Geoffrey Otieno, James J. Owuor, Patrick M. Shem, Simon Adede, Joab O. Onyango, and Austin O. Aluoch

9.1 INTRODUCTION

The pharmaceutical industries in Kenya and Somalia, like many sectors globally, face increasing concerns about chemical security. While chemical safety has been a longstanding focus of the pharmaceutical industry, the deliberate misuse, theft, and diversion of pharmaceutical chemicals for illicit purposes and the potential for weaponization of pharmaceutical-based agents (PBAs) represent a significant and emerging threat. This chapter examines the current state of chemical security within the pharmaceutical industries of these two East African nations, highlighting efforts to enhance security, challenges encountered, and potential opportunities for improvement. The analysis draws significantly from the experiences and reports of the KCS, a leading organization in promoting chemical safety and security in the region, recognizing that its efforts have also extended into Somalia (Cameron, Kuo, & Ho, 2010).

9.1.1 Understanding Chemical Security in the Pharmaceutical Industry

Chemical security refers to the measures and practices designed to prevent the theft, diversion, or misuse of chemicals that could be used for harmful purposes. These chemicals can range from precursors used in the manufacture of pharmaceuticals to toxic substances that could be weaponized. The pharmaceutical industry is particularly vulnerable due to its handling of a wide range of chemicals, some of which are highly regulated because of their potential for misuse (Ta et al., 2020). Chemical security in the pharmaceutical industry is, therefore, critical to preventing accidents, intentional misuse, and environmental contamination.

An incident that highlighted the importance of chemical security in the pharmaceutical sector occurred in 2020, when drugs worth approximately $60 million, including Gemzar, Zyprexa, Prozac, and Cymbalta, were stolen from an Eli Lilly warehouse in the United States. Investigations thereafter revealed that the theft was facilitated by gaps in the chemical security measures at the Eli Lilly facility (Federal Bureau of Investigation (FBI), 2017). Thus, securing the pharmaceutical industry is crucial to prevent access to highly potent pharmaceutical agents, as well as their precursors and derivatives, which could be exploited in a chemical weapon (CW) attack by malicious actors (World Health Organization (WHO), 2020a; Walters, 2015).

In many African nations and other developing regions, where regulatory frameworks and infrastructure are still in development, the risks associated with improper chemical handling are notably significant. Trade in counterfeit medicines and misuse of regulated medicines prevails and poses a serious threat to public safety and security. During Operation Pangea XVI, which ran from 3 to 10 October 2023, the International Criminal Police Organization (INTERPOL) seized potentially dangerous pharmaceuticals that included erectile dysfunction medicines, psychotherapeutic agents, anti-anxiety medicines and stimulants, sex hormones, and gastrointestinal medicines from Mozambique, Qatar, Peru, Ecuador, among other countries (Interpol, 2023).

The lead poisoning crisis in Zamfara State, Nigeria, due to improper handling of chemicals in gold mining, although not directly related to pharmaceuticals, highlights the broader risks of chemical mismanagement. This incident demonstrates how lax chemical security can lead to severe public health consequences, including fatalities (Lo et al., 2010). Traditionally, Africa has not been part of the main drug trafficking routes, but this has started to change over the last three decades, with criminal groups using Africa to smuggle drugs such as cocaine, methamphetamine, and heroine. A new trend typically found in Africa, is the use of nonmedical use of falsified tramadol, an opioid that is not under international control. There is a growing concern about the use of drug mixtures known by names "kush," "karkoubi," and "nyaope." Limited evidence suggests that these concoctions often include a variety of harmful substances such as illicit drugs, pharmaceuticals (like benzodiazepines), alcohol, and solvents (United Nations Office on Drugs and Crime (UNODC), 2024).

9.1.2 Pharmaceuticals, Precursors and PBAs as Chemicals of Concern

The dual-use nature of chemicals necessitates the development of a chemical security framework for the pharmaceutical industry. For example, chlorine is used to manufacture a variety of drugs for cardiovascular, respiratory, gastrointestinal, musculoskeletal, and genitourinary conditions (Fang et al., 2019). Cyanide is also used to make nitrile-containing drugs including anastrazole, gallopamil, and verapamil with many others in different stages of clinical development (Fleming, 2010). Some CW precursors are also used in pharmaceuticals. Triethanolamine, which is used in eardrops, is a blister agent precursor. Phosphorus trichloride, a nerve agent precursor, is used in the synthesis of bisphosphonates used to treat osteoporosis. Pharmaceutical-based agents of concern include opiates and opioids, psychedelics, anticholinergics, Gamma-Aminobutyric Acid (GABA) modulators, neuroleptic anesthetics, α_2-adrenoceptor agonists, and benzodiazepines (Heslop & Blain, 2020).

During chemical security trainings conducted by KCS for the pharmaceutical industry in Kenya, chemicals used in pharmaceutical quality control laboratories, drugs on controlled lists, drug precursors, and Pharmacy-Only Medicines (POMs) such as nonsteroidal anti-inflammatory drugs (NSAIDs) and over-the-counter (OTC) medicines were identified as being vulnerable to theft. A summary of the chemicals, drugs, and agents are shown in Table 9.1. The list is indicative, not exhaustive.

9.1.3 Key Components of Chemical Security

1. *Regulatory frameworks:* Effective chemical security requires robust regulatory frameworks that govern the production, storage, transportation, and disposal of chemicals. These regulations should be enforced by competent authorities to ensure compliance. Additionally, the existing regulatory frameworks primarily address safety and revenue compliance, with minimal emphasis on chemical security.

TABLE 9.1
Examples of Weaponizable Chemicals, Drugs, and Agents in the Pharmaceutical Sector in Kenya

Laboratory Chemicals	Drugs on Controlled Lists	POMs and OTC Drugs
Hydrochloric acid, Sulfuric acid, acetone, formaldehyde, hydrogen peroxide, chloroform, acetic anhydride, toluene, sodium hypochlorite, and methanol	Naloxone, morphine, midazolam, benzodiazepines, opioids and opiates, narcotic agents, and anesthetics, e.g., halothane	*POMs*: NSAIDs, antibiotics, digoxin, ephedrine, warfarin *OTCs*: Ephedrine and NSAIDs like aspirin and ibuprofen

2. *Supply chain security:* The pharmaceutical supply chain is complex, involving multiple stakeholders, including raw material suppliers, manufacturers, distributors, and retailers. Each node in the supply chain presents a potential vulnerability that could be exploited by malicious actors (Walters, 2015).
3. *Physical security:* Facilities that produce, store, or handle chemicals must be secured against unauthorized access. This includes implementing access controls, surveillance systems, and other physical security measures (Organisation for the Prohibition of Chemical Weapons (OPCW), 2021).
4. *Cybersecurity:* With the increasing digitization of the pharmaceutical industry, cybersecurity has become a critical component of chemical security. Cyberattacks on pharmaceutical companies could lead to the theft of sensitive information or the disruption of production processes (Tucker, 2012).
5. *Awareness and training:* Employees at all levels of the pharmaceutical supply chain must be aware of the risks associated with chemical security and be trained to recognize and respond to potential threats (Organisation for the Prohibition of Chemical Weapons (OPCW), 2022). Personnel reliability programs must be established to reduce insider threats.

9.1.4 The Context of Kenya and Somalia

9.1.4.1 Kenya

Kenya is one of the most developed countries in East Africa, with a relatively stable political environment and a growing pharmaceutical industry. However, the country faces several challenges that could impact chemical security:

1. *Regulatory gaps:* While Kenya has established regulatory frameworks for the pharmaceutical industry, enforcement has been inconsistent. Some of chemical control lists and laws in Kenya include the Narcotic Drugs and Psychotropic Substances (Control) Act and the Pharmacy and Poisons Board (PPB) Control list. Regulatory agencies often lack the resources or expertise to effectively monitor and enforce compliance. Additionally, the existing regulatory frameworks primarily address safety and revenue compliance, with minimal emphasis on chemical security (Miller, 2018).
2. *Corruption:* It is a significant issue in Kenya, and it can undermine efforts to maintain chemical security. Bribery and other forms of corruption could allow unauthorized individuals or organizations to gain access to regulated chemicals.
3. *Terrorism:* Kenya has been a target of terrorist attacks, particularly from the Al-Shabaab militant group based in neighboring Somalia. The theft or diversion of chemicals for use in terrorist activities is a real concern (Shem et al., 2021).
4. *Supply chain vulnerabilities:* Kenya's pharmaceutical supply chain is complex and involves multiple stakeholders, including small-scale distributors and retailers. These smaller entities may lack the resources to implement robust security measures, making them potential targets for theft or diversion.

9.1.4.2 Somalia

Somalia presents an even more challenging environment for chemical security. The country has been plagued by decades of civil war, political instability, and weak governance. Key issues include:

1. *Lack of regulatory oversight:* Somalia has limited regulatory framework for the pharmaceutical industry. This lack of oversight creates an environment where chemicals can be easily diverted or misused (Kenya Chemical Society (KCS) and Sandia National Laboratories (SNL), 2022b).

2. *Terrorism and insurgency:* Somalia is a stronghold for the Al-Shabaab militant group, which has been involved in numerous terrorist attacks in the region. The group has shown an interest in acquiring chemicals for use in explosives and other weapons (Okolie-Osemene & Romaniuk, 2024; United Nations Security Council (UNSC), 2023).
3. *Weak infrastructure:* Somalia's infrastructure is underdeveloped, making it difficult to secure the transportation and storage of chemicals. The lack of reliable transportation networks also increases the risk of theft during transit (Payne et al., 2020).
4. *Illicit trade:* Somalia has a thriving black market for various goods, including pharmaceuticals and chemicals. The illicit trade in chemicals is a significant concern, as it provides a pathway for the diversion of regulated substances (Organization for the Prohibition of Chemical Weapons (OPCW), 2025).

9.2 CURRENT STATE OF CHEMICAL SECURITY

Both Kenya and Somalia exhibit critical gaps between the emphasis on chemical safety and the limited awareness of chemical security within pharmaceutical training institutions and the industry.

During a training aimed at introducing chemical security and to raise awareness on weaponization of PBAs to professionals from facilities that conduct research, manufacture, package, or distribute PBAs in Kenya, participants reported limited familiarity with concepts of chemical security. They also had a low level of awareness on weaponization of PBAs as shown in Figure 9.1a. It is evident from Figure 9.1b that the pharmaceutical stakeholders improved their knowledge on the

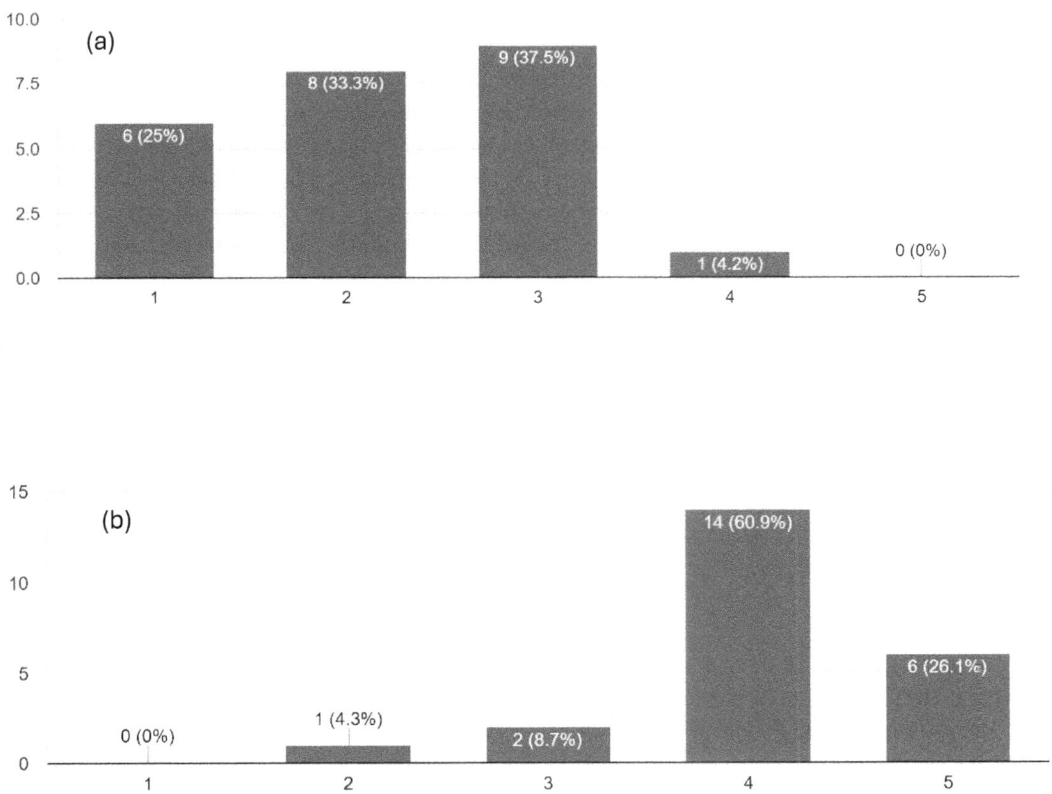

FIGURE 9.1 Level of knowledge of chemical security and weaponization of pharmaceutical-based agents (PBAs) – survey among pharmaceutical stakeholders during a pharmaceutical risk assessment training conducted by KCS in 2022 (a) before the training and (b) after the training. The scale is from **1 (beginner)** to **5 (expert)**.

chemical security risks posed by PBAs, precursors, and derivatives after the training. This shows the importance of engaging stakeholders in the pharmaceutical sector to raise awareness on chemical security. The engagements also build the capacity of the stakeholders to address chemical security gaps and secure PBAs (Kenya Chemical Society (KCS), 2022a).

A similar training for Somali pharmaceutical professionals and stakeholders resulted in a shift in how the pharmaceutical professionals viewed chemical security, leading to a greater awareness of risks associated with PBAs, precursors, and derivatives (Kenya Chemical Society (KCS) and Sandia National Laboratories (SNL), 2022b).

Incorporation of chemical security into the training curricula for pharmacists and pharmaceutical technologists has been proposed as an additional intervention to raise awareness on chemical security and encourage the adoption of chemical security best practices in the pharmaceutical sector. There has been a greater emphasis on chemical safety compared to chemical security in the training programs for pharmacists, pharmaceutical technologist, chemists, chemical technologists, and chemical engineers in Kenya. This has resulted in a chemical security knowledge deficit that contributes to vulnerabilities across the chemical and pharmaceutical supply chain, from storage and handling to distribution and dispensing, increasing the risk of theft, diversion, and malicious use (Kenyan Pharmacy and Poisons Board (PPB), 2017).

The KCS has engaged university leadership in Kenya leading to the incorporation of chemical security principles and practice into chemistry curricula at several universities in Kenya (Shem et al., 2021). Integration of chemical security into the training curricula for pharmacists and pharmaceutical technologists will produce graduates with the knowledge and skills to secure PBAs throughout the supply chain.

9.2.1 Kenya

Reports of chemical theft and sabotage within Kenyan facilities highlight the urgent need to address these vulnerabilities. Some of the chemical security gaps that could lead to the theft or loss of potential biological agents in Kenya include inadequate physical security, insufficient personnel screening both prior to and during employment, and a lack of mechanisms to conduct due diligence for end users of these agents. The informal sectors in Kenya further complicate the issue, often lacking the resources and awareness necessary to adhere to even basic security protocols (Shem et al., 2021). Tackling this complex challenge requires a multi-stakeholder approach (Ministry of Environment and Mineral Resources, 2011). Figure 9.2 shows an exercise for an advanced chemical security training on pharmaceutical prioritization and risk assessment sample carried out by KCS to the pharmaceutical sector stakeholders. In this exercise, the participants used chemical list such as OPCW schedule list and International Narcotics Control Board (INCB) list which are relevant to pharmaceuticals that are used in Kenya to rank chemicals based on security consideration and severity of hazard posed by the pharmaceutical agents. The results of the exercise informed the chemical security mitigation measures to secure the identified pharmaceutical-based agents at facilities.

Key players in Kenya include the following:

- *Pharmaceutical Society of Kenya (PSK):* Responsible for the professional development and ethical conduct of pharmacists.
- *Kenya Pharmaceutical Association (KPA):* Represents the interests of the Pharmaceutical Technologists.
- *PPB:* The regulatory body overseeing the registration and quality control of pharmaceuticals.
- *Ministry of Education:* Plays a crucial role in curriculum development and ensures that chemical security is appropriately integrated into pharmacy training programs.

Collaboration with these bodies is vital for creating national standards and enforcing compliance.

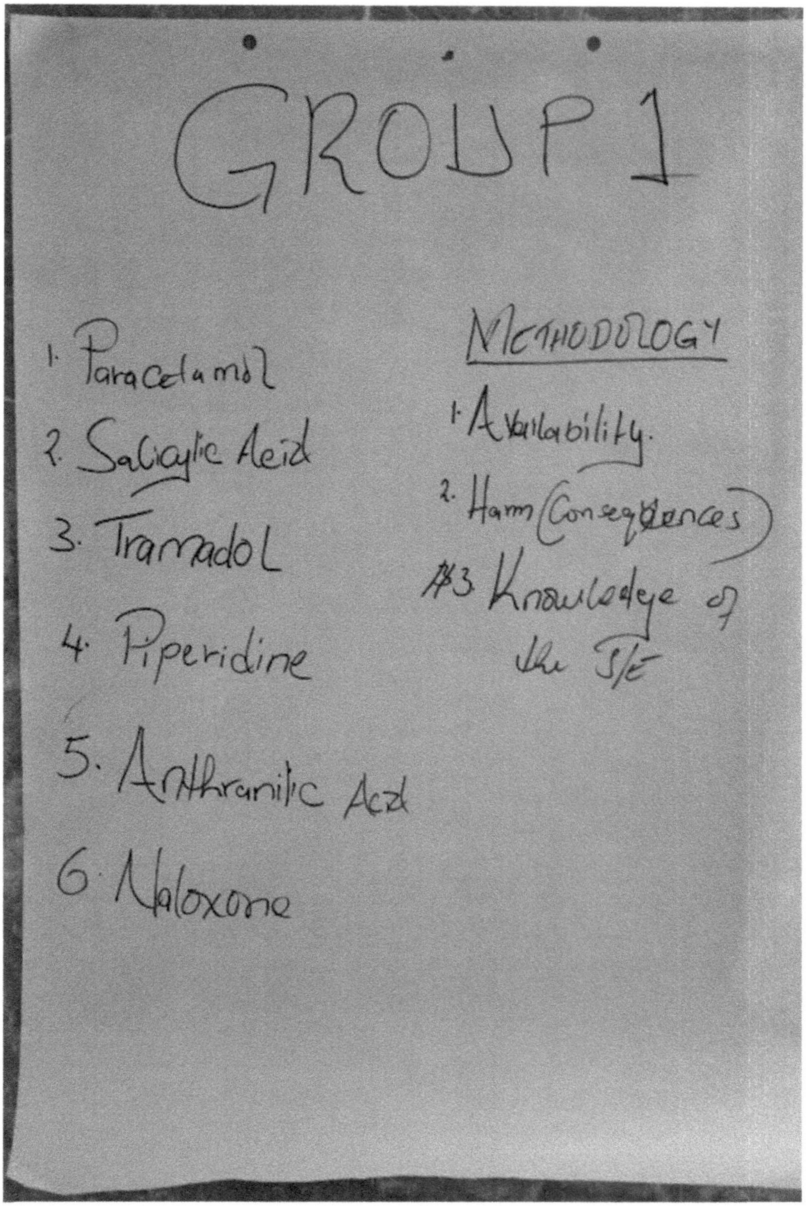

FIGURE 9.2 Knowledge of threats on PBAs by Pharmaceutical Industry in Kenya – exercise on prioritization of PBAs during a workshop on chemical risk assessment for pharmaceutical manufacturers, distributors, and suppliers in Kenya by KCS.

9.2.2 Somalia

The situation in Somalia is more complex due to ongoing political instability and a fragmented governance structure (United Nations Security Council (UNSC), 2023). While a central Ministry of Health exists, its ability to effectively regulate and oversee the pharmaceutical sector is constrained. In Somaliland, a relatively more stable region, similar bodies exist, but their capacity and reach are also limited. Identifying and engaging the relevant pharmaceutical regulatory bodies and professional associations in both Somalia and Somaliland are crucial for building capacity and establishing best

FIGURE 9.3 Somali participants in a workshop for Security of Chemical Hazards of Highly Toxic Agrochemicals and Rudimentary Chemical Materials in Somalia through a Train-the-Trainer Engagement and Outreach Campaigns in March 2023 (Photo courtesy of KCS).

practices. Key players in this landscape include the Somali Chemical Society (SCS), the Somaliland Pharmacy Association (SOPHA), the Somali Pharmacists Association (SPA), the Federal Ministry of Health, the Ministry of Health Development (Somaliland), and the Puntland Ministry of Health (Kenya Chemical Society (KCS) and Sandia National Laboratories (SNL), 2022b).

The KCS is making efforts to build capacity through workshops with Somali partners, as illustrated in Figure 9.3. Pharmaceutical regulatory bodies and professional associations in Somalia and Somaliland are crucial for building such capacity and establishing best practices (Kenya Chemical Society (KCS), 2022b; Kenya Chemical Society (KCS) and Sandia National Laboratories (SNL), 2023). International stakeholders also play a significant role in enhancing chemical security.

The OPCW provides a framework for international cooperation and the development of best practices in this area (Organisation for the Prohibition of Chemical Weapons (OPCW), 2021). The Chemical Security Program (CSP) of the U.S. Department of State has been instrumental in supporting capacity-building initiatives in Kenya, particularly through its partnership with the (KCS) (Shem et al., 2021). The CSP's model of supporting local organizations and adapting best practices to the local context has been replicated and expanded in Somalia, albeit with adjustments to accommodate the distinct political and security realities. A coordinated effort involving all these stakeholders is essential to bridge the chemical security gap in Kenya and Somalia, and mitigate the considerable risks to public health and national security (Ministry of Environment and Mineral Resources, 2011)

9.3 THREATS TO CHEMICAL SECURITY IN KENYA AND SOMALIA

9.3.1 Theft and Diversion

One of the most immediate threats to chemical security in Kenya and Somalia is the theft and diversion of chemicals. In both countries, the lack of robust regulatory oversight and enforcement creates opportunities for criminals to steal chemicals from pharmaceutical facilities or during transportation. These stolen chemicals could then be sold on the black market or used for illicit purposes (Kenya Chemical Society (KCS), 2022a; Kenya Chemical Society (KCS) and Sandia National Laboratories (SNL), 2022b).

Case Study 1: Theft of Pharmaceuticals in Kenya

In 2018, a major theft of pharmaceuticals was reported in Kenya, where a consignment of drugs worth millions of Kenya shillings was stolen from a government medical store. The stolen drugs included antibiotics, antimalarials, and other essential medicines. This incident highlighted the vulnerabilities in the pharmaceutical supply chain and raised concerns about the potential diversion of these drugs to the black market or for use in illicit activities. Various pharmaceutical substances, including Diazepam, Midazolam, Temazepam, Zolpidem, Rohypnol, Ketamine, Zopiclcne, and Scopolamine, have reportedly been used to facilitate criminal activities including theft, sexual assault, human trafficking, or organ theft in Kenya (Bisong & Kemeji, 2024; Mohammed, 2024; Volonnino et al., 2023; Nation Africa, 2022). Among these, the misuse of scopolamine – locally known as *mchele (Kiswahili for rice)* – has become particularly alarming.

The method used with *mchele* involves secretly adding the drug to an unsuspecting person's drink, causing confusion and making them highly susceptible to manipulation (Nation Africa, 2022; Mohammed, 2024). Victims often comply with criminals' demands, willingly handing over valuables, disclosing sensitive information such as Personal Identification Numbers (PINs), or being lured to secluded locations where they can be further exploited. In 2022, a surge in scopolamine-related drugging incidents was reported, particularly in Isiolo County, where traders were targeted and coerced into surrendering their entire stock to criminals. In one disturbing case, a victim, under the influence of the drug, willingly disclosed her M-Pesa PIN, leading to a financial loss of over KSh 60,000 (Mohammed, 2024).

Beyond theft, there have been reports linking scopolamine use to incidents of sexual assault, further highlighting its dangerous and exploitative application. The effects of scopolamine are profound, often leaving victims with little to no recollection of the incident. These drugging incidents are commonly reported in bars, public gatherings, and other social settings where alcohol or beverages are consumed, highlighting the ease with which pharmaceuticals and related substances can be repurposed for criminal activities (Nation Africa, 2022).

Case Study 2: Diversion of Chemicals in Somalia

In Somalia, there have been reports of chemicals being diverted from legitimate supply chains and used in the production of explosives (United Nations Security Council (UNSC), 2023). For instance, in 2016, a United Nations report revealed that Al-Shabaab had acquired chemicals used in the production of improvised explosive devices (IEDs) from the pharmaceutical industry. The report highlighted the ease with which chemicals could be diverted in a country with weak regulatory oversight and a thriving black market (United Nations Security Council (UNSC), 2023; United Nations Security Council (UNSC), 2016).

9.3.2 Terrorism

Terrorism is a significant concern in both Kenya and Somalia, and the pharmaceutical industry could be a target for terrorist groups seeking to acquire chemicals for use in attacks (Kenyan Pharmacy and Poisons Board (PPB), 2017). In Kenya, the Al-Shabaab militant group has carried out several high-profile attacks, including the 2013 Westgate shopping mall attack, the 2015 Garissa University College attack, the 2019 DusitD2 Hotel attack in Nairobi, and the 2020 Manda Bay Airfield attack

(Besenyő & Sinkó, 2024). The group has also been involved in the production of IEDs, which often require chemicals that could be sourced from the pharmaceutical industry (Heslop & Blain, 2020; Volonnino et al., 2023).

Case Study 3: Al-Shabaab's Use of Chemicals in IEDs

Al-Shabaab has been known to use IEDs in its attacks. In 2015, Kenyan security forces intercepted a consignment of chemicals that were being transported to Somalia and were believed to be intended for use in the production of IEDs. The chemicals included nitric acid, sulfuric acid and ammonium nitrate, a common ingredient in explosives, which had been diverted from a legitimate supply chain (United Nations Security Council (UNSC), 2023). In Somalia, the group was involved in the December 2019 attack in Mogadishu, where a truck-borne IED killed at least 82 people and wounded dozens more (Kirechu, 2020). Moreover, the Chair of the Security Council Committee pursuant to resolution 751 (1992) concerning Somalia warned in June 2020 that if current chemicals used for Al-Shabaab's homemade explosives become scarce, the group may seek alternative sources of explosive materials or precursors, possibly through illicit smuggling (United Nations Security Council (UNSC), 2020).

9.3.3 Illicit Trade

The illicit trade in pharmaceuticals and chemicals is a significant problem in both Kenya and Somalia. In Kenya, the black market for pharmaceuticals is fueled by the high cost of legitimate drugs and the presence of counterfeit products. This illicit trade creates opportunities for the diversion of regulated chemicals, which could then be used for harmful purposes (International Criminal Police Organization (INTERPOL), 2025).

Case Study 4: Counterfeit Drugs in Kenya

In 2022, the Kenyan PPB seized a large consignment of counterfeit drugs that were being sold on the black market (Pharmacy and Poisons Board (PPB), 2022a). The counterfeit drugs included antibiotics, antimalarials, and painkillers, some of which contained harmful substances. This seizure highlighted the extent of the illicit trade in pharmaceuticals in Kenya and raised concerns about the potential diversion of chemicals used in the production of counterfeit drugs (Pharmacy and Poisons Board (PPB), 2022b).

Case Study 5: Illicit Trade in Somalia

In Somalia, the illicit trade in pharmaceuticals and chemicals is facilitated by the lack of regulatory oversight and the presence of armed groups (United Nations Office on Drugs and Crime (UNODC), 2025). In 2019, a report by the United Nations Monitoring Group on Somalia and Eritrea revealed that Al-Shabaab was involved in the illicit trade of pharmaceuticals, including the diversion of chemicals used in the production of explosives (United Nations Security Council (UNSC), 2023). The report highlighted the role of the black market in facilitating the diversion of chemicals and the challenges of regulating the industry in a country with weak governance.

9.3.4 Cybersecurity Threats

As the pharmaceutical industry becomes increasingly digitized, cybersecurity threats are becoming a growing concern. In Kenya and Somalia, where cybersecurity infrastructure is underdeveloped, pharmaceutical companies are particularly vulnerable to cyberattacks. These attacks could result in the theft of sensitive information, such as chemical formulas or supply chain data, or the disruption of production processes (Abed, Allain-Ioos, & Shindo, 2024).

Case Study 6: Cyberattacks on a Kenyan Healthcare Sector

Kenyan pharmaceutical companies and hospitals have been targeted in cyberattacks that resulted in the theft of sensitive information, including chemical formulas and supply chain data (Emanuel, 2021). These incidents highlight the vulnerabilities of the pharmaceutical industry to cyberattacks and the potential impact on chemical security.

9.3.5 Environmental and Public Health Risks

The misuse or mishandling of chemicals can have serious environmental and public health consequences. In Kenya and Somalia, where regulatory oversight is weak, there is a risk that chemicals could be improperly stored, transported, or disposed of, leading to environmental contamination or public health crises (Omer, 2024; Shem et al., 2021). This situation is particularly concerning given the potential for widespread toxicity and long-term ecological damage (Bartolo, Azzopardi, & Serracino-Inglott, 2021; Pharmacy and Poisons Board (PPB), 2022b). In terms of security, 77% of facilities in Kenya were reported not to have adequate storage for safekeeping of expired pharmaceutical products, thus exacerbating the security of pharmaceutical agents in the country (Karungamye et al., 2022).

9.4 EFFORTS TO ENHANCE CHEMICAL SECURITY

The KCS, with support from the U.S. Department of State's CSP, OPCW, Sandia National Laboratories (SNL) and other partners has played an instrumental role in bolstering chemical security awareness and practices in Kenya and Somalia. This support has led to several key initiatives, demonstrating a multifaceted approach to addressing chemical security challenges (Shem et al., 2021).

9.4.1 Chemical Security Training

The KCS has developed and delivered numerous chemical security training workshops and hands-on exercises for pharmaceutical professionals. These trainings extend beyond theoretical knowledge, incorporating practical applications through table-top exercises covering crucial aspects of chemical security, including (Kenya Chemical Society (KCS), 2022b; Kenya Chemical Society (KCS) and Sandia National Laboratories (SNL), 2024):

- *Risk and vulnerability assessments:* Identifying potential threats and weaknesses.
- *Robust chemical inventory management systems:* Accurately tracking and accounting for all chemicals.
- *Physical security:* Protecting chemical stores and laboratories.
- *Personnel reliability processes:* Vetting and monitoring those handling sensitive chemicals.
- *Secure chemical transportation:* Ensuring supply chain security.
- *Pharmaceutical emergency response:* Preparing for chemical incidents and accidents in emergency response planning.
- *Dual-use chemicals management:* Addressing their potential for both legitimate and malicious applications.

These capacity-building programs by KCS have trained thousands in Kenya and hundreds from Somalia/Somaliland, including academics, industry professionals, and pharmaceutical professionals.

The tailored curricula have effectively enhanced sector-specific competencies, strengthening institutional capabilities across target fields. The initiative has demonstrably improved technical expertise and professional standards in participating regions.

9.4.2 Curriculum Integration

The KCS in collaboration with its partners in the chemical security sector has engaged different stakeholders in the pharmaceutical sector in both Kenya and Somalia to address gaps in the management of chemicals and controlled substances and to advocate for chemical security education. In Kenya, KCS has actively engaged with universities to integrate chemical security principles into pharmacy and related curricula (Kenya Chemical Society (KCS), 2022b). This has involved collaborative work with university faculty to develop new course modules, revise existing syllabi, and provide training materials for instructors.

In 2022, KCS conducted a workshop on curriculum development for pharmaceutical training institutions in Kenya to mainstream chemical security. The aim of this workshop and other engagements was to equip future pharmacists and related professionals with the knowledge and skills necessary to identify, assess, and mitigate chemical security risks within their professional settings. Several Kenyan universities have successfully integrated these modules into their programs, resulting in graduates who are more aware of and prepared to address chemical security challenges (Kenya Chemical Society (KCS), 2022b).

Similar engagements have been conducted for Somalia in collaboration with SNL and the SCS. An example of these engagements is the training on chemical security risk assessment for Somalia academia and small-to-medium industry conducted in 2021. One major outcome of the training was the recognition by the stakeholders of the need to integrate chemical security training in courses that involve chemicals, for example, chemistry and pharmacy. The participants were provided with the *Chemical Security Handbook* in both English and Somali languages. The handbook is a useful resource for incorporation of chemical security into training curricula as it provides tools and methodologies for identification and prioritization of chemicals (including PBAs) with security concerns.

In addition, a Know-Your-Customer (KYC) training for Somalia pharmaceutical industry stakeholders was conducted in 2022. This workshop targeted universities, research institutions, manufacturers, and distributors of PBAs in Somalia. The process of integrating chemical security into training curricula in Somalia is ongoing. In addition to the capacity-building trainings, several universities across different regions, including the University of Somalia, University of Hargeisa, Juba University of Science & Technology, Afgoi International University, and facilities like the Community Mental Health Care Center (CMHCC) Pharmacy and laboratory facility in Baidoa, Somalia have received funding from the CSP for physical security improvements to secure chemicals and PBAs with security concern.

9.4.3 Outreach and Sensitization Campaigns

The KCS has implemented outreach campaigns in both Kenya and Somalia, focusing on the informal and unregulated pharmaceutical sectors. In both countries, locally trained campaigners work directly with smaller businesses and individuals, raising awareness of chemical security risks and promoting best practices (Kenya Chemical Society (KCS) and Sandia National Laboratories (SNL), 2023). The campaigns were conducted in the informal markets and industries manufacturing, packaging, and distributing pharmaceutical products in the cities of Nairobi and Mombasa (Kenya Chemical Society (KCS), 2023) in Kenya and Mogadishu and Hargeisa (Kenya Chemical Society (KCS) and Sandia National Laboratories (SNL), 2022a).

Figure 9.4 shows a banner utilized during the outreach campaign in Nairobi, which targeted the pharmaceutical sector. Additional materials, such as fliers and posters, were distributed and

FIGURE 9.4 Outreach campaigns to pharmaceutical retailers in Nairobi, Kenya by KCS and supported by CSP.

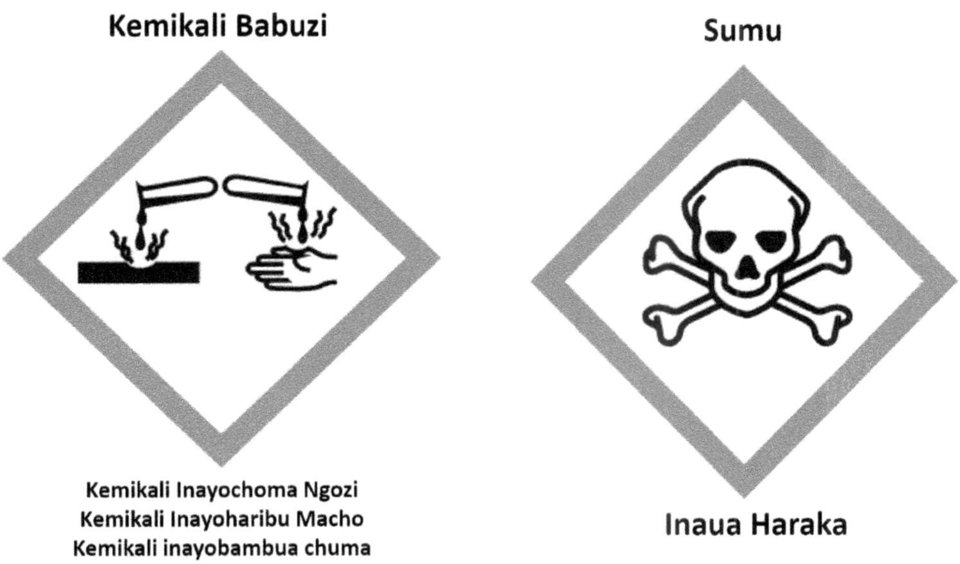

FIGURE 9.5 GHS pictogram for impact of chemical toxicity translated to Kiswahili.

displayed at the visited facilities (Kenya Chemical Society (KCS), 2022a). The initiative successfully heightened stakeholder awareness regarding the security of pharmaceutical-based agents.

Training materials have also been translated into relevant local languages to maximize accessibility. Figure 9.5 shows examples of Globally Harmonized System of Classification and Labelling of Chemicals (GHS) hazard pictogram translated into Kiswahili for distribution to stakeholders during outreach campaigns. Complementing these on-the-ground efforts, the KCS's online e-learning

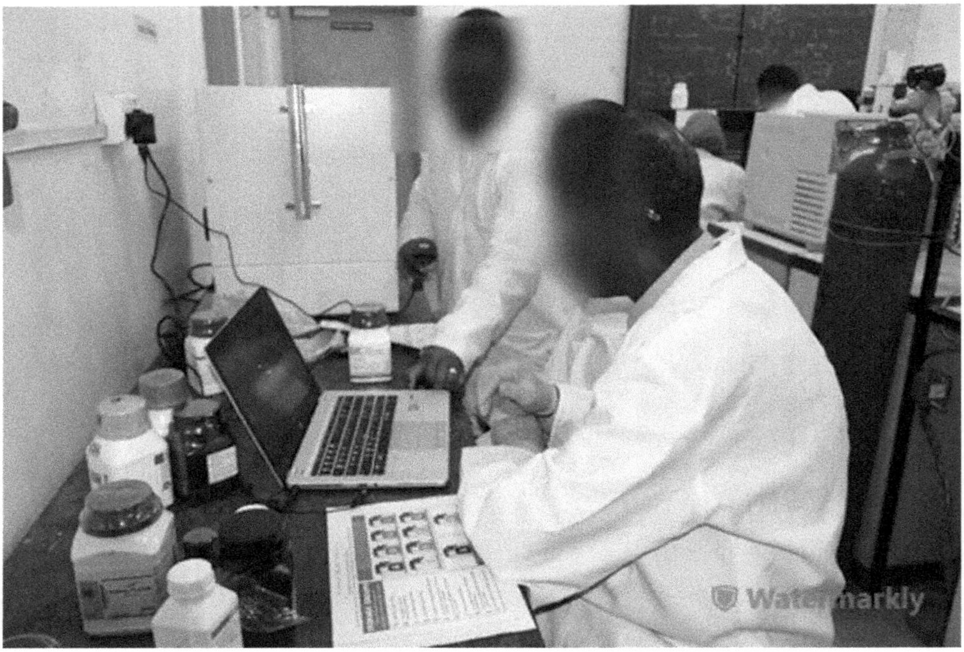

FIGURE 9.6 KCS workshop with CMS software workshop for Somali Academic Staff held in Kenya in 2018.

platform offers flexible access to chemical safety and security training materials, expanding reach across both countries (Kenya Chemical Society (KCS), 2022b).

The scope of direct training in Somalia has been smaller due to ongoing political instability and logistical challenges. To overcome these challenges, KCS has explored strategies to extend its influence through partnerships and training-of-trainers (ToT) programs. KCS has conducted three ToT workshops between 2021 and 2023 to develop capacity for outreach and sensitization campaigns for Somali stakeholders. At least 50 alumni have benefited from the ToT trainings. In addition, through KCS, CSP provided mini-grants for the alumni of the ToT trainings to conduct mini-workshops and door-to-door outreach campaigns in different regions of Somalia in 2021, 2022, and 2024.

For the outreach and sensitization campaigns, informational materials were provided in the Somali language. In 2022, KCS and SNL also conducted two trainings for pharmaceutical industry stakeholders. The training aimed to build sustainable partnerships between industry and government stakeholders to mitigate risks from PBAs, while the second training on KYC for pharmaceutical industry aimed to improve the capacity of the stakeholders to recognize suspicious purchase requests, develop customer vetting strategies, and understand regulations regarding the sale of PBAs and dual-use chemicals. The implementation of chemical inventory management systems (CMS) to improve inventory control has been more extensive in Kenya (Kenya Chemical Society (KCS) and Sandia National Laboratories (SNL), 2024), compared to Somalia.

However, the KCS has made significant efforts to promote the adaptation of CMS in Somalia, for example, at the University of Hargeisa (Kenya Chemical Society (KCS) and Sandia National Laboratories (SNL), 2024). These initiatives are supported by ongoing training and capacity-building efforts as seen in Figure 9.6 and highlight the collaborative approach necessary to address chemical security in the region. Similarly, physical security upgrades, such as installing fume hoods, surveillance systems, and biometric access controls, have primarily been implemented in Kenyan institutions. Commensurate efforts have been made to adapt security enhancements suitable

FIGURE 9.7 Steel door with biometric access control and lockable cabinets, in Somalia. (a) Example of a steel door at Afgoi University. (b) Lockable cabinets at the University of Somalia in 2022 supported by KCS and CSP. Other upgrades included CCTV, alarm systems and sensors lockable cabinets and refrigerators to 5 different institutional laboratories in Somalia with dual-use chemicals.

for the challenging conditions in Somalia (Kenya Chemical Society (KCS) and Sandia National Laboratories (SNL), 2022a). Figure 9.7 shows some of such stated upgrades.

9.5 CHALLENGES AND OPPORTUNITIES FOR INTERVENTION

9.5.1 Challenges to the Adoption of Chemical Security Best Practices in the Pharmaceutical Sector

Challenges to implementing pharmaceutical security measures are pervasive in both countries (Kenya Chemical Society (KCS), 2022a; Kenya Chemical Society (KCS) and Sandia National Laboratories (SNL), 2022b). These challenges include the following:

1. *Limited resources:* Financial constraints hamper the implementation of robust chemical security measures in both Kenya and Somalia. This is especially true for smaller pharmaceutical businesses and training institutions.
2. *Capacity gaps:* The lack of trained personnel in chemical security is a major limitation in both countries. This necessitates a sustained effort in capacity building.
3. *Political instability (Somalia):* Somalia's ongoing political instability poses unique challenges to implementing long-term security programs and data collection. Security concerns make it harder to conduct trainings and risk assessments.
4. *Regulatory gaps:* The absence of comprehensive national standards and regulations on pharmaceutical security hinders consistent implementation of security best practices across both nations.

5. *Informal sector:* The significant informal pharmaceutical sectors in both countries represent a major challenge, as they often lack resources and awareness to implement adequate security practices.

9.5.2 Opportunities for Improvement

Despite the substantial challenges, numerous opportunities exist:

1. *Strengthened regional collaboration:* Increased collaboration between the KCS, governmental agencies, pharmaceutical industry stakeholders, and international organizations can facilitate knowledge sharing, resource allocation, and pharmaceutical security best practice implementation across both Kenya and Somalia.
2. *Development of regional standards:* Creating regional standards and regulations concerning pharmaceutical security would enhance uniformity in practices in Kenya and Somalia.
3. *Targeted support for Somalia:* Specific strategies to address the unique challenges in Somalia are needed, involving targeted training programs, tailored outreach campaigns, and collaboration with local authorities to ensure safety and security of the pharmaceutical-based agents.
4. *Investing in capacity building:* Significant investment in training and education across all levels of the pharmaceutical industry in both countries is critical for enhanced pharmaceutical security in both countries.
5. *Technological integration:* Adopting innovative technologies, including electronic inventory systems and sensor technology, can enhance security capabilities.

9.6 CONCLUSION

Strengthening pharmaceutical security within the pharmaceutical industries of Kenya and Somalia demands sustained commitment from various stakeholders. While the efforts of the KCS provide a strong foundation, continued collaboration, resource allocation, and tailored interventions sensitive to the contexts of each country are essential to mitigate risks related to pharmaceutical chemicals. A holistic approach, including training, improved regulations, technological innovation, and ongoing engagement with all pharmaceutical supply chain is critical for achieving lasting improvements in pharmaceutical security across the region.

10 Regulation of the Handling, Safety, and Security of Herbal Medicines; the Case of China

Lincoln L. Were Munyendo

10.1 INTRODUCTION

Chinese herbal medicine (CHM) occupies a privileged place within that nation's medical institutions. Much more than being merely an assembly of pharmaceutical plants, it's an overall therapy that includes aspects of great interest concerning cultural and historical importance with respect to the philosophical history of China. An Introduction of Herbal Medicine in China will describe the role it plays within this cultural landscape through discussing its historical foundation, philosophical foundation, its movement within contemporary society, and finally its rise globally. In the same vein, this chapter outlines toxic profiles that demand for safe and secure handling.

10.1.1 Traditional CHM: Historical Background

The tale of herbal medicine in China can be drawn back in time, with legendary figures like Shen Nong, the mythical "Divine Farmer," credited with tasting hundreds of herbs and classifying their medicinal properties. Archaeological evidence reveals the use of herbal remedies in China as far back as the Neolithic period, while written records date back to the Shang Dynasty, c.1600–1046 BCE (Liu et al., 2020). Herbal medicine in China, based on the Huangdi Neijing or Yellow Emperor's Inner Canon and the Shen Nong Ben Cao Jing or Shen Nong's Classic of *Materia medica*, systematized knowledge on herbal properties, diagnoses, and treatment principles. These classics laid the foundation for a sophisticated system of medicine that has evolved over millennia.

Emperors and the ruling elite throughout Chinese history realized the importance of herbal medicine in China by setting up imperial medical institutions and sponsoring the compilation of extensive *Materia medicas*. This patronage contributed to the development and refinement of CHM knowledge and practices. For instance, during the Ming Dynasty (1368–1644 CE), Li Shizhen compiled the Bencao Gangmu, also known as the Compendium of *Materia medica*, a monumental work which documented over 1,800 medicinal substances and their applications (Li, 1578/2021). Herbal medicine in China has historically been integrated with other traditional practices like acupuncture, moxibustion, and dietary therapy, forming a holistic approach to health and well-being. This integration is evident in the concept of Traditional Chinese Medicine (TCM), which encompasses various modalities under a unified theoretical framework (Maciocia, 2019).

In the integration of herbal medicine into national healthcare systems, special institutions like TCM hospitals and clinics play a crucial role, offering complementary treatments alongside conventional medical care. These institutions are expected to be regulated to ensure patient safety and quality through standardized practices. Key areas of regulation include practitioner qualifications, dispensing methods, and overall patient safety (Liang et al., 2023; Dubale et al., 2025).

10.1.2 Philosophical Foundations: Harmony and Balance

The TCM is a very holistic natural therapy that applies the philosophical approach of Taoism and emphasizes the fact that everything in this world is all connected, everything has a correspondence, and interdependence exists between all constituents. The concept of Yin and Yang, representing opposing yet complementary forces, is central to CHM. Disease is seen as an imbalance of Yin and Yang, and treatment aims to restore harmony. This principle is reflected in the use of herbs with different properties to address specific imbalances (Flaws, 2019).

Further, herbal medicine practice in China recognizes the flow of vital energy, Qi, through channels called meridians in the body. Herbal medicines are used to regulate the flow of Qi and address imbalances. Other modalities used to influence Qi flow include acupuncture and moxibustion (Cheng et al., 2020). The Five Elements theory (*wood*, *fire*, *earth*, *metal*, *water*) provides a framework for understanding the relations between different organs, emotions, and natural phenomena. Herbal prescriptions are often formulated based on these elemental relationships with the aim of restoring balance within the body (Flaws, 2019).

10.1.3 Contemporary China: Tradition Meets Modernity

Herbal medicine is still vital in healthcare in China despite the rise of Western medicine. It is included in the national healthcare systems, with herbal products and other conventional medical treatments concurrently offered in hospitals and clinics. The herbal products are often more accessible and affordable than Western medicine, particularly in rural areas where access to modern healthcare facilities may be limited (Wang et al., 2021). Herbal products are thus increasingly used for managing chronic diseases like diabetes, hypertension, and arthritis, often in conjunction with conventional treatments. Studies have shown the efficacy of CHM in managing these conditions and improving the quality of life (Li et al., 2020).

Emphasis is also placed on the practice of CHM for preventive healthcare and health promotion, focusing on maintaining balance and strengthening the body's natural defenses. This aligns with the global trend of moving toward preventive medicine and healthy lifestyle choices. Modern scientific research is being carried out to investigate the efficacy and mechanisms of action of CHM, bridging the gap between traditional knowledge and modern science. This is the study that is also confirming the traditional applications of CHM and is highlighting new ones (Chen et al., 2022).

10.1.4 Increasing Recognition and Integration

CHM practice is increasingly gaining recognition and acceptance worldwide as a complementary and alternative medicine modality. The World Health Organization (WHO) has recognized the value of traditional medicine in promoting health and well-being. This paper discusses herbal medicine practice in China. The 2014–2023 WHO Traditional Medicine Strategy supports member states in integrating traditional medicine into their healthcare systems (WHO, 2013). The research on herbal products is also conducted for several other countries to study their applicability for different diseases. This joint collaboration has been developing an in-depth knowledge about the said ethnotherapeutics and their respective benefits that accrue from them (Zhang et al., 2025).

The herbal products of China are being included within the national healthcare system by numerous countries to correlate it with conventional medicine. Integration is often accompanied by the establishment of regulatory frameworks that can ensure the safety and quality of CHM practices. The fast growth in global trade in herbal medicines creates new economic opportunities but also presents challenges. The trade of herbal medicine indicates a requirement for international collaboration and harmonization of regulatory standards (Chen et al., 2021b).

10.1.5 Challenges and Opportunities

While CHM holds great promise, it also faces significant challenges:

Quality Control: Ensuring the quality and consistency of herbal products is crucial in respect to the current global trade context. This requires standardized methods for cultivation, processing, and quality assessment (Li et al., 2022).

Safety: Thorough safety assessments are needed to identify and mitigate potential risks associated with herbal products use. This includes research on herb–drug interactions (HDIs), toxicity, and adverse effects (Wu et al., 2020).

Sustainability: Sustainable harvesting and cultivation practices are required to protect endangered species and preserve biodiversity. This requires responsible sourcing and conservation efforts (IUCN, 2020).

Integration: The effective integration of herbal medicinal products with conventional medicine requires collaboration and understanding between different healthcare systems. This will involve addressing differences in philosophies, practices, and regulatory frameworks (WHO, 2019).

In spite of these challenges, CHM offers considerable opportunities such as the following:

New drug discovery: Herbal products may offer useful leads to new drugs and therapies. A large number of modern drugs are obtained from natural sources, and CHM represents a huge repository of potential candidates (Newman & Cragg, 2020).

Personalized medicine: The holistic application approach of Chinese herbal products can contribute to personalized medicine, tailoring treatments to individual needs. This approach aligns with the growing trend toward personalized healthcare (Polshettiwar et al., 2022).

Global health: Herbal products can play a great role in addressing global health challenges, particularly in developing countries where access to healthcare is limited. The affordability and accessibility of these products make them a valuable resource for primary healthcare and community-based health interventions.

The CHM is a practice with a rich and valuable tradition and a long history of healing. Its philosophical underpinning, emphasizing harmony and balance, appeals to a growing global interest in holistic healthcare. While there are still challenges, the herbal products offer enormous opportunities to improve health and well-being both in China and worldwide. As herbal products continue to evolve and integrate into modern science and healthcare systems, their place in the 21st century and beyond is unquestionable.

10.2 LEGAL FRAMEWORK FOR HERBAL MEDICINES IN CHINA

Chinese herbal products enjoy a priority place in both traditionally rooted and modern oriented medical systems. There is an impending need for laying appropriate law pertaining to safety, effectiveness, and quality assurance within herbal medicine use due to growing utilization and fast increase in popularity even beyond China. This section offers a descriptive overview of the essential elements of legislative instruments, regulatory mechanisms, and systems of enforcement laid down for traditional CHM in People's Republic of China (PRC).

10.2.1 Evolution of CHM Regulation

The regulation of CHM in the PRC has gone through many developments and changes as times, society, politics, science, and technology change. Traditionally, CHM was regulated through practice and knowledge passed from generation to generation among practitioners. However, since the establishment of the PRC in 1949, there have been attempts toward standardization and modernization in the practice of CHM.

In the 1980s and 1990s, further developments in CHM regulation included the promulgation of the Drug Administration Law (DAL) in 1984 and establishment of a State Drug Administration

(SDA, now the National Medical Products Administration, NMPA) in 1998. These steps were to lay a firm foundation for reinforcing quality control, safety, and rational use (Liang et al., 2023).

In the last decade, China has further enhanced the regulatory mechanism for CHM by laying much more emphasis on enhancing safety monitoring, promoting Good Manufacturing Practices (GMP), and integrating CHM into the national healthcare system. This is reflected within the revised DAL (2019) and the newly enacted TCM Law (2017).

10.2.2 Key Legislations Governing Chinese Herbal Medicines

Several key laws and regulations oversee the production, distribution, and use of herbal medicines in China:

10.2.2.1 Drug Administration Law

The DAL is the primary legislation governing all drugs within China, which encompasses CHM. The framework it established regulates the drug registration, manufacturing process, and quality control of drugs. Revised DAL in 2019 strengthened controls around clinical trials, adverse drug reaction reporting, and post-market surveillance (DAL, 2019).

10.2.2.2 TCM Law

This law provides specifically for CHM to promote its inheritance, development, and innovation. Great importance is put on the protection of traditional knowledge, encouragement of scientific research, and integration into the healthcare system (TCM Law, 2017).

10.2.2.3 Pharmacopoeia of the PRC

The Pharmacopoeia represents the official compendium of drug standards in China, including quality specifications for herbal medicines. It contains detailed monographs on individual herbs and herbal formulations that outline identification, properties, and testing methods.

10.2.3 Regulatory Bodies in China

The main regulatory body for CHM in China is the NMPA. Activities of the NMPA related to CHM include the following:

a. *Drug registration:* Studying and approving, for market authorization, new herbal medicines.
b. *GMP certification:* Inspecting and certifying CHM manufacturing facilities to ensure compliance with GMP standards.
c. *Quality control:* Setting quality standards for herbal medicines and conducting post-market surveillance to ensure product quality and safety.
d. *Adverse event monitoring:* Collecting and analyzing data on adverse events associated with CHM use.
e. *Enforcement:* Taking enforcement actions against violations of CHM regulations. Other relevant regulatory bodies include the following:
    - *National Administration of Traditional Chinese Medicine (NATCM):* responsible for promoting the development and inheritance of CHM.
    - *State Administration for Market Regulation (SAMR)*: responsible for market supervision and consumer protection, including regulation of CHM products in the market.

10.2.4 Regulation of Specific Aspects of herbal medicines

The legal framework for CHM in China covers a number of specific aspects:

Cultivation and collection of herbal materials: The regulations encourage Good Agricultural and Collection Practice (GACP) to ensure quality and sustainability in herbal raw materials, and specify guidelines concerning the cultivation, harvesting, and processing of medicinal plants (Chen et al., 2021a).

Manufacturing of herbal products: GMP standards in CHM manufacturing facilities are based on different aspects: hygiene, quality control, documentation, and personnel training. These standards focus on the manufacture of safe and efficient herbal medicines (Liang et al., 2023).

Labeling and packaging: Regulations require clear and accurate labeling of herbal medicines, including information on ingredients, dosage, indications, contraindications, and potential side effects. Packaging must also meet specific standards to protect product quality and prevent contamination (DAL, 2019).

Distribution and sales: Regulations govern the distribution and sale of herbal medicines, including licensing requirements for distributors and retailers. Online sales of CHM are also subjected to specific regulations for consumer safety (SAMR, 2022).

Clinical trials and research: Clinical trials of CHM are subject to ethical review and also receipt of regulatory approval. It provides the regulation to create and promote scientific research on CHM by enhancing its efficacy and safety (TCM Law, 2017).

10.2.5 Enforcement and Penalties

The NMPA and any other relevant authorities may impose penalties for violation or failure to follow the regulations governing CHM products use, which may also involve warnings and rectifications, imposition of fines for violations, and even recall of products from market circulation in cases where the herbal medicines are unsafe or below par.

License suspension and revocation: Manufacturers and distributors, when found guilty of legal and regulatory provisions, may have their licenses suspended or totally revoked altogether.

Criminal liability: In worse conditions where serious violations potentially endanger public health, criminal liability will ensue and punitive action enforced.

10.2.6 Challenges Ahead and Future Directions

Despite the China government having taken significant initiative to control the herbal medicines, difficulties persist (WHO, 2013):

Complex CHMs: The reason herbal formulations represent complex mixtures, and at times raw material variation in standardization and quality control is difficult (Li, Lin, & Han, 2019).

Traditional knowledge: Balancing the protection of traditional knowledge with the need for scientific validation and modernization of CHM practices is an ongoing challenge.

International harmonization: Harmonizing CHM regulations with international standards is crucial for facilitating global trade and collaboration.

Future directions for CHM regulation in China include the following (WHO, 2013):

Strengthening enforcement: Increasing inspections and enforcement actions to ensure compliance with regulations.

Promoting innovation: Encouraging research and development of new CHM products and technologies.

Utilizing technology: Leverage technology, with big data and artificial intelligence (AI), for improvement in safety monitoring and quality control.

International cooperation: There should be cooperation through international organizations and relevant regulating bodies by setting common standards to help ensure responsible use of CHM globally.

The existing legal framework about CHM is of a very dynamic and complex nature. A peculiar feature of such a traditional system of medicine is its amalgamation into the modern system.

While challenges remain, China evidences its commitment to strengthening the regulation of herbal medicines while ensuring safe and effective use. Ongoing modernization of the regulation for herbal medicines, improvement in the quality control, and promotion of scientific research had to be carried out, along with effective ongoing development and international recognition of this health tradition.

10.3 PROCESSING OF CHM PRODUCTS

CHM involves complex processes, from the cultivation and collection of raw materials to the manufacturing and distribution of finished herbal products. Proper handling at each stage is crucial to ensure the quality, safety, and efficacy of these medicines. This section focuses on the key aspects of product handling, highlighting the regulatory requirements, best practices, and emerging technologies that contribute to maintaining the integrity and therapeutic value of herbal products.

10.3.1 Sourcing and Authentication of Herbal Materials

The quality of CHM products starts from the sourcing and authentication of the raw herbal materials through careful selection of the species of plants, proper harvesting techniques, and rigid identification to ensure the identity and quality of the herbs used. It is necessary to accurately identify plant species in order not to get adulteration and ensure correct therapeutic properties. The relevant expertise for this includes botany, besides various techniques involving macroscopic and microscopic examination, chemical analysis, and Deoxyribonucleic Acid (DNA) barcoding (Chen et al., 2022).

The geographical origin of herbal materials can significantly influence their quality and efficacy. In fact, factors such as climate, season, soil conditions, and cultivation practices significantly affect the concentration of active compounds in plants. Regulations and quality standards often specify preferred geographical origins for certain herbs (Zhang et al., 2021). Sustainable harvesting practices are essential to ensure the long-term availability of herbal resources and protect biodiversity. This includes responsible collection methods, avoiding over-harvesting, and promoting cultivation of medicinal plants (IUCN, 2020).

Maintaining traceability of herbal materials throughout the supply chain is crucial for quality control and safety. This involves documenting the species, origin, collection date, processing methods, and storage conditions of each batch of raw materials (Li et al., 2022).

10.3.2 Processing and Storage of Herbal Medicine Products

Proper processing and storage will ensure the quality and effectiveness of herbal medicines, which could easily get adulterated, contaminated, or degraded. Many traditional methods of processing are used, including cleaning, drying, slicing, stir-frying, and steaming of freshly collected samples. These processes may be carried out to enhance the therapeutic properties of herbs or reduce their toxicity. These techniques are passed on through generations and may also be subject to quality control measures (Wang et al., 2021).

Modern technologies, such as freeze-drying, spray-drying, and supercritical fluid extraction, are increasingly used to process herbal materials, offering advantages in terms of efficiency, quality control, and preservation of active compounds (Chen et al., 2022). Proper storage conditions are crucial to maintain the quality and stability of herbal medicines. Factors such as temperature, humidity, light exposure, and packaging can affect the shelf life and efficacy of herbal products. Regulations and guidelines specify appropriate storage conditions for different types of herbal medicines (Liang et al., 2023).

Effective measures for pest control should be applied to avoid the invasion and contamination of stored herbal materials. It may be through natural repellents or controlled atmosphere storage as indicated by Huang et al. (2022b).

10.3.3 Good Agricultural and Collection Practices

The GACP guidelines provide a framework for quality and sustainability in medicinal plants used in the generation of herbal medicine products. These guidelines cover various aspects of cultivation and harvesting, including the following:

Site selection: Selection of appropriate sites for cultivation, taking into consideration factors such as soil, water quality, and climate.

Plant material: High-quality seeds or propagules from authenticated sources.

Cultivation practices: Sustainable cultivation practices, including organic farming methods, to have minimal environmental impacts while ensuring the quality of the products.

Harvesting and postharvest handling: Proper harvesting techniques and postharvest handling procedures to preserve the quality and integrity of herbal materials (Chen et al., 2021a).

Documentation: Recordkeeping should be done, including species cultivated, their origin, and details of processing, cultivation, and harvesting.

10.3.4 GMP for Herbal Medicine Products

The quality and safety of herbal medicine products are assured during manufacturing by GMP standards. The GMP guidelines in China for Herbal Medicines cover all aspects of production, including Facility Design and Hygiene, Equipment and Processes, Quality Control, Personnel Training, and Documentation.

It is important to maintain a clean and hygienic manufacturing environment to prevent contamination. Using appropriate equipment and validated processes usually ensures achievement of consistency in the herbal medicines' product quality. It is therefore key to implementing rigorous quality control measures throughout the manufacturing process, including testing of raw materials, intermediate products, and finished goods (Liang et al., 2023). Training in GMP principles and particular manufacturing procedures will ensure competency and adherence to quality. This is further corroborated by maintaining complete and comprehensive documentation of all manufacturing activities, which include batch records, quality control results, and deviations from procedures.

10.3.5 Use of Modern Technology in Handling Herbal Medicine Products

Therefore, new technologies in herbal medicines are being increasingly put into practice with a view to further efficiency, precision, and traceability for product handling in China. With digital tools and databases managing information regarding herbal materials about origin, processing, and quality control data, the traceability has become better with a view to assuring uniformity (Li et al., 2022).

Advanced techniques of analysis such as DNA barcoding and chemical fingerprinting have been employed for authentication and to check adulteration of herbal materials (Newmaster et al., 2013). More recently, automation and robotics have been introduced in the manufacturing of herbal medicine products to achieve better efficiency, reduce human error, and enhance quality control (Polshettiwar et al., 2022).

AI is also being explored to optimize aspects regarding the handling of herbal medicine products including quality production of herbal materials based on environmental and cultivation data and the identification of optimum parameters in the processing chain (Huang et al., 2023).

10.3.6 Challenges and Future Directions

Despite the improvements in herbal medicines handling, challenges have been realized in certain aspects:

Standardization: Standardized methods of cultivation, processing, and quality control need to be developed across regions and producers to ensure consistency and safety.

Sustainability: The ever-growing demand for herbal medicines requires that plant resources be sourced sustainably and conserved.

Integration of technology: The integration of modern technologies in the practices of traditional CHM should be critically thought through for strengthening validation that will not compromise the integrity of the medicine.

The following therefore entails possibilities that would be explored as future directions in CHM product handling the following:

Strengthening GACP and GMP implementation: Expanding the dissemination and implementation of GACP and GMP guidelines to guarantee the quality and safety of herbal medicines.

Developing traceability systems: Developing robust traceability systems at all levels of the supply chain to trace the origin and processing of herbal materials.

Big data and AI: Use of big data and AI in the optimization of the cultivation, processing, and quality control of CHMs.

Promotion of sustainable practices: Encouragement of sustainable harvesting and cultivation practices that do not harm biodiversity and ensure long-term availability of resources.

Proper handling of herbal medicine products is critical to maintaining their quality, safety, and effectiveness. This requires great attention to detail at every stage, from raw materials sourcing and authentication to the manufacture and packaging of finished products. By adhering to regulatory requirements, implementing best practices, and embracing modern technologies, the herbal medicine products industry can ensure continued access to high-quality herbal medicines that promote health and well-being.

10.4 SAFETY OF CHINESE HERBAL MEDICINES: A MULTIFACETED APPROACH

The CHM products represent a rich repertoire of natural remedies for a variety of health conditions. Ensuring the safety of herbal medicine products cannot be overemphasized in the face of increased global use and integration into conventional medicine. This section presents the multifaceted approach to CHM products safety, with key highlights on assessment of safety and toxicity, adverse event reporting, HDIs, and considerations for vulnerable populations.

Evaluating the safety of herbal medicine products presents unique challenges due to its complex nature. The CHM often involves multicomponent formulations with multiple herbs, each containing numerous bioactive compounds. Understanding the interactions between these components and their combined effects on the human body is crucial for safety assessment (Wu et al., 2020).

The quality and composition of herbal materials can be highly variable due to factors like species, geographical origin, cultivation practices, and processing methods. This variability may affect the safety and efficiency of CHM products (Li et al., 2022). The gap between traditional knowledge and modern scientific methods also applies to comprehensive safety evaluation. Whereas traditional use thus forms a valuable basis, claims about safety and risk estimations need rigorous scientific investigation (Wu et al., 2020).

10.4.1 Toxicity and Security of Chinese Herbal Products

The majority of TCMs are toxic in nature, and can be harmful on intentional or accidental use. Toxicities are typically well documented in TCM theory, and hazards are neutralized and therapeutic effects utilized by precise preparation, dosing, and concomitant use with other herbs (Cai et al., 2019; Xie, 2011).

Some of the best-known toxic constituents contained in most Chinese herbs include the following:

- *Alkaloids:* Present in *Aconitum* species (aconitine, mesaconitine, hypaconitine), *Datura* species (hyoscyamine), and others, with effects on the nervous and cardiovascular systems (Cai et al., 2019; Australian Prescriber, 2003).

- *Aristolochic acids:* Present in *Aristolochia* species (Guan Mu Tong, Ma Dou Ling, Qing Mu Xiang, Tian Xian Teng, Guang Fang Ji, Xi Xin), which have been shown to be nephrotoxic (poisonous to the kidneys) and to cause urothelial carcinoma. These have been alerted against by regulatory agencies (Australian Prescriber, 2003; Xie, 2011).
- *Diterpenoid esters:* Found in *Aconitum* species, to which neurotoxic and cardiotoxic activities may be attributed (Cai et al., 2019).
- *Ephedrine and pseudoephedrine:* Contained in *Ephedra* (Ma Huang), which can lead to cardiovascular disease (Australian Prescriber, 2003; Xie, 2011).
- *Pyrrolizidine alkaloids:* Present in certain herbs such as Comfrey (*Symphytum officinale*), are accountable for inflicting liver injury (veno-occlusive disease) (Xie, 2011).

One of the most significant safety and security issues with CHMs is their adulteration with other substances. Undeclared chemical compounds such as warfarin, alprazolam, diethylstilbesterol, non-steroidal anti-inflammatory agents such as diclofenac, and weight loss or erectile dysfunction medication have been discovered, which resulted in severe side effects.

Contamination with heavy metals like arsenic, lead, cadmium, and mercury has been documented (Australian Prescriber, 2003). Pesticides and microorganisms also cause health risks.

Accidental or intentional substitution errors during dispensing or manufacture can result in the substitution of toxic herbs for intended ones, causing serious complications (e.g., *Tupistra chinensis* for *Panax notoginseng*, or *Datura metel* for *Campsis grandiflora*) (Cai et al., 2019).

Chinese medicinal herbs can interact with Western drugs, either modifying efficacy or enhancing toxicity. For instance, *Ephedra* will interact with cardiovascular and sympathomimetic medications, and certain herbs will interact with anticoagulants.

10.4.2 Profiles of Some Very Poisonous Herbs

10.4.2.1 Aconitum Species (Fu Zi, Chuan Wu, Cao Wu)

The *Aconitum* species contain aconitine, mesaconitine, and hypaconitine, which are powerful neurotoxins and cardiotoxins, and can produce rapid development of symptoms such as nervous system stimulation followed by depression, arrhythmias, and potentially fatal cardiac arrest (Cai et al., 2019; Australian Prescriber, 2003; Xie, 2011). The classical uses for *Aconitum* species are for "internal warming," for pain and illness associated with Yang deficiency. They need careful processing and dosage (Australian Prescriber, 2003).

10.4.2.2 *Aristolochia* Species (Guan Mu Tong, Guang Fang Ji, etc.)

They contain aristolochic acids, which are both nephrotoxic and carcinogenic, causing aristolochic acid nephropathy and urothelial cancer risk. Effects may be cumulative and delayed (Australian Prescriber, 2003). They have been used historically for various conditions but, being toxic, their use is now generally prohibited or severely limited in all countries. Replacement by sounding-alike but nontoxic herbs is required (Australian Prescriber, 2003).

10.4.2.3 *Ephedra* (Ma Huang)

Contents of these species include ephedrine and pseudoephedrine, and other sympathomimetic amines. These may lead to cardiovascular problems such as rapid heartbeat, elevated blood pressure, and even stroke or heart attack, particularly in those with pre-existing heart conditions or when used with other stimulants (Australian Prescriber, 2003; Xie, 2011). It is utilized for respiratory conditions like asthma and nasal decongestion, as well as for weight loss in some preparations. It is controlled for use today in most parts of the world.

10.4.2.4 *Croton* (Ba Dou)

These species contain croton oil, which is a strong irritant and purgative. The oil can lead to violent gastrointestinal disturbance, such as vomiting, diarrhea, and abdominal pain, is poisonous when taken in very large quantities, and is classically used as a powerful purgative in obstinate constipation (Xie, 2011). It needs to be carefully dosed and is often taken in processed forms to minimize toxicity.

10.4.2.5 *Datura* Species (Yang Jin Hua, Zi Hua Man Tuo Luo)

These species contain anticholinergic alkaloids such as hyoscyamine, atropine, and scopolamine. They can produce a variety of symptoms such as dry mouth, blurred vision, dilated pupils, tachycardia, confusion, hallucinations, and even coma (Australian Prescriber, 2003). They are employed in the alleviation of pain and respiratory ailments in very precise and controlled situations because of their powerful actions.

The key considerations that would enable mitigation of the adverse effects include the following:

- *Processing:* Conventional processing techniques (e.g., soaking, boiling) are usually helpful in minimizing the toxicity of certain herbs such as *Aconitum*. Poor preparation may result in side effects (Australian Prescriber, 2003; Cai et al., 2019).
- *Dosage:* The toxicity of most herbs is dose-related. Going beyond the recommended dosages can produce unwanted effects.
- *Individual sensitivity:* People may possess different sensitivities to herbal preparations.
- *Misdiagnosis and inappropriate use:* Utilization of herbs for conditions they are not prescribed for or without proper TCM diagnosis can cause damage.

It is important to consult experienced practitioners of TCM who possess extensive knowledge about herbal properties, possible toxicities, and guidelines for safe use. Additionally, in most countries, there are regulatory agencies that oversee the safety of herbal products, and it is recommended to stay informed about warnings and recalls that are likely to be posted.

10.4.3 Safety and Toxicity Assessment

Safety is a cardinal attribute of any medication. The safety and toxicity assessment will form the cornerstone of the protection of public health and further ensure responsible use of herbal medicines (Zhao et al., 2024). It is a task at many levels when it comes to integrating traditional knowledge with modern scientific methodologies (Balkrishna et al., 2024). The use of herbal medicine products has a long history in China, with traditional knowledge that has passed down through generations, providing valuable insights into the safety profile of many herbs. However, it is essential to critically evaluate this knowledge and complement it with rigorous scientific investigation (Li et al., 2023).

Preclinical studies, including in vitro and in vivo experiments, are conducted to assess the potential toxicity of herbal medicines. These studies evaluate various parameters such as acute and chronic toxicity, genotoxicity, and reproductive toxicity. Modern technologies, including *omics* approaches, like genomics, transcriptomics, and metabolomics, are increasingly used to gain a deeper understanding of the biological effects of herbal compounds (Huang et al., 2022c).

Clinical trials are, therefore, crucial in the assessment and evaluation of the safety and effectiveness of CHM in humans. Clinical trials follow strict ethical guidelines, monitoring adverse events and potential drug interactions. Well-designed clinical trials yield quality evidence necessary for the safe and effective use of CHM (Du et al., 2024).

After a CHM product has been approved into the market, post-market surveillance systems monitor its safety profile within the real-world setting. This will include data collection from adverse events that have been reported by health professionals and patients. It is this constant monitoring that detects long-term and rare adverse effects which might not have been captured in the clinical trials conducted as stipulated by the NMPA rules (Liang et al., 2023).

10.4.4 Adverse Event Reporting

Establishment of a strong adverse event reporting system is vital for identifying and responding to possible safety issues arising from the use of CHM products. China has established a national Adverse Drug Reactions (ADR) monitoring system that provides mechanisms both for healthcare professionals to report ADRs and for patients to report the same. Such a system would allow collation and analysis of ADR data associated with all drugs, including CHM products. It involves the analysis of collected ADR data to identify safety signals and trends (Cheng et al., 2024). This information is used for updating safety guidelines, revising product labeling, and taking regulatory action if required.

Challenges in ADR reporting involve underreporting, incomplete information, and a problem in ascertaining causality. Efforts are underway to improve ADR reporting through education and training of healthcare professionals, public awareness campaigns, and the use of technology to facilitate reporting (Wu et al., 2020).

10.4.5 Herb–Drug Interactions

Herbal products contain a complex of biologically active compounds. Therefore, their potential to elicit occurrence of HDIs is even more probable than for drug–drug interactions. Such interactions can occur pharmacokinetically and pharmacodynamically (Suroowan, Abdallah, & Mahomoodally, 2021).

HDIs are a huge concern in the safety of CHM products (Czigle et al., 2023) owing to the fact that many people administer herbal medicines together with conventional drugs. The HDIs may occur through various mechanisms, including pharmacokinetic interactions (affecting drug absorption, distribution, metabolism, and excretion) and pharmacodynamic interactions (the pharmacological effects of drugs). Research has identified a number of herbs and drugs commonly implicated in HDIs. For instance St. John's Wort (*Hypericum perforatum*) interacts with a variety of medications, including antidepressants, anticoagulants, and immunosuppressants (Zhou et al., 2021).

It has been said that assessment of HDI risks requires deep understanding of the pharmacokinetic and pharmacodynamic properties of both herbal medicine and convectional drugs. Health professionals should be informed about possible HDIs and should give the necessary guidance to the patients appropriately (Hu et al., 2021).

10.4.6 Vulnerable Populations

Pregnant women, children, and elderly individuals tend to be more prone to the adverse action of herbal medicines. Many herbs have not been adequately studied for safety during pregnancy and lactation. Caution is, therefore, advised; hence, healthcare professionals should be consulted before using CHM during these periods (Wang et al., 2014; Zhang et al., 2022). Children may have more different sensitivities to herbal medicines compared to adults; therefore, dosage adjustments and careful monitoring are necessary when using CHM in these age-sets (Huang et al., 2014).

Age-related physiological and pharmacological changes and altered drug metabolism in the aged can also impact response to herbal medicine. The state of polypharmacy or consumption of a variety of medications and herbal medicines simultaneously among aged individuals increases further possibilities of HDI conditions or herbal–drug interactions (Xin et al., 2020).

10.4.7 Promotion of the Safe Use of Herbal Medicine Products

Any of the several ways below can be adopted to ensure people do safely use herbal medicines:

Quality control and standardization: Strict quality control and standardization of herbal products should be implemented to reduce the risk of contamination and adulteration, and to ensuring consistency among the products (Wang et al., 2023; Edo et al., 2024).

Education and training: Safety and responsible use of CHM must be taught to all health professionals and the public, showing potential risks, interactions between herbs and drugs, and adequate dosage (Sile et al., 2023).

Pharmacovigilance: Strengthening the pharmacovigilance system to monitor and report adverse events associated with CHM use is critically important as a means of identifying and mitigating potential safety issues (Choudhury et al., 2023).

Research and innovation: Further research on safety and efficacy, particularly for complex formulations and multi-herb combinations, should be continued to expand the knowledge base and inform clinical practice (Balkrishna et al., 2024).

Ensuring safety is an ongoing and complex process regarding CHM. By integrating traditional knowledge with modern science, promoting research and innovation, and strengthening the pharmacovigilance systems, the CHM community has been in a position to continually improve the safety and effectiveness of this valuable healthcare tradition for the benefit of people worldwide (Balkrishna et al., 2024).

10.5 SAFETY OF CHM PRODUCTS: PROTECTING AUTHENTICITY, SUSTAINABILITY, AND HERITAGE

The concept of security in CHM products is multivariate and designed to protect these precious resources from a wide array of threats such as adulteration, counterfeiting, unsustainable harvesting, and intellectual property (IP) infringement. This section discusses the major dimensions of CHM security by highlighting regulatory measures, technological developments, and international collaborations that will help safeguard authenticity, sustainability, and heritage related to this ancient healing tradition.

10.5.1 Preventing Adulteration and Counterfeiting

Adulteration and counterfeiting pose significant threats to the safety and efficacy of CHM products, with potentially deleterious consequences for public health and a loss of confidence in this medicine system (Raj et al., 2025). China has taken some stern steps to overcome these:

According to the DAL (2019) and other relevant regulations, the manufacture and sale of adulterated or counterfeit herbal medicines are prohibited. The laws have strict penalties for violators, including fines, product recalls, and criminal prosecution.

Quality control in the CHM supply chain, from cultivation and harvesting to processing and manufacturing, is stringent. Authentication techniques, such as macroscopic and microscopic examination, chemical analysis, and DNA barcoding, are used to verify the identity and purity of herbal materials (Chen et al., 2022).

Traceability systems are being implemented to track the movement of herbal materials and products throughout the supply chain. This helps to identify the origin of products, detect potential adulteration or counterfeiting, and facilitate product recalls if necessary (Li et al., 2022).

Advanced technologies, including blockchain, are under investigation for improving supply chain security against counterfeiting. Blockchain can provide a secure and transparent record of transactions that would guarantee the authenticity and origin of herbal products (Yang et al., 2023).

10.5.2 Protecting Endangered Species

The utilization of endangered species in CHM has brought about concerns in the conservation of biodiversity and ethical sourcing of herbal materials. China has taken steps to address these concerns, including the following:

Legal protection: The Wildlife Protection Law of the PRC (2022 Version) and other regulations are formulated so as to protect or rescue rare and endangered species, preserve biodiversity and ecological balance, advance the construction of an ecological civilization, and promote a symbiotic coexistence with nature. They laws prohibit the illegal trade in and use of endangered species in CHM, and provide for strict penalties against violators, including fines and imprisonment (CITES, 2022).

Sustainable sourcing: Such regulations encourage sustainable sourcing, mainly toward encouraging the cultivation of medicinal plants in place of endangered species, research support in finding alternative substitutes, and methods of sustainable harvest practices in general. For example, this means being aligned with the International Union for Conservation of Nature (IUCN) rules of 2020 (Marcelino et al., 2023).

International collaborations: China works with international organizations and other countries in the fight against the illegal wildlife trade and protection of endangered species used in CHM (Jiao, Yeophantong, & Lee, 2021). This includes participation in international agreements such as the Convention on International Trade in Endangered Species of Wild Fauna and Flora, commonly known as CITES 1 (CITES, 2022).

Public awareness: Public awareness programs are carried out to educate consumers on the need for the protection of endangered species and the use of sustainable CHM products.

10.5.3 IP Protection

Traditional knowledge related to CHM products is an important asset that needs protection against misappropriation and exploitation. China has established ways of protecting traditional knowledge of CHM, including documentation and registration of traditional knowledge, community-based conservation, and benefit-sharing arrangements (WIPO, 2021).

Patents are granted for the protection of the rights of new inventions and innovations, including the development of new herbal formulations, extraction methods, and drug delivery systems in CHM (Xu & Wang 2023). It encourages research and development in the field by protecting IP rights.

Geographical Information Systems can also serve to protect both the reputation and quality of herbal products linked to geographical origins. By so doing, it aids the conservation of distinctive features and traditional knowledge corresponding to a specific region or origin (WTO, 2019). Trade secret protection can be applicable to confidential information pertaining to formulations or manufacturing processes relating to CHM formulations (Wunsch, 1980; Varadarajan, 2014).

10.5.4 Challenges and Future Directions

Despite concerted efforts toward the security of CHMs, the following challenges remain:

Enforcement: Effective enforcement of laws, rules, and regulations is critical in preventing adulteration, counterfeiting, and illegal trade in endangered species.

Supply chain complexity: Multiple actors, diverse sources of raw materials, and the general intricacy of the CHM supply chain all make traceability and quality control very difficult to master, let alone implementation.

Balancing tradition and modernity: How to balance protection of traditional knowledge with the need for scientific validation/modernization of practices is a critical challenge.

The following would be considered the future directions for CHM security:

Regulatory framework strengthening: Continuously updating and strengthening regulations in addressing emerging challenges, such as online sales and the use of new technologies in CHM.

Enhancing traceability and authentication: Developing and implementing more sophisticated traceability systems and authentication technologies to combat adulteration and counterfeiting (Fernando et al., 2024; Maritano et al., 2024).

Promotion of sustainable sourcing: Giving more strength to the promotion of sustainable sourcing of herbal materials, including support for the cultivation of medicinal plants and protection of endangered species (Marcelino et al., 2023).

International collaborations: Improve international collaborations on cross-border issues such as illegal wildlife trade and IP infringement.

China continues to strengthen and forte her CHM security by focusing on three key areas: *authenticity protection*, *sustainability*, and *heritage protection*. To safeguard this resource, China is focusing on building strong regulatory frameworks to enhance the practice and apply advanced technology in practice (Lu & Ahmad, 2023). Moreover, it requires further work at different levels to assure integrity and credibility nationally and internationally.

10.6 REGULATION OF HERBAL MEDICINE PRODUCTS FOR LOCAL AND EXPORT MARKETS: NAVIGATING THE COMPLEXITIES

Herbal medicine products are increasingly in demand both in the domestic and international markets, offering both opportunities and challenges in their regulation. Ensuring the safety, quality, and efficacy of herbal medicine products in various markets requires profound knowledge of the regulatory landscape, international standards, and market demands (Liang et al., 2023). The present section considers the problems of CHM products regulation for local and export markets by major considerations, challenges, and the strategies for this complex issue.

10.6.1 Domestic Regulation of Herbal Medicine Products in China

China has established a wide range of regulatory frameworks for herbal medicine products in its domestic market to ensure safety and quality for its citizens. Generally, the legal framework for CHM in China is provided by the DAL (2019) and TCM Law (2017). These laws outline requirements for drug manufacturing, quality control, labeling, registration, and sales. The NMPA is the primary regulatory body responsible for overseeing CHM in China. It sets quality standards, conducts inspections, and enforces regulations to ensure the safety and efficacy of herbal medicines (Liang et al., 2023).

GACP guidelines promote sustainable and quality-assured cultivation and collection of herbal raw materials. They cover aspects such as site selection, plant material, cultivation practices, harvesting, and postharvest handling (Chen et al., 2021a). The GMP standards for CHM manufacturing facilities ensure the quality and safety of herbal products during production. These standards cover various aspects, including facility design, hygiene, equipment, quality control, and personnel training (Liang et al., 2023).

Herbal medicine products are incorporated into the national healthcare system, with special TCM hospitals and clinics offering herbal treatments together with conventional medical care. Such institutions are supposed to be regulated in terms of practitioner qualifications, dispensing practices, and patient safety (Dubale et al., 2025).

10.6.2 Exporting Herbal Products: Meeting International Standards

The export of CHM products is even more difficult because firms must follow the regulations and standards of the importing countries. Each country has its own set of regulations regarding herbal medicines, including requirements for registration, labeling standards, and quality control parameters. Firms exporting CHM must do their homework in understanding the laws and regulations of target markets.

International organizations such as the WHO and ISO have, over the years, developed guidelines and standards that relate to herbal medicines. Indeed, meeting such standards would facilitate the acceptance and registration of CHM products in international markets (WHO, 2013; ISO, 2021).

Many herbal medicine products require heavy documentation to be exported, such as a certificate of analysis, free sale certificate, and GMP compliance certificate. Third-party certification by independent organizations may also be required by companies for international acceptance of their products markets (WHO, 2013). The labeling and packaging of exported CHM products should conform to the laws and regulations of the importing country. These could include, among other things, meeting specific labeling standards and providing information in the local language.

10.6.3 Challenges in Exporting Herbal Medicine Products

Some of the challenges arise in the exportation of CHM products:

Regulatory divergence: Diversity in the regulatory environment of different countries may pose complexities and raise compliance costs for exporters.

Language and cultural barriers: Language and cultural barriers to understanding CHM may be problematic in marketing and promotion of herbal products in the international market.

Scientific evidence: Some countries may require sound scientific evidence to back up the efficacy and safety of CHM products, which may be hard to provide for traditional medicines with long histories of use but limited clinical research.

Protection of traditional knowledge: Protecting traditional knowledge associated with CHM while navigating IP regulations in different countries can be complex.

10.6.4 Strategies for Successful Exportation

The companies willing to export herbal medicine products can successfully do so by following various strategies, including the following:

Market research: Doing thorough market research to understand the specific regulatory requirements, market demand, and consumer preferences of the target country.

Collaboration with local partners: Partnerships with local importers, agents, distributors, and/or research institutions may provide better insight and facilitate access to the market.

Investment in quality and standardization: Establishment of appropriate quality control measures and standardization procedures to ensure the quality of the products and meet international standards.

Creation of scientific evidence: Clinical trials and research are done to build scientific evidence that supports the efficacy and safety of the CHM products.

IP protection: Efforts toward the protection of IP rights associated with CHM, such as patents and trademarks.

Brand reputation: Build brand reputation by ensuring consistent quality, ethical sourcing, and transparency in communications.

10.6.5 Harmonization Efforts and International Cooperation

International harmonization of the regulations for herbal medicine products is important to facilitate global trade and ensure the safe and responsible use of CHM worldwide. The WHO Traditional Medicine Strategy (2014–2023) encourages member states for integrating traditional medicine including CHM into the healthcare system and also calls for international cooperation in regulation and research (WHO, 2013). An International Pharmacopoeia is under development by the WHO to which monographs on herbal medicines are being added, hence contributing to the harmonization of quality standards (WHO, 2019).

China is actively pursuing bilateral and multilateral agreements with other countries on the regulation, research, and trade of CHMs (Chen et al., 2021b). A very promising future is envisaged, both in the local and export markets concerning CHM, impelled by the increasing global demand, scientific developments, and supportive regulatory environments. Key trends include the following:

Growing consumer interest: The rise in consumer interest in natural health products and traditional medicine propels the demand for CHMs globally.

Scientific validation: More scientific research is being conducted to prove the efficacy and safety of CHM, thus increasing its credibility and acceptability.

Integration with modern medicine: CHM is increasingly combined with conventional medicines to provide complementary approaches to health.

Technological advancements: New technologies, including AI and big data, are being applied to improve quality control, safety monitoring, and drug discovery in CHM.

The regulation of herbal medicine products for both local and export markets is a daunting task but an absolute necessity. China has already done a lot in building up a very good regulatory framework for its domestic market and is actively involved in international harmonization (Hock, Wah, & Kian, 2020). Meeting the challenges and grasping the opportunities, the industry of CHM can only continue to serve global health and well-being.

10.7 FUTURE OF HERBAL MEDICINES: ADDRESSING CHALLENGES AND AVAILING OPPORTUNITIES

Traditional medicine is at the crossroads between its millennial tradition and modernization, globalization, and scientific advance. The present section looks to the prospect for herbal medicine products in China through an evaluation of challenges and embracing opportunities that lie ahead for this ongoing effort of modernization with its essence and international harmonization for global reach and acceptance. The prospect for herbal medicinal products is bright, with continuous modernization of regulations, encouragement of research, and integration of CHM into the national healthcare system. Key areas of focus include the following:

Personalized CHM: To apply advanced genomics and diagnostics to deliver personalized treatment by CHM to every individual, keeping in view their needs and characteristics.

Integration with modern medicine: Permit further integration of CHM with conventional medicine in the way of holistic and patient-oriented care.

Global leadership in CHM: China will lead in global development and promotion while contributing to international research, standardization, and regulation pertaining to CHM.

Sustainable development: To ensure that the CHM industry is developed in a sustainable manner, necessary measures shall be taken to promote responsible sourcing, cultivation, and manufacturing practices.

After a long journey, CHM has arrived at a fork in the road. The complex tradition of regulation and modernization continues to present various problems. However, China is embracing policy measures aimed at enhancing safety, efficacy, and national and international acceptance of CHM. On new scientific developments, improved international cooperation, and sustainable practices, the position of CHM is better than ever to expand its role in healthcare both in China and globally.

11 Industrial Effluent and Chemical Insecurity in the Indian Subcontinent

Yasodha K. Janapati

In developing regions, industrialization and urbanization strain land and water resources, while using industrial waste in agriculture disrupts soil nutrients, leading to serious hazards. The Indian subcontinent, for example, faces serious problems with chemical insecurity and industrial effluent discharge, mostly as a result of industrial expansion and a lack of proper waste treatment infrastructure (Jahan & Singh, 2023). Water bodies are contaminated by untreated industrial effluents that include heavy and dangerous chemicals, endangering aquatic life as well as human health and the environment. The problem is made worse by inadequate treatment facilities and lax enforcement of laws (Mukherjee et al., 2015).

Industrial waste exists in gaseous, liquid, and solid states. However, industrial effluent is broadly defined as liquid waste (treated or untreated) discharged from industrial operations into the environment, characterized by a complex mixture of organic and inorganic constituents that pose significant risks to the ecosystem and human health.

Industrial pollution from mining, chemical production, and waste management releases various contaminants like solvents, fertilizers, and metals. Urban development worsens this, leading to environmental degradation and health issues. The presence of heavy metals in sewage presents a lower risk compared to their release into ecosystems; however, the discharge of untreated wastewater introduces significant contamination issues for drinking water and agricultural produce, thereby augmenting health hazards (Tariq & Mushtaq, 2023; Singh, Kumar, & Chandra, 2022). These pollutants have long-lasting effects triggering a chain reaction that spreads contamination and disease.

In the Indian subcontinent, this effluent typically exhibits wide-ranging physicochemical parameters (pH, temperature, suspended/dissolved solids, oxygen demand), a spectrum of heavy metals (both essential and non-essential), nutrients (nitrogen, phosphorus), oils, surfactants, and dyes, as well as emerging contaminants (ECs) such as pharmaceuticals and personal care products (PCPs) (Agarwal & Kumar, 2024; Sharma et al., 2022c).

Regulatory standards in India set permissible ranges (e.g., pH 5.5–9.0) to mitigate discharge impacts, yet many industrial estates still release effluents exceeding these limits (Schellenberg et al., 2020). Sector-specific profiles—textile mills, smelters, petroleum refineries, food processors, and pharmaceutical manufacturers—contribute distinct pollutant signatures, intensifying chemical insecurity across surface and groundwater resources. ECs such as endocrine disruptors, antibiotics, and microplastics further complicate treatment processes and risk assessments, underscoring the urgent need for advanced monitoring and control strategies (Khan et al., 2023; Lellis et al., 2019; Li et al., 2024; Murtaza et al., 2010).

11.1 SOURCES OF INDUSTRIAL POLLUTION

A wide range of industries contribute to industrial pollution in the Indian subcontinent, each of which releases a different profile of contaminants into air, water, and soil. The textile and apparel sector, especially in Bangladesh and India, is a major contributor, producing large amounts of sludge that contains microplastics and wastewater that is heavy with dyes (Anindita Saha et al., 2023). Chromium, sulfides, and organic load are released by tanning and leather processing facilities in Kanpur, Jakarta, and Dhaka, overwhelming local treatment systems (Fadilah et al., 2023). High levels of chemical and biochemical oxygen needs, nutrients, and suspended particles are released by pulp and paper mills, sugar manufacturers, and food processors (Anindita Saha et al., 2023).

Concerns regarding antibiotic resistance and long-term toxicity are raised by the solvents, active pharmaceutical components, and heavy metals released by chemical, petrochemical, and pharmaceutical plants (Godman et al., 2022). Lead, nickel, arsenic, and other hazardous substances are released into soils and waterways by mining, metallurgical, and metal-finishing processes (Rampley et al., 2020). Particulate matter (PM), hydrocarbon-rich effluents, and sulfur and nitrogen oxides are released by fossil fuel-based thermal power plants and petroleum refineries. Additional sources of particulates, fluoride, ammonia, and per- and polyfluoroalkyl substances include shipbreaking yards, cement kilns, and fertilizer facilities. These sources present serious risks to human health and the environment (Uddin & Jeong, 2021).

11.1.1 Textile and Garment Manufacturing

India: Although all integrated textile units, such as facilities that produce cotton, wool, polyester, and clothing, are subject to the Central Pollution Control Board (CPCB) effluent rules, many mills, nevertheless, discharge high levels of BOD/COD (biochemical oxygen demand)/(chemical oxygen demand) and color units into rivers (Khan & Malik, 2014; Azanaw et al., 2022).

Bangladesh: It is the second-largest exporter of clothing in the world; the industry produces an estimated 577,000 tons of garbage a year, much of which is unsafely released by unofficial workshops (Mia et al., 2019).

Pakistan: The country's textile industry continues to employ the most people, but it is also the biggest cause of water pollution due to dyes, surfactants, and heavy metals (Chatha, Asgher, & Iqbal, 2017).

11.1.2 Leather Tanning and Processing

Kanpur, India: One of the world's largest urban pollution burdens is caused by the poisonous effluent released by hundreds of leather tanneries that burn coal and employ chromium salts (Balakrishnan et al., 2019).

Dhaka, Bangladesh: Untreated wastewater frequently avoids municipal treatment, and dyeing and tanning industries are significant sources of sulfide, phenolic, and chromium slats (Anindita Saha et al., 2023).

Pakistan: Sulfides, chromium, and nitrogenous pollutants released by the leather industry, which is centered in the Karachi and Lahore areas, contaminate canals and groundwater (Syed et al., 2010).

11.1.3 Paper, Pulp, and Agro-Based Sectors

India: Mills in Punjab, Odisha, and West Bengal release black liquor that is rich in lignin, which raises the COD and color of rivers (Ruba et al., 2021). Located in Uttar Pradesh, Punjab, and Maharashtra of India and in Pakistan, sugar mills increase BOD and nutrient loads by releasing biodegradable organics, soil nutrients, and vinasse (Lahori et al., 2024). High-strength effluent containing proteins, lipids, and carbohydrates is produced by fruit, vegetable, and dairy farms and is frequently released with little pretreatment (Uddin & Jeong, 2021).

Pakistan: In Pakistan, integrated mills generate approximately 205.6 m^3 of wastewater per ton of product, with a significant portion of this wastewater being discharged into drains such as Hudiara. These mills utilize wheat straw, bagasse, and refuse paper. The environmental standards are exceeded by the high levels of suspended particulate (~431.8 kg/ton) and BOD of ~726 mg/L in this effluent (Akhtar, 2013). Small-scale agro-based mills produce approximately 150–200 m^3 of effluent per ton of paper. The "black liquor" they produce is a byproduct that contains inorganic compounds, chlorophenols, and fiber residues. It exhibits high COD, BOD, suspended solids, and intense coloration (Zulfiqar, 2025).

11.1.4 Chemical, Petrochemical, and Pharmaceutical Industries

India: Solvents, phenols, chlorides, and sulfates are produced in the effluents of red-category industries, which include organic chemicals, dyes, caustic soda, and petroleum refining in India (Kato & Kansha, 2024). Antimicrobial resistance hotspots surrounding significant active pharmaceutical ingredients (APIs) clusters in Maharashtra and Gujarat are exacerbated by APIs and antibiotic residues in effluent (Anjum et al., 2024).

Pakistan: Surface waters are contaminated by nitrates, phosphates, and persistent organic pollutants released by chemical factories in Punjab and Sindh (Azizullah et al., 2011).

11.1.5 Metal Finishing, Mining, and Metallurgy

India: Soils and groundwater close to industrial estates are contaminated by arsenic, lead, nickel, and fluoride released by nonferrous smelters and metallurgical facilities (Angon et al., 2024).

Pakistan: Wastewater from industrial estates near Lahore is contaminated with heavy metals, and risk quotients show ecological damage (Iqbal et al., 2024).

11.1.6 Petroleum Refineries and Energy Production

India: Ash slurry, hot water, and flue gas desulfurization effluent, including sulfates and heavy metals, are released by coal-fired power plants (Meawad, Bojinova, & Pelovski, 2010).

Hydrocarbons, phenols, oily sludge, and nitrates are among the contaminants in wastewaters from petroleum refining that are controlled under the Central Pollution Control Boards Schedule I, yet often surpass discharge limitations (Selvam et al., 2022).

Pakistan: Fly ash contains toxic elements, including lead (Pb), arsenic (As), cadmium (Cd), and chromium (Cr), which are deposited in the air, water, and sediment. In regions near coal-fired power facilities in Pakistan, over 90% of cabbage samples exceeded Pb limits, and more than 30% exceeded the limitations that affect human health and ecosystems (Luqman et al., 2024).

Smog, respiratory ailments, and environmental degradation are all exacerbated by the emissions of sulfur dioxide (SO_2), nitrogen oxides (NO_x), volatile organic compounds (VOCs), and PM from refineries. Refining produces effluent that is contaminated with oil, heavy metals (such as lead, zinc, and copper), polycyclic aromatic hydrocarbons (PAHs), and VOCs. These can pose significant health risks and pollute aquatic ecosystems without proper treatment, such as oil–water separators or biological systems (Khan et al., 2024).

11.1.7 Cement Plants

India: Cement manufacturers are substantial source of air pollution. Particulates and fluorides are released by kilns. Wastewater from cooling and dust suppression has a high alkalinity and suspended solids content, causing respiratory, skin, and eye health issues. Respiratory maladies, including asthma, chronic bronchitis, cardiovascular issues, and in some cases, lung cancer, have been associated with exposure to cement-related PM, particularly in regions like Kashmir. Deforestation has been prevalent in ecologically sensitive regions, such as the Jaintia Hills of Meghalaya, as a result of the construction of cement plants. Local communities have reported that mining has caused the water to turn red, and airborne ash is causing children to become unwell (Lamare & Singh, 2016).

One of the industrial facilities in India that uses the most water is the fertilizer plant. They use 4.55–12.73 m^3 of water per ton of urea on average, which is much more than what is recommended globally. The depletion of resources is exacerbated by certain public sector institutions that are dependent on groundwater. In comparison to the Bureau of Indian Standards (BIS) limit of 0.5 ppm, studies reveal startlingly high quantities of ammoniacal nitrogen in groundwater close to fertilizer factories, ranging from 0.51 to 93.5 ppm. This suggests severe nitrogen pollution. Ammonia, total Kjeldahl nitrogen, nitrates, cyanide, and other contaminants such as diammonium phosphate are present in high concentrations in wastewater from the production of urea. Near 14 plants, over half of the samples did not fulfill the MOEF&CC discharge standards (Madjar, Vasile Scăeţeanu, & Sandu, 2024).

Pakistan: High concentrations of $PM_{2.5}$ and PM_{10} are emitted by cement plants in Hattar, Haripur, Pakistan, which exceed the World Health Organization (WHO) guidelines in certain facilities. PM_{10} was measured at 1,552 μg/m^3 and $PM_{2.5}$ at 7,867.5 μg/m^3 in specific areas of the Haripur plant. These levels have the potential to impact both laborers and the surrounding community significantly (Bărbulescu & Hosen, 2025).

CO_2, NO_x, SO_2, and fine contaminants are all released during the manufacturing of cement. These pollutants are associated with premature mortality, cardiovascular diseases, respiratory maladies (e.g., asthma, bronchitis), and reduced cognitive performance. The systemic effects and profound lung penetration of fine particulate matter ($PM_{2.5}$) render them particularly harmful (Bărbulescu & Hosen, 2025).

11.1.8 Marine Industries and Shipbreaking

India: The world's largest ship breaking yard, the Alang Ship Breaking Yard, dismantles a substantial portion of retired global vessels by beaching them during high tides and painstakingly dismantling them as the tide recedes.

Hazardous substances, such as asbestos, polychlorinated biphenyls (PCBs), hydrocarbons, antifouling residues, and oil, are released into coastal environments during the dismantling process. These substances include heavy metals (lead, cadmium, chromium, mercury), and oil. This contamination has a detrimental impact on the quality of coastal water, sediments, and adjacent ecosystems (Singh & Frank, 2025). Additionally, pollutants are disseminated throughout the region during monsoons, which negatively impact mangroves, agriculture, marine biodiversity, and local fisheries. India has signed the Convention, which promotes the dismantling of dry-docks and the implementation of pollution containment measures such as effluent facilities and waste storage. Metal contamination at Alang has been mitigated by certain enhancements, including improved containment and cemented cutting areas.

More than 100 yards along Bangladesh's coast experience release of PCBs, asbestos, heavy metals, and oil into the Bay of Bengal (Rabbi & Rahman, 2017).

11.1.8.1 Composition of Industrial Effluent

Industrial effluent comprises liquid waste discharged from industrial activities, characterized by a complex mixture of physicochemical parameters, inorganic and organic constituents, and ECs. Key physicochemical parameters include pH, temperature, total suspended solids (TSS), total dissolved solids, BOD, and COD (Angon et al., 2024). Inorganic constituents such as heavy metals (e.g., cadmium, lead, chromium, nickel) and major anions (chloride, sulfate, phosphate) are routinely detected at concentrations exceeding environmental safety thresholds.

Organic pollutants such as oils, greases, phenolic compounds, and synthetic dyes contribute to toxicity and resist conventional biodegradation (Khan et al., 2023). ECs, such as pharmaceuticals, PCPs, endocrine-disrupting chemicals (EDCs), and microplastics, are increasingly found in treated effluents, posing long-term ecological and human-health risks (Li et al., 2024). Table 11.1 presents a summary of the main physicochemical constituents typically found in untreated industrial effluents, along with their usual concentration ranges.

TABLE 11.1
Typical Physicochemical and Heavy Metal Composition of Raw Industrial Effluents

Parameter	Typical Range	Unit	References
pH	5.5–9.0	–	Singh et al. (2024a)
Temperature	≤40	°C	Singh et al. (2024a)
Total suspended solids (TSS)	63.7–243.5	mg/L	Mary Sadler et al. (2022)
Total dissolved solids (TDS)	23.3–1,000; (influent) 472–555	mg/L	Nirgude, Shukla, & Venkatachalam (2013)
Biochemical oxygen demand (BOD_5)	456.6–1,931.5	mg/L	Nirgude, Shukla, & Venkatachalam (2013)
Chemical oxygen demand (COD)	585.4–1,967.5	mg/L	Mary Sadler et al. (2022)
Oil and grease	7.6–21.6	mg/L	Nirgude, Shukla, & Venkatachalam (2013)
Total phenolics	Up to 1.2	mg/L	Singh et al. (2024a)
Chloride (Cl^-)	Up to 600	mg/L	Singh et al. (2024a)
Sulfate (SO_4^{2-})	Up to 1,000	mg/L	Singh et al. (2024a)
Heavy metals (Zn, Cu, Ni, Pb, Cr, Hg, Cd)	Zn ≤ 2; Cu ≤ 0.2; Ni ≤ 3; Pb ≤ 0.1; Cr^{6+} ≤ 0.1; Cr_{tot} ≤ 1; Hg ≤ 0.01; Cd ≤ 0.1	mg/L	Singh et al. (2024a)

11.1.8.2 Regulatory Frameworks Governing Industrial Waste Disposal

To control industrial waste disposal, enforce pollution restrictions, and guarantee environmentally sound management, the Indian subcontinent uses a multilayered legal framework that includes national statutes, subsidiary regulations, and institutional mandates (Gupta et al., 2023). The Central and State Pollution Control Boards in India implement the Water (Prevention & Control of Pollution) Act, 1974, the Air Act, 1981, and sector-specific regulations (such as the Hazardous Waste Management, Handling & Transboundary Movement Rules, 1989, amended 2008) in addition to the Environment (Protection) Act, 1986, which provides overall authority for pollution control (Grandhi, Dagwar, & Dutta, 2024).

The Pakistan Environmental Protection Act of 1997, which created the Pakistan Environmental Protection Agency and gives it the authority to define and enforce National Environmental Quality Standards (NEQS) for industrial emissions and effluents, serves as the main foundation for Pakistan (Luken, 2009).

The Department of Environment oversees industrial discharges in Bangladesh under the Environment Conservation Act of 1995 and the Environment Conservation Rules of 1997, which establish emission and effluent standards (Wahidul Alam, 2023).

While the South Asian Association for Regional Cooperation (SAARC) Convention on Cooperation for the Environment (2010) and its 2019–2030 roadmap promote cooperative waste management and resource-circulation strategies, the Basel Convention (1989) binds subcontinental nations to control transboundary movements of hazardous wastes (Gamaralalage et al., 2019).

11.2 IMPACT OF INDUSTRIAL EFFLUENT

Industrial effluent discharges pose serious dangers to ecosystems and human health, because they contaminate agricultural soils and food chains, damage terrestrial vegetation and forest cover, introduce huge loads of organic and inorganic pollutants into important rivers, and seep into aquifers (Singh et al., 2025; Afrad et al., 2020).

11.2.1 Contamination of Major Rivers

More than 150 highly polluting enterprises discharge their effluents into the Ganges River, frequently violating CPCB discharge regulations and resulting in high levels of heavy metals (Cd, Pb, and Cr) and BOD that often exceed 40 mg/L (Singh et al., 2025). Seasonal increases in phosphates associated with textile and chemical effluents and COD (>150 mg/L) are observed in Delhi's Yamuna, resulting in continuous eutrophication and foam formation (Sharma et al., 2024b). Untreated industrial and municipal wastewater in Pakistan pollutes the Indus, contributing nitrates, phosphates, and pesticides at up to 10 mg/L and increasing downstream public health hazards (Pappas, 2011). Despite its massive flow, the Brahmaputra receives more than 179 million liters of untreated wastewater per day, increasing turbidity and pollutant loads that endanger riparian biodiversity (Barbulescu et al., 2021).

11.2.2 Groundwater Contamination

In areas like Kanpur and the Bengal basin, industrial effluents from industries like tanning, textiles, and chemical production transport heavy metals like arsenic, chromium, and lead into shallow aquifers, frequently surpassing WHO drinking water requirements (e.g., arsenic > 10 μg/L) (Shankar, Shanker, & Shikha, 2014). Hexavalent chromium (Cr VI) concentrations in groundwater wells have been found to reach 0.05 mg/L in studies conducted in the Kanpur industrial zone, exceeding the US Environmental Protection Agency (USEPA) limit contamination level of 0.01 mg/L (Sharma et al., 2012).

Over 68,000 km^2 of the Bengal delta region, which includes West Bengal and Bangladesh, are heavily contaminated with arsenic, with aquifer concentrations ranging from 50 to 300 μg/L (Shaji et al., 2021). In Punjab and Gujarat, it has been reported that organic solvents, phenols, and nitrates from fertilizer and petrochemical plants have seeped into aquifers, causing increased nitrate levels (>50 mg/L) linked to the risk of methemoglobinemia (Taloor et al., 2024). According to a national evaluation, there is heavy metal contamination in groundwater in 718 districts of India, with numerous industrial belts having levels of lead, cadmium, and chromium over allowable limits. Long-term contamination has caused heavy metals to bioaccumulate in nearby crops that are irrigated with contaminated groundwater, endangering rural communities' health and neurological development (Angon et al., 2024).

11.2.3 Aquifer Depletion

Between August 2002 and October 2008, India's groundwater reserves decreased at an average annual rate of 4 cm, or around 109 km^3 of water, or double the volume of the Bhakra Dam reservoir (Rodell, Velicogna, & Famiglietti, 2009).

More than 17% of India's evaluated groundwater blocks are categorized as overexploited according to the Central Groundwater Board, with water-table decreases surpassing 1 m/year in regions like Punjab and Haryana. In the Indo-Gangetic plains, increased irrigation and diminishing recharge have caused accelerated groundwater storage losses, according to satellite-based Gravity Recovery and Climate Experiment (GRACE) data from 2002 to 2017 (Seka et al., 2022).

Because warmer temperatures increase evapotranspiration and irrigation demand, resulting in seasonal shortages and jeopardized water security, climate change has made depletion rates even worse (Bhattarai et al., 2023).

In metropolitan areas like Ahmedabad and portions of Delhi, land subsidence brought on by aquifer overdraft has damaged infrastructure and made areas more vulnerable to flooding. Smallholder farmers are frequently forced to abandon wells or rely on costly surface-water alternatives as a result of groundwater depletion, which demands deeper drilling and increases extraction costs and energy consumption (Gautam, Chandra, & Henry, 2024).

11.2.4 Effects on Vegetation and Forest Cover

Heavy metals that are hazardous to plants build up in industrial wastewater, which severely inhibit seed germination, lowers plant growth parameters, and reduces agricultural yields (Ali et al., 2015). It has been demonstrated that the high concentrations of mercury (Hg) and cadmium (Cd) in the wastewater released by the JK Paper industry in Rayagada change the physicochemical characteristics of river water downstream, resulting in a significant reduction in the biomass and species richness of riparian flora within one kilometer of the discharge point (Tripathy, Dixit, & Panigrahi, 2022). Chlorophyll content and photosynthetic ability were significantly reduced. The chlorophyll content in lettuce (*Lactuca sativa*) and onion (*Allium cepa* L.) leaves was significantly reduced after exposure to industrial effluents with elevated zinc contents (78 mg/L), suggesting compromised plant physiological activities (Kravtsova et al., 2024).

An estimated 0.5% of tropical forest cover is lost each year as a result of extensive forest clearing for industrial development, such as building factories and mining operations. These have also caused habitat fragmentation and the local extirpation of sensitive plant species in impacted areas (Giljum et al., 2022). By interfering with nutrient cycling and microbial communities necessary for forest regeneration, heavy metals and persistent organic pollutants from industrial discharges build up in forest soils, preventing seedling establishment and decreasing species diversity (Roshni et al , 2023).

11.2.5 Effects on Agriculture and Food Chain Contamination

In India, irrigation with industrial wastewater has significantly contaminated food crops and agricultural soils, introducing heavy metals into the food chain and creating major health hazards. Long-term irrigation with industrial wastewater has notably caused accumulation of heavy metals, including cadmium (Cd), lead (Pb), and nickel (Ni), in soils. These metals are then absorbed by crops, contaminating edible plant portions (Rashid et al., 2023).

Health Risks from Contaminated Produce: Eating vegetables and grains that are grown in fields that are irrigated by wastewater has been associated with higher levels of heavy metals in human diets, such as rice. The Food and Agriculture Organization (FAO) and the WHO define the levels of heavy metals allowed in human diets, and higher levels than these raise the risk of chronic health problems (Afrad et al., 2020). In addition to directly harming human health, heavy metals in crops cause bioaccumulation in the food chain, affecting animals and other living things that consume these tainted plants. The results highlight the critical necessity for efficient wastewater treatment and management techniques to stop more contamination of agricultural areas and guarantee food safety (Deshmukh, Amolic, & Mutkule, 2025).

11.2.6 Impact of Industrial Effluent on Health

Numerous harmful health effects result from the complex mixture of toxicants borne by industrial effluents (heavy metals and organic compounds) from metallurgical, battery-making, and tanning facilities. Chronic contact increases the risk of cancer (liver, lung, and breast), neurodegenerative diseases, cardiovascular disease, and reproductive developmental abnormalities and can result in severe gastrointestinal distress, renal failure, and encephalopathy (Oladimeji et al., 2024). Acute exposures show up as neurological symptoms, respiratory distress, and gastrointestinal irritation. Persistent pollutants' bioaccumulation and bio-magnification increase long-term toxicity, especially for susceptible populations like children and communities that depend on tainted water sources, and they also fuel periodic outbreaks of waterborne diseases (Ali et al., 2025).

When released into rivers untreated, industrial solvents (such as benzene and trichloroethylene) cause bone-marrow suppression, liver damage, mucosal irritation, and disorientation. People living close to industrial areas have higher rates of liver, lung, gallbladder, and breast cancers when they are exposed to carcinogenic effluent elements, including heavy metals, PAHs, and PCBs regularly (du Plessis et al., 2023). Dose–response connections are highlighted by longitudinal cohort studies, where cancer is correlated with cumulative pollution load (Baek, Park, & Kwak, 2022).

In downstream populations, bacterial and protozoan pathogens (such as *Vibrio cholerae*, *Giardia lamblia*, and Cryptosporidium) proliferate due to inadequate treatment of industrial effluent, causing outbreaks of typhoid, hepatitis A, diarrhea, and dysentery (Ali et al., 2025). Children, expectant mothers, and those with weakened immune systems are disproportionately affected by effluent-related diseases because of their increased water intake, developing organ systems, and impaired detoxifying processes. Environmental justice concerns and the need for focused public health measures are highlighted by epidemiological surveys that show increased illness prevalence and enhanced biomarkers of exposure (such as blood lead and urinary arsenic) in industrial belt locations (Saravanakumar et al., 2022).

11.3 SUSTAINABLE SOLUTIONS AND FUTURE PERSPECTIVES

An integrated strategy encompassing source reduction, resource recovery, sophisticated treatment, strong governance, and regional cooperation is necessary for industrial sustainability on the Indian subcontinent. There is implementation of advanced biological and physicochemical treatment to achieve zero-liquid discharge (ZLD); the use of digital tools, artificial intelligence/Internet of Things (AI/IoT) for real-time process optimization; and strengthening of policy frameworks and public–private partnerships to enforce compliance and promote innovation. Besides, the integration of green chemistry and cleaner-production principles at the design stage, and the scaling of circular economy and industrial symbiosis models to valorize wastes are all important strategies (Castegnaro & Sansone, 2012).

By using safer reagents, atom-economy processes, and waste-prevention design, green chemistry reduces harmful inputs and byproducts, which benefits industry and the environment (Unnikrishnan & Hegde, 2007).

Cleaner production combines material substitution, energy efficiency, and process optimization to lower effluent loads; case studies in pulp and paper mills in India (like the Raval mill) show that staff training and cleaner-production audits reduce effluent COD and TSS by 40%–60%. Furthermore, several uses of green chemistry, including substituting bio-based solvents for chlorinated solvents in pharmaceuticals, have reduced the production of hazardous waste in pilot plants by as much as 70% (Bennett, 2024).

Frameworks for the circular economy encourage closed-loop resource flows, converting waste streams into raw materials for additional operations. According to reviews, India can recover valuable resources (metals, polymers, and organics) from agricultural wastes and industrial trash, generating new sources of income while lessening its impact on the environment (Imran et al., 2025). Pilot projects in Gujarat and Maharashtra have demonstrated up to 20% energy savings and 15% water usage reductions across clusters using Kalundborg-style industrial symbiosis, in which waste heat or byproducts from one facility are converted into inputs for another (Mallawaarachchi et al., 2023).

Phytoremediation using *Typha* spp. (*Narrowleaf cattail*) has sequestered more than 50% of the chromium from tanning wastewater, whereas constructed wetlands and microbial bioaugmentation have eliminated 70%–85% of colors and phenolics in textile effluents in South Asian experimental studies. These low-cost, low-energy systems are particularly well suited for small- and medium-sized businesses with little funding (Ugrina & Milojković, 2024).

Under ideal circumstances, advanced oxidation processes (AOPs) (ozonation, Fenton's reagent, and UV/H_2O_2) generate hydroxyl radicals that break down resistant organics (dyes, medications, and EDCs) with >90% elimination; combining AOPs with biological post-treatment significantly lowers costs and sludge production (Ugrina & Milojković, 2024). Mylan has successfully developed ZLD programs in India's API clusters, integrating thermal brine concentrators, reverse osmosis, and ultrafiltration to achieve nearly zero effluent discharge and over 98% water reuse (Liao et al., 2025).

Real-time monitoring of effluent parameters, predictive maintenance of treatment units, and automated chemical dosage management are made possible by AI and IoT systems, which can increase treatment efficiency by up to 25% while using less energy (Chen et al., 2025).

11.4 CONCLUSION

Industrial effluent discharge in the Indian subcontinent presents a serious and growing threat to environmental and public health, largely driven by rapid industrialization, inadequate treatment infrastructure, and weak regulatory enforcement. Key industries—including textiles, leather tanning, pharmaceuticals, and petrochemicals—release complex mixtures of heavy metals, organic pollutants, and ECs that contaminate rivers, groundwater, soils, and the food chain. These discharges have led to severe degradation of major water bodies, groundwater depletion, reduced agricultural productivity, forest ecosystem damage, and heightened human health risks, including cancer and developmental disorders. Addressing this crisis demands urgent investment in advanced waste treatment technologies, stricter enforcement of environmental regulations, cross-border cooperation, and sustainable industrial practices to safeguard water resources, biodiversity, and public health for future generations.

12 Management and Regulation of Chemical Laboratory Practice in India

Yasodha K. Janapati

12.1 INTRODUCTION

Laboratories play a crucial role in the realm of scientific research, enabling a wide range of activities. The importance of health and safety measures in chemical laboratories is paramount. Countries must implement robust policies to maximize advantages, minimize potential hazards, and uphold quality standards. In the Indian subcontinent, chemical laboratories are crucial to industry, research, and education, as they foster creativity and quality assurance across various fields. These laboratories must run safely and effectively to ensure human wellness, environmental preservation, and scientific accuracy.

The creation of vaccines, medicine formulation, and quality testing all rely heavily on laboratories (Sharma et al., 2024a). Indian laboratories such as the Indian Council of Medical Research – National Institute of Virology (ICMR-NIV) contributed to antiviral research and diagnostic kits during the COVID-19 epidemic, highlighting their significance (Kumar et al., 2021a). From petrochemicals to agrochemicals, laboratories guarantee product safety, quality, and adherence to ISO 17025 and other standards. The $220 billion Indian chemical industry in 2023 depends on laboratories for research and development as well as process optimization (FICCI, 2023). Laboratories conduct pollutant analyses, track the quality of the air and water, and create remediation plans. Standards are enforced by Central Pollution Control Board (CPCB)-mandated laboratories under the Water (Prevention & Control of Pollution) Act, 1974 (CPCB, 2025b). By examining poisonous materials, drugs, and explosives, forensic laboratories assist law enforcement organizations in criminal investigations (Basu, 2021).

Although the modern industrial laboratory appears more orderly and refined than earlier iterations, many of the foundational safety challenges remain unresolved.

12.1.1 Chemical Laboratory Practices in India

Chemical facilities present significant hazards, thereby requiring the implementation of stringent protocols. An analysis of the safety culture within these environments could enhance training programs in laboratories located in Sri Lanka's Western Province (Samaranayake, Nishadya, & Jayasundara, 2021).

India is establishing robust benchmarks for chemical laboratories by evaluating global regulations and safety protocols. It is advisable for the State Department to formulate a customized regulatory framework that aligns with the specific needs of the nation. The categorization of reagents and chemicals is based on their safety conditions and required equipment. This method complies with national standards established by the Bureau of Indian Standards (BIS) under IS1641 and meets international criteria set by the International Organization for Standardization (ISO) (Korde et al., 2024; Rossete & Ribeiro, 2021). The management of chemical and laboratory waste in India is regulated by the Ministry of Environment, Forest and Climate Change (MOEFCC) and the Basel Convention's Article 5.

Numerous regulations dictate waste disposal protocols, highlighting the importance of training initiatives for effective laboratory operations. Undergraduates must receive training that aligns with these standards, combined with a solid scientific foundation (Freese et al., 2024). Training is essential for safe lab operations and reducing accident risks. Untrained personnel in chemical labs pose significant hazards. Thorough risk assessments are critical for prevention, following LRM principles. National and international guidelines provide frameworks for compliant practices (Mourya et al., 2017).

The Indian Chemical Society (ICS) is dedicated to the progression of the field of Chemistry, guided by ethical standards that prioritize the meticulous documentation of experimental findings. This practice is essential for fostering trust and integrity within the discipline. In this context, the OECD Principles of Good Laboratory Practices are significant benchmarks that inform the establishment of regulatory standards for Good Laboratory Practices (GLP) across diverse settings.

The management of chemical laboratory operations in India requires the establishment of national standards that align with international benchmarks. The National Accreditation Board for Testing and Calibration Laboratories (NABL) is responsible for defining these standards, which emphasize safety, quality, and ethical practices. Adhering to IS/ISO/IEC 17025 is crucial for obtaining accurate results, while the implementation of operational criteria enhances the reliability of laboratory practices (Balaram & Satyanarayanan, 2022). Accreditation by NABL is a validation of competence, which is vital across all industry sectors.

The rapid advancement of chemical sciences in India causes a talent drain from academia, as researchers prioritize career growth over lab safety. This trend prompts society to invest more in health and environmental regulations, which can hinder research because they often lag resource development. Enhanced training programs could minimize the need for intricate laws (Kumar et al., 2024; Fatemi, Dehdashti, & Jannati, 2022).

12.1.2 Important Chemical Laboratories in Research in India

Organizations such as the Council of Scientific & Industrial Research (CSIR), Indian Institute of Chemical Technology (IICT) in Hyderabad (founded in 1944), and the CSIR National Chemical Laboratory (NCL) in Pune (founded in 1950) spearhead multidisciplinary chemical science research, ranging from organic synthesis and catalysis to biochemistry and materials engineering, which propels innovation in a variety of industries. The importance of chemistry laboratories in converting natural product research into therapeutic medicines is demonstrated by the Central Drug Research Institute (CDRI) in Lucknow and the Indian Institute of Chemical Technology (IICT) in Hyderabad, which historically led the way in drug design and development derived from traditional treatments (such as ormeloxifene and artemether); they also focus on biomedical research and pharmacological work. National Institute of Pharmaceutical Education & Research (NIPER) focuses on pharmaceutical sciences (Differding, 2017; Singh, Chawla, & Kapoor, 1985).

Defense Research & Development Organization (DRDO) Laboratories has multiple laboratories in 50 different locations which focus on defense research such as aeronautics, armaments, electronics, combat vehicles, missiles, and naval systems. Located in Trombay, Mumbai, the Bhabha Atomic Research Centre (BARC) is the leading multidisciplinary nuclear research facility in India. In addition to its work in biotechnology, health, and agriculture, BARC supports India's nuclear program by doing research in nuclear physics, physical sciences, chemical engineering, and materials sciences, The Indian Institute of Science (IISc) is situated in Bengaluru, India engages in research in a wide range of scientific fields such as materials science, computer science, brain research, nanoscience and engineering, and hypersonics. Table 12.1 shows some of the important chemical laboratories in India, along with their location, year of establishment, and key research areas in brief.

TABLE 12.1
Important Chemical Laboratories in Research in India

Research Institute	Location	Established	Type of Research Work
Indian Institute of Science (IISc)	Bengaluru, Karnataka	1909	Nanotechnology, materials science, drug discovery
Council of Scientific & Industrial Research (CSIR) Laboratories	Multiple locations	1942	Diverse chemical research (pharma, petrochemicals, environment)
CSIR-National Chemical Laboratory (NCL)	Pune, Maharashtra	1950	Polymers, catalysis, green chemistry
CSIR-Indian Institute of Chemical Technology (IICT)	Hyderabad, Telangana	1944	Medicinal chemistry, agrochemicals, energy
Bhabha Atomic Research Centre (BARC)	Mumbai, Maharashtra	1954	Nuclear chemistry, radiopharmaceuticals
Tata Institute of Fundamental Research (TIFR)	Mumbai, Maharashtra	1945	Chemical physics, nanomaterials
Central Drug Research Institute (CDRI)	Lucknow, Uttar Pradesh	1951	Drug development, pharmacology
Indian Institute of Chemical Biology (IICB)	Kolkata, West Bengal	1935	Biochemistry, drug design
Defense Research & Development Organization (DRDO) Laboratories	Multiple locations	1958	Defense research in high-energy materials, explosives, aeronautics, combat vehicles, naval systems propellants and naval systems
National Institute of Pharmaceutical Education & Research (NIPER)	Multiple locations (Mohali, Ahmedabad, Hyderabad, Kolkata, Guwahati, Hajipur, Raebareli)	1998	All the campus focuses on excellence in advance research and education in pharmaceutical sciences, etc.
Central Leather Research Institute (CLRI)	Chennai, Tamil Nadu	1948	Leather chemistry, waste management
Indian Institutes of Technology (IITs) – Chemistry Departments	Multiple locations	Various (1950s–)	Catalysis, electrochemistry, organic synthesis
National Environmental Engineering Research Institute (NEERI)	Nagpur, Maharashtra	1958	Environmental chemistry, pollution control
Central Scientific Instruments Organisation (CSIO)	Chandigarh	1959	Analytical chemistry, instrumentation
Institute of Chemical Technology (ICT)	Mumbai, Maharashtra	1933	Industrial chemistry, biofuels

12.1.3 Importance of Chemical Industries in India

The chemical industry is one of the largest and most diversified sectors in India. It plays a crucial role in the country's economy, industrial growth, and public welfare. Some of the important chemical industries in India, along with their location, year of establishment, and key research areas, are shown in Table 12.2 (IBEF, 2025a; FICCI, 2023; IBEF, 2025b).

TABLE 12.2
Important Chemical Industries in Research in India

Sector	Importance in India	Research Institute	Location	Est.	Type of Research Work
Pharmaceuticals	• India supplies 60% of global vaccines • Third largest drug producer by volume	CDRI (Central Drug Research Institute)	Lucknow, UP	1951	Drug discovery, antimicrobial resistance, and traditional medicine research
Agrochemicals	• Supports agricultural productivity • Exports worth $3.5 billion	IARI (Indian Agricultural Research Institute)	New Delhi	1905	Pesticide formulations, soil chemistry, sustainable farming
Petrochemicals	• Contributes 3% to India's GDP (PIB, 2023) • Reliance & GAIL (India) Limited are key players	IIP (Indian Institute of Petroleum)	Dehradun, Uttarakhand	1960	Fuel technology, refinery processes, alternative energy
Textile chemicals	• India is the second largest textile exporter • Dyes and auxiliaries market: .1B	SITRA (South India Textile Research Association)	Coimbatore, Tamil Nadu	1956	Eco-friendly dyes, textile processing chemicals
Specialty chemicals	• Fastest-growing sector (CAGR 12%) • Used in electronics, cosmetics, etc.	NCL (National Chemical Laboratory)	Pune, Maharashtra	1950	Polymers, nanomaterials, green chemistry
Biotechnology	• $100 billion industry by 2025 • Vaccines, enzymes, biofuels	CCMB (Centre for Cellular and Molecular Biology)	Hyderabad, Telangana	1977	Genetic engineering, biopharmaceuticals, disease diagnostics

In FY2023–2024, the Indian pharmaceutical market was estimated to be worth between USD 50 and USD 55 billion. The domestic pharmaceutical market, which refers to internal consumption, was valued at approximately USD 41 billion in 2021. It is anticipated to increase to USD 65 billion by 2024 and USD 130 billion by 2030. Some reports claim that the overall objectives of the Indian pharmaceutical, biotech, and medical technology sectors are much higher. For instance, biotech aspires for USD 300 billion by 2030; and medications, pharma for USD 130 billion (Shekhar & Rai, 2025). India is one of the world's biggest producers of vaccines, giving it the name "vaccine factory". It provides a significant portion of the world's need for vaccines against measles, DPT, BCG, and other diseases. For instance, more than half of the world's demand for different vaccines and around 60% of the market for several inexpensive or WHO-prequalified vaccines. Exports account for two-thirds of vaccine production (Akhtar, 2013).

India has one of the largest pharmaceutical ecosystems in the world, with over 10,000 manufacturing facilities and over 3,000 registered pharmaceutical businesses. Table 12.3 lists India's top ten pharmaceutical industries along with their location, areas of expertise, and significant contributions:

Agrochemicals: India produces pesticides, fertilizers, and crop protection products annually, making it the world's fourth-largest producer of agrochemicals (Pandey, Joshi, & Kumar, 2020). With centers in Gujarat and Maharashtra, petrochemicals—the backbone of India's industrial economy worth USD 40 billion—provide the raw materials for fuels, synthetic fibers, and plastics (Shalu et al., 2023). India is a significant global provider of textile chemicals due to its sizable textile industry, which fuels demand for dyes, finishing agents, and auxiliaries. Specialty chemicals, which include paints, coatings, adhesives, and additives with significant export potential, are one

TABLE 12.3
Pharmaceutical Industries in India with Their Location, Specialization, and Major Contributions

Company	Headquarters	Specialization	Key Contribution
Sun Pharmaceutical Industries Ltd.	Mumbai	Generics, specialty formulations, APIs	Largest Indian pharma company; strong U.S. and global market presence
Dr. Reddy's Laboratories	Hyderabad	Generics, biosimilars, APIs	Affordable medicines globally; strong in United States, Russia, Europe
Cipla Ltd.	Mumbai	Respiratory, antiretrovirals, oncology	Pioneered low-cost HIV/AIDS drugs, major impact in Africa
Lupin Ltd.	Mumbai	Cardiovascular, diabetology, respiratory	Major exporter to United States and Japan; strong generic portfolio
Aurobindo Pharma	Hyderabad	Generics, APIs, antiretrovirals	One of the largest suppliers to U.S. and EU markets
Zydus Lifesciences (Cadila Healthcare)	Ahmedabad	Vaccines, biosimilars, generics	Developed ZyCoV-D, India's first indigenous COVID-19 DNA vaccine
Glenmark Pharmaceuticals	Mumbai	Respiratory, dermatology, oncology	Strong pipeline of novel molecules and biosimilars
Biocon Ltd.	Bengaluru	Biopharmaceuticals, insulin, biosimilars	India's leading biologics firm; global insulin and biosimilars supplier
Serum Institute of India (SII)	Pune	Vaccines	World's largest vaccine manufacturer; supplies >50% of global childhood vaccines
Bharat Biotech	Hyderabad	Vaccines, biologics	Developer of Covaxin, India's first indigenous COVID-19 vaccine

of India's fastest-growing market categories with estimated compound annual growth rate (CAGR) of ~11%–12%. The biotechnology sector in India, which includes biopharma, bio-agriculture, and bio-services, is estimated to be worth over USD 80 billion in 2024 and is expected to grow to USD 300 billion by 2030 (Vandy, Mohanty, & Mohapatra, 2025).

12.1.4 History and Evolution of Laboratory Regulations in India

The legal framework for laboratories in India has changed from the early drug-law mandates and standard setting to a complex network that includes sector-specific monitoring, internationally aligned practices: Globally Harmonized System of Classification and Labelling of Chemicals (GHS), GLP, NABL accreditation, and lifecycle chemical management.

12.1.4.1 Early Beginnings: Pre-1940 and 1940s

The Indian Pharmacopoeia, a national statute for drug regulation, and the establishment of central and state drug laboratories were all suggested by the Drug Enquiry Committee (Chopra Committee, 1931), opening the door for institutional management of standards and testing (ELABSS, 2025;

Ministry of Health & Family Welfare, 2021). As a result, India's import, production, distribution, and sale of pharmaceuticals and cosmetics were formally regulated by the Drugs and Cosmetics Act (1940) to regulate the import, production, distribution, and sale of drugs and cosmetics. Later in 1945, rules and regulations were formulated that included requirements for ingredient disclosures, labeling, and drug safety testing as prescribed by Central Drugs Standard Control Organization (CDSCO) (Pawar, Gore, & Palekar, 2023).

12.1.4.2 Standardization and Early Accreditation: 1940s–1980s

The BIS, which was founded in 1946 as the Indian Standards Institution (ISI), is responsible for establishing technical standards, particularly those pertaining to chemicals and labeling (Bureau of Indian Standards, n.d.). India's early standardized framework was established in large part by Lal Chand Verman, the first Director-General of ISI, and it had an impact on laboratory regulations and procedures. The Environment (Protection) Act, 1986, became the overarching legislation controlling hazardous waste management and chemical safety in 1986–1987. Safety thresholds, major accident hazard protocols, and handling standards were enforced by the Manufacture, Storage, and Import of Hazardous Chemicals Rules (1989, modified in 1994 and 2000) (Gupta, Nair & Singh, 2013).

12.1.4.3 Recent Reforms: ICMSR and Chemical Safety Integration (2021–Present)

The MoEFCC introduced the Indian Chemicals Management and Safety Rules (ICMSR) in March 2021 to unify disparate chemical regulations into a single lifecycle-based framework. Similar to EU REACH (Registration, Evaluation, Authorization, and Restriction of Chemicals) systems, they demand consent, registration, hazard communication, safety evaluations, and limitations for priority compounds.

Although small and medium enterprises (SMEs) cite capacity and cost issues, industry polls as of July 2025 indicate strong support for ICMSR (≈80%) and the adoption of GHS-based categorization and labeling. Furthermore, there are thoughts in BIS of expanding the regulatory scope of laboratory identification and testing by adding 58 additional compounds, such as acetic acid, formaldehyde, and benzene, to its required list of controlled substances (Paudyal, 2025; Kumra, 2025).

12.2 NATIONAL LAWS AND REGULATIONS GOVERNING CHEMICAL LABS OF INDIA

Chemical laboratory regulations in India are complex and encompass several key laws and regulations. The Government of India's regulatory framework for chemical production and importation is defined by key legislation, especially the Chemical Accidents (Emergency Planning, Preparedness and Response) Rules of 1996 under the Environment (Protection) Act of 1986. This law requires personal protective equipment (PPE) in labs and classifies lab waste as "hazardous" per the Hazardous Wastes (Management and Handling) Rules of 1989, reflecting the commitment to safety and environmental protection (Mourya et al., 2017).

The Environment (Protection) Act of 1986 is the main piece of legislation and offers a comprehensive framework for managing dangerous substances and protecting the environment. The Hazardous Chemicals (Manufacture, Storage and Import) Rules, 1989, and the Bio-Medical Waste (Management and Handling) Rules, 1998 are other important rules. Additionally, the Chemical Management and Safety Rules (CMSR), which are being implemented in India, are anticipated to resemble the REACH framework in the EU. Key national regulations and laws governing chemical laboratories in India are given in Table 12.4 (Kumar, Maurya, & Henly, 2021; Kumar et al., 2023).

TABLE 12.4
National Regulations and Laws Governing Chemical Laboratories in India

Regulation/Law	Year of Implementation	Governing Body	Importance
Factories Act, 1948	1948 (amended multiple times)	Ministry of Labour & Employment	Ensures workplace safety, including chemical laboratories; mandates proper handling, storage, and disposal of hazardous chemicals
Hazardous Waste (Management, Handling & Trans-boundary Movement) Rules, 2016	2016 (amended in 2019)	Ministry of Environment, Forest and Climate Change (MoEFCC)	Regulates disposal and management of hazardous waste from laboratories to prevent environmental pollution
Chemical Accidents (Emergency Planning, Preparedness, and Response) Rules, 1996	1996	MoEFCC	Ensures laboratories have emergency plans for chemical spills, leaks, or accidents to protect workers and the public
Indian Standard (IS) Codes for Laboratory Safety (e.g., IS 14433, IS 9523)	Various	Bureau of Indian Standards (BIS)	Provides guidelines for laboratory design, ventilation, chemical storage, and personal protective equipment (PPE)
Environmental Protection Act, 1986	1986	MoEFCC	Regulates emissions and waste disposal from laboratories to prevent environmental harm
Occupational Safety, Health and Working Conditions Code, 2020	2020 (replaced parts of Factories Act)	Ministry of Labour & Employment	Enhances workplace safety, including chemical exposure limits, PPE, and worker training in laboratories
Narcotic Drugs and Psychotropic Substances (NDPS) Act, 1985	1985	Narcotics Control Bureau (NCB)	Controls storage and use of regulated chemicals (precursors) in laboratories to prevent misuse
Explosives Act, 1884 & Explosive Substances Rules, 2008	1884 (updated 2008)	Ministry of Commerce & Industry	Regulates handling and storage of explosive or highly reactive chemicals in laboratories
Hazardous and Other Wastes Rules, 2016	2016 (amended 2019, 2023)	MoEFCC	Applies to laboratories handling biomedical/clinical waste; mandates proper segregation and disposal
Drugs and Cosmetics Act, 1940 & Rules, 1945	1940 (amended multiple times)	Central Drugs Standard Control Organization (CDSCO)	Regulates laboratories manufacturing/testing pharmaceuticals and cosmetics for safety and quality
Poison Act, 1919	1919	State Governments	Controls sale and storage of poisonous chemicals used in laboratories
Disaster Management Act, 2005	2005	National Disaster Management Authority (NDMA)	Requires laboratories to have disaster management plans for chemical leaks/spills

12.3 THE ROLE OF REGULATORY AGENCIES GOVERNING CHEMICAL LABS IN INDIA

The goal of Indian regulatory bodies that oversee chemical laboratories is to guarantee quality, safety, and compliance in the pharmaceutical and chemical sectors. The Department of Chemicals and Petrochemicals (DCP), which is in charge of strategy and planning for the chemical industry, and CDSCO, which is the main regulating body for medicines, are important organizations. These organizations control manufacturing, set standards, and manage both the import and export of chemicals and associated goods.

The Indian Pharmacopoeia Commission (IPC) operates as an autonomous entity under the auspices of the Ministry of Health & Family Welfare, tasked with the formulation of pharmaceutical standards that govern drug production in India. In parallel, the CDSCO serves as the primary regulatory agency overseeing pharmaceutical products within the nation (Pawar, Gore, & Palekar, 2023).

A legislative framework allows the Indian government to oversee the quality of pharmaceuticals, cosmetics, diagnostics, and medical devices via licensing, inspection, and market surveillance. The Department of Chemical and Petrochemicals from the Ministry of Chemicals and Fertilizers manages this oversight, establishing pricing policies for bulk drugs and enhancing the chemical and petrochemical sector. The National Pharmaceutical Pricing Authority (NPPA), created in 1997, regulates prices of controlled bulk drugs, enforces pricing regulations, and ensures access to essential medications while compiling data on production, export, import, and pricing (Mourya et al., 2017).

The Ministry of Health & Family Welfare oversees policy formulation, strategic planning, research, disease management, and export-import policies. In contrast, the Ministry of Chemicals and Fertilizers focuses on chemical sector regulation, including drug manufacturing. Multiple entities, such as the Department of Industrial Policy and Promotion, manage regulatory oversight on patents and pharmaceutical exports, with the CDSCO collaborating on licensing and quality assurance (Kumar, Maurya, & Henly, 2021).

The regulation of clinical trials in India follows Rules 122 A–E of the Drugs and Cosmetics Rules, along with Schedule Y, the Good Clinical Practice Guidelines from CDSCO, and ICMR's Ethical Guidelines for Biomedical Research. Rule 122 DA defines clinical trials on new drugs as studies evaluating safety or efficacy in humans, requiring prior authorization. Rule 122 E categorizes new drugs, including new molecular entities, modifications of existing drugs, and fixed-dose combinations with at least one new ingredient (Mourya et al., 2017; Imran et al., 2013; Kumar, Maurya, & Henly, 2021). Table 12.5 shows the key regulatory bodies overseeing chemical laboratories in India (Kumar, Maurya, & Henly, 2021; Kumar et al., 2023).

International standards also play a vital role in regulating chemical laboratories in India. These include ISO 45001 for occupational health and safety, and ISO 17025 for testing and calibration requirements. ISO 45001 provides a framework for effective safety management, helping organizations create safer environments, manage hazardous chemicals, and prioritize health. ISO 17025 outlines essential criteria for laboratory competence, impartiality, and operational consistency, integrating quality control principles relevant to chemical testing across industries, government, and academia. This standard also addresses sampling processes critical for assessing chemical hazards and managing hazardous materials.

Storing chemicals requires following regulations and best practices. The Comprehensive Environmental Response, Compensation, and Liability Act (CERCLA) mandates immediate action during hazardous spills, including notifying authorities about environmental risks. Laboratories must also adhere to national safety standards and codes for regulatory compliance (Mourya et al., 2017).

12.4 SAFETY AND WASTE MANAGEMENT IN CHEMICAL LABS IN INDIA

Chemical laboratories in India produce environmental pollutants like solvents, heavy metals, and pathogens, requiring careful disposal to prevent contamination. Waste is collected in central, labeled areas following the GHS. Mercury, cyanide, and heavy metals have specific disposal procedures, and must never enter drainage systems.

TABLE 12.5
National Regulatory Bodies Overseeing Chemical Laboratories in India

Regulatory Agency	Year Established	Headquarters	Key Role in Regulating Chemical Laboratories
Central Drugs Standard Control Organization (CDSCO)	1940 (under Drugs and Cosmetics Act)	New Delhi	Regulates drug testing laboratories, ensures GLP/GMP compliance, and approves pharmaceutical and cosmetic testing
Indian Pharmacopoeia Commission (IPC)	2006 (as autonomous body)	Ghaziabad, UP	Sets drug standards, accredits laboratories for pharmacopoeial testing
Food Safety and Standards Authority of India (FSSAI)	2006 (FSS Act)	New Delhi	Accredits food testing laboratories for contaminants, additives, and pesticide residues
Central Pollution Control Board (CPCB)	1974 (under Water Act)	New Delhi	Certifies environmental laboratories for air/water pollution, hazardous waste, and industrial effluent testing
Bureau of Indian Standards (BIS)	1986 (replacing ISI)	New Delhi	Accredits laboratories for chemical quality testing (ISO/IEC 17025), sets industrial chemical standards
National Accreditation Board for Testing and Calibration Laboratories (NABL)	2000 (under QCI)	New Delhi	Provides ISO 17025 accreditation to chemical laboratories across sectors (pharmaceutical, food, environment, etc.)
Directorate General of Factory Advice Service & Labour Institutes (DGFASLI)	1948 (Factories Act)	Mumbai	Regulates industrial hygiene laboratories for workplace chemical exposure limits
Narcotics Control Bureau (NCB)	1986 (NDPS Act)	New Delhi	Oversees narcotics and psychotropic substance testing laboratories
Central Forensic Science Laboratories (CFSLs)	Established in phases (e.g., CFSL Delhi in 1957)	Multiple (Delhi, Kolkata, etc.)	Conducts forensic chemical analysis for criminal investigations
Quality Council of India (QCI)	1997	New Delhi	Promotes quality standards, oversees NABL accreditation for laboratories

Safety and management of waste in chemical laboratories are essential for safeguarding the environment, public health, and laboratory personnel. These practices are regulated by national regulations, institutional policies, and international safety standards in India. In order to guarantee the safety of personnel and environmental preservation, chemical laboratories in India are required to comply with rigorous safety and waste management protocols.

The management of liquid hazardous waste and laboratory effluents adheres to the CPCB regulations under the Water Requirements Regulations of 2010. Wastewater assessment analyzes pH, color, turbidity, fluoride, heavy metals, chemical oxygen demand, biological oxygen demand, total suspended solids, and total dissolved solids to ensure responsible disposal methods. The CPCB oversees hazardous waste management and licensing per the Hazardous Waste (Management and Handling) Rules (2016). Research identifies effective chemical waste treatment methods. Compliant facilities manage various waste types as chemical laboratories contribute notably to the waste stream, requiring specific control strategies (Mourya et al., 2017).

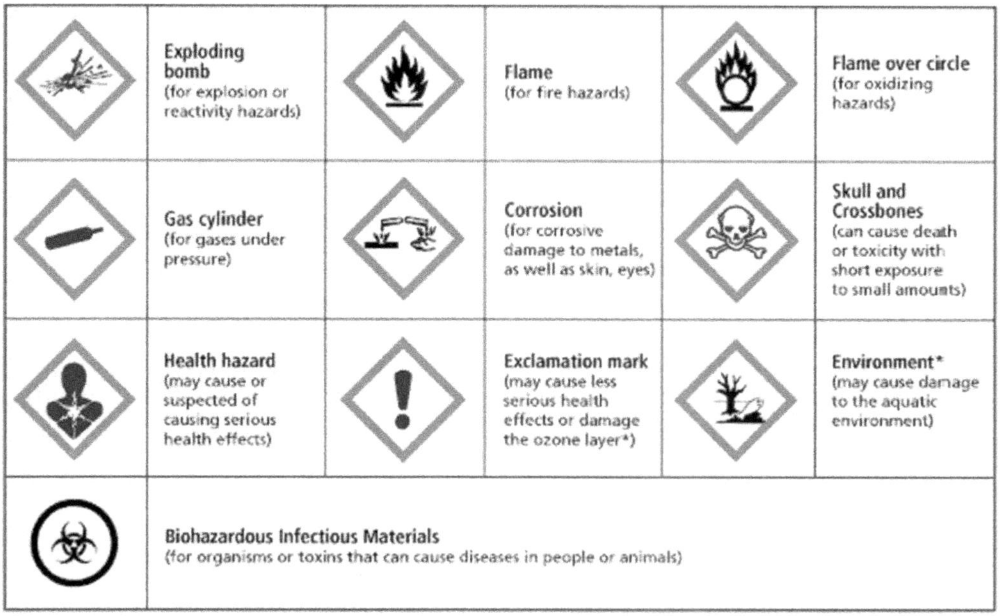

FIGURE 12.1 Pictorial representation of chemical hazards under the GHS.

12.4.1 Chemical Safety Protocols

Chemical safety protocols are procedures and measures that are implemented to reduce the risks associated with the management of hazardous chemicals (Abedsoltan & Shiflett, 2024). These protocols are typically implemented in Indian laboratories and consist of the following:

Risk assessment: Identification of hazardous chemicals (toxic, reactive, corrosive, and combustible). Assessment of exposure risks and implementation of control measures.

PPE: The use of safety goggles, face shields, laboratory coats, and mittens is mandatory. The provision of appropriate training on the use and maintenance of PPE.

Safe handling and storage: Segregation of incompatible compounds (e.g., acids and bases). Utilization of fume chambers for volatile or toxic substances. Labeling and secure storage of compounds with Material Safety Data Sheets.

Emergency preparedness: Availability of fire extinguishers, safety showers, and eyewash stations. Attending consistent training sessions regarding evacuation procedures, incident response, and first aid.

Regulatory compliance: Compliance with agency guidelines, including the Occupational Safety and Health Administration (OSHA-India equivalent), Factory Act 1948, and Chemical Accidents (Emergency Planning, Preparedness and Response) Rules, 1996. GHS is an acronym for the Globally Harmonized System of Classification and Labeling of Chemicals. It is an internationally recognized system that is intended to standardize the classification and dissemination of chemical hazards. The GHS establishes a standardized structure for hazard communication, guaranteeing that information regarding hazardous chemicals is easily accessible and comprehensible, irrespective of the location of their manufacture, transportation, or use as shown in Figure 12.1.

12.4.2 Handling and Disposal of Chemical Waste

It is imperative to ensure that chemical residue is handled and disposed of properly in order to prevent health hazards and environmental pollution. Laboratory protocols in India are governed by the

Hazardous and Other Wastes (Management and Trans-boundary Movement) Rules, 2016 (Dehal, Vaidya, & Kumar, 2022; CPCB, 2025a).

Waste segregation: Dividing waste into biomedical, non-biomedical, and hazardous categories. The utilization of color-coded containers for waste collection (e.g., yellow for biohazard, red for flammable waste).

Collection and storage: Temporary storage in containers that are labeled and leakproof. Preventing the amalgamation of incompatible substances.

12.4.2.1 Methods of Treatment

In neutralization for waste that is either acidic or basic, oxidizers and reducers undergo chemical treatment. Incineration is employed for the disposal of toxic debris and organic solvents. Biodegradation for specific biodegradable chemicals. Disposal through authorized waste management vendors. Adherence to the regulations of the CPCB and the State Pollution Control Boards (SPCBs). Ongoing maintenance of records regarding the maintenance and disposal of waste, periodic audits and reporting to regulatory authorities (Karthikeyan et al., 2018; Kumar et al., 2017).

12.5 COMPLIANCE, AUDITS, AND ETHICAL CONSIDERATIONS OF CHEMICAL LABS IN INDIA

Compliance with safety, environmental, and ethical regulations is a fundamental requirement for chemical laboratories in India. These laboratories must undergo regular audits, adhere to legal frameworks, and uphold ethical standards in research and testing to maintain credibility, ensure safety, and protect the environment. Here is a summary in Table 12.6, covering the regulatory agencies, audit frequency, and penalties for noncompliance in chemical laboratories in India.

TABLE 12.6
The Regulatory Framework for Audits and Penalties

Regulatory Agency	Scope of Oversight	Audit/Inspection Frequency	Penalties for Noncompliance
Central Pollution Control Board (CPCB)	Hazardous waste management, pollution control, environmental compliance	Annual/Periodic and surprise inspections	Fines, operational stoppage, and prosecution under the Environment (Protection) Act, 1986
State Pollution Control Boards (SPCBs)	State-level monitoring of waste disposal and emissions	Periodic & surprise checks	Monetary penalties, revocation of permits, legal action
Directorate of Industrial Safety and Health (DISH)	Occupational health and safety in industrial and laboratory setups	Annual or as needed	Fines, shutdown orders, criminal liability for accidents caused by negligence
Ministry of Environment, Forest and Climate Change (MoEFCC)	Implementation of environmental protection laws	Compliance reviews based on reports	Suspension of licenses, penalties, and project halts
Institutional Safety Committees (internal)	Laboratory safety, waste handling, and emergency preparedness	Monthly to quarterly internal audits	Internal warnings, retraining, temporary halts in laboratory operations
Third-Party Certification Bodies (ISO auditors)	Compliance with ISO 14001 (environmental) and ISO 45001 (safety) standards	Annually for certification renewal	Loss of certification, reputational damage, restrictions on collaborations and funding

12.6 FUTURE PERSPECTIVES AND INNOVATIONS IN LAB MANAGEMENT IN INDIA

12.6.1 Laboratory Process Automation and Digitalization

Integrating laboratory management systems poses a significant challenge for laboratories in developing countries like India. Case studies showcasing their use and potential difficulties are essential for laboratory staff and managers. Laboratory administration is being transformed by the integration of automation and digital technologies. Automated analytical instruments, robotics, Laboratory Information Management Systems (LIMS), and AI-powered data analysis improve reproducibility, minimize human error, and increase efficiency. The transition to digital laboratories (Lab 4.0) facilitates real-time monitoring, remote operations, and predictive maintenance, thereby optimizing workflow and reducing operational costs (Saxena, 2024; Gholamizadeh et al., 2023; Mutlu, Altuntas, & Dereli, 2023).

12.6.2 Implementation of Green Chemistry Practices

The primary objectives of green chemistry are the reduction of hazardous substances, the reduction of pollution, and the utilization of sustainable resources. In order to mitigate their environmental impact, laboratories are increasingly employing energy-efficient apparatus, waste recycling protocols, and eco-friendly solvents. It is probable that the future will witness a more extensive implementation of green technologies, including the integration of renewable energy and solvent-free reactions, to ensure that laboratory operations are in accordance with global sustainability objectives (Bennett, 2024; Kapoor & Mehendale, 2021).

12.6.3 Enhancing Public Awareness and Policy Enforcement

The promotion of secure laboratory practices will be contingent upon the effective enforcement of stricter regulatory frameworks. Future developments may encompass enhanced penalties for violations, digital audit trails, and real-time compliance monitoring. A culture of responsibility will be cultivated among researchers, institutions, and the general public through public awareness campaigns regarding environmental safety and chemical hazards (Veleva et al., 2018).

12.6.4 The Contribution of Industry and Academia to the Improvement of Laboratory Safety

It is imperative to foster collaboration between academia and industry to enhance laboratory safety. Academic institutions can incorporate comprehensive safety education into their curricula, while industries can offer practical training, resources, and innovations. The development of safer chemicals, improved protective equipment, and advanced risk assessment instruments can be achieved through collaborative research initiatives. This collaboration will guarantee that laboratory safety standards are consistently enhanced on a global scale (Ezenwa et al., 2022).

Part 3

Design and Construction of Affordable Quality Laboratories

13 Design Features of an Affordable, Quality Laboratory

Philip Mark Othigo

13.1 INTRODUCTION

The effectiveness of a laboratory is significantly influenced by its architectural and functional design (Roberson et al., 2013). An inadequate physical environment can negatively impact a laboratory's operational effectiveness, whereas a thoughtfully constructed facility promotes optimal performance. Critical factors encompass illumination, air circulation, and essential utilities. Additionally, design considerations such as spatial arrangement and workflow processes play significant roles in enhancing lab efficiency. Laboratories vary in size, from large central ones to smaller units, often facing financial constraints while needing to uphold high-quality standards. This chapter highlights key features of cost-effective lab designs that sustain quality and operational needs, offering insights for creating economical, high-quality laboratory spaces (Kamau, 2013).

The design and construction of laboratories are crucial for operational efficiency, safety, and costs. It involves a cohesive framework covering construction methods, material selection, safety guidelines, and mechanical systems facilitation (Roberson et al., 2013). Various laboratory design and construction principles, along with relevant standards, have been implemented to achieve key objectives, to (1) create an affordable laboratory from common materials, (2) ensure high health and safety standards for personnel, (3) facilitate sample assessment from prevalent matrices, and (4) adopt a strategy that minimizes costs while maintaining performance and quality.

13.2 COST-EFFECTIVE LABORATORY DESIGN

Developing a cost-effective, high-quality laboratory requires the integration of functionality, safety, sustainability, and financial viability. Functionality involves selecting suitable equipment, structuring environments, and optimizing operations for efficiency. These are fundamental for workflows to balance precise design with quick execution. Flexibility is vital, utilizing modular equipment and versatile furniture to lower renovation costs, while safety measures protect personnel and reduce liability by meeting regulatory standards. Sustainability efforts aim to lessen environmental impacts through energy-efficient upgrades and material reuse, aligned with the Long Life, Loose Fit, Low Energy (L3) design philosophy, underpinning operational needs, occupant-flexible, regulatory standards, and financial viability of labs (Mosteiro-Romero et al., 2023). These principles are elaborated in the next sections.

13.2.1 Functionality

The core design principle in lab operations is functionality which warrants careful alignment of activities and equipment with the available spatial constraints. While workflow considerations are frequently underestimated, the development of workflow and relationship diagrams during the planning phase can significantly enhance space utilization by minimizing unnecessary movement and improving material flow. The strategic positioning of high-load equipment is crucial for operational efficiency. Depending on specific tasks, the incorporation of full-height workstations may be essential. Additionally, open-plan office setups featuring lab benches along the periphery contribute to increased flexibility, optimized storage solutions, and heightened productivity (Kaur, Downey, & Choate, 2018).

The operational sequence in a clinical laboratory is categorized into three distinct phases: pre-analytical, analytical, and post-analytical. This workflow can be visually represented in a circular diagram illustrating the movement of personnel and materials. Space limitations play a crucial role in the arrangement of laboratory layouts (Eggert et al., 2020). Within the context of a Molecular Microbiology laboratory, the critical procedures encompass sample drop-off, subsequent processing, and final disposal. The presence of sufficient equipment and sampling stations is essential to mitigate any potential delays, while ensuring that all setups adhere to established safety protocols. Furthermore, effective strategic planning necessitates both adaptability and comprehensive evaluation prior to the installation of any laboratory equipment.

A well-structured laboratory design houses adequate storage solutions for reagents, instruments, glassware, and waste materials (Eggert et al., 2020). Additionally, strategic allocation of bench space, as well as the closeness of a reporting station to the specimen area, significantly improves operational efficiency. Assessing current and future equipment needs is crucial for scalability.

13.2.2 Flexible Adaptability

Flexibility greatly improves adaptability and integration of new technologies, aligns with sustainable architecture, and is key across various goals. A laboratory environment must adapt to workload fluctuations and technological advancements, as adaptability is crucial for the growth of smaller labs. Flexible design elements enhance operational efficiency, allowing labs to swiftly address changing stakeholder demands and manage unexpected workload shifts. Easy renovation support is essential for updating finishes, improving ergonomics, or transitioning diagnostic functions (Doshi & Clay, 2017; Saha et al., 2024). Modular furnishings are key, with suppliers offering preconfigured safety cabinets. Laboratory configurations, including workbenches and cabinetry, should be based on operational needs, enabling various layouts like corridor, island, or peninsula designs.

13.2.3 Safety in Affordability

An affordable laboratory is attainable while still ensuring safety, efficiency, comfort, durability, and adaptability. Evaluating the human, environmental, and economic impacts of lab incidents reinforces this goal. Designs must concomitantly consider personnel safety and the future well-being of clients or stakeholders consuming its outputs. However, despite established guidelines, uncertainties regarding the criteria for safety and cost-effectiveness remain challenging, as minor factors and rushed decisions can cause major financial and safety issues.

Creating affordable labs goes beyond cost-cutting to involve careful planning of layout, equipment, materials, and processes, necessitating no compromises in features. Safety, like the other core lab design principles, must therefore be prioritized above cost. The facility's design must thoroughly understand and integrate the requirements of testing and relevant processes.

13.2.4 Sustainable Labs

Sustainability seeks to maintain ecosystems over the long term, particularly within the construction sector, by mitigating health and environmental threats through the effective management of energy, water, and material resources. Laboratory designers play a vital role in reducing potential hazards and safeguarding personnel. Clients delineate specific safety requirements, enabling managers to implement checklists for effective budget management. Strategic design can lead to a reduction in material handling expenses by 10%–30%. The integration of energy-efficient technologies and sustainable materials is imperative for diminishing environmental repercussions and improving the long-term operational efficiency of laboratory infrastructures (Roberson et al., 2013).

The design of a laboratory must prioritize reducing environmental impacts from its construction and operation. Specific construction techniques, effective site work, and appropriate materials and equipment are crucial for achieving sustainability goals. Sustainable laboratories enhance safety and health for occupants and the community, conserve resources, generate less waste, and minimize pollutant emissions, including greenhouse gases. They utilize efficient chemical scrubbing, optimize natural lighting and ventilation, and incorporate recycled or renewable materials, which are vital for sustainability, particularly under high occupancy loads affecting Building Research Establishment Environmental Assessment Model (BREEAM) scores.

13.2.5 Financial Viability of a Laboratory Project

Establishing a cost-effective, high-quality laboratory requires rigorous financial management. Creating and supervising budgets for design, construction, equipment, and operational costs is essential. Managing these budgets involves assessing expenditures and generating financial records, with regular updates and reports for donors and stakeholders (Avivar, 2012).

Government and donor organizations typically finance the procurement of equipment, yet this support frequently suffers from issues related to affordability and oversight. The assistance provided tends to prioritize the preferences of donors rather than the actual needs of laboratories. This approach raises concerns about the potential for acquiring inappropriate equipment and neglecting to consider recurring expenses, such as travel costs for suppliers. Optimal lab operation requires careful financial resource allocation for reagents, consumables, spare parts, and maintenance. A systematic procurement approach is vital to reduce shortages and wastes (Roberson et al., 2013).

Optimal lab design balances cost, safety, environmental concerns, and functionality. Design choices shouldn't be driven solely by finances; instead, focus on blending components for quality and affordability. Professionals are refining budgeting skills to address increasing demands for resources and equipment.

Integrating budget planning in the early stages of laboratory design is essential for aligning with architectural and engineering considerations. This ensures budgetary constraints and objectives are relevant in later phases, aiding effective design oversight. The budget plays a significant role in the design process, though it is not the only consideration. Emphasizing cost-efficiency may negatively impact quality and safety, as it is essential for the laboratory to fulfill all its designated functions. At the outset, the budget acts as a hopeful point of reference.

Financial assistance is crucial for the establishment of high-quality laboratory facilities; however, challenges in securing funding often arise from various limitations. Numerous organizations are available to offer financial support. It is essential for grant proposals to highlight the societal advantages of the research, as funding bodies tend to prioritize initiatives that demonstrate potential for long-term positive effects.

13.2.6 Emerging Laboratory Designs

The integration of technologies enhances laboratory design and workflows, boosting productivity and quality while cutting waste and costs. This change significantly impacts the construction of research and educational facilities. The LIMS automate tasks such as data analysis and inventory management, but many labs still rely on manual processes. Anticipated advancements in laboratory design indicate better equipment integration and technology use. Open-plan layouts, plug-and-play systems, automation, microanalysis, digital tools, organic design, sustainability, and modular configurations can improve operations and reduce costs.

Laboratories are required to incorporate advanced technology in order to comply with changing standards. Implementing effective strategies is essential to meet these demands. Facilities with lower equipment density are well-positioned to integrate new devices seamlessly, while open layouts facilitate adaptable workspaces and the use of mobile equipment. The concept of plug-and-play laboratories significantly increases flexibility by providing various options for specialized equipment (Weldon, 2012).

Automation boosts quality and capacity, reducing spatial needs. Microanalytical techniques enhance safety and lower facility demands. Digital collaboration connects remote teams for efficient equipment sharing. An organic design integrates functionality with space use, while sustainability initiatives target energy reduction and increased recycling. Modular design enables flexible space reconfiguration with minimal disruption (Doshi & Clay, 2017).

13.3 LABORATORY CONSTRUCTION PROCESS AND CRITERIA

13.3.1 Programming

Before any design work is commenced in the laboratory design process, the stakeholders (architect, project engineer, laboratory consultant, the owner's administrators, and laboratory users) develop a building program, a document that describes and quantifies the goals of a particular project. The program functions to assess the number and type of personnel who inhabit the building, the functions to be housed within the space, and the inter-relationships between the aforementioned functions (Roberson et al., 2013). The importance of the program cannot be gainsaid as it ensures sufficient space provisions for its occupants as well as uses, ascertains safety, and guarantees the client's budget and needs are met.

The program document may be organized into three components based on the objectives of the client's project team:

1. *Conceptual program:* often used to source funds, a conceptual program is used to show how achievable the construction or renovation project may be. It represents the net usable area or gross area for each generic space type, i.e., the laboratory, laboratory support, office and administration, personnel support, and building support.
2. *Outline program:* used for further fund-raising and as a tool for acquiring additional researchers, this document lists and quantifies the numbers and areas of specific room types.
3. *Detailed functional program:* this document allows the client to evaluate the layout that is developed by the architects and engineers as well as estimating the construction cost of the project. This document contains architectural, mechanical, electrical, plumbing, IT, and fire safety performance for all functions that the building is meant to accommodate. Additionally, this document identifies potential safety hazard areas such as those that use toxic and pathogenic materials and processes and the procedures to be followed in waste removal for sensitive materials.

13.3.2 Program Document Procedure

There are sequential steps to be followed in developing a program document. This section discusses briefly what each step entails.

Step 1: Analysis of existing facility occupancy

This step involves investigation, through observation and data recording, of occupancy patterns from current work area of the intended occupants. This may include population density, major equipment used in the laboratory, processes conducted in the laboratory, linear dimensions of lab benches, number of chemical fume hoods and their distribution within the lab, and how hazardous materials and waste are handled. This step helps make the client and the intended user more aware of their current use of space and informs them of what their goals for a renovated facility of new building could be.

Step 2: Special analyses and studies

For factors such as threat and security, site location, and environmental impact, additional expert knowledge is sought by the stakeholders, to better identify the possible obstacles that may arise in developing the project and suggest possible remedies to each obstacle.

Step 3: Focus group discussions (FGDs) and stakeholder interviews

Through workshops and FGDs, the program project team conducts interviews with laboratory stakeholders for information on current population numbers, the functions of specific spaces and their forecast for future capacities. Stakeholders may include department heads, principal investigators, administrative heads, laboratory managers, health and safety professionals, hygiene officers, material handling personnel, and other potential users and individuals involved in operating the proposed building. This step is important as it highlights the critical health, safety, environmental, and operations information, as well as hidden assumptions that might be overlooked by the program project team (Berezin et al., 2023).

Step 4: Net or revised net area standards

Here, the design team determines the required area by calculating area standards from prevailing laboratory population density through analysis of current and more preferred laboratory occupancy patterns, assessing the extent to which existing conditions and net assignable areas are adequate. Thereafter, the team sets realistic but safe area goals for the new development. Factors such as bench length, shared equipment walk length, computational station length, length of fume hoods, hazardous waste storage area, and linear feet of sink are instrumental in establishing standards per full-time equivalent.

Step 5: Room type list

Here, the project team notes down all proposed spaces in the intended facility. These may be broadly grouped into laboratories, laboratory support facilities, administration, personnel support facility, and building support.

By planning for future adaptation of the facility, the non-laboratory spaces can be designed in such a way that enables them to be converted into laboratories. This would require specification on the non-laboratory area that should be engineered for future conversion into laboratories.

Step 6: Laboratory performance criteria data

Extremely useful to the building engineers, this stage involves gathering comprehensive information on the performance criteria of all room types by the design team from laboratory personnel on their current and expected facilities. This data provides relevant scope and quantity information for the entire design team, as well as ensuring that all requirements are met in the design documents, and guarantees quality assurance during construction.

Step 7: Room type diagrams and net area estimates

At this stage, the design team estimates the net areas of each room type by generating options on laboratory module dimensions, configurations, and net area, and testing the estimated area by developing simple diagrams of all room types using the module template. Once a building module is generated by the designers and the most appropriate dimensions chosen, an estimated Net Assignable Square Foot/Meter of the laboratory can be determined. This is a vital exercise which eventually determines the building's structural grid layout (through the established dimensions of the laboratory module).

Step 8: Quantities of room types

By evaluating the number of functions and room types required by the laboratory personnel, the building layout and its influence on access to shared facilities, and making decisions on centralized and proprietary facilities, the program team develops an estimated number of each room type, i.e., the number of laboratories and support rooms, that will be hosted in the development or renovated facility.

Step 9: Building net and gross area calculations

The team determines the net assignable laboratory area on each floor, which is then converted into gross area using the most appropriate net to gross ratio which may vary for different laboratory facilities. This step is critical in estimating construction costs and the design process following the completion of the program.

Step 10: Spatial relationships of functions

The architects and engineers are informed of the importance of various relationships between the room types within the facility. This may be with regard to hierarchy, material handling concerns, health and safety considerations, proximity of support spaces to laboratory spaces, and building mechanical service distribution. This will help the drawing team in developing conceptual layouts and plan for future adaptation of the facility.

Step 11: Basis of design

This is a written document that provides all participants involved with the project a baseline, narrative, and overview of how the final design of the HVAC/Electrical/Plumbing and Fire equipment systems, and integrations related to the project have been selected and designed against the project requirements of the client. It is instrumental in developing narrative description on the proposed project as well as estimating the cost through a summary of major scope components and building systems.

Step 12: Cost model or estimate of construction and project cost

Professional cost estimators may be brought in to analyze the program description and develop a construction estimate which may be ±25%. A cost estimate is an important planning tool that ensures the success of the project.

13.3.3 Schematic Design

At this stage, the designers will work to develop the building configuration with regard to the laboratory module, identify where spaces in each floor will be located and how each floor will relate to one another (for multi-story laboratory facilities), develop the exterior image of the laboratory, and conceptualize the individual engineering system based on the laboratory module. They will also embark on reviewing the predicted project cost as well as reviewing the associated codes.

13.3.3.1 Design Development

This stage involves finalizing design decisions with regard to site context, building facade, and casework layouts for each laboratory space. Detailed floor plan layouts for each laboratory space, including service requirements, and systems and materials specifications, as well as coordination of engineering systems with the architectural plans, should be made concrete at this point.

It is also important that consultations are held with the end user to come up with and agree on adjacencies between rooms, layout schemes, and building finishes, as well as casework selection.

13.3.3.2 Construction Documents

Upon this phase, the architects and engineers develop working drawings and all specifications that are essential to developing the facility. Construction documents comprise drawings and specifications meant for bidding and construction and a final cost estimate.

13.3.3.2.1 Bid Phase

The construction documents will be issued for bidding at this stage. Once construction bids are received and reviewed, a contract will be awarded.

13.3.3.3 Construction Administration

This step involves the review, coordination, and approval of shop drawings. Regular site visits by the architects ensures that construction is done in accordance to the construction documents. The duration of construction will depend on the size and complexity of the project.

13.3.4 Laboratory Construction Considerations

13.3.4.1 Structural Considerations

There are several factors that determine the structural system of a laboratory building, such as the following:

1. Coordination of the structure with the laboratory planning module
2. Slab thickness and effective floor-to-floor height

3. Ability to create penetrations for laboratory services in the initial design and through the life of the building
4. Potential for vertical or horizontal expansion
5. Superimposed loads
6. Vibration criteria
7. Cost.

13.3.4.1.1 Structural Grid

The structural system of the laboratory is often determined by the labcratory planning module. It is common that the structural grid width is equivalent to two basic laboratory modules as in Figure 13.1, while its length is determined by the basic planning requirements, cost-effectiveness, and functional requirements of the structural system.

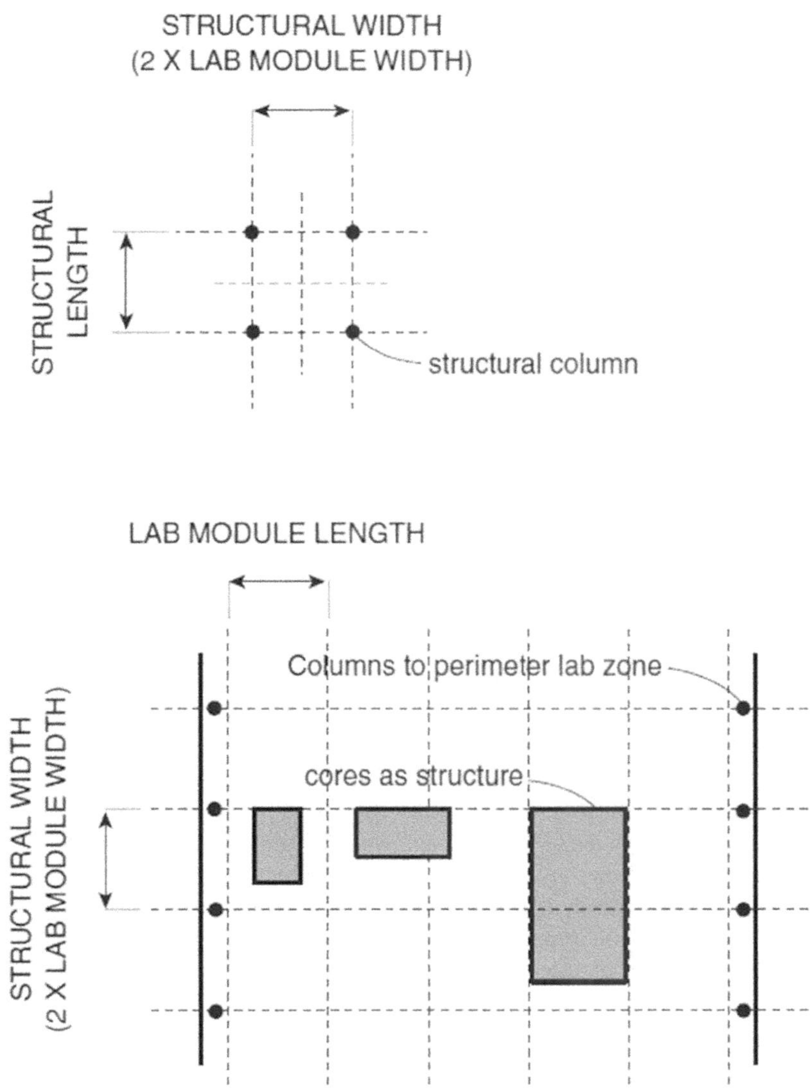

FIGURE 13.1 The building's structural grid derived from the laboratory planning module. For maximum flexibility, columns should be integrated or placed outside the grid (Metric Handbook: Planning and Design Data (Buxton, 2021).)

Columns should be incorporated into the design (within built-out walls, cores, or perimeter of the laboratory zone) such that it forms an unobstructed work space. The structural system should also be organized in a manner that facilitates extension and reconfiguration of the layout.

It is common practice that structural engineers will offset the structural grid from the module grids to prevent structural complications that may arise from floor beams and drain pipes. Offsetting the two grids will also prevent installation of drainage sleeves on structural members. Designers should ensure that stinks and equipment drain away from structural members where possible.

The depth of structural members such as beams should be done in such a way that it provides adequate space above finished laboratory ceilings for the installation of ducts, pipes, and electrical services. As consequence, designers should provide ample floor-to-floor heights to enable easy installation of mechanical systems, reduce cost, and increase future flexibility (Doshi & Clay, 2017).

Special consideration is required for spaces that contain equipment that are unusually heavy, or those that are either sensitive to vibration or produce significant vibrations. Data on such equipment should be considered during the building programming process so that their structural design can be established at the very start of the design process.

13.3.4.1.2 Vibrations

Increased structural mass combined with building foundation stability can help reduce structure-borne vibrations that are both internally and externally generated within the laboratory building. Use of active vibration suppression measures and special vibration dampening and (or) mounting devices can be used to compliment the aforementioned steps. Moreover, careful organization of the building layout, especially location of mechanical rooms, elevators, and utility distribution pathways, as well as good site location, can help reduce vibration in spaces that are highly sensitive to them. This should be ultimately coupled with careful detailing of machine, duct and pipe mountings, use of local vibration isolating devices, and structurally independent platforms at spaces that host sensitive equipment.

13.4 REGULATION OF LABORATORY STANDARDS

Laboratories are required to comply with established safety protocols and to engage with relevant authorities to obtain necessary licensing (Steere, 1973). For a clinical lab, conducting a biosafety risk assessment is essential for the effective implementation of safety measures during the design phase (Jayamani et al., 2022). The precautions necessary are contingent upon the designated biosafety level, and is determined by the associated infection risks.

Laboratories play a vital role in healthcare, research, and industry globally. However, challenges like limited access and poor management hinder growth, especially in Central Asia. The high costs of establishing labs require a focus on cost-effective, high-quality solutions. Key considerations include equipment selection, material quality, facility design, and qualified staff recruitment. This section examines these crucial areas (Roberson et al., 2013).

Designers encounter obstacles in the development of high-quality, economically viable laboratory models by examining numerous variables. The development process adheres to budgetary constraints, emphasizing factors such as layout design, workflow efficiency, the availability of skilled personnel, and compliance with import regulations. While efforts to reduce costs may impact quality, a thorough evaluation of architectural designs can facilitate the preservation of affordability. Moreover, strategic investments are capable of providing potential savings in the long run, while also ensuring the safety and functional integrity of operations.

It is crucial to comply with local safety standards and regulations, especially in the absence of explicit guidelines. Various resources, including literature and checklists, can facilitate adherence to these regulations, although compliance dynamics differ based on the laboratory's intended purpose, occupancy classification, and construction norms (Guide et al., 2001).

Laboratory safety regulations are vital for the design, operation, and maintenance of labs, protecting employees, the public, and the environment. Various safety organizations create, provide, and enforce these important protocols. Organizations such as EPA, OSHA, ISO, American National Standards Institute (ANSI), National Fire Protection Association (NFPA), and the European Agency for the Environment, Economy, and Sustainability, issue comprehensive laboratory standards, codes, regulations, and guidelines.

The NFPA and OSHA focus on hazardous materials, EPA for waste management, ANSI for ergonomics, European directives for construction, and ISO for medical lab criteria. Evaluating hazardous activities and their risks impacts laboratory processes. Mechanical systems, like plumbing and electrical, must comply with safety regulations (Ikranbegiin et al., 2019).

13.5 SPECIALIZED LABORATORY AREAS

13.5.1 Biosafety Level Laboratories

13.5.1.1 Biosafety Level 2 (BL-2) Laboratories

These laboratory spaces need 100% outside air. Exposed surfaces within this space should be smooth and easy to maintain, while caulking at furniture to wall seams improves cleanability. Where there's need for future flexibility, the laboratory should be designed for BL-3 standards as it is much easier to convert it into a BL-2 as compared to converting a BL-2 room into a BL-3 room.

13.5.1.2 Biosafety Level 3 (BL-3) Laboratories

These laboratory spaces are designed to contain the agents used within them. As a rule, these spaces should be accessible through a controlled entrance. Walls, floors, and ceilings should be finished in liquid-resistant finishes, while work surfaces require impervious surfaces that are resistant to acids, bases, and organic solvents. The space and all the furniture it hosts should be easy to clean in order to avoid contamination. An automatic handwashing sink should be placed near the exit. All windows to these spaces should be sealed shut.

13.5.1.3 Biosafety Level 4 (BL-4) Laboratories

Designed to limit personnel exposure and prevent the spread of BL-4 organisms to the outdoor environment, these laboratory spaces demand special attention to its containment requirements. Personnel can be protected through protective gear, while the environment can be protected through use of HEPA filters at the exhaust systems. Laboratory organisms can be placed in Class 3 BSC or in contained caging systems. The designers should provide for chemical showers at both means of egress that will also allow decontamination of equipment. The laboratory spaces should have a contained room, stainless-steel duct work and cabinetry, column-free spaces, and minimal penetrations in the wall surface. Other support rooms such as showers, washrooms, suit storage, and water fountain facilities should also be taken into consideration by the design team.

13.5.2 Clean Rooms

Often involved in manufacture of miniature components, these are spaces in which several parameters such as temperature, room pressure, and humidity are controlled, monitored, and maintained. This is done to ensure such a room has a lower level of airborne particulate contamination than normal.

A clean room should incorporate air locks or vestibules that are kept at a slightly lower pressure than the clean room itself, which should be under positive air pressure to avoid ingress of contaminants. Having several small clean room chambers is better at temperature regulation as opposed to having one large clean room facility.

Surfaces within these rooms should be easily cleanable, nonabrasive, and chip-resistant. This could include high-pressure phenolic laminate on steel or aluminum panels.

13.5.2.1 Environmental Rooms; Cold Rooms and Warm Rooms

When prefabricated facilities are used, they offer superior performance as compared to built-in facilities. These rooms require heated triple-pane observation windows with an option for rubber flaps and freeze-proof, self-closing doors with magnetic gaskets. It is worth noting that only one door should be provided for environmental rooms, unless the area exceeds 37 sqm. All surfaces within these facilities should be vapor-proof. Finally, it is recommended to have several small chambers with their own separate power systems with emergency backup than having one large room.

13.5.2.2 Tissue Culture Suites

These spaces should be under positive pressure in relation to its adjacent spaces. The flooring should be seamless and easy to clean. The doors to these spaces should comprise an in-swinging, self-closing door with a panic bar on the exterior for easy opening.

13.5.2.3 Radiation Laboratories

Equipment that produce radiation and radioactive materials in these facilities demand the use of special shielding. Shielding could be done by placing cobalt and cesium irradiators in separate small rooms that are linked with lead or constructed from dense concrete walls, ceilings, and floors to shield the source. As these facilities are heavy on the structure, they call for special consideration during structural design.

13.5.3 Specialized Equipment and Equipment Spaces

This section looks at laboratories for equipment that can produce magnetic fields and those that are sensitive to interference. As a rule, sensitive equipment must be kept away from steel columns, magnetized doors, and other metal equipment. Considerations for some of these equipment are briefly stated below:

13.5.3.1 Nuclear Magnetic Resonance Apparatus

These equipment should be located on the lowest floor level and UPS have a rigid floor slab and structure to minimize vibrations. They should be located as far away as possible from elevators, and mechanical equipment and steel objects should be kept outside the safety zone.

13.5.3.2 Electron Microscope Suite

The dimensions of this room will depend on the size and number of electron microscopes needed. They should also be stable to minimize vibration and where necessary, vibration dampening may be used.

13.5.3.3 Magnetic Resonance Image Suite

Metal objects should be kept away from this space. All electric conductors within this space require electrical filters. A clearance of 3 m should be maintained for refilling of liquid helium and nitrogen that is needed to cool the magnet.

13.5.3.4 Lasers

It is often desired that rooms with lasers have no glazing, and should have the ability to be blacked out.

13.5.3.5 Mass Spectrometry

Rooms with this equipment should be isolated so as to deal with vibration. Noise control measures should also be taken into consideration when designing this space.

13.5.4 Safety Showers and Eye Washes

Safety showers should be easily identifiable and accessible. The maximum distance to any safety shower should be about 30 m with a distance of 23 m being the recommended length. Floor drains in the shower may result in contamination within the drain or a leak into the floor below and should therefore be discouraged. Mopping the shower is better as it identifies what elements were on the individual. Care should be taken to keep safety showers away from electric sources.

Each laboratory should be equipped with an eye wash and a body wash at least on one sink. They should be supplied with multi-stream cross flow of potable water.

Eyewash and safety showers are often located along corridors with visual and physical proximity to laboratory exits. It is necessary to have anti-slip flooring in these spaces.

13.6 MATERIAL SELECTION CRITERIA

The process of material selection in construction necessitates identifying materials that are not only cost-efficient but also exhibit durability and sustainability, particularly for applications such as furniture, wall systems, and flooring. In environments subjected to high foot traffic, such as laboratories, opting for cost-effective materials at the outset can result in suboptimal performance and subsequent unforeseen maintenance expenses. Therefore, a thoughtful selection process is crucial, as it can significantly minimize long-term maintenance requirements and enhance overall operational efficiency.

Using cost-efficient materials can greatly reduce the financial burden of equipment and infrastructure maintenance. It's essential to consider cost-effectiveness, longevity, and environmental impact. For example, moisture-resistant materials can prevent mold and mildew growth. Recent advancements in material technology indicate promising new alternatives are likely to emerge soon.

In laboratory design, the aspect of durability plays a fundamental role, as it directly affects maintenance decisions regarding material and equipment selection. It is essential to comprehensively analyze the origins of damage, which are frequently associated with human interactions. Designers are advised to choose resilient materials that can withstand significant usage and rigorous cleaning procedures. Additionally, the reliability of equipment significantly influences operational expenditures; investing in trustworthy instruments tends to be more economical in the long run compared to opting for costly and unreliable options.

Environmental impact is crucial in laboratory design; hence, planning must include sustainable principles and Lean operations, as materials handling significantly affects operational costs. Goals of optimized material efficiency include improving workflow, minimizing walking distances, enhancing visibility, and managing inventory. Maintaining indoor air quality is also vital due to high ventilation demands in labs (Alali, Beheiry, & Atabay, 2024).

Increased air exchange rates alone do not guarantee satisfactory indoor air quality. The presence of contaminants, inadequate filtration systems, and humidity levels can adversely affect workplace environments, ultimately contributing to increased absenteeism and diminished productivity. Adhering to the ASHRAE 62-2001 standards is crucial for promoting health, while the implementation of sustainable outdoor air intake methods within HVAC systems is also of paramount importance (Persily, 2015).

13.6.1 Walls

High-grade water-soluble latex paints are suitable for spaces with low to moderate chemical use. Latex paints off-gas volatile organic compounds (VOCs) during the painting and drying phase which may have significant effect on the health of the building occupant (Table 13.1).

TABLE 13.1
Laboratory Wall Material Criteria

Material	Cost	Advantages	Disadvantages
Solvent-based coatings	Low	• Durability • Ease of application • Excellent coverage • Water resistance	• Fairly high VOC emittance
Latex coatings	Low	• Breathable (prevents accumulation of moister) • Low VOC emittance • Ease of maintenance	• Low durability • Cleanability is reduced • Has a fairly short life span
High-performance coating systems	Medium	• Durability • Easy to clean • Low VOC content	1. **Application and Technical Limitations** • Stringent Surface Preparation • Complex Application Process • Sensitive Curing Conditions • Difficult Repairs • Long Curing Times 2. **Environmental and Safety Hazards** • Volatile Organic Compounds (VOCs) • Hazardous Ingredients 3. **Performance Weaknesses** • Brittle Behavior • Poor Thermal Shock Resistance • UV Degradation • Slippery Surfaces 4. **Economic and Practical Constraints** • High Initial Cost • High Operator Skill Required • Under-film Corrosion
Fiberglass-reinforced panels	High	• Durability • Ease of installation • Easy to clean • Low maintenance	• Poor sound attenuation • Difficult to repair
PVC panels	High	• Durability	• Difficulty to repair • Poor sound attenuation
Reinforced epoxy wall coatings	High	• Durability • Easy to clean • High chemical resistance • Water resistance	• Moisture Sensitivity and Delamination • UV Sensitivity (Yellowing) • Intense Installation Process • Strong Odors and VOCs • Brittleness and Scratching • Difficult and Costly Repairs • Long Curing Time • Weight and Application Complexity

Where chemicals are prone to produce stains and where surfaces need to be regularly cleaned to remove particulate contamination, then alkyd, solvent-based paints and industrial coatings comprising epoxies, resins, polyurethane, elastomeric membranes, and fiberglass coatings are recommended. These finishes off-gas VOCs during the painting period, but their ease of cleaning reduces the frequency of paint reapplication, thereby reducing the number of times the finish will emit VOCs. Added ventilation may be required during installation and curing to mitigate the effects of the VOCs.

13.6.2 Floor Materials

13.6.2.1 Concrete

Concrete floors should be sealed with industrial coatings which are resistant to constant abuse such as abrasion, scraping, impact, and rolling loads as well as stains from oils and strong acids. The method of application will strongly determine the floor's resistance to damage. This could be either mortar applied, broadcast, or slurry applied. In some cases, the method of application will allow the coating to go up the wall forming wainscots or cove bases which make cleaning way easier. Concrete floors could also be covered with terrazzo topped with vinyl sheets, vinyl composition tiles, linoleum tiles, and quarry tiles (Table 13.2).

13.6.2.2 Terrazzo

By nature, terrazzo is made up of naturally occurring stone aggregates, glass, or recycled plastic bound by cement or epoxy-based binders. It has high durability and low VOC emissions.

13.6.2.3 Vinyl Sheet

These are appropriate for spaces with low-risk spillage of aggressive, staining chemicals. Their durability is greatly determined by its installation methods, moderate use, and low abuse of the floor surface. Vinyl sheets are water-resistant and easy to clean and repair in case of damage.

13.6.2.4 Vinyl Composition Tiles

This type of floor finish is easy to install and maintain. However, as tiles aren't sealed, they are susceptible to chemical leakage into lower floors and are relatively harder to clean as joints collect dirt. They are also prone to damage along the joint edges making them trip hazards.

TABLE 13.2
Laboratory Floor Material Selection Criteria

Material	Cost	Advantages	Disadvantages
Linoleum tiles	Medium	• Durability	• Less resistant to chemicals and abrasives • Difficult to clean • Limited use to dry areas as they are not water-resistant
Vinyl composite tiles	Low	• Durability • Low VOC formulations • Low installation and life cycle cost • Easy to repair, replace, and maintain	• Susceptible to impact damage • Has many joints that can collect bacteria; hence needs application of sealers and polishes (for high traffic potential and stain-prone areas)
Rubber flooring	Medium	• Durable • Low VOC formulations • Slip-resistant • Walking comfort • Wears fairly well • Damage-resistant to chemicals • Easy to repair • Easy to maintain	• Poor impact resistance
Concrete flooring	Low	• Durable • Ease of installation • Damage and stain-resistant • Easy to repair • Easy to maintain	• Poor chemical resistance • Uncomfortable to walk on

13.6.2.5 Linoleum

Made of petrochemicals, linoleum is prone to stains or is difficult to keep clean. It is therefore not recommended for laboratory use.

13.6.2.6 Quarry Tiles

It is not common for contemporary laboratories to employ use of quarry tiles. Being heavy in nature, they add significant dead load on the structural members. Cracks around joints make them trip hazards while they are potential slip hazards when wet. They are also significantly harder to clean than other finishes discussed above.

13.6.3 Ceiling Material Selection

Ceilings seldom experience spills and stains from chemicals and therefore have a different criterion in its finishing. Suspended ceilings with good acoustic properties are often used to reduce noise levels within the laboratory. Nevertheless, acoustic finished are fitted with fibrous materials that shed particulates over time. When these surfaces are lifted, particulates may dump down into the laboratory causing contamination. Alternative finishes that may reduce airborne participate contamination include painted, solid and seamless gypsum boards, sealed metal panels and sealed fiberglass panels. Access doors to ceilings must be sealed to reduce chances of dust and debris escaping into the laboratory space (Tables 13.3 and 13.4).

TABLE 13.3
Laboratory Ceiling Selection Criteria

Material	Cost	Pros	Cons
Fiberglass ceiling	High	• Excellent acoustic performance • Durability	• Particulate contamination during installation and removal • VOC formation
Mineral fiber	High	• sound absorption • fire resistance, and • light reflection	• susceptibility to moisture (leading to sagging/staining) • potential respiratory irritation during handling
Vinyl-wrapped ceiling	Medium	• Durability • Humidity (moisture)-resistant • Easy to clean • Easy to maintain	• VOC formation • Warps in extreme temperatures
Mylar-faced tiles	Medium	• High Light Reflectance • Moisture and Water Resistance • Easy to Clean • Durability due to resistance to abrasion and heat • Improved Acoustics (with fiberglass core) • Aesthetic Versatility	• Aesthetics Not for All Styles • Shows Fingerprints and Scratches • Requires Precision Installation • Potential for Glare • Cost: Mylar-faced options can be more expensive than standard ceiling tiles • Limited Thermal Insulation

TABLE 13.4
Laboratory Casework Selection: Pros and Cons

Material	Where Applicable	Advantages	Disadvantages
Wood	Often used in academic institutions. They are chosen for their aesthetic qualities and ability to handle chemical spillage.	• Easy to repair • Quieter when using as compared to steel casework	• Quite flammable • Warps due to water and chemical damage
Steel	Used in labs where spills, corrosive agents and heavy use commonplace.	• Resistant to chemical spillage • Resistant to abuse	• Noisier than wood • Easy to dent • Prone to corrosion • More difficult to repair than wood

13.7 CASES OF AFFORDABLE LABORATORY PROJECTS

Cost-effective, high-quality laboratories illustrate that spatial and financial constraints can be addressed. Collaborating with local staff and tailored solutions are crucial for future projects. Analyzing current facilities shows effective and economical designs. Ecuadorian labs exemplify successful quality standards. Research offers insights into organizational growth and quality management, as seen in the Centro Académico de Talentos (CAT) lab in Cuenca, Ecuador, which highlights how quality standards enhance outcomes and sustainability.

The lab showcases diligent procedural documentation, recordkeeping, and inventory management to enhance equipment maintenance. Insights from CAT demonstrate the value of an effective lab layout, fair practices, and following quality standards for financial health. Conversely, two laboratories at the neighboring University of Azuay faced issues after neglecting quality management, leading to their eventual closure despite the initial funding (Kaur, Downey, & Choate, 2018).

Setting up laboratories in developing countries, particularly in regions impacted by conflict or disasters, presents distinct obstacles that demand dependable and cost-effective solutions. It is essential that efforts to reduce costs do not sacrifice safety or quality standards. When concentrating on diagnostic laboratories, the importance of efficient design cannot be overstated, as it directly influences both operational effectiveness and financial outlay. Emphasizing functional priorities and spatial organization can significantly improve quality and cost-efficiency, while also addressing often neglected considerations.

To establish a laboratory, it is imperative to prioritize safety protocols, quality assurance, and financial sustainability. Effective administration involves strategic planning, clear roles, and optimal resource allocation. Key operational processes include selecting, installing, and maintaining equipment. Standardized record management and testing protocols are vital for consistency and accuracy. Information management practices support data collection while ensuring confidentiality (Berezin et al., 2023). Performance evaluation through quality assurance and customer feedback drives continuous improvement. Moreover, well-equipped facilities significantly enhance reliable testing outcomes and ensure safety compliance.

Construction failures stemming from design flaws have major consequences, underscoring lapses during the design stage. Failures encountered in laboratory design bring attention to four primary principles of safety and cost-effectiveness. Neglecting these principles exacerbates the difficulties faced during construction. While teams may effectively manage simpler projects, there is a tendency to incorrectly apply these methods to more intricate laboratory designs. This misapplication can lead to a phenomenon known as "kitchen-sink syndrome," where trivial details obscure essential design objectives. Consequently, this contributes to the creation of hazardous laboratory conditions. Furthermore, the inclusion of BSCs and fume hoods complicates regulatory controls, affecting biosafety policies and escalating expenses, which result in laboratory environments that are challenging to maintain.

13.8 CONCLUSION

The economic viability of a laboratory is fundamentally linked to its design framework. Four core principles govern the architectural blueprint, ensuring that it adequately incorporates environmental factors and operational efficiency of equipment while remaining within financial constraints.

The design of a space must prioritize functionality, ensuring organized operations. This involves choosing finishes, furnishings, and equipment aligned with the intended purpose, avoiding unnecessary embellishments. Additionally, the design should allow flexibility for future modifications and repurposing, reducing extensive reconfiguration and costs. Such adaptability broadens equipment usage and accommodates various laboratory setups, leading to cost-efficiency and improved facility utilization. The design should comply with established safety standards to decrease the likelihood of injuries, concurrently facilitating effective training and management processes. Additionally, sustainability is critical, which can be achieved through the optimal utilization of resources, enhanced recycling practices, and the minimization of maintenance expenses.

14 General Design and Layout of a Laboratory

Philip Mark Othigo and Lincoln L. Were Munyendo

14.1 INTRODUCTION TO LABORATORY DESIGN AND LAYOUT

A science laboratory is a critical teaching infrastructure needed to sufficiently inculcate practical skills to learners in the fields of Science, Technology, Engineering and Mathematics (STEM). Careful planning, consultations, design and construction can enable schools and higher learning institutions to create infrastructural layouts with plenty of flexibility and space for independent study as well as collaborative group work.

The design adopted needs to ensure that students have a clear view of the teacher and provides the teaching staff with shorter lines of communication and greater control over their learning environments plus flexibility to easily switch between practical and theory lessons (Bae & Martin, 2023).

14.2 SPACE OPTIMIZATION IN LABORATORY DESIGN

A laboratory should have space that could be re-developed in emerging diversified designs to accommodate open shelving for books, bags and other personal items, as well as concealed storage to house equipment, glassware, reagents and specimen. More importantly, designs should be integrated to enable seamless, temporary conversion of the laboratory space into a classroom design, allowing surfaces to be kept clean and clutter-free. Services, including gas and electric outlets, usually need to be located on the front face of each bench, away from the work surface, ensuring they are easy to access but tamper-proof.

The design of laboratory layouts plays a crucial role in optimizing operational efficiency. Research laboratories demand specialized configurations and workflows tailored to their distinct requirements for consistency and transparency in processes. However, operational and institutional challenges hinder efficient space utilization, worsened by inadequate frameworks. Analyzing these issues helps identify root causes and develop solutions (Jia et al., 2023).

Recently there has been a phenomenal global move toward emphasizing the value of STEM subjects. The LMIC countries do not have an accompanying school capital funding to enable them to actualize the desires. This has, therefore, orchestrated huge gaps in the quality of secondary schools' science laboratories. It is noteworthy that these laboratories are overwhelmingly unsuitable, low tech and unfit to inspire learners to excel in STEM subjects. Not surprisingly, therefore, there is a high level of dissatisfaction among teachers of the STEM courses, who have loudly indicated that they are not confident about pupil safety during their science lessons. This is definitely a ticking bomb waiting to go off and needs immediate arrest through careful management.

It has been known since time past that teachers often adjust their teaching style to suit their learning space. In teaching and learning of practical subjects, the settings require practical spaces that enable staff and students to get hands-on. This is achievable by specific considerations during the design, construction and furnishing of laboratories.

A general layout strategy to achieve space optimization for teacher and learner ownership would be adopting a dual-function laboratory benches approach, with one side for theory and the other for practical work (see Figure 14.1). This results in much more circulation space. Meanwhile since all

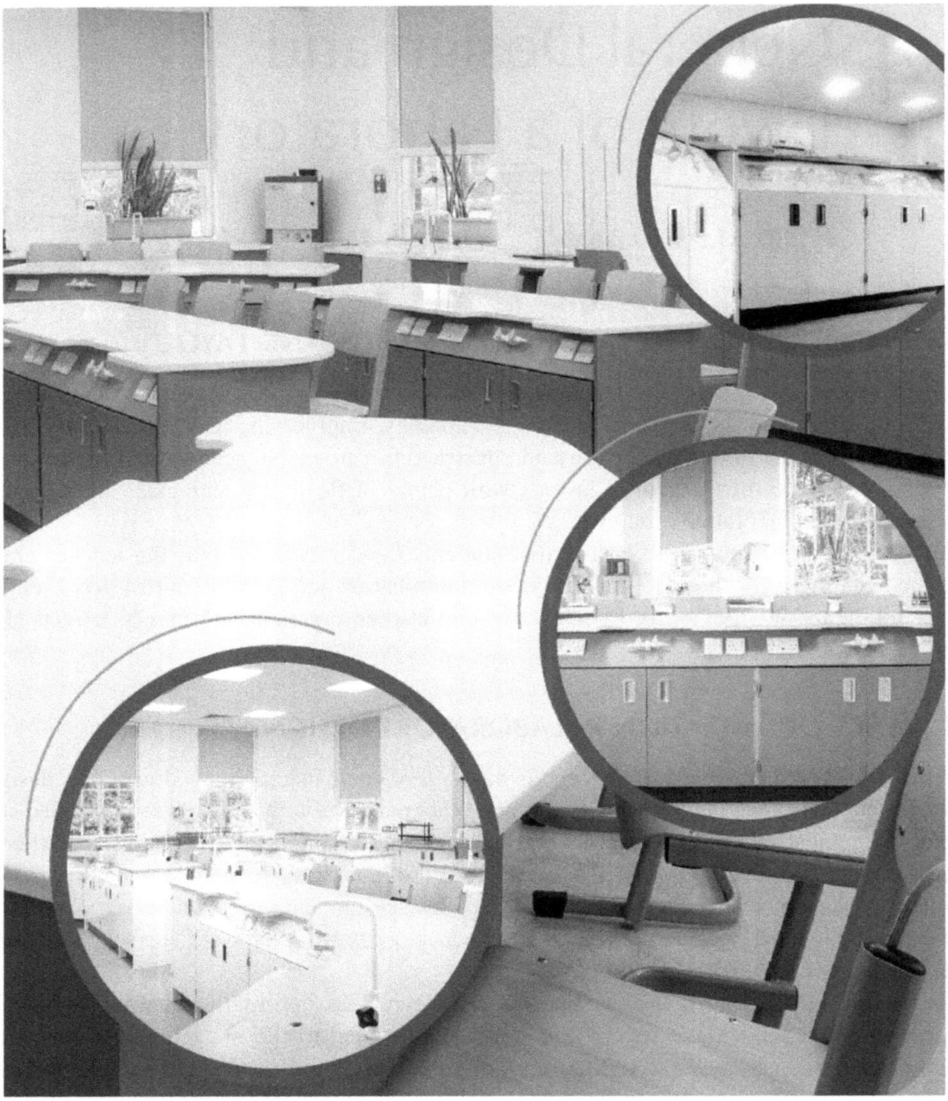

FIGURE 14.1 Sample dual-function laboratory benches.

the students are seated, they can see and hear the instructor throughout the session, and similarly the instructor can see and hear the students very well. This gives a much more flexible learning space that promotes ownership of the learning process as compared to when the benches and tables are laid out so learners sit behind each other in rows, which makes it hard for those at the back to see and hear what was being taught. On the other part the instructors see all students, thus enabling identification of those struggling so that they can easily get round the laboratory room to help them so no one falls behind.

Since ownership manifests greatly in learners' attitudes, the operationalization of the dual-function laboratory benches (see Figure 14.2) results in increased students' interests and desire to study science as they really look forward to practical sessions. In addition to the students feeling more confident in their ability to carry out practicals, the teachers also enhance their proficiency in planning and coordinating practical work, knowing they got the facilities to achieve what is required in a space that really works and inspires.

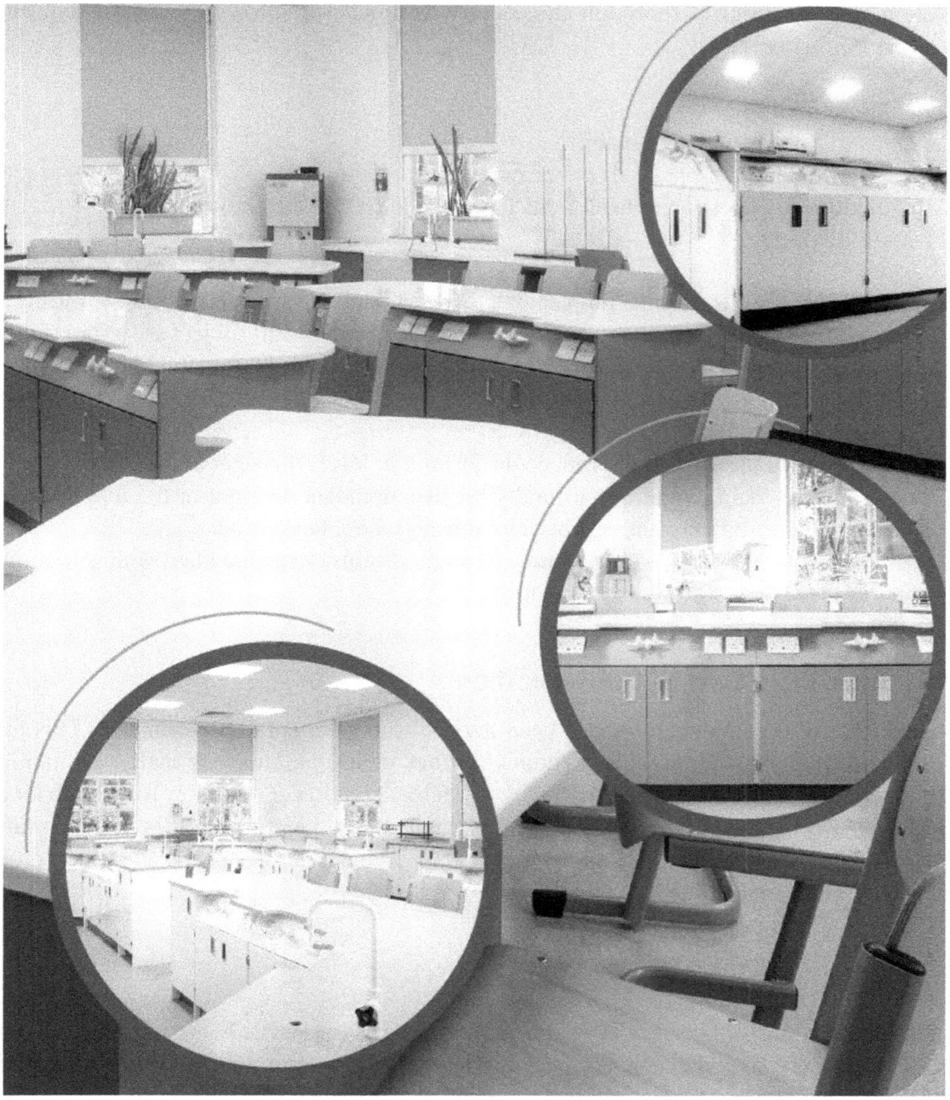

FIGURE 14.2 Operationalized dual-function laboratory benches.

Our next discussion is on the following desirable design outcomes that are easily attainable by institutions generally, despite funding inequities:

- Ventilations for quality air circulation
- Natural lighting
- Ideal rapid lines of communication
- Users' health and safety
- Optimized storage
- Sustained technology.

14.2.1 Ventilations for Quality Air Circulation

Quality ventilation is most essential in settings for scientific learning as it protects every user of the space from short-term and long-term exposure to toxic substances. Identifying the best ventilation systems requires thorough considerations of set variables. The crucial requirements include the following:

- Indoor air quality
- Energy efficiency and sustainability
- Underwritten legislation
- Regulations in educational institutions for new and refurbished buildings
- Natural lighting.

Natural light is best but is usually supplemented by artificial lighting from Light-Emitting Diode bulbs. This has the potential to greatly help in improved learning in all subject areas and is therefore a key consideration for the perfect science laboratory layout. Natural light could be maximized in a science lab design, mainly by orientation, since

- north-facing windows tend to produce uniform light levels throughout the day;
- east- and west-facing windows normally receive abundant daylight in the fore and afternoon, respectively, hence the risk of glare during school hours; and
- south-facing windows, however, usually require curtains or special glass with solar film as window panes to control the degree of light.

14.2.2 Ideal Rapid Lines of Communication

Shortened lines of communication have been noticed to revolutionize performance. This must be emphasized during science laboratory room's settings of layout. Conventionally, positioning the teacher on the long wall rather than the short wall enables creation of a shallow room with long rows of benches or workstations. The consequence of this is that no learner shall be ever too far away from the teacher which in turn goes a long way in improvement of all aspects of learning processes from behavior and concentration to attainment.

14.2.3 Users' Health and Safety

A serious hazard in school science laboratories manifests as limited or constricted access. The layout of the laboratory should enable teachers and learners to move around the room with minimum potential for accidents, particularly when conducting experiments. Some or all of the following attributes should be implemented to assure safety:

- A good-quality nonslip flooring that is durable, safe, stain-resistant and easy to maintain should be installed in laboratories to minimize the likelihood of accidents.
- Work surfaces should be heat-, scratch- and impact-resistant and able to withstand daily use in order to aid in decontamination.
- Electricity and gas services should be of appropriate design to facilitate rapid shut off if required to ensure capability of being turned off easily if there is a leak or drop in pressure.
- Fume cupboards (FCs) for evacuating harmful fumes, vapors or gases during the time for conducting experiments are very vital. This should be given much attention during laboratory construction and also regularly maintained.

14.2.4 Optimized Storage

Intelligent storage, when incorporated into laboratory designs, tend to provide solutions in even the most constrained of spaces and is easily achievable by simply ensuring it is in-built into the specification.

The suggested strategies for optimization of storage (Roberson et al., 2013) may, among others, include the following:

- Locating storage solutions such as lockers in corridors, so long as it does not limit access.
- Addition of under desk storage in the room itself with integrated cupboards plus pigeon-hole units to store coats and bags.
- Hiding away resources such as interactive screens and power cables, seamlessly, for example, by incorporating sliding write on whiteboards that can be used to conceal test results or reveal answers to equations.

14.2.5 Technology, Automation and Robotics

The diversity of tasks within, especially clinical, laboratories demands the use of flexible devices and software, a characteristic that traditional automation systems do not possess. The implementation of tailored solutions can result in significant expenses, which could impact laboratory budgets and necessitate the establishment of enhanced infrastructure for cell monitoring. Automation systems have the potential to enhance the quality of care while simultaneously reducing operational costs. Mobile robots play a key role by handling non-value-added tasks, such as the transportation of samples and the replacement of nutrient media, effectively employing their mobile bases, robotic arms, storage capabilities and navigation technologies (Kleine-Wechelmann et al., 2023).

Technology is currently an extremely important feature of any high-quality science initiative, including in laboratory setup. Educational technology tools are used appropriately and effectively in science laboratory sessions to actively stimulate students' engagement for knowledge construction to improve their thinking and problem-solving skills. Use of technology in laboratories comes in various forms, and their use could be through incorporation of the following:

- Modern interactive white boards
- Screens or projectors
- Data collection and analysis software
- Digital microscopes
- High-power projector or interactive TV
- Laboratory simulation.

14.3 ARCHITECTURAL CONCEPTS IN LABORATORY DESIGN

14.3.1 Laboratory Layout for Better Efficiency

It is essential that the planning and design of laboratories take into consideration both the present best practices and the flexibility to accommodate future needs and changes. A well-planned laboratory will ensure safe and secure work environments for its users; future-proof to make it more adaptable to the rapid advancements in science and technology; and achieve sustainability by reducing energy demands, minimizing waste and creating healthy and productive environments.

An optimal layout enhances the performance of unit operations. The design of a laboratory is based on three fundamental principles: (1) *Workflow:* Arrange areas to facilitate the seamless movement of materials and tasks; (2) *Safety:* Guarantee accessibility to safety gear while reducing potential hazards; (3) *Space Utilization:* Maximize the effective use of space to accommodate both personnel and equipment efficiently. An added benefit to efficient laboratory layouts is its ability to attract and retain personnel as well as enhanced interaction and collaboration which is vital for the development of new inventions, cures and scientific progress. Exchanging tools, equipment and workspace is vital for effective collaborations (Zhi & Lien, 2024).

In designing a laboratory layout, it is important for the planners and design team to examine the extent to which their design will influence or affect the space users. Some of the elements they may observe are the users' access to views and natural light, preference to cellular laboratories over open laboratories (Sharma et al., 2019) and the desired nature of interaction and collaboration between laboratory personnel.

14.3.2 Laboratory Types

Laboratories can be distinctively grouped into three main categories which are determined by the primary function and end goal of each group. These groups consist of *private-sector laboratories*, whose primary goal is to bring innovation to market; *government laboratories*, which are primarily concerned with research that are geared toward regulation for public interest; and *academic laboratories*, which are largely fashioned for teaching, experimental and research purposes (Roberson et al., 2013).

14.3.2.1 Private-Sector Laboratories

The private-sector laboratories ***function*** to discover, innovate and ultimately make profits from their research by bringing their products and services to the market.

14.3.2.1.1 Spatial Requirements

One of the primary goals of a private-sector laboratory is to attract and retain the best scientists in their respective fields. Additionally, improving productivity and efficiency of their employees is essential (Roberson et al., 2013). These facilities will therefore have additional spaces such as day-care centers, fitness facilities, walking trails and cafeteria, to name a few. As the end goal is to bring their products to market, researchers work closely with the marketing teams who will need their offices in close proximity to the research facility.

14.3.2.1.2 Special Considerations

It is worth looking into start-ups, which are a more common form of private-sector laboratories these days. Acknowledging that, at formation, most of such enterprise have very little capital to design their own facilities, they may either choose to lease existing facilities and customize them to fit their needs; or they could opt for a design-built system, where much of the investment comes from the developer. In case of the latter, the developer may choose to construct a more generic laboratory in case the initial start-up fails, so that the facility can, in turn, serve future research studies.

14.3.2.2 Government Laboratories

The ***function*** of government laboratories often focuses solely on research, having very few or no teaching laboratories, and is often geared toward the interest of the public. They may also test the research findings conducted by private-sector laboratories to ascertain that various products are fit for the general public, for instance, how the Food and Drug Administration (FDA) in the USA approves drugs developed by various pharmaceutical companies. Such government laboratories are often used as legal tools to provide evidential support like in the case of the Government Chemist and Public Health Laboratories in Kenya. They are also used for sectoral regulation like the Dairy Board of Kenya Laboratories and the National Quality Control Laboratories (NQCL) which are used to sample, assess and qualify the standards of milk and medicines, respectively, in the Kenyan market.

The NQCL is a legal body set up to support the Pharmacy and Poisons Board (PPB) in regulating the quality of medicines at registration and post-market surveillance through a process called pharmacovigilance.

14.3.2.2.1 *Special Considerations*

Security is of vital importance in these facilities to ensure only authorized personnel have access to respective government departments. Security personnel, cameras and a card access system are typically used in these facilities.

Government laboratories should promote innovation and creativity by promoting interaction among scientists. This could be through efficient communication systems and collaborative spaces. They should be both flexible and adaptable while offering security, safety, affordability and energy efficiency (Weldon, 2012). As there is a goal to retain their employees, the spaces should also promote the quality of life of its inhabitants.

14.3.2.3 Academic Laboratories

These facilities ***function*** mostly by offering instructions to students as well as conducting scientific experiments and research.

14.3.2.3.1 *Spatial Requirements*

Academic laboratories vary greatly in regard to the discipline, course level, equipment, laboratory type and the degree of required flexibility (Weldon, 2012). Thus, the requirements for biology, physics and chemistry laboratories as well as the needs for undergraduate, secondary and primary school laboratories may vary. However, the usual standard requirements for academic laboratories include the following:

- Mobile and flexible casework and furniture that facilitate variety of teaching and learning scenarios as well as allowing both research and teaching to be conducted in the same space.
- There should be sufficient storage facilities for equipment and students' belongings. Overhead storage should, however, be avoided to offer better sight lines for teaching and learning.
- The laboratory spaces should be placed adjacent to prep, storages and equipment supply to allow for efficiency of the teaching laboratory.
- Flexible workspaces.
- Adequate bench space for equipment and instruments.
- Adequate number of fume hoods and biosafety cabinets (BSCs) placed along the perimeter of the lab space.
- Write up stations where students can document observations and results from their various experiments.

  Biology laboratories: These are wet labs. In addition to the aforementioned considerations, it is necessary for biology labs to be designed with a level of flexibility to accommodate a range of biological research, from general biology to molecular genetics and biochemistry (Weldon, 2012). To be included in the design of this facility is support spaces such as vivarium facilities, tissue culture areas, incubators, growth chambers, dark rooms, refrigerators and freezers.

  Pharmaceutical laboratories: These are a mixture of wet and dry labs that play a critical role in the research and development as well as the testing of new drugs, encompassing a range of specialized disciplines including Pharmaceutics, Pharmacology, Toxicology, Pharmacognosy, Biotechnology, Microbiology, Analytical Chemistry and Quality Control.

  The architectural configuration of such laboratories necessitates systematic arrangements for intricate machinery that conform to established safety regulations (Roberson et al., 2013). A significant challenge lies in achieving a harmonious balance between spatial utilization and the delineation of functional areas for workflows, storage and contamination mitigation. Strategic design and the positioning of equipment play a pivotal role in improving operational efficiency, safety and ease of access. The incorporation of adaptable workspaces is essential for accommodating the evolving requirements of research, thereby optimizing laboratory settings for the development and production of novel pharmaceutical products in accordance with safety and quality standards (Avivar, 2012).

Physics laboratories: These are mostly dry labs. Some of the key considerations for these spaces include noise and vibration mitigation, magnetic shielding and support for computer and telecommunication. Due to the bulky equipment used in these spaces, special consideration on their floor loads needs to be addressed.

Chemistry laboratories: These are wet labs. Attention is to be directed toward the mechanical work needed in within these rooms. They include supply of piped gases, heavy electrical and data infrastructure for instruments and ventilation requirements.

Engineering laboratories: These are often dry labs which are often larger due to the customized setups and equipment that would be far too large to fit in a standard laboratory. Space provision should take into account the size of apparatus used within these spaces. This should extend to the calculation of floor loads as well as the sizing of openings such as doors.

Computer laboratories: These require extensive cabling network and management, and should ensure electrical and data outlets are readily accessible on the floor, walls and overhead. Raised flooring will allow for better cable management along the floor, facilitate under floor cooling and shield the space from electromagnetic fields. The light requirements of these spaces should also be specialized to reduce glare.

Geology laboratories: These are often dry labs. They require ample space for display of maps and rocks. Some facilities would require fume hoods, water supply and drainage services for experiments as well as hydraulic lifts to transfer heavy rocks from place to place.

14.3.3 Architectural Considerations

14.3.3.1 Ceiling Height

The vertical dimensions of a given laboratory space is chosen based upon the requirements of all stakeholders (Roberson et al., 2013) and the needs of the equipment, lighting and building services that are meant to be incorporated into the space. Generally, the minimum ceiling height for laboratories should not be less than 2,700 mm.

14.3.3.2 Laboratory Doors

The recommended minimum door width should be 910 mm. However, using double doors with active and passive panels may be useful. A 900 mm active leaf can be paired with a 300 mm inactive leaf to bring the total width of the opening to 1,200 mm. This will facilitate better flow of lab users and equipment.

14.3.3.3 Aisles

As it will be discussed later, a minimum aisle width of 1,500 mm is recommended in the lab to permit a person to pass behind a researcher while they're working. Aisles wider than 1,800 mm should be discouraged, as these are susceptible to clutter with researchers using them to store carts and equipment.

14.3.3.4 Overhangs

Use of overhangs can help shade windows and improve the quality of natural light entering the interior space of the facility. It can also help in generating an appealing design for the building's exterior image.

14.3.3.5 Glazing

Exterior windows should be equipped with a thermal break and an insulating section between the inner and outer sections of the frames. Frames should be made of wood or fiberglass to give the glazing a better thermal performance.

14.3.3.6 Roof and Walls

In order to reflect light and heat, the selected roofing should be light-colored with high-albedo coating. To prevent air leakages, all electrical outlets and plumbing, and wire penetrations into the building should be sealed.

14.3.4 Laboratory Configurations and Layouts

Laboratory configurations can be categorized into three distinct types: open, modular and cellular. Open layouts provide shared work areas that enhance flexibility among users. In contrast, modular layouts incorporate movable furniture arrangements that facilitate both individual separation and community interactions. Cellular layouts, on the other hand, isolate personnel to perform designated tasks efficiently (Sharma et al., 2019). The choice of an appropriate layout is contingent upon several factors, including workflow optimization, safety requirements and spatial constraints (Tanutomo & Octavia, 2016; Sharma et al., 2019).

The contemporary approach on designing laboratories employs an open layout which promotes team-based work. This layout facilitates communication between researchers and allows for flexibility for future needs. Researchers share more than space; they share bench space and equipment, which helps to reduce cost of the research.

Closed laboratories may still be required for certain research and equipment (see Table 14.1). These may include nuclear magnetic resonance equipment, tissue culture labs and darkrooms, to name a few. There are cases where a researcher would demand a closed space for his research and equipment, requiring a closed space as well. The use of a laboratory module when designing the laboratory space can allow temporary partitions to be installed and removed at will so as to provide flexibility in creating closed labs in an open layout. This would demand a comprehensive design of the building services to incorporate this change.

14.3.4.1 Design of Laboratory Exterior

Typically, a building's outward appearance is influenced by the site context in which it finds itself, the architectural and cultural context, and the client's vision and objectives. For laboratories, the physical organization and design of offices, lab spaces and other major mechanical services, as well as the size, number and architectural expression of exhaust stacks, play an important role in determining the image of the facility's exterior (Roberson et al., 2013; Weldon, 2012).

TABLE 14.1
Pros and Cons of Open vs. Closed Laboratory Layouts

Layout Type	Advantages	Disadvantages
Open laboratory layout	• Cheaper to construct as it requires fewer partitions • Improves efficiency of space per square meter • Accommodates more casework and equipment • Allows for team-based research	• Less secure than closed labs
Closed laboratory layout	• Allows for tighter security • Better containment	• More expensive to construct • Reduces communication and interaction between researchers

14.3.4.2 Physical Organization and Design

The physical organization of different laboratory spaces can have an impact on the outward image of the facility. When offices, for instance, are consolidated into a large space, then their exterior walls can be more sculptural and incorporate more glazing. In order to receive natural light, many researchers prefer their offices to be placed along the exterior wall.

The lighting needed for the interior spaces within the laboratory facility will also determine the amount of glazing used on its façades. Of course, other factors such as budget and the client's objectives will play a part in making this design decision. Glazing should be elevated between 900 and 2,900 mm to allow wall space for furniture, casework and equipment.

The building's appearance can also be influenced by the location of mechanical equipment within the laboratory building. This will have a considerable impact on the massing of the building. Mechanical equipment could be located at the roof, in the basement, in mechanical rooms on each floor and within the interstitial space. Hiding equipment using screen wall can enhance the aesthetic of the building as well as lower the construction costs, but the building equipment might be susceptible to weather and human damage.

14.3.4.3 User-Centered Designs

User-centered design is a design philosophy and systematic approach focused on considering the needs, desires and constraints of end users in every design phase. This process functions as a comprehensive, multistage methodology that systematically examines user interactions while also corroborating assumptions via empirical testing. When implemented in laboratory environments, this approach emphasizes the essential requirements of laboratory personnel, focusing on effective communication, flexibility, support for performance, overall well-being and ergonomics (Doshi & Clay, 2017; Roberson et al., 2013).

Ergonomics focuses on workplace efficiency, prioritizing user comfort and safety by designing components that address physical, physiological and psychological needs. This approach involves optimizing thermal and acoustic conditions and enhancing tools and layouts. Understanding human capabilities and workplace dynamics is crucial, with principles like "fitting the task to the person" guiding design. Comfort encompasses both physical and perceived aspects, influenced by conditions and personal perceptions. Quantitative comfort assessment considers pain absence, support and environment. Discomfort reduces efficiency, so ergonomics aims to improve work environments, enhance operator–machine interactions, minimize health risks and accidents and promote staff well-being (Saha et al., 2024).

14.3.4.4 Expression of Exhaust Stacks

Choosing on whether to have individual or bundled exhaust stacks also determines the image of the laboratory building.

Individual stacks: Each fume hood in the building has its own dedicated exhaust stack, for example, perchloric and radioactive hoods. The main advantage of this design choice is that it eliminates the chance of cross-contamination between the exhausts.

Bundled stacks could either bundle individual stacks or the use of a manifold exhaust system. Architects and designers have used these elements to develop unique architectural forms that enrich the skyline of the areas where they are built.

14.3.5 Utilization Metrics for Laboratory Space

The role of space in laboratory settings is critical, as it directly influences operational expenditures. Analyzing space utilization can enhance the effectiveness of the facility and ensure compliance with budgetary constraints. Key performance indicators serve to quantify space efficiency and underscore areas ripe for improvement. These indicators facilitate the identification of inefficiencies, establish performance benchmarks and promote advancements in space management strategies (Kiechle & Arcenas, 2016). Four specific metrics contribute to the assessment of space usage within projects:

a. *Laboratory-net-to-gross ratio (Lab–Net/Brutto):* The gross-to-net assignable square footage ratio in a laboratory project quantifies the allocation of space designated for laboratory functions, encompassing both the laboratory areas and their associated support spaces.
b. *Laboratory–bench ratio (Lab–Bench Ratio):* The ratio indicates the relationship between the area designated for bench lab work and the overall space available, and encompasses workstations and fume hoods, but does not account for auxiliary areas.
c. *Laboratory–equipment ratio (Lab–Equip Ratio):* The metric indicates the ratio of laboratory space allocated to significant instruments in relation to the overall laboratory area. This encompasses areas for substantial, fixed equipment such as centrifuges and spectrophotometers, as well as the necessary space for their operation, maintenance and the storage of smaller instruments.
d. *Laboratory–table ratio (Lab–Table Ratio):* This metric assesses the ratio of laboratory space designated for bench work relative to that assigned to major instruments. It reflects the balance between workstations and equipment within the laboratory arrangement. An elevated ratio signifies an increased allocation of space for bench-related tasks, whereas a diminished ratio indicates a greater allocation for equipment usage.

In U.S. governmental entities, the laboratory-net-to-gross ratio typically falls between 0 55 and 0.65. A ratio exceeding 0.65 could reflect a constraint in available office space, while a ratio below 0.55 may suggest an abundance of space or opportunities for expansion. Furthermore, the laboratory–bench ratio usually lies within the range of 0.25–0.50, where elevated values indicate a greater allocation of bench space compared to office space. Conversely, a ratio lower than 0.25 implies a higher availability of space for alternative purposes.

14.4 LABORATORY LAYOUT

14.4.1 Laboratory Module

A laboratory module is the basic unit for planning, design and organization of laboratories. By definition, a laboratory module is a three-dimensional planning unit composed of a specific floor space (often determined by the bench spacing which in turn is determined by the anthropomorphic dimensions of the laboratory worker and his/her workspace). An ideal module should contain suitable space for partitions, benches, floor standing equipment and extract devices and aisles that minimize circulation conflicts/safety hazards.

The module depth is determined by the bench depth and the aisle width. The length is often determined by the unit size of chosen laboratory furniture, requirements for free standing equipment and the number of scientists that will occupy the space. A typical laboratory module width is between 3,300 and 3,500 mm, and it considers a bench depth of between 1,200 and 1,800 mm with a minimum aisle width of 1,500 mm and a service spine of 400 mm (see Figure 14.3). For generic laboratories, a module of 3,300 mm wide is preferred, as it gives ample space for a bench and a FC in either wall of the module.

Considerations of the type of work and the general arrangement of equipment can facilitate a more economic layout. Priority should, however, be given to the flexibility of the space to accommodate future practices and equipment as well as the ease by which the laboratory can be expanded.

14.4.2 Combining Modules

14.4.2.1 Two-Directional Module

By employing the standard width and depth of modules, casework can be organized in either direction (see Figure 14.4). This arrangement is beneficial in the sense that it is more user friendly than the basic lab module, and it can accommodate different lengths of run for casework. Mobile casework will enable lab users to rearrange the space to fit different types of research within the space.

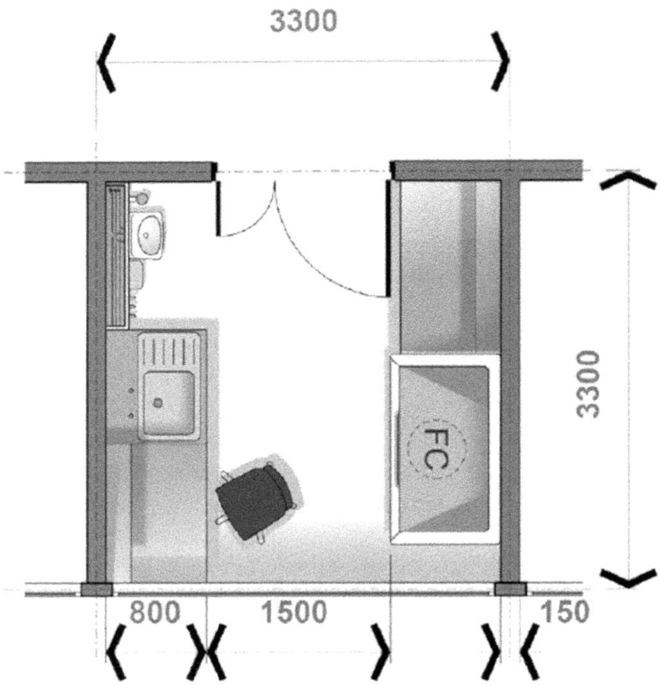

FIGURE 14.3 Basic laboratory module. (*Metric Handbook: Planning and Design Data* (Buxton, 2021).)

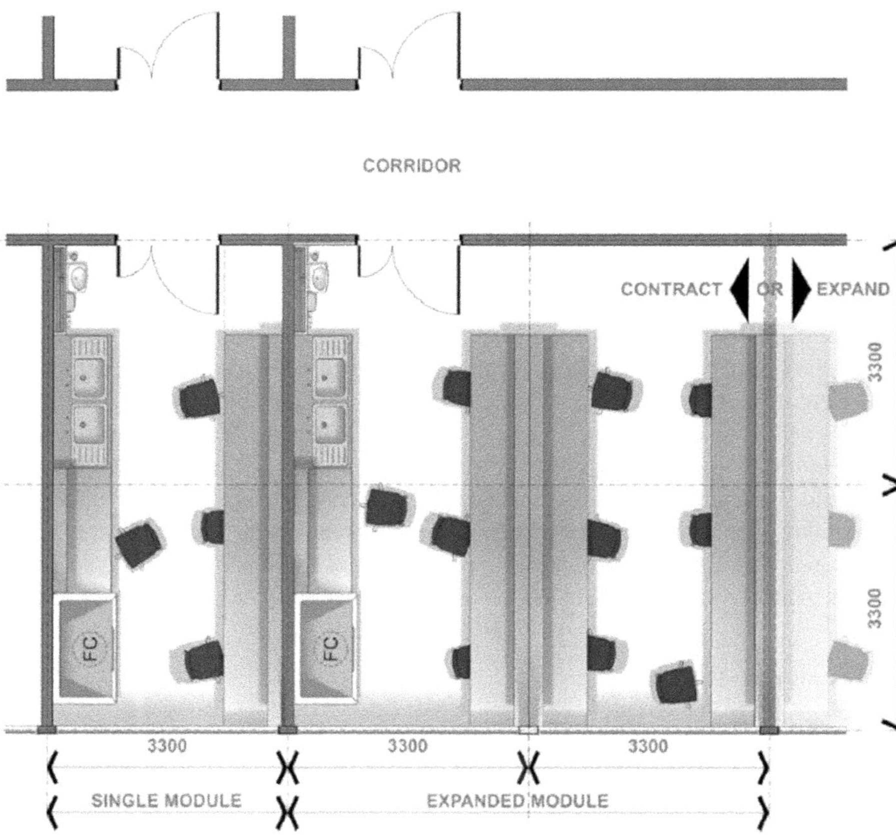

FIGURE 14.4 Flexible, adaptable and expandable two-directional module (can expand and contract on either side, hence flexible and adaptable). (*Metric Handbook: Planning and Design Data* (Buxton, 2021).)

14.4.2.2 Three-Directional Modules

The basic module, or the two-directional module is combined with any laboratory corridor arrangement for each floor of the building to give rise to three-directional modules (see Figure 14.5). While creating a three-dimensional module, the basic or two-directional module must be well defined; proper coordination of all vertical risers (such as fire stairs, restrooms, elevators, and shafts) and coordination of mechanical, electrical and plumbing system with the ceiling work must be established.

Three-dimensional module enables the designer to easily plan for change in corridor configurations on different floors of the laboratory facility, allowing for easier renovations. It also allows him/her to be more responsive to the program needs of the lab users on each floor. While this design option is recommended for most facilities, it requires proper coordination and design thought in the initial design.

14.4.2.3 Circulation and Access

The importance of proper and efficient circulation systems within laboratories cannot be understated as it guarantees safety of its personnel. Safety, in this case, is concerned with ensuring emergency egress for building occupants as well as allowing emergency personnel such as fire fighters, police and hazardous materials response officers, to easily access the building and its internal parts.

An efficient circulation system allows for easy movement of the laboratory users, material, waste and equipment; reduces susceptibility to accidents while allowing for ease of escape from hazards; and accommodates desired relationships between primary, secondary and tertiary spaces.

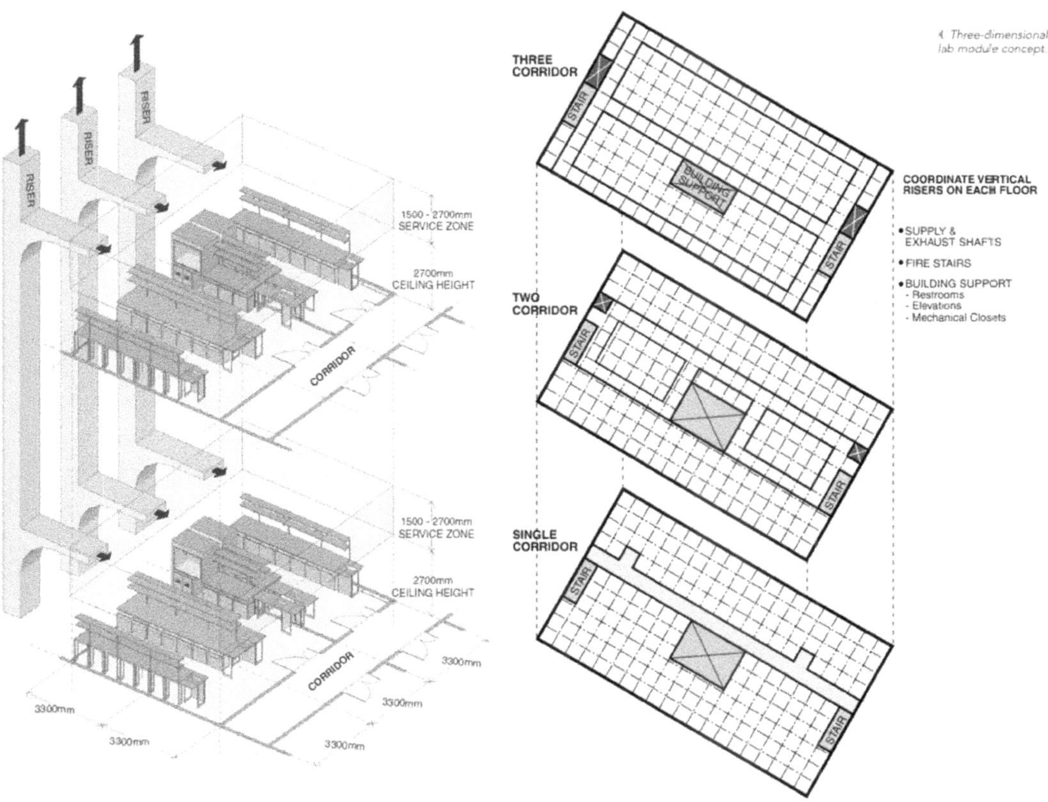

FIGURE 14.5 Varying corridor configuration on each floor. (*Metric Handbook: Planning and Design Data* (Buxton, 2021).

There are four main laboratory corridor configuration layouts: *single corridor configuration, internal loop corridor configuration, perimeter loop corridor* and a *grid corridor system* which are discussed below.

14.4.2.4 Single Corridor Configuration

This layout configuration comprises a single corridor traversing through the center of the building. Laboratory buildings with this configuration generally have emergency egress doors or fire stairs at the end of each corridor to facilitate easy escape from fire.

Single corridor configurations (Figure 14.6) are characterized by their high efficiency and increased occupant interaction. Unfortunately, they lack service corridors and have reduced visual access to the building's exterior. It is also quite difficult to get a wide floor plate with this configuration type.

14.4.2.5 Internal Loop Corridor

An internal loop corridor (Figure 14.7) manifests as a path that wraps around the building's center core spaces, providing access to laboratories and support spaces on two or four sides of the building. They are double loaded, which increases their efficiency, and it enables use of wider floor plates. However, occupant interaction in this type of layout is considerably lowered. Furthermore, it is increasingly difficult to achieve natural lighting within these spaces, and they lack service corridors.

14.4.2.6 Perimeter Loop Corridor

This comprises a circulation path that wraps around the building along the exterior walls on at least two sides. Presence of large windows within this configuration type can create for a good office space by widening the corridor.

Perimeter loop systems (Figure 14.8) allow for service corridors, have plenty of natural light and enable use of wide floor plates. They have, however, low efficiency and reduced occupant interaction. This configuration also reduces natural light within the interior spaces.

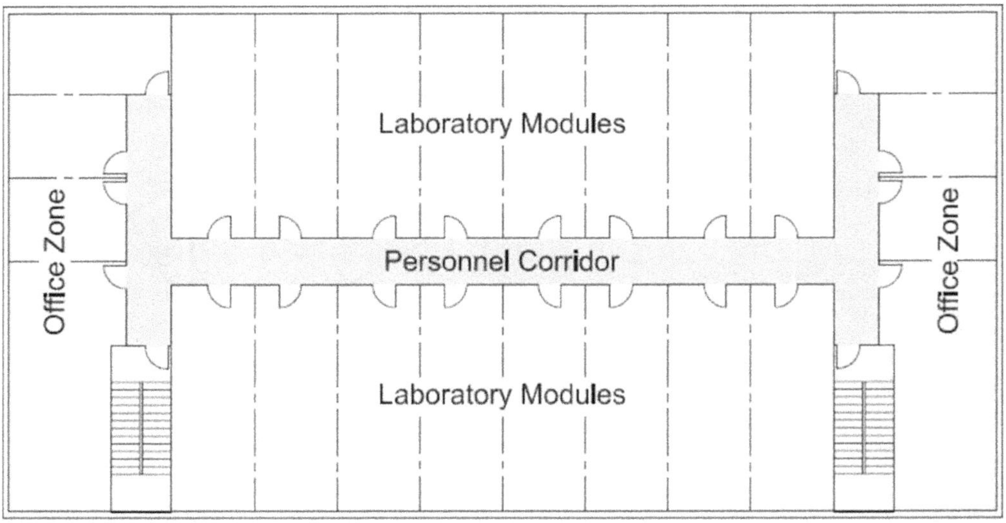

FIGURE 14.6 Single center corridor, building layout. (*Metric Handbook: Planning and Design Data*) (Buxton, 2021).

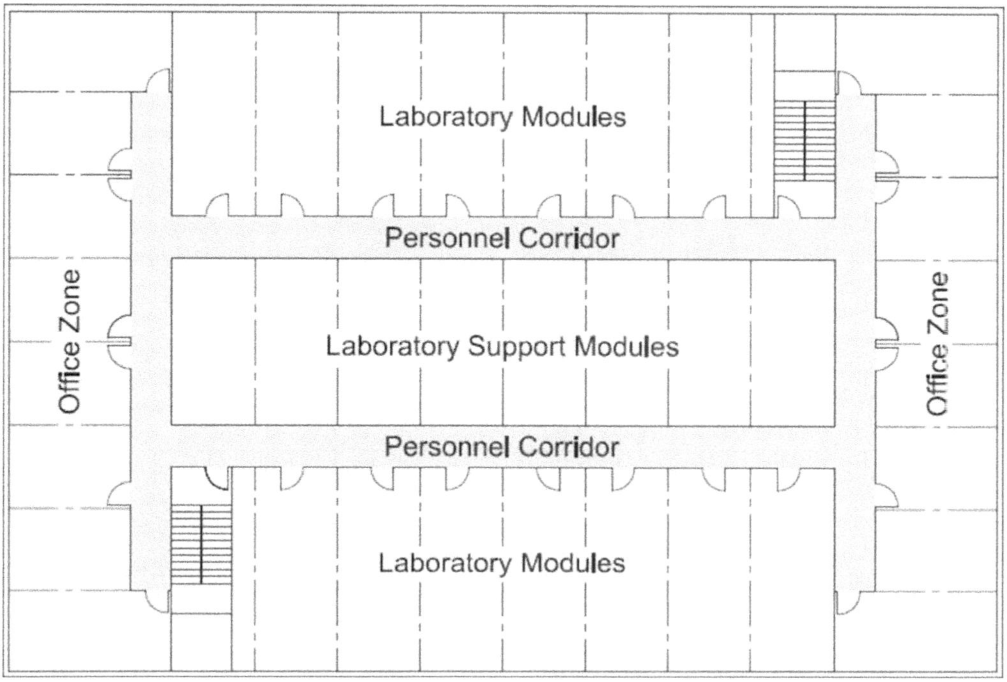

FIGURE 14.7 Internal loop corridor with core, building layout. (*Guidelines for Laboratory Design* (DiBerardinis, 2013).)

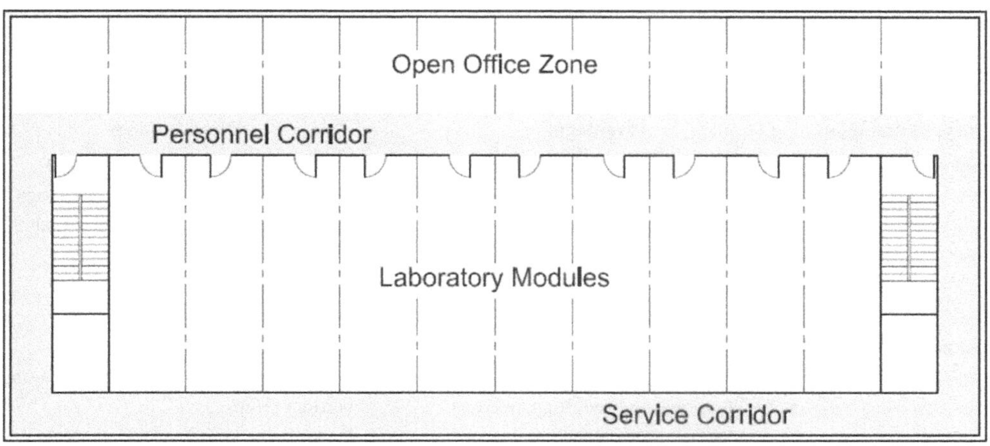

FIGURE 14.8 Perimeter loop corridor with core, building layout. (*Guidelines for Laboratory Design* (DiBerardinis, 2013).)

14.4.2.7 Grid Corridor System

A grid corridor system (Figure 14.9) may be used in cases where the interior spaces and functions of the building do not require windows for access to natural light or views to the exterior. While this system facilitates easy access to all parts of the laboratory, they need good signage to prevent new users from losing their way due to its network of corridors.

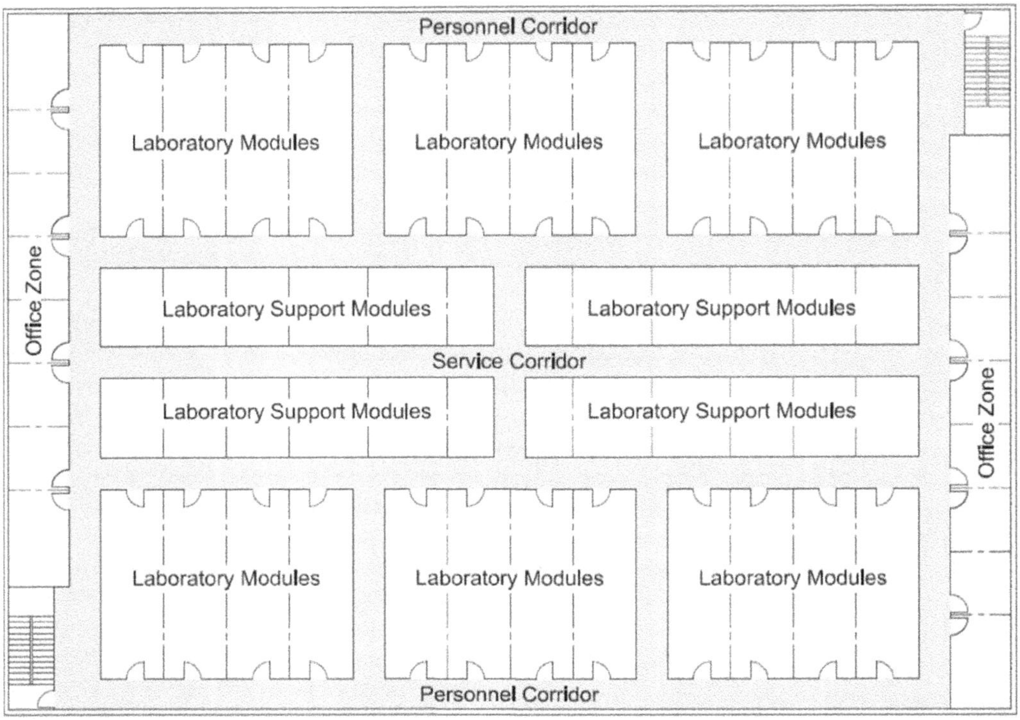

FIGURE 14.9 Grid corridor system, building layout. (*Source*: *Guidelines for Laboratory Design* (DiBerardinis, 2013).)

TABLE 14.2
Advantages and Disadvantages of Different Lab Layout Configurations

Configuration Type	Advantages	Disadvantages
Single corridor	• Building net-gross-ratio is 60% or greater. • Better communication opportunity as the corridor acts as a main street.	• Limits the width of the building and consequently, restricts floor plan design. • Difficult to achieve natural light when the corridor is interior. • Likelihood for material/people flow crossover and conflict.
Multiple corridors	• The floor plan is wider allowing for more layout options. • Ability to separate people and material circulation within the facility. • Allow for "ghost" corridors which facilitate better movement within the facility.	• Separates laboratory users. • Less efficient and more costly than a single corridor setup. • Presence of "ghost" corridors might compromise security.

14.4.2.8 Service Corridor

Often found in perimeter loop layouts and grid corridor systems, service corridors (Table 14.2) are restricted access pathways that are used by laboratory personnel.

14.4.3 Laboratory Aisle Dimensions

As a standard, an aisle should be 1,500 mm wide to adequately allow a laboratory user to comfortably push a laboratory cart between persons working at benches or equipment on either side of the

TABLE 14.3
Aisle and Corridor Widths

Circulation Space	Dimensions in mm
Aisles running along the ends of benches providing circulation within a laboratory	1,200–1,500[a]
Similar aisles, where there is no separate corridor for general circulation outside the laboratory	1,800–2,000[b]
General circulation corridors	1,500–2,000[a]

Source: *Metric Handbook: Planning and Design Data* (Buxton, 2021).

[a] Allowance must be made when the aisle may be partially obstructed, for example by staff using bench-end sinks, by doors opening into the aisle, by equipment and by fume cupboards and microbiological safety cabinets.

[b] Where possible and particularly when corridors are relatively long, a width of 1,800–2,000 mm is preferable, for both functional and aesthetic reasons.

TABLE 14.4
Recommended Aisle Distances in Laboratories

Minimum Distances	Dimensions in mm
Between front of bench or workstation and a facing wall, other furniture or equipment or pedestrian route (with one person at bench)	1,000[a]–1,400[a]
Between benches, furniture or equipment without workspaces either side allowing passage of one person at a time	1,000[a]
Between two people back to back but no need for a third person to pass between front of facing benches, workstations or equipment where people work allowing one of the people to pass behind the other	1,400[a]
Between two people back to back where space is required for a third person to pass between front of facing benches, workstations or equipment where people work allowing a third person to pass behind the others	1,450[a]–1,650 (1,800 exceptionally)

Source: *Metric Handbook: Planning and Design Data* (Buxton, 2021).

[a] 1,200 mm is required for Disability Discrimination Act (DDA) compliance and 1,500 mm (subject to risk assessment) when opposite a fume cupboard or microbiological safety cabinet.

aisle (Tables 14.3 and 14.4; Figure 14.10). This dimension also caters for access for people with disabilities. Aisle widths of less than 1,500 mm should only be used in instances where there is limited access, and where adjacent support rooms are often unoccupied such as store rooms.

In analytical chemistry laboratories and instrument laboratories that require utility access, aisles demand a length of about 600–1,000 mm. Aisle lengths of 1,800 mm are acceptable in research, testing and development labs, but dimensions larger than that may compromise safety as laboratory personnel would be tempted to place their equipment and other objects along the path. In case of emergency, these obstacles would endanger the safety and security of those seeking to escape. In certain cases, however, where bulky equipment need to be moved or assembled, such as chemical pilot plants, as well as engineering and physics laboratories, wider aisles are desirable.

To better facilitate escape from the laboratory, aisles should be oriented along the direction of the egress. This can be accompanied by floor signage and emergency egress lights to assist users to find their way. The accompanying figures illustrate the different aisle widths in a laboratory.

14.4.3.1 Escape

To facilitate escape in case of fire or other danger, the maximum distance to be traveled within the laboratory should be 18 m if it's in one direction only, and 45 m where travel is in more than one

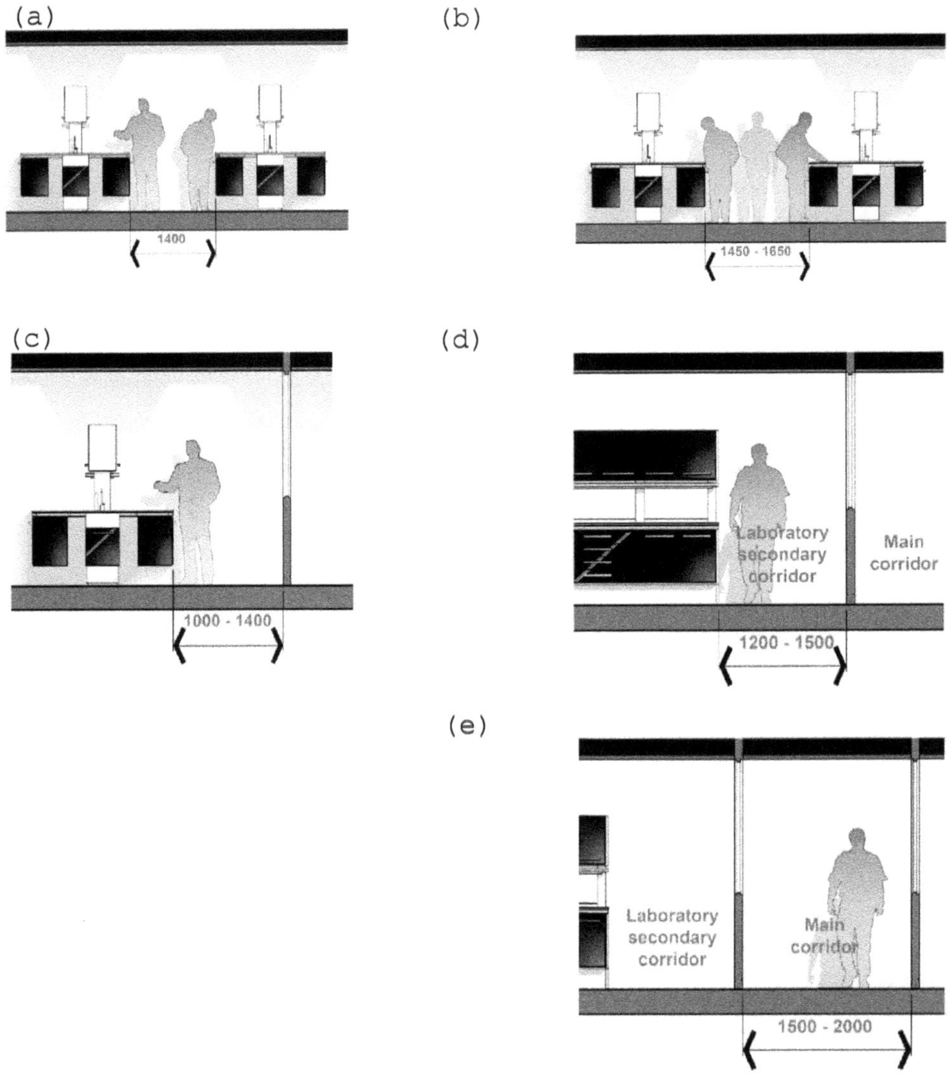

FIGURE 14.10 Aisle distances: (a) bench and wall/equipment, single throughway; (b) two workers back-to-back, single throughway; (c) two workers back-to-back, multiple throughways; (d) Internal laboratory dimensions; (e) main corridor dimensions. (*Metric Handbook: Planning and Design Data* (Buxton, 2021).)

direction (see Figure 14.11). Fire escape routes should be clear of all health hazard equipment such as a FC and gas cylinders.

14.4.4 Distribution of Mechanical Equipment and Services

The design team should develop a distribution network for the laboratory that is flexible enough to cater for future additions of exhaust chemical fume hoods and special exhaust systems. Additionally, the service distribution system should be installed in a way that does not interfere with the laboratory functions or activities.

A typical service distribution system comprises a vertical shaft of adequate dimensions that contain electric conduit, piped utilities and ventilation ducts, as well as a horizontal distribution system on each floor. Vertical shafts may be centrally placed or may be distributed at each module and commonly occupy from 1% to 10% of net usable area on a typical laboratory floor.

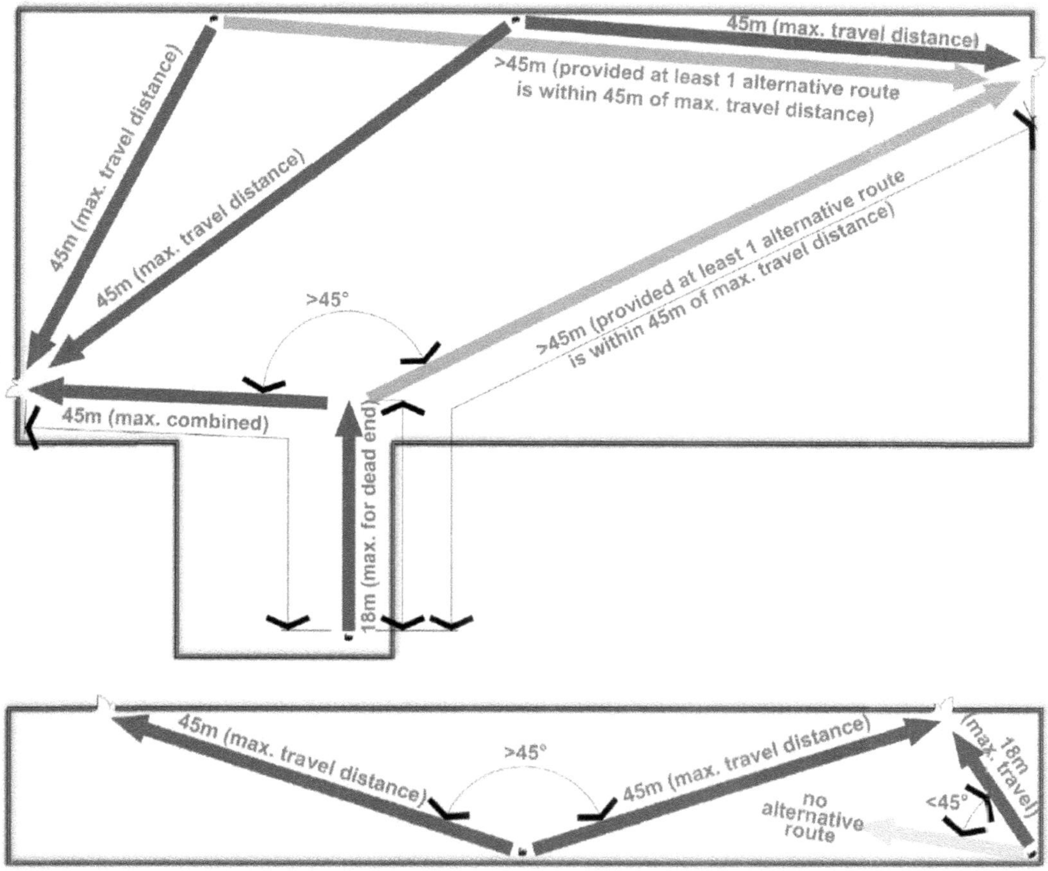

FIGURE 14.11 Design for laboratory escape. (*Metric Handbook: Planning and Design Data* (Buxton, 2021).)

For horizontal distribution systems, adequate floor-to-floor height is needed. The recommended floor height in laboratory, therefore, ranges between 4 and 5 m. In renovation projects where this cannot be achieved, exposed soffit fixed services could work. Horizontal distribution systems should be connected to vertical risers which extend to the top of the building to discharge contaminated exhaust air way above the roof surface. To prevent increase of static pressure in ducts, vertical continuity from floor to floor is vital in both central vertical shafts and individual riser shafts.

Primary air supply into multistory laboratory buildings depends on both vertical shafts and horizontal duct systems. An unobstructed space above the ceiling will facilitate horizontal supply and exhaust air ducts to run through the laboratory space. This may be achieved by interstitial floors which allow for easy maintenance of utilities and facilitation of air distribution.

14.4.5 Types of Distribution Systems

14.4.5.1 Central Utility Shafts in Laboratory Service Cores

In this system, shafts are located within a core area on the laboratory floor in buildings with an internal loop corridor configuration (see Figure 14.12). The core, in these buildings, contains laboratory support spaces, vertical circulation and building services. This system is suited for mid- and high-rise laboratory buildings.

Its main advantage is its flexibility while the long distance to be covered by horizontal ducts being its main disadvantage. Note that long distance horizontal ducts may need to be larger to cater for static pressure gains.

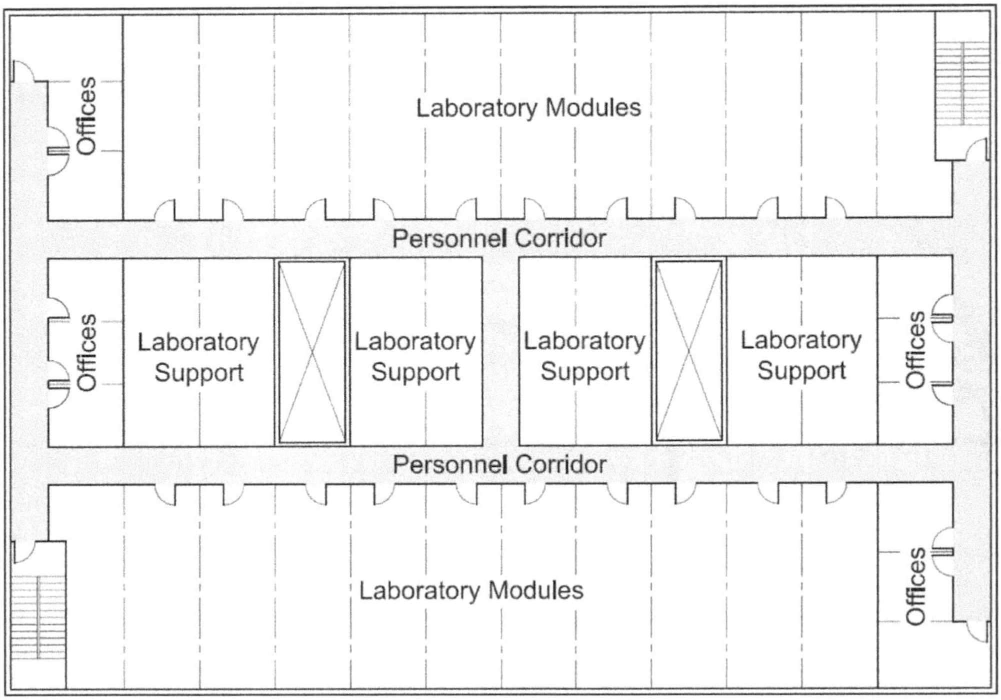

FIGURE 14.12 Central utility (mechanical, electrical and plumbing) shafts located in central core zone. building layout. (*Guidelines for Laboratory Design* (DiBerardinis, 2013).)

14.4.5.2 Centralized Utility Shafts

These systems (Figure 14.13) promote high building efficiency, as they produce a relatively low ratio of utility shaft area relative to net story area. However, there is reduced access to service these shafts, as maintenance personnel would need to access at least one laboratory on every floor, which may be undesirable to laboratory activities. To mitigate this limitation, having a finger corridor to the vertical shaft may reduce interference, as maintenance workers can access it using a separate route.

Chemical hoods can be placed adjacent to the service shaft or in a line at the back of the laboratory. The layouts for all pipes and ducts for a cluster of up to four laboratories can be accessed from one of any of the central utility shafts, which allows for variation in module length.

14.4.5.3 Utility Shafts at Ends of Buildings

This system (Figure 14.14) is desirable in conditions in which both supply air and exhaust air are distributed in continuous horizontal rounds along both sides of the building. An increase in distance between the end shafts reduces air volume which makes for efficient layout of ducts. In situations where the cross-section area in ducts remains constant, then flexibility in future additions and (or changes of exhaust devices and supply air volume requirements is easily achieved. Additionally, this system allows for continuous laboratory floors, as no major shafts interfere with laboratory layouts, adjacencies or cores.

Its major disadvantage is its lack of aesthetic, as it reduces views to the exterior. In long building layouts, end-located shafts are not as effective as centralized shafts due to pressure loss which is a a result of the long distance the ducts have to go.

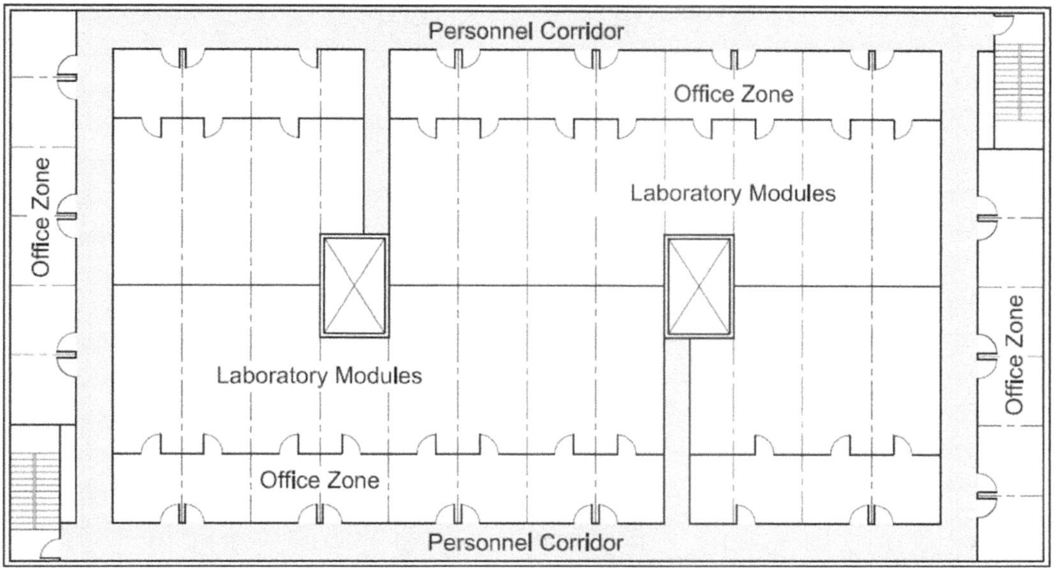

FIGURE 14.13 Central utility (mechanical, electrical and plumbing) shafts located in center with access corridor, building layout. (*Guidelines for Laboratory Design* (DiBerardinis, 2013).)

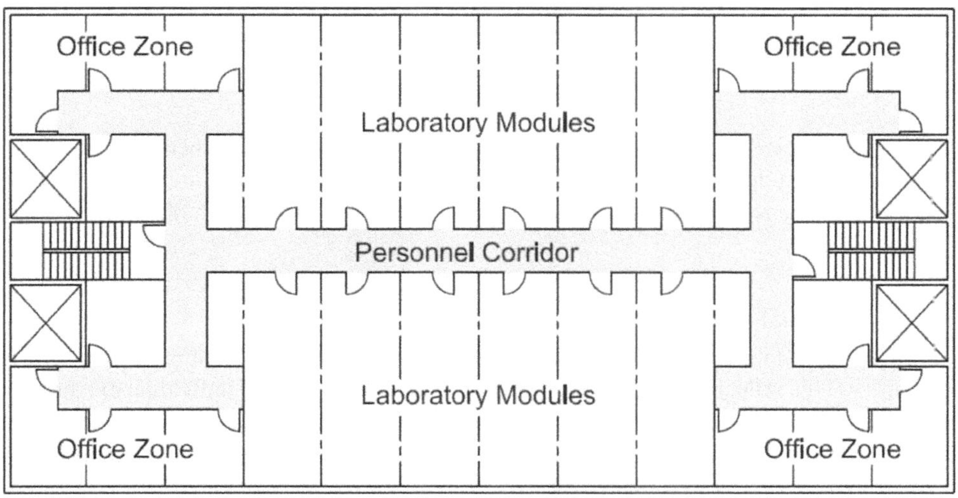

FIGURE 14.14 Central utility (mechanical, electrical and plumbing) shafts located at ends, building layout. (*Guidelines for Laboratory Design* (DiBerardinis, 2013).

14.4.5.4 Individual Utility Shafts Adjacent to Service/Public Corridors

In this layout (Figure 14.15), laboratories are often interior spaces (in buildings with loop corridor configurations) which limits the exposure to natural light. However, where appropriate, this set up allows for repairs to be done in the corridor, outside the laboratory space, thereby increasing work efficiency within the laboratory space (Roberson et al., 2013). It also allows for expansive areas of glazing on the exterior wall, as they cause no interference with the building façade.

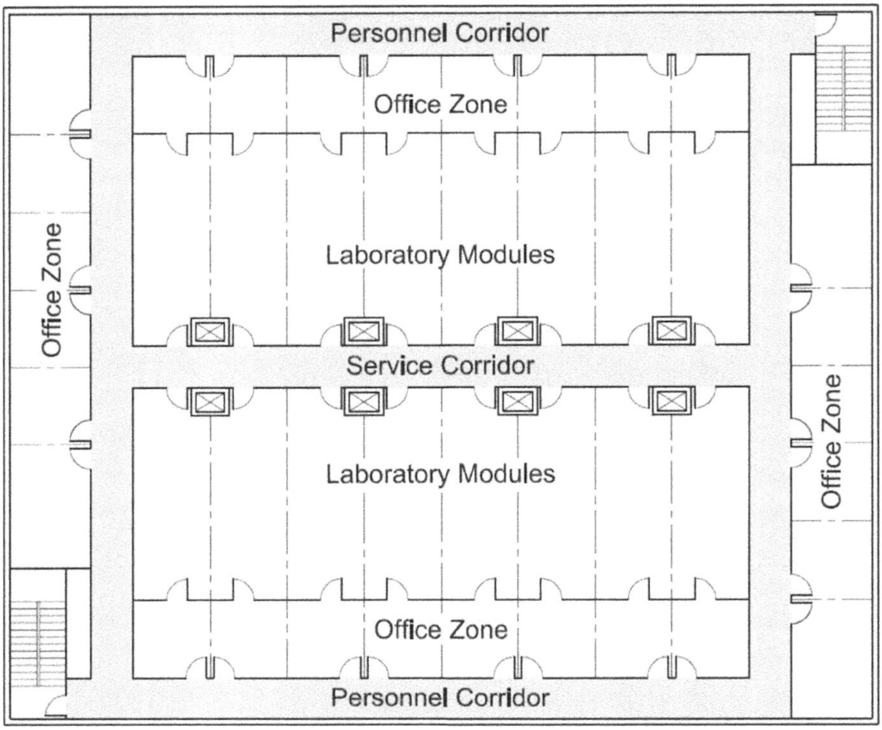

FIGURE 14.15 Central utility (mechanical, electrical and plumbing) shafts located on service corridor, building layout. (*Guidelines for Laboratory Design* (DiBerardinis, 2013).

Unfortunately, this setup requires long horizontal runs to connect the chemical fume hoods, or other special exhaust devices, to the riser in the shaft. Long horizontal runs are susceptible to corrosion from acid condensation than vertical ducts. This layout will also limit building flexibility when changes are needed.

14.4.5.5 Individual Utility Shafts at Exterior Wall

This configuration (Figure 14.16) allows fume hoods to be located adjacent to shafts at the rear of the laboratory. This reduces the horizontal runs for duct connections to individual exhaust risers in the shaft, they keep the primary exits safe and they allow data entry stations to be located along the corridor, away from potential hazards. Unfortunately, this layout means that maintenance work is done inside the laboratory, which may interrupt scientific work or pose a health and safety risk for lab users.

14.4.5.6 Interstitial Floors

As shown in Figure 14.17, these are structural floors located between occupied floors and are used by construction and maintenance workers to access mechanical services such as duct work, plumbing and fire pipes, electric and IT distribution conduit and cables, and controls for all utility systems. They offer flexibility and ease of maintenance without interfering with laboratory operations.

The minimum height of 2 m is required for these spaces, with the highest zones being reserved for electrical and IT cables, while wet utilities occupy the lower zones of the floor for safety reasons.

Interstitial floors increase the total gross area of the building by 50% of the actual floor area, and they require the same number of floors, exits and circulation paths as occupied floors. An additional construction cost is also associated with this floor system.

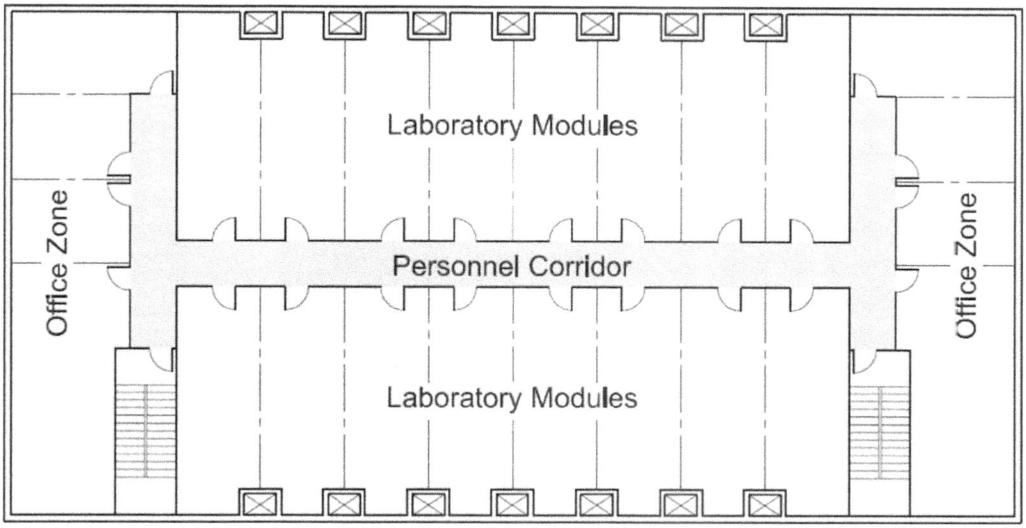

FIGURE 14.16 Central utility (mechanical, electrical and plumbing) shafts located on exterior walls of lab modules, building layout. (*Guidelines for Laboratory Design* (DiBerardinis, 2013).

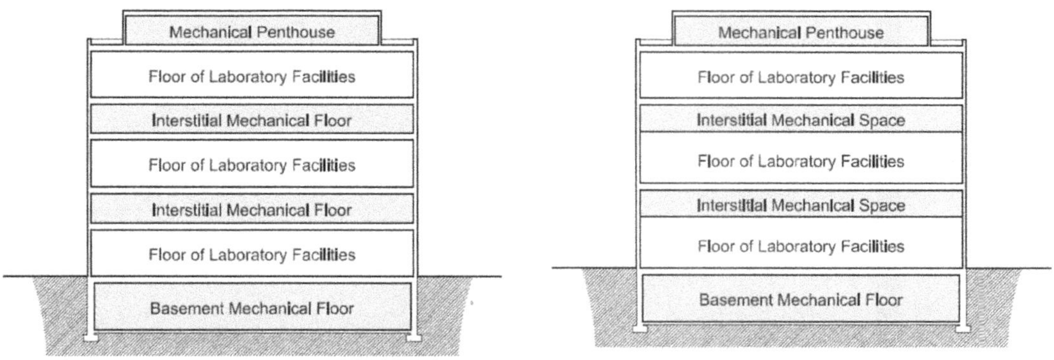

FIGURE 14.17 Diagram of laboratory building with (a) interstitial floors, and (b) interstitial spaces between floors. (*Guidelines for Laboratory Design* (DiBerardinis et al., 2013).

Alternatively, a metal frame platform, called a catwalk, can form an interstitial space above the usable laboratory area where maintenance workers can easily access control systems and equipment. Each interstitial ceiling should be provided with a set of stairs for safe egress and access.

14.4.6 Utility Distribution to Work Benches

The delivery option chosen to supply utilities to laboratory benches has a significant influence on the configuration and flexibility of the laboratory to future renovations and/or reconfigurations. The various options include an overhead service carrier (Figure 14.18) which may be fixed or flexible, perimeter service distribution (Figure 14.19) which is fixed on the laboratory walls, and an underfloor service distribution (Figure 14.20).

Overhead service carriers, hung from the underside of the structural floor, provide a point of access for services that run above the ceiling. Flexibility can be effectively achieved in this method by providing flexible service delivery to point of use through movable ceiling pods and bench-mounted masts (which are connected to a ceiling services infrastructure through a flexible

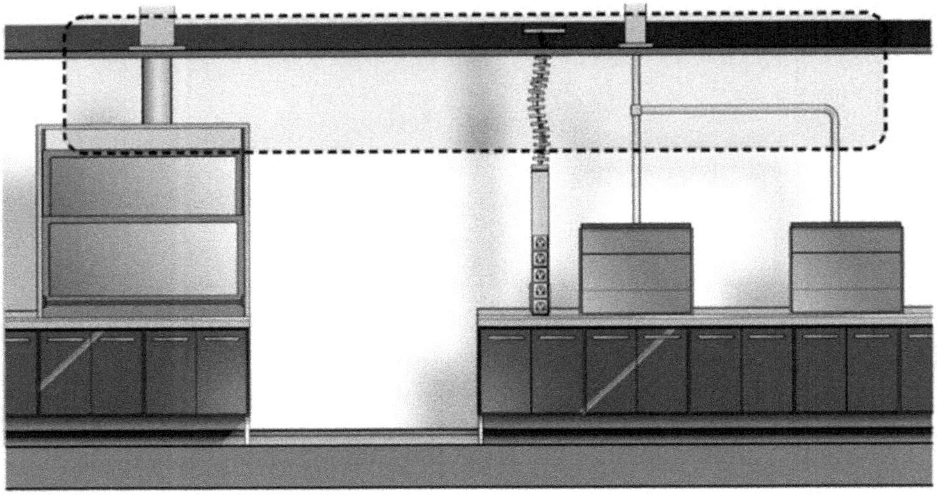

FIGURE 14.18 Overhead flexible services distribution. (*Metric Handbook: Planning and Design Data* (Buxton, 2021).)

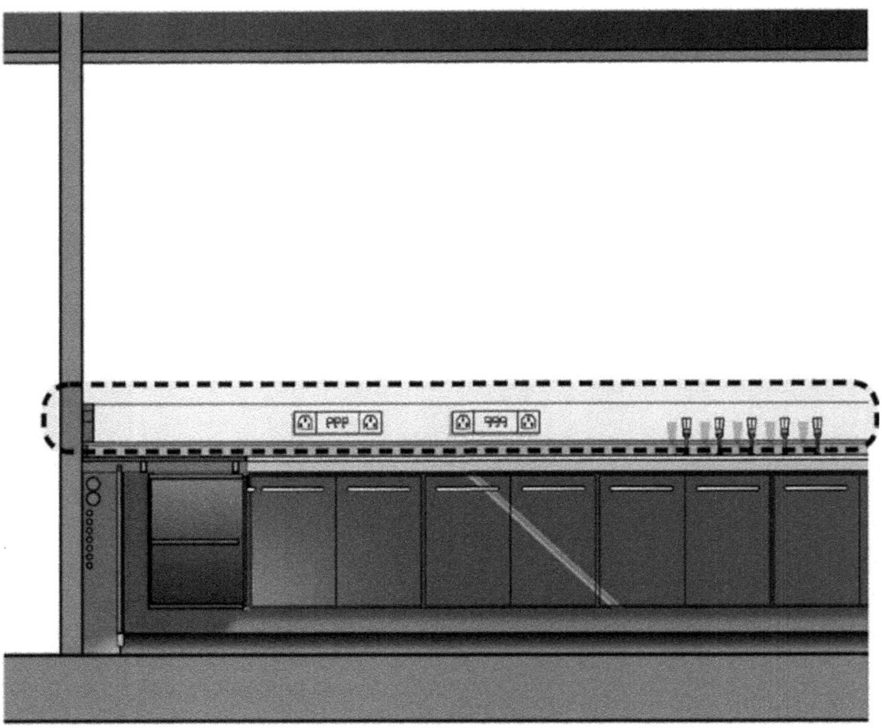

FIGURE 14.19 Perimeter services distribution through walls and trunking. (*Metric Handbook: Planning and Design Data* (Buxton, 2021).)

hose and flying electrical leads). Fully movable services within the ceiling should, therefore, be integrated within the ceiling as well as the laboratory furniture and equipment for this method to work. Overhead service distribution hosts communication outlets, light fixtures and service fixtures for process piping and exhaust apparatus.

Underfloor service distribution involves having floor-mounted docking stations or floor-mounted bollards that accommodate all services as well as a sink, which isn't hosted by the overhead system.

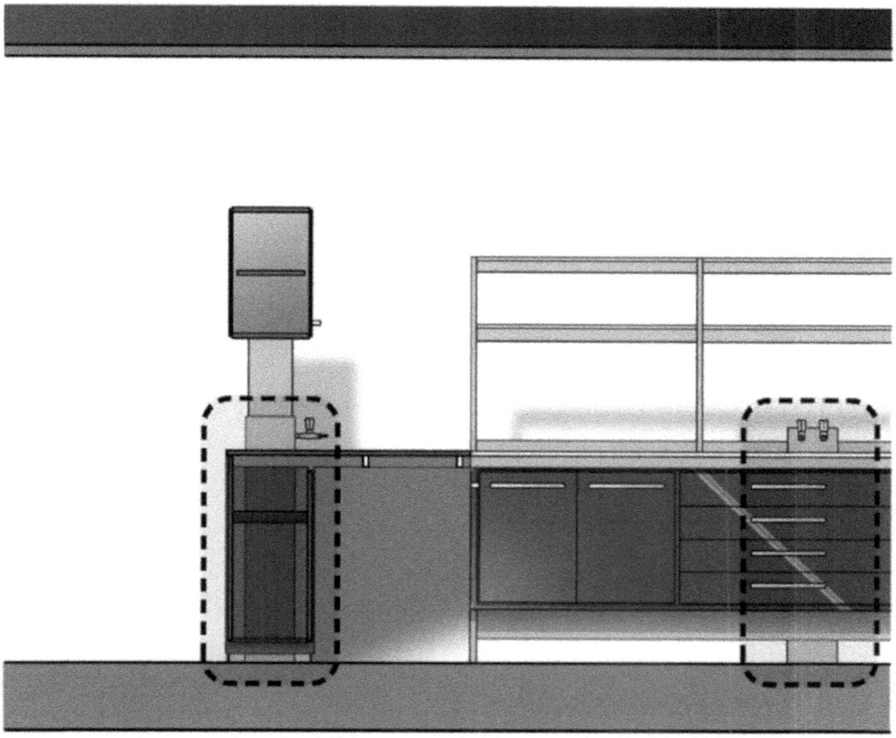

FIGURE 14.20 Distribution of services through floor-mounted bollards. (*Metric Handbook: Planning and Design Data* (Buxton, 2021).)

Standard distribution for power and data outlets are 450 mm centers/bench and equipment run and 900 mm centers/bench and equipment run, respectively. For health and safety reasons, sockets should not be placed near sinks and cleaners should have their sockets well demarcated.

Piped services should be provided to each laboratory on a modular basis to increase flexibility of the space and reduce the costs of remodeling and retrofitting the laboratory. Each laboratory needs its own separate shut-off valve for emergency shutdowns and repairs. Shut-off valves should be accessible and placed in a consistent area in all laboratories.

14.4.7 Mechanical Systems

14.4.7.1 Electric Systems

Electric cabling in laboratories is done through cable trays and panel boxes. Due to the need for constant maintenance for electric and data wires, ease of access is very important. For this reason, electric cables are located in a race way for easier maintenance. Cable trays can be found along corridors, below the ceiling, at the same level as the ceiling or just above it.

Panel boxes, on the other hand, are located in zones on each floor or at the entry alcove to each lab. Panel boxes require an adequate number of extra breakers to accommodate future needs. Power outlets in instrument usage areas should be located at 900 mm center to center.

14.4.7.2 Lighting Systems

There are a number of factors a designer must consider when choosing appropriate lighting for a laboratory. These factors include the quantity and quality of light, direction of light, light source color, the ability to accurately render colors, contrast, light uniformity and surface reflectance.

Other considerations such as the laboratory environment, work to be carried out within the laboratory, the physical properties of each space and the opportunities and constraints of various lighting systems can also aid the designer in picking the most appropriate lighting for the lab interiors.

Proper laboratory lighting enables the user to perform a variety of tasks quickly and accurately without visual strain or discomfort. It should also promote a sense of well-being, enhance productivity and foster creativity within the workspace. Thus, a good lighting system in a laboratory is one that reduces glare, is aesthetically and visually appealing, is affordable and is easy to maintain.

Factors such as user expectation, illumination levels and uniformity of light are key issues to be addressed by laboratory designers.

14.4.7.2.1 User Expectation

Visual comfort is an important factor, as they directly impact upon the researcher's productivity, accuracy and mood. Minimal glare and uniform lighting levels must be achieved throughout the laboratory space.

14.4.7.2.2 Illumination Levels

By taking account of both luminance and reflectance and light dispersing properties of surface materials within the space, the lighting designer is able to provide for appropriate levels of light within the laboratory.

14.4.7.2.3 Uniformity

Uniform lighting can be defined as one that reduces glare, extreme contrast and harsh shadows. The purpose of uniform light is to reduce the amount of adaptation the user's eye must endure when changing tasks and increase productivity while reducing eye strain.

Horizontal luminance uniformity allows for high levels of task mobility and flexibility while vertical luminance uniformity enables the user to perform three-dimensional tasks at multiple levels without experiencing harsh contrast and shadows from one work area to the other. Uniform ceiling luminance, on the other hand, can work to reduce extreme contrast and glare while increasing the overall perception of brightness and the space of a room.

14.4.7.3 Light Distribution

Light in laboratories can either be distributed through indirect/direct distribution or through direct distribution.

a. *Indirect/direct distribution:* This method of distribution typically involves the use of fluorescent sources for spaces with minimum ceiling heights of 2,850 mm. Luminaires are located such that they provide uniform ceiling luminance while concentrating the direct light component into the work surface.
b. *Direct distribution:* Direct distribution is often chosen in cases where the ceiling height is lower than 2,850 mm, where there is a high equipment density whereby users only use the space for experimental preparation and clean-up and where the ability to suspend luminaires is limited. Direct distribution method offers better control of glare, reduces shadowing and increases uniformity. Lensed fluorescent luminaire and parabolic fluorescent luminaire are the two common types of luminaires used in direct distribution method.

14.4.7.4 Luminaire Distribution

Luminaires in laboratories can be placed either parallel or perpendicularly to work benches. Where parallel, the luminaire should be placed along the edge of the bench to allow for easy access and maintenance through the aisle. Where placed perpendicularly to work benches, then they should extend 300–450 mm beyond the edge of the bench. It is imperative that the luminaire is placed to illuminate the work surface and not cast shadows onto it.

The lighting designer will consider all factors before choosing a suitable distribution method. These include the ease and costs of maintenance, dimensions of the lab module, equipment layouts and the selected luminaire.

14.4.7.5 Special Considerations

Lighting design for areas where team and individual seating arrangements are constantly changing; indirect lighting offers the most appropriate solution. With luminaires hanging 450–720 mm below the ceiling, ceiling materials should have a minimum reflectivity of 80% and a minimum of 65% for major walls.

To increase efficiency, separate switches are preferred. This could work well for light fixtures adjacent to windows, where daylight can adequately light the work surface. Multiple ballast switches could also be used in this scenario to only switch on lights required in selected rows.

14.4.7.6 Plumbing Systems

14.4.7.6.1 Floor Drains, Roof Drains and Sprinklers

Reducing the number of floor drains within a project will ensure less leakages occur in the lower floors, as well as less interference with the construction system of the building. Horizontal piping from a drain should flow directly to the outside wall. In cases where this cannot be achieved, however, an interior chase via a "wet" column will do. This should be channeled down the building to the building-wide drain system. Drains and horizontal piping must be well coordinated with the beams and interior walls in situations where the drains are adjacent to the columns.

For flexibility, laboratory renovations and future-proofing, the plumbing risers should be located on a modular basis. The length of the pipe should not be too great and there should be adequate space to accommodate the slope without interfering with the ductwork, other piping or wiring.

Roof drains are often installed alongside fume hoods or plumbing risers in a common chase which is located along the corridor or on the exterior wall at most structural bays. The vertical riser of the roof should conform to the laboratory module at the beginning of the project.

Piping for sprinklers and other wet piping should be placed below the duct work and above electrical systems. Sprinklers may be placed below the ceiling or on the underside of the structure (where no ceilings are present) to cover the entire volume of the space below.

14.4.7.7 Laboratory Piping Systems

Process piping can be installed on racks, either vertically or horizontally. Horizontal racks result in less interference with other engineering systems. A proper laboratory piping system should achieve flexibility for easy renovation and modification, provide appropriate plumbing system for each laboratory with regard to the building program, minimize energy usage, cater for the client's budget and have shut-off valves that are legible and easily accessible. Shut-off valves could be located on valve boxes inside the laboratory or at the entry alcove, behind casework at the rack or in scenarios where the piping comes into the bench.

14.4.7.8 Laboratory Waste and Vents

There are a number of ways in which waste neutralization is done inside laboratories. They include passive neutralization (with and without monitoring), active neutralization with monitoring and passive and active neutralization with monitoring.

Hot water systems in laboratories can be affected through heat tracing, recirculation, and the use of local water heaters.

14.4.7.9 Piped Gases

Gases can be supplied within the laboratories in the following ways:

- *Through a central location:* This allows gas to be supplied along vast areas. It should, however, make financial sense.
- *Local closets:* These serve multiple laboratories. It is important that they are fire-rated so that they adhere to the building codes, and they also require an auto-switchover. Contractors are required to install the piping.
- *Locating the piped gases in the laboratory, near the door:* Piping serves multiple benches in this instance and may not require auto-switch valves. It is also easy to achieve flexibility with this setup. Nonetheless, they do take more program space.

Where cylinders are used, they should be anchored to a wall, a wall rack or a bench. Provisions should be made with regard to the storage of incoming cylinders and water cylinders at the receiving docks.

14.4.7.10 Water Supply in Laboratories

Before we look at the methods of water supply in laboratories, it is important to know the types of water supplied into a laboratory. They are as follows:

1. *Chilled water*, which is used to cool special equipment
2. *Potable water*, used in work areas and restrooms
3. *Domestic water*, used at drinking fountains, eyewashes, break rooms, etc.
4. *High-purity water*, used for numerous research processes. High-purity water may subsequently be divided into three types, depending on the work the researcher sets out to do. *Type 1 water* is used for critical laboratory applications such as organic and inorganic analysis; *type 2 water* is used in reagent preparation and glassware rinsing, while *type 3 water* is used for qualitative procedures, glassware washing and preliminary rinsing.

Like gas, water could be supplied through a *central system*, with water piped throughout the building, or a *local point-of-use system* at the sink.

14.4.7.11 Water Purification Systems

There are a number of ways in which laboratory water is purified before use in conducting research and experiments. This may depend on the local variations that make up the water quality as well as the end use requirements. Water purification methods include distillation, deionization, reverse osmosis (RO), filtration and photo-oxidation. While each method has its pros and cons in relation to energy demands, cost, demand and use, they are often used together when there is need to produce the highest quality water.

14.4.7.11.1 Distillation

Distillation is used to produce types II and III water by removing all contamination from the raw water, except microorganisms, and dissolved gases whose boiling points are below 100°C. In this process, raw water is fed into a boiler where it is heated to steam. This steam is then channeled into a condenser where it is cooled, reverting to its previous liquid state, free of any impurities. To increase efficiency, the water used to cool the condenser is redirected back into the boiler.

It is worth noting that this process has a lot of energy and water demands. Some 3 kW of energy is required to produce about 4 L of purified water, while only 6% of purified water is obtained from the raw water. As this process is not an on-demand system, water is stored for future use, which may result in contamination of the purified water.

14.4.7.11.2 Deionization

Deionization involves removal of ions from the feed water, and it is the only purifying method that is capable of achieving the resistivity requirements of high-purity pure water (type I water). Feed water is passed over a synthetic resin at high pressure where the cations react with the resin to form hydrogen while the anions release hydroxyl ions which later combine to form water molecules.

This process demands pretreatment, monitoring and maintenance of the ion exchange beds which may deteriorate over time or play host to microbial growth and release of particulates which eventually leads to inconsistent water performance. Contaminants such as microorganisms, pyrogens and bacteria are not removed in this system.

14.4.7.11.3 Filtration

Filtration is used as a pretreatment system before the water is further processed. It is often used to remove colloids, enzymes, microorganisms, particles and endotoxins. Filters of different pore size ratings are used to remove particles above their rated pore size in accordance with the filter's efficiency or its beta ratio[1] at a given rate of flow. The rated pore size can vary between 0.2 and 10 μm. Maintenance is fairly easy in this system as only the filters need to be replaced. Failure to replace them, however, may lead to clogging of the filter which may spread contamination. It's worth noting that filtration does not remove organic and inorganic molecular substances.

14.4.7.11.4 Microporous Filtration

Using membranes of about 0.2 μm, filtration method is able to filter out bacteria and particulate matter like resin fragments from ion exchange columns.

14.4.7.11.5 Ultrafiltration

This type of filtration uses filters of typically 1–10 nm and is capable of removing pyrogens and other organic molecules with a molecular weight greater than 10,000 Dalton (Da).

14.4.7.11.6 Activated Carbon Filters

These are capable of removing chlorine and chloramine through adsorption. Dissolved organic matter can also be removed through this process.

14.4.7.11.7 Reverse Osmosis

In this process, raw water is passed at high pressure (of about 60 psi) over a RO membrane. The RO membranes are able to filter out bacteria, pyrogens and inorganic and organic solids, but have very little control over dissolved gases.

Feed water might have chlorine, mineral deposits and hard particles that could damage the RO membrane. It is therefore prudent to pretreat the feed water before using RO. Water in this method is often stored due to the slow nature of the process.

14.4.7.11.8 Photo-oxidation

Used often in systems requiring ultrapure/type I water, photo-oxidation is responsible for the removal of trace organisms at 185 nm as well as inactivating microorganisms at 245 nm emitted by the UV light. The oxidation of trace organisms results in purified water with low total organic carbon but also leads to decreased water resistivity as a result of the ionized contaminants (such as the carbon dioxide which is released in the process) which would require more downstream treatment to resolve.

14.4.7.11.9 Electro-deionization

This system combines both electrodialysis and ion exchange to eliminate ions from purified feed water. This system constantly and automatically regenerates itself. However, it does not remove organisms, particles, pyrogens and bacteria, but it may be less susceptible to microbial contamination as compared to the deionization process.

14.4.8 Fume Hoods

14.4.8.1 Fume Cupboards

FCs are local exhaust ventilation systems in laboratories, whose primary function is the control of exposure to hazardous airborne substances (such as hazardous gases, dusts, mists, aerosols and vapors) in laboratories. Figure 14.21 shows the FCs which also act as physical barriers between reactions and the laboratory, protecting the user against inhalation exposure, chemical spills, run-away reactions and fires.

Placement of FCs in laboratories should be done in such a way that it guarantees the safety of the user working with his chemicals/aerosols. To ensure a safe working environment between the FC and its operator, designers should consider locating it away from areas with high traffic patterns, doors and air supply/exhaust grilles that may interrupt airflow patterns (Figures 14.22 and 14.23). It would also be judicious for the design team to provide an alternative means of escape in case of fire or explosion.

Where flexibility is needed, recirculating and relocatable FCs with flexible and variable ducting are best suited. They also promote low energy consumption within the laboratory.

14.4.8.2 Biosafety Cabinets

A BSC, also commonly referred to as microbiological safety cabinet (MSC), is an enclosed workspace with a ventilated hood that is designed to specifically handle organisms, infectious agents, processes and particles that can potentially contaminate and harm other projects, personnel or the environment.

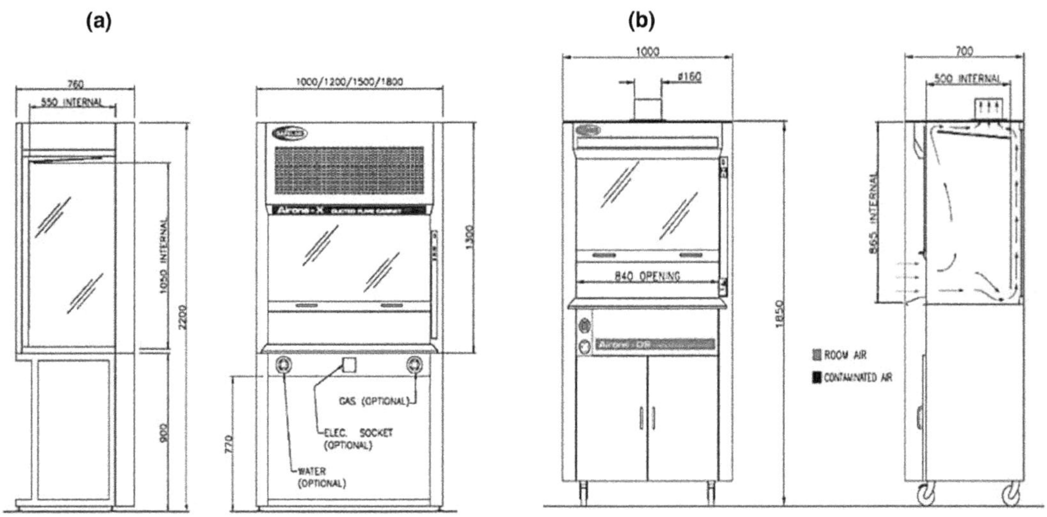

FIGURE 14.21 (a) Ducted fume cupboard. (b) Mobile fume cupboard.

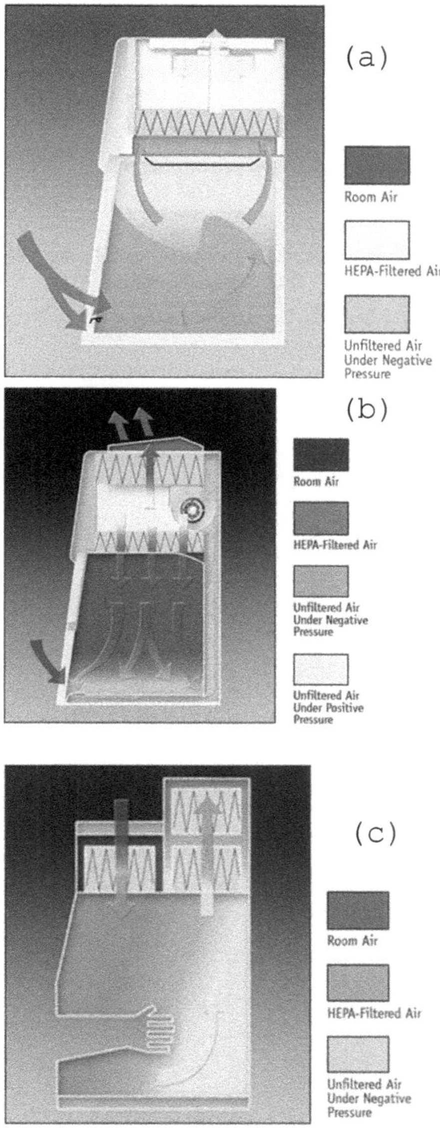

FIGURE 14.22 Fume cupboard minimum distances to avoid disturbances. (*Metric Handbook: Planning and Design Data* (Buxton, 2021).)

Like FCs, BSCs (Figure 14.24) should be placed away from high traffic areas, doors and air supply/exhaust grilles that may interrupt airflow patterns. There are three classes of BSCs and they are described below:

14.4.8.2.1 Class 1 Microbiological Safety Cabinets

These provide protection to the operator and the environment only, while the sample material in the cabinet is not offered the same level of protection. It works by exhausting all extracted air through a high-efficiency particulate absorbing (HEPA) air filter to the atmosphere. These cabinets are susceptible to cross-contamination within the laboratory which may affect experiment consistency. As a result, they are hardly ever used in laboratories.

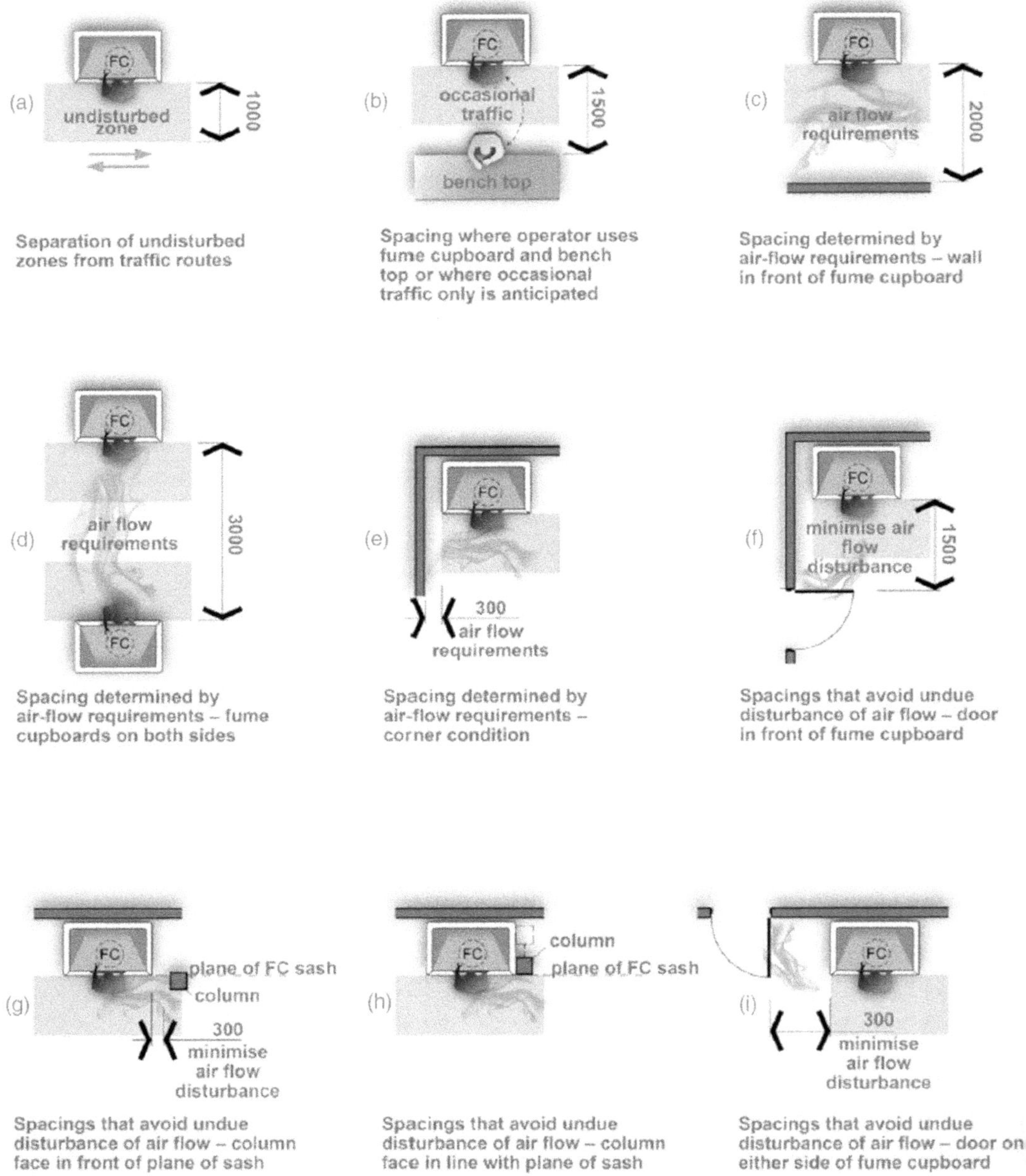

FIGURE 14.23 Fume cupboard arrangements minimizing disruptions from other lab users. (*Metric Handbook: Planning and Design Data* (Buxton, 2021).)

14.4.8.2.2 *Class 2 Microbiological Safety Cabinets*

Class 2 MSCs provide protection to the user, the environment and the sample material in question. An advanced inflow air curtain provides protection for the user, while the sample material is protected from dirty air in the laboratory through the use of a constant stream of supply filtered air.

14.4.8.2.3 *Class 3 Microbiological Safety Cabinets*

Specified for works with the most lethal biological hazards, Class 3 MSCs provide maximum protection for its users and sample specimen. The user operates his specimen through an additional

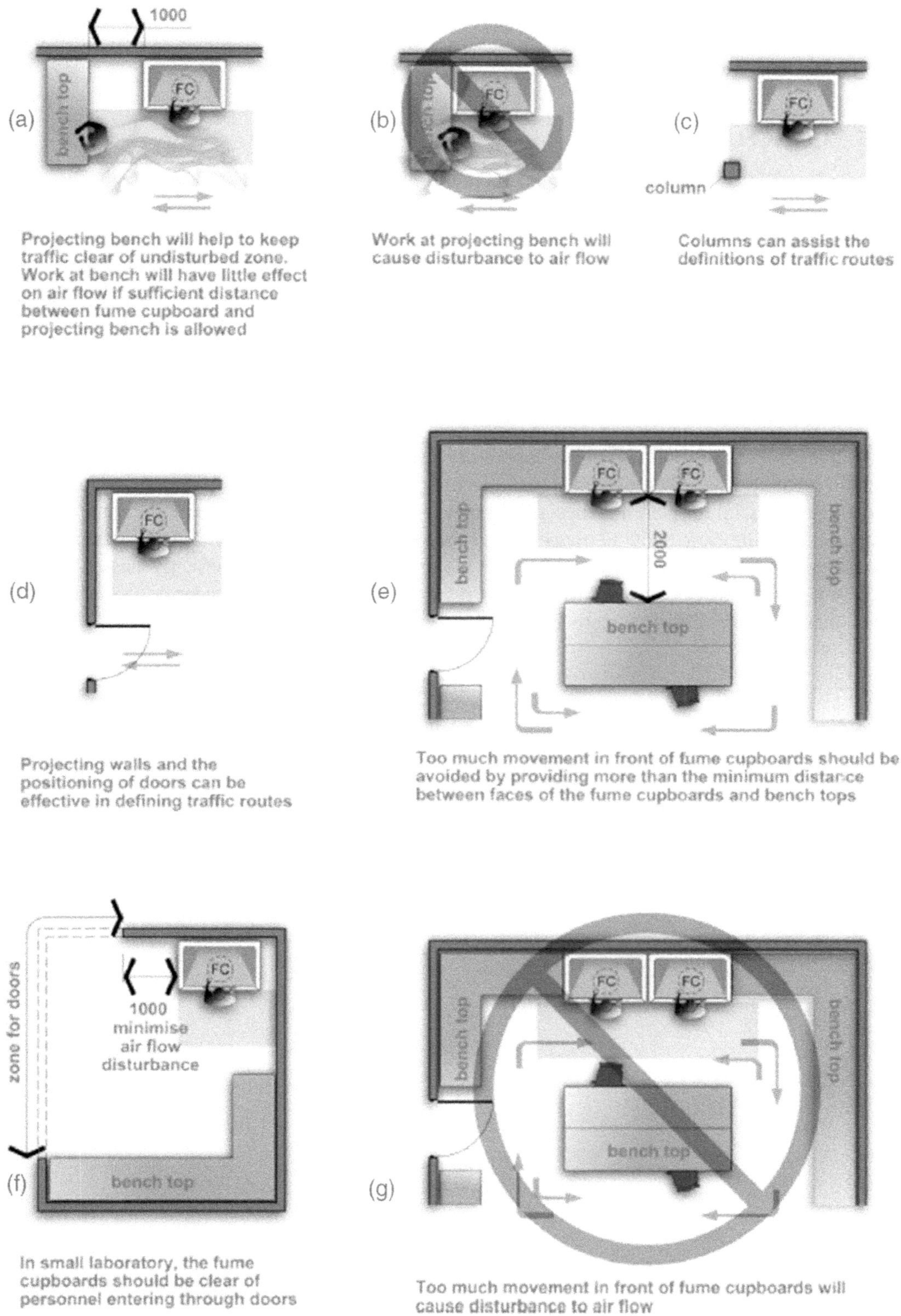

FIGURE 14.24 (a) Class 1 microbiological safety cabinet. (b) Class 2 microbiological safety cabinet. (c) Class 3 microbiological safety cabinet. (https://www.labconco.com.)

barrier, commonly through glove ports, while the specimen is protected by a stream of supply filtered air that is kept at a low pressure within the cabinet relative to the ambient environment while the exhaust air is vented out through a HEPA filter to the environment.

14.5 CONCLUSION

The configuration of a laboratory is fundamentally important for operational efficiency, as a lack of organization can hinder established methodologies. Engagement with User Groups is essential for defining objectives. Following the selection of a site, architects and engineers analyze flow diagrams with the aim of enhancing both construction processes and operational functionality. The integration of sustainable design principles and lean methodologies is pivotal for cost management and the optimization of material handling processes. The layout must be designed to facilitate workflow, reduce unnecessary movement, increase visibility, eliminate redundancies and promote automation, all while effectively managing inventory levels. Maintaining an accurate inventory of equipment and chemicals is crucial for ensuring safety and compliance with biosafety standards, particularly in light of anticipated future expansions of automated systems.

At another level, the optimization of laboratory space utilization is essential due to the increasing need for adaptability and cost-efficiency in teaching and research facilities. Understanding laboratory usage is crucial for planning, design and management. Use of metrics can enhance and reduce costs in evaluating data. Such metrics include space occupancy rate, equipment utilization rate and laboratory turnover rate, all of which can provide insights into usage and effectiveness. Validation of these metrics can be achieved through surveys across various clinical, academic and research lab types, demonstrating effective methods for quantifying space efficiency.

NOTE

1 The ratio of the number of particles of a minimum given size upstream of the filter to the number of particles of the same size and larger found downstream.

Part 4

Cost-Saving Laboratory Use

15 Laboratory Operations and Management

Sarah A. Otieno and Apollo Odhiambo Maima

15.1 INTRODUCTION

This chapter presents a discussion on the management of laboratory operations. The objective of management of laboratory operations is to organize activities across various settings and to guide the laboratory personnel to efficiently deliver their assigned duties using limited resources and time. The process of management entails the acquisition of the needed resources, designing of workflows, managing laboratory sessions, and overseeing the daily operations of the lab to ensure maximum effectiveness and efficiency is achieved. Effective lab oversight is vital for success, as it reduces costs, enhances turnaround times, and boosts safety and motivation. Key operations include equipment maintenance, diagnostic tests, quality control (QC), research support, administrative tasks, staff training, and cost reporting (Agily et al., 2022; Roy, Mandal, & Das, 2025).

Laboratories are classified into clinical, research, and industrial types based on their functions. Clinical laboratories evaluate human and animal health through diagnostic tests, while research laboratories conduct experiments to expand knowledge and must adhere to strict publication criteria. Industrial laboratories play a vital role in manufacturing, covering everything from product development and design, to QC, sales, and product launches.

Effective management of laboratory operations is crucial, including control over equipment purchase, use, cleaning, calibration, and maintenance, which involve cost analysis. Detailed documentation of maintenance history and adherence to standard operating procedures (SOPs) is essential for consistent performance, reduced downtime, and safety (Roy, Mandal, & Das, 2025; Mohzana et al., 2023).

For the most part, lab management entails laboratory care activities such as purchasing equipment, maintaining instruments, restocking consumables, scheduling, giving technical advice and keeping records of various laboratory activities and incidences. A considerable aspect of the laboratory management process is to ensure that the lab condition and the personnel adhere to the set contemporary standards of laboratory operation.

15.2 QUALITY MANAGEMENT IN THE LABORATORY

Quality assurance in laboratories requires standards for material handling, equipment use, hazard management, and personnel protocols. QC on the other hand is vital for precision in analytical procedures and waste management, emphasizing instrument calibration and maintenance. Structured administration is necessary for reliability. International quality assurance standards, such as International Organization for Standardization (ISO) 9000 and ISO 17025, play a crucial role in establishing quality benchmarks. In the USA, the Clinical Laboratory Improvement Amendments (CLIA) are indispensable for clinical laboratories, underlining the significance of regulatory compliance assessments and laboratory accreditation in upholding quality standards (Bekmirzayevich & Bakhodirovich, 2024; Pillai & Fox, 2025).

Laboratories are essential in sectors like healthcare, forensics, and environmental management, focusing on improving performance and precision. Integrated processes are vital for achieving goals, with quality and systematic protocols being able to lower errors and boost productivity

(Aslan, 2018). Regulation of laboratory quality is grounded in specific standards, rules, and sectoral legislation, supplemented by QC line operations (Mandal, Ansari, & Aswal, 2021).

Laboratory quality assurance plays a critical role in establishing reliable evaluations by implementing systematic procedures throughout the pre-examination, examination, and post-examination stages. Effective management of laboratory equipment is essential, necessitating regular cleaning, calibration, and meticulous documentation to ensure adherence to compliance standards.

Standardized workflows play a critical role in ensuring quality assurance within laboratory settings. The implementation of documented procedures enhances operational efficiency and guarantees the consistency of results. An effective quality assurance framework is characterized by its capacity for oversight and the initiation of corrective measures when analytical outcomes surpass established tolerance thresholds (Kaur, Downey, & Choate, 2018).

15.3 CONTROL OF LABORATORY EQUIPMENT USE

Most equipment-related hazards in labs are those emanating from electric gadgets that work with compressed gases, and appliances for high or low pressures and temperatures. Other physical threats include electromagnetic radiation from lasers and radiofrequency generating gadgets. Apparently ordinary threats like overflows from water-cooled devices, accidents with rotating appliances and machines or tools for cutting and drilling, loud sound, falls, trips, stumbles, lifting, and mediocre ergonomics are the major causes of laboratory accidents and injuries. Noticeably, hand injuries are the most common in laboratories, hence the need to wear appropriate gloves when handling laboratory equipment to safeguard against electrical, heat, and chemical burns; cuts; pricks; and perforations.

To safely use and work with lab chemicals, some of which are hazardous, requires proper use of apparatus and equipment. Usage of laboratory equipment and apparatus is governed by frameworks with operational protocols and regular inspections for reliability. Equipment use focuses on cleaning, calibration, repair, and refurbishment, all activities requiring careful documentation (Roy, Mandal, & Das, 2025). Many of the mishaps and accidents that occur in the lab are attributable to improper use or maintenance of the equipment. The next sections of the chapter are therefore dedicated to the discussion of prudent practices for handling routinely used lab equipment.

15.3.1 Systematic Device Wash

Cleanliness of laboratory equipment is one of the easiest, most affordable and most obvious ways to keep the lab in greatest shape but it's surprisingly more overlooked than most people may imagine (Roy, Mandal, & Das, 2025; Rethorst, 2015). The following should be done:

- Wipe clean all equipment external surfaces on a daily basis.
- Conduct a weekly thorough cleaning of all equipment.
- Conduct a regular thorough cleaning of microscopes using a 70:30 mixtures of ether and alcohol to ensure that they are clean enough to yield most accurate results.
- Refer to the manual or laboratory manager on any specific cleaning procedures for specified equipment requiring cleaning. A hematology analyzer, for example, normally needs checking by an engineer every 6 months, a weekly analyzer surface cleaning, and a haematology control (HC) control check monthly.
- Outsource cleaning of sensitive equipment to qualified, prequalified third-party firms, to make the exercise professional and cost-effective.

Abiding by the above simple cleaning steps keeps equipment in optimum condition and enables the lab to run smoothly. The process of cleaning and sanitation of equipment includes the thorough cleaning of tools, containers, racks, surfaces, and glassware to ensure the validity and reliability of research outcomes (Moura et al., 2025). External surfaces should be wiped clean on a daily basis while scheduling a thorough equipment cleaning to be conducted weekly.

15.3.2 Systematic Device Calibration

Laboratory equipment require regular calibration for it to continue giving accurate data. Inaccurate data could easily lead to invalidation of experiments. Further, satisfactory calibration can enhance safety in the lab in the process of using hazardous chemicals, hence the importance of carrying out the procedure regularly. Calibration is a crucial step in the systematic management of laboratory equipment and needs careful documentation for consistent performance and operational reliability, integrity, and functionality over time (Braga, Infusino, & Panteghini, 2015).

There are a number of firms that are licensed or accredited to offer calibration services at the right frequency and to the desired standards. It is always important to prepare a list of the lab equipment indicating the most appropriate requirement of each, from simple preventative maintenance to more advanced validation of accuracy. Equipment that are undergoing preventative maintenance particularly require regular calibration in order to optimize the lab condition.

15.3.3 Lab Equipment Renovation

Often, lab equipment and apparatus deteriorate in condition and stop working, and hence may need renovation, repairs, refurbishment, or replacement. It is normal practice that before faulty equipment or apparatus are replaced, they should first get restored by repairing or replacing the faulty parts. The evaluation of repair documentation provides insight into the operational state. Effective organization of schedules and management of spare components are critical for the prompt resolution of technical challenges.

Replacement of equipment components must follow original specifications, ideally sourcing parts from the original manufacturer. Refurbishment goes beyond maintenance to restore instruments to their original condition, including repainting and replacing worn parts. Entire chambers may also need replacement.

This is also a cost-containment approach for the lab and, like the maintenance of a motor vehicle, often works perfectly. In fact, through such procedures and more so for larger ones, the device can be maintained, updated, and its productive lifespan increased. Lab Managers through experience get to know the parts of equipment that wear out faster than others and that when these parts are repaired or replaced in a timely manner it is possible to prevent worse problems or functional failure. In water purification systems, for example, filters are often replaced severally before the whole machine is finally written off and replaced.

15.3.4 Equipment Refurbishment

When equipment malfunctions, repairing may be useful though if it still works (except, may be, not as well as when it was new) refurbishment could be a better option. To refurbish older components of the device, the procedure below should be followed:

- The equipment should be carefully dismantled.
- Each part of the equipment should be thoroughly cleaned.
- If need be, the components should be polished.
- All the joints and moving parts should be relubricated.
- If parts are showing signs of aging or misfit, this preventative stage could be the right time to replace them.
- The equipment should be properly and tightly reassembled.

It is important to try and understand the roles and fits of the components as they are dismantled, as this helps in the reassembly process. If professionally done, the process above can make used equipment to look and function as new.

15.3.5 Quality Renewal of Lab Equipment

Even with the most diligent daily cleans, systematic calibrations, repairs, and refurbishments of devices, time comes when replacements must be procured. In such instances, the temptation to go for cheaper models, in the name of economics, can be pretty high. It is important to be forewarned that this is often a route to false economy as the adage "cheap is expensive" goes. Often, the cheaper devices lack sufficient integrity and quality, as their parts tend to be more prone to wear and tear and therefore less likely to cover the journey. It follows then that the choice of high-quality lab equipment over cheaper alternatives will offer longer durability, particularly for frequently used devices.

Equipment selection and management are crucial in laboratories, covering purchasing, inspection, installation, usage, maintenance, and retirement. Proper equipment choice ensures accurate results, while neglect can cause liabilities. Ongoing maintenance, timely calibration, waste management, and planning uphold quality standards and ensure the functionality of lab equipment (Kaur, Downey, & Choate, 2018).

It is noteworthy that parts for high-quality lab devices are usually easier to find, as the manufacturers of these devices ensure they are available in the market, even if through international purchase. This makes it an easy exercise to clean and refurbish the high-quality equipment. Further, if one needs to use an equipment for a short time, it might be wiser to seek for and hire a high-quality device instead of buying a poor-quality equivalent.

15.3.6 Equipment Maintenance Catalog

Maintaining laboratory equipment requires organized documentation of specifications, procedures, and servicing directives, including operational limits, maintenance schedules, and safety protocols. It covers disassembly, inspection, cleaning, repair, reassembly, and testing for effective maintenance. In compliance with Good Laboratory Practice, it's necessary to keep an elementary housekeeping catalog (checklist) in the lab to guide the staff and to serve as a quick reference for routine and daily duties in the lab like cleaning devices and around the lab, countertops, sweeping floors, disinfecting and calibration.

The maintenance catalog enumerates equipment with unique identifiers for documentation and evaluation. Each listing features manufacturer specs, catalog numbers, equipment types, calibration and verification protocols, maintenance schedules, and service status. The catalog and procedures are regularly updated to incorporate new assets, retire obsolete equipment, and resolve recurring issues identified through analysis. The maintenance catalog principle has evolved from managing scientific equipment to overseeing machinery, tools, and personal assets, focusing on standardized and cost-efficient operations. Some of the items on the catalog may include the following:

a. *Use of quality personal protective equipment*

   The condition, integrity, and quality of protective gear must be assessed at the start of any project or lab exercise, and the same maintained throughout. It is important that lab users regularly check and ensure eye and face protection, gloves, footwear, and hearing aid devices for any depreciation.

b. *Handwashing stations should be fully functional*

   Handwashing sinks should be hygienic, while antiseptic soap dispensers and towels should be easily accessible. It's important to ensure the sinks don't get clogged and if they do, the must be unclogged immediately to reduce the risk of logjams and contamination.

c. *Emergency stations to be maintained*

Sustain emergency eyewash and shower facilities in compliance with the given directives and always make time to activate pieces in all the facilities several times a week to guarantee effective functioning. Organize for regular facility inspections by trained and qualified staff.

d. *State of lab equipment to be checked routinely*

It is strongly recommended that at the start of any new project, the required devices be assessed for cleanliness and functionality, while at the end of the project, the devices should be cleaned to meet specified standards.

e. *Records of lab consumables to be properly kept*

It is crucial to maintain proper accounting records of all consumable items in the lab, including leftovers of projects. Especially key is the records of items that require cold storage (be it in refrigerators or freezers) which include such critical information on the items' source and expiry dates. Meticulous rotation makes inventory preservation and prompt disposal of expired items much easier.

f. *Use of appropriate glass cleaning techniques*

It is important to wash laboratory glassware as soon as possible after use. If, however, diligent cleaning is not possible immediately, they should be left to soak in water before washing; otherwise, it may become impossible to remove the residue if left uncleaned for long. Soaking can also be done for several hours in acid water (1% nitric acid or hydrochloric acid solutions) before washing. Brushes used should have handles made of materials that don't scratch glass surface, like wood and plastic.

If glassware has become too cloudy or dirty or contains dried organic matter, it must be washed with chromic acid solution and then rinsed with the cleaning solution or filled and allowed to stand. For how long it is allowed to stand will be determined by the level of contamination on the glassware; moderately dirty glassware may need only a few minutes. If dirtier, it would need to be kept overnight. Certain hard types of dirt may need nitric acid or sulfuric acid to clean, though these are very corrosive.

If glassware has been washed for sensitive or immediate use, it must be dried by rinsing 2–3 times with acetone which will remove all the water and evaporate fast. Acetone is completely miscible with water and hence dilutes and clears any residual water from the glassware, followed by vaporization to dryness. This approach is preferred to blowing air into glassware to dry it.

Routine lab glassware washing and drying has now been made much easier by use of glass washer-disinfector machines with hot-air drying capacities and produce water of varying temperatures. It is incumbent upon the Lab Manager to ensure the staff are empowered with requisite skills on the application of different manual cleaning techniques and washing products like detergents.

g. *Disposal of broken glass*

A labeled, puncture-resistant garbage bin should be assigned for *non-contaminated* broken glass and sharp waste items. Preferably this should be a sealed, stout pasteboard carton lined with a plastic bag, located at an easily accessible position, for use by all staff. This kind of bin must never be allowed to fill to capacity; it should be emptied whenever the bin is half-full to reduce the risk of injury to staff from its contents.

h. *Assessment of emergency equipment*

The laboratory fire safety and sprinkler networks should be frequently assessed for functionality. Well-kitted first aid boxes should be kept in visible and easily accessible areas of the lab, while fire extinguishers should be frequently assessed and checked for full charge and proper storage.

15.4 OPERATIONAL GUIDES FOR INSTITUTIONAL LABORATORIES

15.4.1 Staff Development and Evaluation

Laboratory operations focus on structured processes to enhance efficiency. Daily tasks and methodologies are organized for optimal performance, aligning with operational objectives.

Adequate staff training is vital for efficient lab operations, allowing personnel to manage equipment, follow quality assurance protocols, and address any operational challenges. Supportive policies enhance worker welfare and create a positive environment. Job descriptions must align responsibilities with competencies, while development programs boost knowledge. Performance evaluations should connect tasks with expertise through training sessions, workshops, and seminars (Mocerino et al., 2009). Whereas general education offers foundational theories, skills training focuses on practical procedures and familiarity with instruments and QC for accuracy.

While engaged, it is invaluable to conduct performance assessments to evaluate all regular employees within six months of hiring and annually thereafter. These entail Supervisors meeting with staff using written tools and providing copies of the assessments. The meetings should clarify job duties, highlight strengths, and set growth objectives. Employees can add comments and confirm completion with their signature. Evaluations are not grievance processes, and negative information in HR files is added only with employees' awareness. They should also be allowed to participate in political activities while at work.

15.4.2 Roles of Key Laboratory Stakeholders and Personnel

Operations of a lab of any level can be a hard task, and it is often a question of how many personnel should be included in the operational process. It should be taken into account that too many personalities and levels of operational management can create confusion that may lead to delays in terms of efficiency in decision-making and delivery. It is, therefore, important to know and have a good understanding of the people to include based on their roles that can bring effective workflows and higher operational output.

It is important that key stakeholders as well as investors are in the know while designing a new lab. Depending on the size of the lab and the kind of work it is meant to undertake, it is a noble idea to provide key stakeholders with regular updates concerning the operations so that they can provide feedback on a regular basis when needed. Most of the key stakeholders are not normally involved in the day-to-day running of the lab; rather, they usually interact at a higher level. As such, stakeholders usually like information to be brief, concise, and very relevant. They therefore tend to look at the organization from profit view of the investment.

Laboratory workers should be trained on equipment use, safety, and SOPs of equipment used in areas of their responsibility. The entire workforce should be responsible for ensuring compliance to quality that pertains to the work in the labs. Each lab should strictly follow its institutional or organizational requirements for maintenance of equipment.

When chemicals or equipment are received from suppliers, labs should keep track of procurement records, including the list of samples that should be included in the shipment. Laboratory staff should thoroughly check the purchase forms with the equipment or chemicals received to ascertain that there are no inconsistencies. If any chemical or equipment is damaged, then the head of the lab should be notified immediately.

The lab protocols are adopted as policy in research facilities or institutions of higher learning and thus their implementation is expected to be done by all lab personnel. The main responsibility for implementation of lab protocols solely lies with the heads of labs, and it is their responsibility to ensure that labs are guided by culture of safety, security, and best practice. This is an important aspect of ensuring that the reputation of a lab is guaranteed at all times. The institutional head is ultimately mandated with the responsibility of implementing policies and at times may delegate some of the responsibilities where and when necessary. The chief Lab Manager is responsible for the daily oversight of lab practice. The safety officer is responsible for periodic random inspections and other breaches of work protocols may be reported to the chief Lab Manager.

Whereas institutions endeavor to limit risks posed to the health and safety of persons who visit, work, or learn in the labs under its control, it nonetheless recognizes that certain key stakeholders can, more so than others, have a significant impact on establishing, implementing, and maintaining safety standards within the labs. The sections below outline the roles played by key lab stakeholders.

15.4.3 The Principal or Chief Executive Officer or Director

Based on the type and size of the organization, the principal, chief executive officer, or directors are normally charged with initiation of projects based on emerging needs, acquisition, and diversification. These individuals normally develop a basic scope and budget for the organization team. They, therefore, look for fulfillment of mission and vision of the organization and will mainly focus on return on investment in terms of results.

A manager should ensure that communication, collaboration, and meetings are comprehensively done with the right people at the right time. This will assist greatly in ensuring best results. The principal/executive officer should be able to deliver regular progress updates. Predetermined channels of communication to all stakeholders should be used. Take advantage of the opportunities to celebrate milestones and to highlight the success criteria established during the planning phase. Management success depends on adequate planning and cross-collaboration, which calls for input from and comprehensive communication with all the stakeholders from the top to the bottom of the organization. Communication and involvement at every level of the organization will greatly avoid common hindrances that may result from inadequate communication.

15.4.4 The Finance Manager

The role of the finance manager is to build a viable scope with the lab team so as to ensure that there are smooth operations. The manager will often work with the organization work force to ensure that the operations of the organization are streamlined in accordance with budgetary allocations so as not to incur losses or wastages. The finance manager will more often than not work as part of the management personnel to assist with return on investment. Some of the key roles to be chaperoned by the finance manager include budget consolidation, costing, and revenue generation.

A structured budgeting approach gives other managers insights into financial inflow and outflow. By overseeing the budget, managers can provide diagnostic services cost-effectively while optimizing funds. Thus, a solid grasp of budgeting and financial principles is crucial for laboratory managers (Finkler, Calabrese, & Smith, 2025; Eccher et al., 2023).

Cost analysis is essential for laboratory management, evaluating financial performance and identifying inefficiencies. It supports corrective actions and budgeting, allowing managers to project financial scenarios and prepare for unforeseen events. Effective financial management depends on good recordkeeping and information sharing. Continuous cost control involves assessing operational impacts linked to costs and comparing these to actual spending. Budgeting enhances financial accountability among staff, raising awareness of their costs (Eccher et al., 2023).

Funding sources are essential for laboratory operations, enabling salary payments, supply acquisition, and equipment maintenance. Evaluating financial management includes budgeting, accounting, and costing systems. Budgeting compares income to spending, while accounting records all transactions. The costing system determines test pricing and categorizes costs. The Level 1 Budget estimates specific activity costs, with adjustments made after funding denials. The Level 2 Budget analyzes yearly costs to sustain services, using historical data adjusted for inflation (Finkler et al., 2025).

15.4.5 Head of Department

In certain administrative structures, especially in teaching institutions, the Head of Department (HOD) sits at the apex of the governance structure of the Institution Labs, assisted by the Lab Manager or Chief Technologist. He/she is the Accounting Officer for all lab expenditure. In particular, the HOD will be expected to

- be the overall overseer for the Safety and Operations of the Labs, creating a professionally enabling environment for teaching, research and self-sustainability;
- communicate the Institution teaching and research policy issues to the teaching and technical staff;
- receive and act on issues regarding operations of the teaching and Technical team, including disciplinary cases, from the Lab Manager;
- oversee the procurement and maintenance of stocks of materials and consumables on advice from the Lab Manager;
- carry out performance reviews of the technical staff and other HR functions as necessary on the advice of the Lab Manager;
- oversee the orientation and induction process for new students;
- in discharging the above functions, the HOD will be the link between the School and the Institution's Management structures and accountable to the Chief Executive Officer or his Deputy.

15.4.6 Laboratory Manager of a Research Laboratory

As the principal investigator's right-hand person, the Lab Manager is a critical player in ensuring a lab's success. Day-to-day activities require a multifarious skill set and an ability to seamlessly switch gears at the drop of a hat. Although specific responsibilities vary between institutions, lab managers are generally master multitaskers, performing bench work, training personnel, maintaining existing lab equipment, contributing to grant submissions, overseeing safety permit applications and inspections, and establishing relationships with vendors. Managing lab purchases and juggling finances are among the top priorities, and as the world of research reagents continues to expand, lab managers are faced with the growing challenge of cost-efficiently acquiring high-quality reagents for the lab. The specific role of the laboratory manager is discussed below.

Lab Manager position can come in many forms. In some establishments, a Lab Manager may be termed Head of Science or laboratory operation coordinator. This is because a Lab Manager's tasks can be strategic, tactical, or operational, depending on the size and the nature of the lab-affiliated organization. In academia, for example, a principal investigator may handle the strategic aspect of lab management, while a senior technician or postdoctoral researcher oversees the tactical and operational tasks. In a larger industrial lab, however, strategic, tactical, and operational duties can split between two or more staff who interact with lab personnel and report to the organization's executives.

15.4.6.1 Strategic Roles

Lab managers who conduct the strategic duties of lab management are highly experienced. They are responsible for the direction of the lab, and they make certain that the lab operation fits in with the lab visions and missions in the long term. Examples of lab managers' strategic roles are as follows:

- Setting lab objectives
- Designing long-term plans

- Overseeing lab operations and relevant regulations
- Assigning tasks, monitoring and evaluating staff progress, performance, and customer satisfaction
- Developing and administering the budget.

15.4.6.2 Reviewing Regulatory Requirements

Apart from these roles, senior lab managers in some organizations may supervise, give technical advice, and contribute to certain executive duties. For instance, senior lab managers may advise on the desired qualifications and technical background in the staff recruitment process and assist in budgeting and procurement plans when drafting funding proposals or business plans.

15.4.6.3 Tactical and Operational Roles

From a tactical and operational standpoint, lab managers provide an implementation system for the overall strategic plan. They facilitate the day-to-day operation in the lab by setting up schemes and timeframes for each job, including personnel involved in each job, and the extent of involvement. This can be broadly categorized into the following forms:

15.4.6.4 Internal Work Processes

Lab managers are responsible for several internal work processes that are crucial for the day-to-day functioning of the lab. Some examples of internal work processes are as follows:

- Inventory management
- Shared equipment reservation
- Equipment maintenance
- Resources acquisition and management
- Waste management
- Information management
- Regulatory adherence.

Many day-to-day works must be performed under certain regulations. In this regard, lab managers must ensure that the lab operations comply with relevant regulations without compromising working hours or work output. To achieve this, lab managers may do one of the following:

15.4.6.5 Design Workflows and Set Up Working Areas

In this regard, streamlined workflows are essential, because they allow lab personnel to easily comply with regulating protocols, and they also support the controlling procedure. In particular, a Lab Manager can decide to set up in a single working area for experiments using a toxic substance, which must be disposed of in a specific manner. In this working area, the substance can be stored, utilized, and discarded, which will ensure that the use and disposal of the substance is contained in a single working area and ease the regulatory surveilling procedure, while keeping additional tasks for the purpose of regulatory compliance to a minimum.

15.4.6.6 Mentoring

Lab managers often take on the duty of mentoring, especially in the form of lab orientation. This is to inform new lab personnel of the rules they must abide by when working in the lab, and to make them aware of the associated health risks and how to minimize them.

15.4.6.7 Paperwork

Apart from active lab duties, lab managers are usually assigned some form of paperwork, especially those involving procurement and regulatory requirements. This varies between laboratories, topic of study, and lab settings. Paperwork ranges from gathering invoices or proof of purchase, to drafting, reviewing, and filing regulatory papers. In the case of regulatory paperwork, oftentimes, lab managers still have to collect data, keep records, and report them as per legal compliance even after the authority has processed and approved them.

Ultimately, the duties of a Lab Manager are to ensure that staff focus on their tasks, without having to sacrifice their attention or their time for bureaucratic and/or nonscientific works.

15.4.6.8 Maximizing Lab Efficiency

Minimizing the cost and time spent searching for suitable products are priorities in order to maximize lab efficiency and productivity. Scientists today are fortunate to have an abundance and a variety of reagents available, but sifting through the assortment to select just one quickly becomes daunting. The process of paring down a long list of comparable products is not straightforward, and the fact that not all products perform equally well complicates the decision further (Gajdzik & Wycislik, 2012). Seasoned scientists are well aware that substandard reagents are risky and can compromise the outcome of a study. Searching for a product, performing the experiment, troubleshooting, searching for a new product should the original one not work as expected, and repeating the experiment cost the lab money and cost the staff time that would be better spent elsewhere. Therefore, in order to maximize efficiency, it is in the best interest of the lab to consciously source high-quality products.

An increasing concern in the life science field is that of reagent quality. Researchers often take for granted that a product will perform as marketed, but some, in fact, perform below par. Research using commercial antibodies suggests that while many perform as intended, some do not work as expected, and users are cautioned to rigorously test their research antibodies prior to use.

Upon generating a list of possibilities, the researcher must differentiate the quality of a product manufactured by company X versus that of similar products manufactured by company Y, company Z, and so on. A seemingly quicker alternative to this approach is to blindly purchase any reagent from the list. While bench work may begin sooner, the researcher runs the risk of wasting a great deal of time and money having to troubleshoot a problematic reagent.

Another limitation with which scientists are faced is that the uses of a product are dictated by what the manufacturer has tested. For example, a rabbit antihuman CD31 antibody tested by its manufacturer for flow cytometry is naturally suited for detecting human CD31 by flow cytometry. However, might it also be useful to a scientist looking to determine CD31 expression by immunohistochemistry? Knowing whether an unconventional use or application has already been tested, successfully or unsuccessfully, is significant information that would undoubtedly impact the design of an experiment.

A great deal of effort and resources go into designing and carrying out a study. To improve productivity and to avoid losses in time and money, it is key to be able to identify a product with a proven performance record. The challenge then is to find useful information about product quality and performance in order to make an informed decision.

15.4.6.9 Identifying Information Resources

High-quality reagents are central to experimental success, and information about a product helps determine its suitability for a particular study. As a result of continual lab turnover, key information can sometimes get lost in old notebooks. Knowing that a previous researcher spent two months, troubleshooting three different antibodies before finding one that performed as expected is invaluable information and can quickly set the next lab member on the right track. Too often, however, labs lack a formal repository to store such key information. In other instances, the lab may be moving into a new area of research in which it has little experience. The learning curve may be steep due to inexperience with study-related reagents unless there is a resource—a colleague, a citation, or other—from which to gain insight. To avoid squandering lab resources trying to reinvent the wheel, it is imperative that experiences be shared among members of the lab and others in the same field of study.

A Lab Manager not only offers firsthand research experience but also has seen the mistakes and successes of past members of the lab who have since moved on. Other colleagues are also great resources and may be able to suggest a suitable reagent they have used. They may be able to provide guidance and technical details based on their experiences with the product. Of course, the insights they are able to offer are limited by their familiarity with the area of study.

Scientific literature represents another vital source of information worth investigating. Papers that reference a product of interest provide pertinent experimental details and, better yet, provide concrete proof from independent scientists that a product works and produces publication-quality data. A caveat, however, is that final figures represent several repeated experiments and do not necessarily indicate how easy or difficult it is to optimize and work with the product of interest.

15.4.7 Laboratory Manager of a Non-research Laboratory

Based on the size and type of lab, the role of the Lab Manager may include leading the lab staff as well as ensuring that quality is not compromised by ensuring that the lab produces great results. But, overall, the Lab Manager is the Principal Assistant and Advisor to the HOD in the management of the labs. In particular, the Lab Manager is expected to

- advise the HOD on the policies to guide on lab operations, growth, and development;
- communicate any issues regarding the operation of the technical team to the HOD, where these cannot be resolved within the team;
- answer general enquiries from students and staff, where these cannot be met directly from a member of the technical team;
- supervise the instruction of students in the preparation of materials and the use of lab equipment;
- ensure practical requirements and lab written methods are developed, reviewed, and tested before sessions are delivered to students;
- lead the technical team in the labs, providing support for teaching and research activities and preparing regular reports to the HOD;
- act as line manager for the technical team, ensuring smooth, efficient, and flexible operation of the team, leading regular meetings, delegating tasks;
- respond to student requests for support in practical sessions and where appropriate with projects and equipment, noting to promote development but not dependency;
- manage inventory budget, providing support and guidance on the use of materials and equipment for teaching and research;
- make collaborative decisions with academic staff and the technical team regarding preparation for lab sessions;
- initiate proposals and contribute to collaborative decisions with academic staff regarding, for example, the purchase of new capital equipment;
- work under the general guidance of the HOD, planning and organizing the workload of the technical team to meet the needs of staff, students, and researchers in the labs;
- supervise the technical team to ensure efficient use of labs, stores, and teaching resources;
- be consulted for advice on standard day-to-day problems and complex technical problems to ensure success;
- maintain knowledge and awareness of product developments in a wide range of relevant lab discipline areas;
- ensure the lab activities of the faculty are fully compliant with all Health, Safety, and Environmental requirements and are fully risk assessed, and ensuring that students, academic staff, and members of the technical team work safely;
- ensure risk assessments are in place, and to provide advice on the safe use of equipment and facilities;

- develop the technical team into an effective, proactive, and energized support team;
- supervise the orientation and induction process for new students;
- provide instruction, support, and guidance on the use of lab and workshop equipment and techniques as necessary;
- have knowledge of Health, Safety, and Environmental legislation as it is applied to the lab or workshop environment;
- set a clear research agenda for the labs; and
- design mechanisms for long-term self-sustainability of the labs.

15.4.8 Researchers and Research Supervisors

Researchers (including Research Supervisors of Masters' and PhD students) are the primary persons involved in research projects in the Institution, its Schools or Research Centers, and as such have a responsibility to implement measures to ensure the safety of all those associated with the research. In particular researchers are required to ensure that

- any research that requires Ethical and Biomedical Research approvals has been issued with one;
- approval has been given by the HOD for the research to proceed;
- they fulfill their responsibilities as outlined in the Research Code of Practice Policy and Procedure of the Institution;
- staff and students associated with the research are made aware of their responsibilities;
- they conduct formal risk assessments on all work that is included as part of the research program and ensure that all control measures are implemented;
- staff and students associated with the research have received the appropriate lab safety induction and training to enable them undertake their work safely;
- staff and students associated with the research are formally advised that failure to comply with this Manual may result in disciplinary action being taken by the Institution including expulsion from the use of the labs;
- staff and students associated with the research have access to and wear personal protective clothing required to undertake their work safely;
- all incidents, hazards, and near-miss incidents are reported to the Lab Technologist in-charge using the relevant forms issued by the Institution, and a copy sent to the Lab Manager within 12 hours of the incident;
- staff and students associated with the research are formally advised that unauthorized experimentation is strictly forbidden and any new work must not proceed until a formal risk assessment is conducted and approval issued;
- staff and students associated with the research are formally advised that they are required to take personal responsibility for ensuring their own safety and the safety of others.

15.4.9 Technologists and Technicians

These technical staff are the primary custodians of the lab resources (specimen, references, equipment, chemicals, etc.), standards, and systems. They are encouraged to consult with other technical staff in the same field because this may enrich their knowledge of current trends and challenges in the market. Other functions of the lab technical staff based on the type and size of the lab may include the following:

- Restrict access to and use of the labs to authorized persons only.
- Ensure keys to the labs are not allowed into any other persons' hands except the Institution Security, for security reasons.

- Keep clean records of access and use of the labs, including the equipment and consumables used.
- Monitor compliance with the guides within their work area(s) and report instances of non-compliance to their immediate supervisors for action.
- Arrange for practical classes and assist faculty in demonstrations during practical sessions with students.
- Ensure sufficient stocking; organize for stock taking in the chemicals and consumables stores and forward any requirements to the Lab Manager to plan for procurement.
- Operate and maintain equipment in their respective labs.
- Prepare SOPs, Manuals, and Signing in forms for students, staff, researchers, and visitors to their respective labs.
- Establish, monitor and maintain appropriate levels of hygiene and housekeeping within labs and preparation rooms.
- Prepare weekly, monthly, semester, and annual Lab Use Records for submission to the Lab Manager.
- Prepare weekly, monthly, semester, and annual Risk Assessment Reports for submission to the Lab Manager.
- Prepare monthly, semester, and annual Equipment Serviceability Reports for submission to the Lab Manager.
- Assist academics in implementing the risk control measures they have outlined in their course/unit risk assessments.
- In consultation with the section heads, conduct risk assessments on the work that they perform in the lab (e.g., equipment setup, practical class preparations, waste disposal, etc.).
- Regularly check, test, and document the serviceability of specific lab emergency equipment not under the control of Capital Works & Facilities.
- Oversee the proper Segregation, Storage, and Disposal of Wastes.
- Clean, prepare, and isolate lab equipment prior to handover for maintenance to ensure that the maintenance work is carried out safely by a person(s) other than technical staff.
- Monitor the serviceability of fixtures, portable equipment, and apparatus and facilitate repair and/or replacement as required.
- Ensure that all chemicals and hazardous substances are stored, labeled, and used in accordance with the law and advice contained in material safety data sheets (MSDS).
- Ensure MSDS are current, legible, readily available, and retained appropriately.
- Ensure all lab safety equipment (fixed and portable) is properly installed and remains in a serviceable condition.
- Remain abreast of any legislative or industry changes that may materially affect health and safety management within a lab setting.
- Report all incidents, hazards, and near-miss incidents to the Lab Technologist in-charge using the relevant forms issued by the Institution, and a copy sent to the Lab Manager within 12 hours of the incident.
- Prominently display and maintain emergency management information in each lab, including the essential telephone numbers of
    1. County Fire Service,
    2. Independent Fire Service,
    3. Ambulance Services,
    4. Institution Security,
    5. Hospital Emergency Department, and
    6. Police Service.
- Attend and contribute to meetings of subject working groups as necessary.
- Communicate with suppliers and service engineers.
- Deliver a high-quality standard of service.

- Contribute to the induction and development of new staff as needed.
- Contribute to the orientation and induction process for new students.
- Provide technical advice and information to enable students to achieve their potential.
- Develop systems to ensure facilities and equipment are maintained to the necessary standards.
- Participate in material analysis and research.
- Provide advice to students and colleagues where necessary to allow them to make decisions concerning practical work.
- Maintain and update lab notebooks to ensure compliance test specifications and procedures.

15.4.10 College and Undergraduate Students

College and undergraduate students' practicals represent a major proportion of the work undertaken in Institution Labs. It is essential, therefore, that these students

- are fully conversant with these operational guides, any safety procedures that apply to their practical work in particular and the lab in general;
- dress appropriately for lab sessions, failure to which the Technologist in-charge may refuse access or report to the relevant authorities;
- understand the risk assessments that are relevant to their practical work and are able to implement the risk control measures that are acceptable to the Institution;
- adhere to the safe work practices at all times;
- have completed appropriate Lab Safety inductions, and have passed Lab Induction Tests, prior to commencing lab work;
- adhere to acceptable housekeeping standards;
- correctly use the provided safety equipment;
- understand that unsatisfactory behavior will be dealt with as a disciplinary matter;
- report all accidents to the Instructor or Technologist responsible for their practical so that an appropriate investigation can be carried out;
- do not perform unauthorized experiments in the labs.

15.4.11 Postgraduate Students

Masters' and PhD students who conduct their research work in INSTITUTION labs and associated facilities are required to ensure that

- any research that requires Ethical and Biomedical Research approval has been issued with one;
- approval has been given by the Lab Manager, for the research to proceed;
- they fulfill their responsibilities as outlined in the Research Code of Practice Policy and Procedure of INSTITUTION;
- they are fully conversant with the contents of this Manual and any safety procedures that apply to their research work in particular and the lab in general;
- carry out formal risk assessments in conjunction with their research supervisor prior to commencing work in a lab or associated facility;
- understand the risk assessments that are relevant to their research work and are able to implement the risk control measures that are acceptable to the Institution;
- do not undertake any new research work in a lab or associated facility that is not included in a current risk assessment, without consultation with their supervisor and completion of a formal risk assessment;

- adhere to the safe work practices at all times;
- have completed appropriate Lab Safety Induction Training, and have passed Lab Induction Tests, prior to commencing their research work;
- undertake all necessary training and instruction relevant to their work;
- adhere to acceptable housekeeping standards;
- correctly use and maintain the safety equipment provided;
- understand that unsatisfactory behavior will be dealt with as a disciplinary matter;
- report all incidents, hazards, and "near-miss" incidents to the Lab Technologist in-charge using the relevant forms issued by INSTITUTION, and a copy sent to the Lab Manager within 12 hours of the incident.

15.4.12 Contractors

Contractors (cleaners, suppliers, repair, and maintenance technicians) are required to

- strictly adhere to any conditions or requirements imposed by the Institution in contracts, agreements, scopes of work, specifications, variations, permits to work, risk assessments, work method statements, or that may be part of any workplace induction program, orientation, inspection, handover, or be required under legislation;
- comply with any reasonable direction given by lab staff in the interests of health, safety, and welfare; immediately report any incident or accident or any unexpected occurrence to lab staff.

15.4.13 Visitors

All visitors are required to comply with any reasonable directions that may be given by lab staff in the interests of promoting and maintaining health, safety, and well-being within the labs and facilities.

A visitor is any person who is not permanently authorized by the Institution to work or learn in the lab or facility concerned and who has not received appropriate lab induction training. Visitors may include but not necessarily be limited to other INSTITUTION staff, visiting academics, students, clients, contractors, and members of the public.

The following groups of visitors shall not be allowed into the labs:

1. Children or minors, except organized school parties on pre-arranged study tours
2. Animals and animal pets
3. Intoxicated or drunk persons, visitors, students, and staff (including lab staff and faculty)
4. Persons acting in a manner likely to cause a breach of the peace
5. Persons suspected to have mental instability.

Technologists must immediately call Security in to action whenever this clause is breached.

15.4.14 Breaches of Security to be Reported

It is expected that when all protocols are adhered to, then infringements should be rare occurrences. However, whenever they occur, all such breaches of lab security (including suspected breaches) should be reported as follows:

1. When the chief lab officer identifies an infringement, then he or she should bring it to the notice of the personnel responsible with immediate effect. A reminder should be issued detailing the significance of compliance with the laid down protocols. The direct manager

of the worker involved has to be informed that an infringement has occurred and the identity of the lab worker responsible for the infringement shall be made known to them. It is then expected that this will prompt a discussion in the relevant organizational group concerning the importance of the protocols.

2. In the event of occurrence of a second infringement, then the chief lab officer should act as detailed above, but a written warning should also be issued to the worker responsible for the infringement and copied to institutional head.
3. The institutional head will then deal with any further infringements and may therefore impose appropriate sanctions that may include suspension from the lab.
4. Where an event is certified to be serious, then a written report is required and involvement of the safety officer should be immediate.

15.4.14.1 Categories of Breaches

1. Minor breaches

   These include not wearing lab coats, safety glasses, eating in the lab, and unsafe or incorrect operation of equipment and handling of chemicals.
2. Major breaches

   Major breaches may be related to safety, and this includes breaches of scientific ethics, ignoring laid down protocols and unwarranted behavior toward fellow workers.

15.5 RECORDS AND INFORMATION MANAGEMENT

Information management and recordkeeping play a pivotal role in the functioning of laboratory operations, as they are essential for maintaining safety through the retention of critical information pertaining to experiments, personnel stocks, and equipment. The integrity of information is contingent upon its accuracy over the entirety of its life cycle, coupled with the need for accessibility during audits. Electronic laboratory notebooks (ELNs) offer a compact alternative to traditional recordkeeping, facilitating researchers in the efficient analysis and sharing of findings (Berezin et al., 2023).

The integrity of data is a fundamental aspect in laboratory settings. The validity of research findings and the safety of operations are contingent upon the processes of accurate data generation, systematic backup, proper archiving, and secure sharing methods, all of which are necessary to safeguard against potential tampering (Berezin et al., 2023). The Minimum Information for Biological and Biomedical Investigations (MIBBI) guidelines conceptualize the laboratory notebook as an electronic document marked by time stamps, which facilitates the replication of data analysis.

An ELN enhances data management by offering improved and structured accessibility and representation, while simultaneously ensuring the integrity of information through unchangeable records that track all interactions with the data.

Recent developments in web-based ELN platforms have enhanced user accessibility and compatibility with simpler devices. Customizable systems are crucial for effective electronic recordkeeping. Integrating workflows involves considering user availability and reducing IT support. Nonetheless, ELN adoption in basic science labs remains low, around 4%, due to difficulties in managing diverse data formats and file types (Vandendorpe et al., 2024; Chen et al., 2024b).

Data management and recordkeeping are essential for labs, ensuring data integrity and fulfilling commercial and ethical aims. The ELNs are crucial for data generation and discussions, emphasizing reproducibility in diagnostics, research, and development.

15.6 REGULATORY STANDARDS AND ACCREDITATION

Laboratories must follow standards like ISO 15189 and USA's CLIA of 1988, which define quality criteria. Accreditation processes vary by region, and systematic measurement and monitoring are vital for compliance and operational efficiency. Laboratories undergo accreditation through systematic evaluations, audits, and assessments (Franić et al., 2020). Internal audits assess operations through planning and feedback, with regular updates to patient safety protocols. Staff roles and growth opportunities are clear and supported by performance reviews. Continuous education improves managerial and technical skills. Clear communication of reference intervals is crucial for effective testing, while test results are approved after verification, with critical values released autonomously (Ilinca et al., 2023; Khattar & Nehme, 2024). Better information access enhances efficiency and compliance with accreditation standards.

The ISO standards provide a crucial framework for managing laboratory activities, ensuring QC and efficiency. ISO sets important standards across industries, including ISO 9000 for quality management. ISO/IEC 17025 outlines the competencies required for testing and calibration labs. Previously named ISO/IEC Guide 25 or EN 45001, it defines criteria for labs to demonstrate technical competence in these areas (Panagiotidou et al., 2025). Adhering to these standards aligns processes with client and community quality expectations, focusing on management strategies, methodologies, and client relations to help labs regulate practices and manage resources effectively (Erkaboev et al., 2023; Bennari et al., 2025).

The CLIA create a regulatory framework for clinical laboratory testing in the USA, requiring labs to obtain CLIA certification to ensure quality. Waived tests need a Certificate of Waiver, while more complex tests require a Certificate of Compliance. The Centers for Medicare & Medicaid Services (CMS) oversees inspections, CDC sets technical standards, and FDA classifies tests and develops kits (Bennett et al., 2025).

15.7 THE FUTURE OF LABORATORY MANAGEMENT

Implementing sustainable laboratory practices significantly lessens the environmental impacts of reagents and energy use. Global initiatives stress waste management and energy efficiency. Key strategies include classifying waste, separately collecting hazardous materials like mercury and silver, and promoting recycling. Proper electronic waste disposal and solvent recovery highlight reduction, reuse, and recycling principles. Furthermore, energy-efficient lighting and eco-friendly supplies are increasingly adopted (Berezin et al., 2023).

Laboratory activities are evolving through digitalization and globalization, propelling scientific advancement. Automation is essential in diagnostics and research, using open-source frameworks to link instruments. Future lab management will blend digital intelligence with biosensor technologies. Tackling complex challenges requires interdisciplinary strategies, new Research and Technology Organizations, and innovative collaborations that involve social sciences and humanities (Khankhoje, 2023).

Automation and artificial intelligence (AI) are vital in laboratory management, enabling complex experiments with reduced human involvement and generating vast data essential for biological research in pharmaceuticals and biotechnology. Effective automation requires management support regarding training, resources, and funding. Laboratory leaders must promote automation, inform upper management about its advantages and limitations, and monitor emerging technologies while fostering internal support.

The AI offers valuable support for laboratory managers in equipment maintenance, results management, and decision-making. Algorithms consistently extract quantitative data from instruments, enhancing accuracy through diluted sample analysis. In chemical warfare agent testing, AI plays a crucial role in decision-making, interpreting results, and selecting optimal confirmatory tests amid various analytical methods (Odunayo, 2024). Open-source AI systems enable managers to tailor algorithms for reporting aid and detect common lab issues.

Integrated platforms enhance laboratory operations by optimizing layouts, inventory control, workflow management, and simulating experimental conditions, leading to better decision-making and reduced risk. Currently, simulations rely on equipment data, but adding sensing modules can improve precision. Future advancements in reinforcement learning will boost system adaptability. Additionally, robotics integration and modular designs will ensure compatibility across scientific systems, promoting improved management practices (Marescotti et al., 2021).

The implementation of automation in virology laboratories improves the efficiency of manual tasks while ensuring their proper execution. Advanced IT boosts data analysis capabilities across testing platforms. Laboratory Information Systems (LIS) have evolved from data storage to direct analyzer connections, optimizing administrative operations. Today's LIS are vital for laboratory organization and departmental integration (Zubanov et al., 2025; Shah, Shah, & Jbeen, 2024).

15.8 CONCLUSION

The management of laboratories plays a crucial role in advancing clinical, industrial, and scientific endeavors. The incorporation of integrated systems enhances operational efficiency, thereby streamlining data generation in an increasingly automated environment. Well-structured protocols are instrumental in directing various administrative functions, including budgeting, inventory control, human resource management, and employee training. Additionally, these protocols supervise analytical methodologies and information technology frameworks, ensuring consistent performance and dependable data outcomes.

Laboratories engaged in research and analysis play a crucial role in bridging technical and scientific gaps that have an impact on competitive advantages. Maintaining high standards of reliability and quality is imperative for protecting these advantages. To effectively confront modern challenges, it is necessary to evolve from inflexible structures to more agile frameworks that can swiftly adapt to shifting objectives and technological advancements. Effective management is fundamental to the ongoing maintenance and sustainability of laboratory infrastructures.

16 Management of Laboratory Inventories and Waste

Apollo Odhiambo Maima and Sarah A. Otieno

16.1 MANAGEMENT OF INVENTORIES

Inventory management is an essential ingredient of the quality management system as the laboratory requires an uninterrupted availability of reagents, supplies and services. Any shortages, however briefly, are likely to disrupt services that depend on full or proper laboratory operations. In setting up an inventory control system, various factors need consideration. The design of the system should be such that the conditions of all supplies and reagents can be monitored closely, available quantities ascertained and the system alerted when reorder levels are reached for initiation of procurement procedures. Assiduous control of inventory helps to reduce waste and financial loss due to poor storage, expiry or obsoleteness of reagents and supplies. Institution of an inventory management system, therefore, assures that reagents and supplies are ever available as required; high-quality reagents are acquired cost-effectively; and reagents and supplies are not lost to spoilage or expiry before use.

It is noteworthy that the lifespan of reagents can stretch from a couple of weeks to a few years; hence, it pays to keenly monitor their expiration dates to avoid using or losing expired reagents. On the other hand, overstocking is too wasteful and expensive. Another issue requiring much attention is receipt of donation of equipment and supplies; they must meet the laboratory's set standards and operational needs. If that is not the case, then the Laboratory Manager may opt to reject such donations, albeit diplomatically to ensure opportunities for future donations are not jeopardized. Donors are not expected to use labs as dumping grounds for what is not in sound shape, spoilt or expired elsewhere!

16.1.1 Implementation of an Inventory Control Program

The steps below (and summarized in Figure 16.1) are a good guide in the implementation of a workable inventory control program:

- Appropriately assigning duties and responsibilities.
- A process is needed for assessing the laboratory needs for materials. A list should be compiled of all the tests carried out in the laboratory, and all the supplies and reagents required for each test determined.
- Institution of the minimum stock needed for a defined period.
- Development of required forms and logs.
- Institution of a system for receipt, inspection and storage of supplies.
- Maintenance of an inventory system in all warehouses for all chemicals, reagents and other consumables.

16.1.2 Quantification

Quantification is a method for assessing or calculating how much of an item is required for a defined period, hence how much to be ordered. In empirical science, **quantification** (or **quantitation**) involves counting and measuring that maps human sense observations and experiences into quantities.

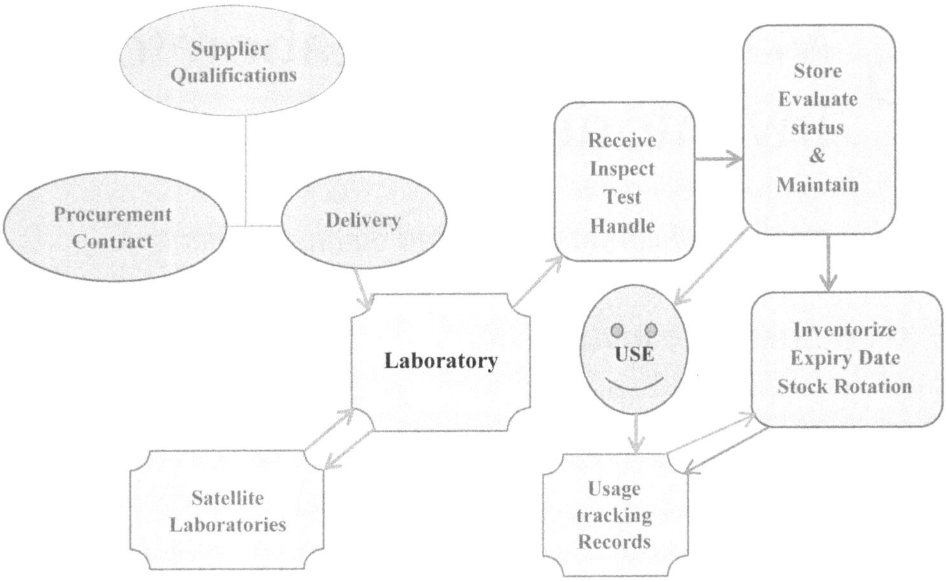

FIGURE 16.1 Essential elements of an inventory management system.

It is an important part of an effective inventory management system. Proper quantification ensures critical supplies are always available; prevents overstocking, hence wastage; helps in gauging annual demand and budget; allows for better, more informed management decisions; and in monitoring and evaluation of the inventory management process. The most commonly used quantification methods are consumption-based and morbidity-based quantification methods.

Consumption-based quantification is undoubtedly the more commonly used between the two methods, being the method of choice for almost all non-medical labs. This method relies a lot on the experience of a laboratory over time, draws on actual consumption and, therefore, requires a number of considerations including wasted, expired and spoilt quantities. For management and planning purposes, it is important to note the items that may have been out of stock for a while, enough to interfere with operations, implying that the last supply was insufficient, or that the wastage or expiry is higher than foreseen.

Morbidity-based quantification method calls for the laboratory to consider the actual number of cases, illnesses and health states that necessitate laboratory testing. In the planning phase the laboratory should be able to foretell and estimate as closely as possible the episodes. By use of standard guidelines for diagnosis and treatment, and assuming health clinicians fully comply with these guidelines, can help to approximate the number of laboratory tests to be performed. A robust morbidity-based quantification method is more precise than the quantification by consumption method, provided the data are correct.

16.1.3 Procurement of Goods and Services in the Laboratory

The specifications of the required products should be given to the procurement department of the organization. Once a detailed profile, specification and quantities of a product have been given the procurement process can then proceed. As much as the normal procurement process is used in the selection of the correct suppliers, the Laboratory Manager should be involved in quality evaluation, to select a supplier that offers best quality. Once a supplier has been shortlisted for a given group of products, the institution can be often engaged to supply the needed products.

TABLE 16.1
Summary of the Most Commonly Used Procurement Methods (WHO, 2021)

Procurement Method	Requirement	Market Conditions	Appeal/Request	Competition
Request for quotation (RFQ)	Goods	Competitive market place/ prequalified	Request for quote from prequalified vendors or public tendering	Global, regional or local
Invitation to bids (ITB)	Goods/Services	Competitive market place/ prequalified	Formal invitation to prequalified vendors or public tendering	Global, regional or local
Request for proposal (RFP)	Services	Competitive market place/ prequalified	Formal invitation to prequalified vendors or public tendering	Global, regional or local
Direct procurement/ contracting	Goods/Services	Emergency cases Time critical activities Standardization/calibration of equipment Nonavailability of goods or services in local or regional market	Direct invitation or negotiation	No competition

There are various methods for purchasing equipment, supplies and services by different labs. These include the use of request for quotation (RFQ) method, invitation to bids (ITB), request for proposal (RFP) and direct procurement/contracting. In some jurisdictions, there is a national procurement system with a central stores/institution that receives, stores/handles and distributes items directly to the labs. Still, many organizations and institutions rely on donors for the procurement of their equipment, supplies, reagents and services (WHO, 2021). These methods are briefly described in the next paragraphs and summarized in Table 16.1.

A **RFQ** is an unofficial invitation to present a quotation for a required item. Most organizations set an upper limit on how much money can be spent using this method since it tends to be abused. Suppliers are usually given at least 7 working days to respond to RFQs. Prices, specifications and other commercial terms and conditions, that often include samples, are requested, and award is usually made to the most economical and technically admissible offer.

An **ITB** is usually associated with clearly and succinctly defined criteria, including estimated procurement value. Besides, ITB could be accompanied by requirement of public bid opening if the value goes beyond a certain defined value. Below that value, there would be no requirement for public bid opening. An ITB presupposes that the bidder and her required product(s) meet set technical criteria; hence, price is the only consideration; an award is made to the lowest evaluated and responsive bidder. Suppliers are usually given 15 working days or more to respond, subject to how complicated the criteria are.

A **RFP** is a formal request to present a proposal usually linked with criteria for services rather than goods, which cannot be clearly or succinctly defined. A number of factors form the criteria for evaluation, cost being just one of them. Award is made to the responsive qualified bidder whose bid considerably fulfills the criteria given in the requisition documents and is evaluated to be the lowest in price. Suppliers are usually given 15 working days or more to respond, subject to how complicated the criteria are. In **direct procurement/contracting**, procurement personnel may buy goods or services directly from a vendor without subjecting them to competition. Direct procurement is also called **single-sourcing** and is a rarely applied method except when it is deemed to be in the best interest of an institution.

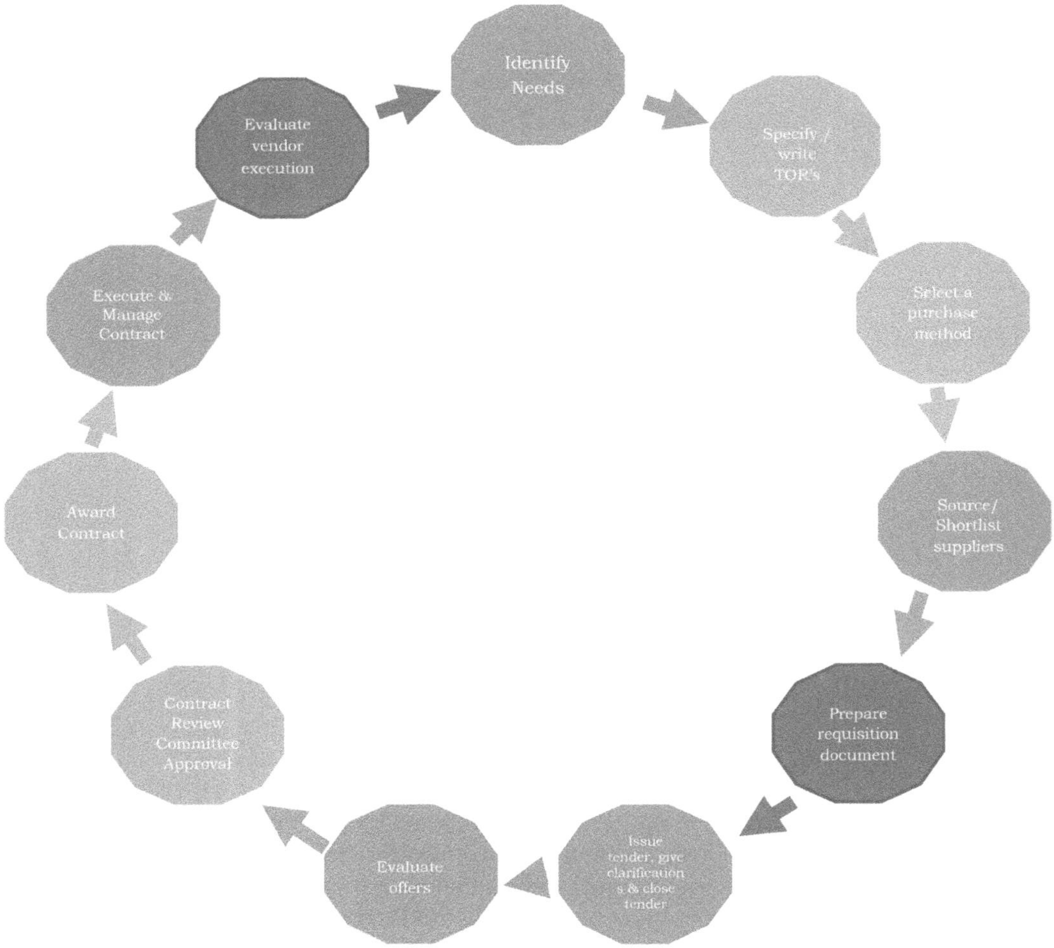

FIGURE 16.2 Flowchart of the procedure of a procurement process.

The procedures of the procurement process (outlined in Figure 16.2) are conducted in strict adherence to the institution's financial rules and regulations, and to the home country's laws, rules and regulations for public institutions. In most cases, contracts are awarded through competitive mechanisms with bids obtained through local or international tenders or through prequalified suppliers/vendors (which may be individuals, companies or organizations) for specialized items. In certain special circumstances, an institution may turn to vendors prequalified by the central government, local government or quasi government agencies so long as it is allowable by law or regulations. This kind of process is called a prequalification program whose primary goal is to ensure quality is prioritized. It is achieved through regulator assessment and inspection activities.

Laboratories should carefully consider vendors' and manufacturers' qualifications, product specifications and methods of transport. They should thus predetermine the basis for supplies or items to be procured, source for lowest prices, and consider too the qualifications and integrity of the vendors. Laboratories that are supplied with consumables from their central government stores should interact with the stores operating staff to ensure those objectives are met. Meanwhile, consideration should be made for the pros and cons of procuring "branded" versus "generic" products, as sometimes there is no value addition in using branded products, despite their very high costs. Often, it helps to compare notes with other labs when evaluating quality, reliability of supply and cost. It is equally important to evaluate vendors' after-sales service and response.

16.1.4 Procurement of Laboratory Consumables

Laboratory consumables are single-use, nonreusable items for tests and reactions bought for daily use. They may include test-specific items as microscope slides and cover slips. Some, like bleach, distilled water, alcohol, masks and gloves transcend all lab services and are grouped as general lab consumables.

Quality is essential in the procurement of lab consumables. Unfortunately, what is found within the market is not always quality, safe and appropriate. This is a major issue as the myriad choices and products in the international and local markets complicates the task of access to the appropriate consumables required for daily lab use. The cornerstone for procurement of consumables should, therefore, be the evaluation of obtainable market alternatives for quality, value and convenience. Whereas procurement processes can vary, certain rules and regulations shall remain cardinal in the purchase of consumables.

The lab needs to set some standard operating procedure for the procurement and supply chain of laboratory supplies, as this would help in reducing unwarranted requisitioning. Possession of proper safety data sheets before procurement of supplies assures correct handling and storage of the materials upon reaching the premises.

The sourcing for these consumables can be done strategically so as to create various openings for buyers. Most lab buyers have a primarily transaction-based approach and lose out on cost-saving openings due to lack of exposure to functional statistics on consumables procurement market.

Consumable dealers in the Sub-Saharan Africa have modest bargaining power due to low pressure and competition from substitutes and new entrants. This makes it inevitable for sellers to get the pricing and pricing model right, to stay afloat. On the other hand, labs can benchmark their favorable pricing models for supplies with the larger market and identify the potential to cost-cut. Some of the most common pricing models for lab supplies are volume-based, unit-based and cost-plus pricings. Every model presents its own prime advantages and add-ons in distinct situations. Each institution or lab entity should, therefore, determine the model that best fits their operations and associate supplier capabilities with the pricing models.

16.1.5 Purchase of Hazardous Chemicals and Reagents

Hazardous chemicals often need to be purchased by a laboratory for the purpose of either teaching or carrying out research. The procurement process of such hazardous chemicals is fairly distinct and the process dictated by the procuring department considering issues like cost and academic need.

Ordering decisions are usually made by the head of the procuring department, or the lab manager. Lab requirements presented by the manager (American Chemical Society, 2016) would usually not be rejected by the Head of Department as the manager knows best the needs of the lab. The two should work harmoniously to assure safe procurement of dangerous chemicals, after identifying the lab requirements. Authority to procure should only be based on the evidence for the laboratory requirements. Ordinarily, the purchase and restocking of hazardous chemicals and reagents are dependent on satisfying all storage and compatibility requirements.

16.1.6 Procurement of Laboratory Apparatus and Equipment

Laboratory apparatus and, especially, equipment are generally very costly; hence, procuring them is a major investment for any lab. It is important that the multiple facets of the procurement process are evaluated to maximize the lab return on investment. This is certainly a herculean task to carry out. The procuring entity (buyer) can lean on the manufacturer for advice and guidance through the process and decision-making. This approach is, however, based on the assumption that the manufacturer or her agent will have the buyers' best interests in mind and won't take advantage of the trust to sell a product that will not serve the buyer's purpose. It is hence necessary that the lab has reliable, genuine, authentic, trustworthy and technically apt suppliers.

In addition to the general considerations before procurement discussed earlier, we now look at some of the important factors to take in to account on arriving at the apparatus and equipment to procure for the lab. The following 6 factors will help labs to streamline their decision-making process and substantially reduce the buying cycle.

Availability: Laboratorians always want their supply orders delivered immediately, including those of equipment. This works well only where the products are in stock and available off the shelves for prompt delivery. For low- and medium-income countries, lab equipment are mostly imported from industrialized countries and the required items could take more than 10 weeks to be delivered. Where products are not available locally and must be imported, suppliers may be asked to quote lead times to be considered alongside technical and financial bids. Then the laboratorians can compromise on lead time versus price.

After-Sales Service is crucial for proper running of, especially, specialized equipment. Often, laboratorians need extra coaching on equipment use and maintenance service, making it necessary to include this aspect into the sales contract of such equipment. It would be reassuring to know that the equipment supplier would provide support if users required any assistance with using the equipment.

Warranty is a written guarantee, provided to the buyer of an equipment by its manufacturer, pledging to repair or replace it partially or wholly as the situation demands, within a given time-frame. There are many variations in the elements covered by warranties, usually depending on individual manufacturers and regulations at play in the country of origin. Despite the variations in the type of support pledged, warranties are extremely important when procuring lab equipment. Some aspects of warranty stand out, as every lab would be expected to put premium value on them. **Length of warranty** period should be considered carefully since in some cases it counts from the time of purchase. This would disadvantage the lab, especially, if the item is to be imported, a process whose lead time is often more than 10 weeks. Ideally, the counting should start from the point of receipt. Still, in some cases, the equipment needs to be registered before a warranty can issue.

Labs should keenly study the **factors that would void the warranty** so as not to be caught flat-footed. But it is noteworthy that warranties are assured only for as long as the equipment is used as directed (in the product manual or catalog). It would be devastating to send an equipment back for a warranty claim just to have it sent back by the manufacturer, with the advice that the warranty is void.

A lab is expected to place a **warranty claim** with the company that manufactured the equipment, and not the seller, since they are the ones with all the data and records in their system regarding the product. They are able to more promptly deal with the warranty issue. Besides, the warranty document is only legally binding to the manufacturer and not the seller. The laboratory leadership should therefore be very clear on the process of acquiring a warranty. That includes the location for presenting the warranty, when and how to acquire new products, plus all the costs for processing the warranty, should be known before making purchase payment.

The laboratorian also needs to evaluate upfront the **workability of the warranty claim process**. It may for example involve sending the equipment back to the manufacturer, including paying for freighting, or other additional costs. The "fine prints" of the warranty document always state whether the manufacturer will repair the equipment, replace it or give a refund. It is better to have a more workable warranty claim process, free of hassles. Often manufacturers offer **extended warranties** but at extra charges. Whether the extension is a worthwhile investment depends on the cost of the equipment and the related extended warranty charge. It is equally important to consider the coverage of the extended warranty, and how significantly it differs from the standard warranty. By the time the lab management decides to pay for extended warranty, all these considerations should have been made and it is deemed worthwhile, not just under pressure!

In summary, the laboratory management should strive to work with suppliers who offer stress-free warranties from whom the replacement products, or parts thereof, are delivered to clients at no extra cost. Assessment of the presence of extended warranties is also a vital consideration.

Other factors to consider when procuring lab equipment include usage, servicing and cost.

On equipment requiring regular **servicing** or **calibration**, the lab needs to decide whether these will be done internally by the maintenance staff or externally by more specialized technical staff. Some manufacturers have trained and approved specialist technicians found within defined localities or regions of the world. They are especially trained to service, maintain and calibrate the manufacturers' tools, apparatus and equipment. Local suppliers should always be able to help the lab in reaching these technicians and plan for the works. There are also local, regional or global organizations whose major forte is to service, maintain and calibrate equipment on consultancy basis. In many countries, calibration is regulated by law and the organizations that carry out calibrations have to be approved and registered by the governments in whose jurisdictions they operate. Often, warranty is voided if an equipment is serviced, maintained or calibrated by an unapproved technician, including internal staff, or at inappropriate intervals.

Equipment usage is a crucial factor that should be placed at the center of the procurement decision-making process, yet is often ignored. Quality and reliability aspects are more critical in more frequently used equipment than one that is seldom used. The more frequently an equipment is used, the more these aspects become important. Usage must therefore be appropriately considered before procurement is made.

Before using an equipment price in determining whether it should be procured, there are other related considerations to be made. The more relevant cost aspect that should considered is called the total cost of ownership (TCO), which in the useful life of equipment could be composed of disparate items such as costs for installation, training, operations, maintenance, calibration and user efficiency. The TCO may be weighed up against total benefits of ownership (TBO) which considers what benefits the lab will accrue by procuring the equipment. Depending on the nature and cost of equipment proposed for procurement, TCO and TBO may or may not be evaluated; the cost and time spent on such evaluation may not be worth it for a low value equipment. Either way, the lab needs to have in place a protocol for TCO/TBO analysis as required for equipment procurement over a given currency sum (S).

16.1.7 Practical Steps in Equipment Procurement

There are many areas of potential errors in the process of purchasing equipment, more significantly by inexperienced buyers. That is why we have found it necessary to add this section even after what looks like an exhaustive section above on equipment procurement.

Getting equipment procurement right must never be a matter of trial and error as they are very expensive. The lab manager hence needs a lot of prior knowledge of both the specified equipment and the supplier track record. Hopefully, the chosen supplier is trusted enough to help the laboratorian through the process and selflessly support their decision-making process. The laboratory management should look for suppliers that they can trust to provide wholesome guidance bereft of unwarranted product push. The steps briefly mentioned below can form a good guide on precautious and streamlined approach to the process and can significantly reduce the buying cycle (Ezzelle et al., 2008).

16.1.7.1 Ascertainment of Equipment Need

It is crucial to ascertain the lab need for a piece of equipment before requisitioning for it. It does not help the lab to buy an equipment to be brought and then hardly used. Where a specific brand or model is favored, that should be justified by the functional features desired or essential for the lab's teaching or research functions. The required features need to be listed before deciding on the brand, not vice versa.

16.1.7.2 Choosing from Available Supplier Options

Equipment supplies may be from the manufacturers, third-party (agency) suppliers or online market sites. Manufacturers mostly sell equipment, which should come with warranties and service contracts. Manufacturers also often sell discontinued, used or demo models at discounted rates (Seery, Agustian, & Zhang, 2019). The greatest advantage of procuring directly from a manufacturer is that the organization has significant knowledge and technical competencies of their instruments and can provide the same and reliable support.

Most third-party suppliers trade in both used and new equipment, and in all sorts of brands and models, with the advantage that a buyer can compare specifications and costs of different instruments that perform similar functions in the lab. Third-party suppliers also often provide warranties and service contracts with purchase.

Equipment can also be bought from online markets or auctions. The major disadvantage of this avenue is the lack of warranty to accompany the equipment, though one can also land some very good deals. The costs of buying online tend to be relatively cheaper than buying directly from the manufacturers or third-party suppliers. Lab managers should therefore always conduct considerable investigations and research to determine the best option for obtaining the equipment.

16.1.7.3 Ensuring Demonstrations Are Conducted

Before concluding the procurement process for an equipment, it is important to ensure a demonstration is conducted of the equipment and its use. The potential users will thus have the chance to see how the equipment works and learn how to use it firsthand. After the demonstrations, it would be possible to tell how easy or difficult it is to use the equipment or others. It is really like having to take a new car on a test drive; it is that important! Requiring demonstrations of equipment before buying is a crucial practice not only for new equipment, but even for used ones too. Perception of success or failure of demonstrations will determine how much trust can be bestowed upon the supplier. Demonstrations also bring out any, if at all, special steps required to get the equipment to work optimally, before considerations for purchase.

16.1.7.4 Negotiation with Suppliers

Upon receiving quotes from various suppliers, one can use the same to negotiate with a view to paying lower than the listed prices. Negotiations can take the same format used in car bazaars or livestock markets. If, for example, one supplier offers a lower price, that may be used to negotiate with the other suppliers for lower prices. If there is time, the new prices can then be used for another round of negotiations by telling the suppliers with higher prices how much lower you have been offered elsewhere.

The above method can be used for either used or new equipment, to extract lower prices, longer warranties or better service contracts. For this approach to be successful, it is better to get at least three competitive supplier bids.

16.1.7.5 Enquiring about Service Contracts

On expiry of warranties, there is still a realistic possibility that something could still go wrong with the instrument, whether new or used. That is where service contracts come in, procured as an optional add-on. Instead of a contractor being paid to repair malfunctioned instrument, a service contract offers flexible payment of monthly fees regardless of whether the equipment malfunctions or not. Of course, this can be money-saving if the equipment develops multiple problems but a loss if nothing malfunctions. Therefore, a service contract is more useful in cases of used equipment.

In assessing service contracts, the first thing to consider is the equipment has an existing manufacturer's warranty and its validity period. Next to consider is the cost of the contract and the coverage period. It is crucial to read all contracts, especially the fine prints, before appending any signature on them to ensure they clearly capture what was discussed and agreed on with the supplier representative.

16.1.7.6 Consider Supplier Reviews and Referrals

On getting prepared to procure an instrument from a specified supplier, it is helpful to research for information on both the instrument and supplier. Reviews on the suppliers should shed light on their previous performance, for example if other users have faced any difficulties dealing with them or with similar instruments. If feedbacks on a supplier has a high rating and recommendation, the confidence to deal with them gets better.

If from the feedback from other customers, the supplier was reliable and timely in addressing concerns, then this gives them a higher suitability rating and will be helpful in deciding whether to procure a service contract or not. Similarly, the lab manager can make enquiries from neighboring labs, and other colleagues experienced with similar equipment, for referrals on where to procure from. Some organizations have contracted suppliers who supply them with equipment at discounted prices, and it may be advisable to contact them before making the final procurement decision.

16.1.7.7 Determine the Costs Associated with the Equipment

It is one thing to buy a piece of equipment and a totally different one to operate it! As an equipment is being bought, it is extremely important to know the consumables it would use, and what they cost per month. While a second-hand equipment may save money at the beginning, it can be expensive to run if its consumables are costly or not readily available. In the long run, therefore, the equipment may turn out to be more expensive.

It is equally important to find out about the possible replacement parts and their costs for specific brands or models. These variables can be used to help determine which one of two or more different pieces of equipment the lab should spend money on. It is advantageous to have the ability to fix an equipment at a lower cost whenever it malfunctions, even if it initially costed slightly more at procurement.

16.1.8 Laboratory Supplies Registration and Control

Instituting a suitable supplies registration and control system is an invaluable step in inventory management, and must be detailed but clear enough for all the responsible staff to follow. Established instruments for stock control include standardized forms, card systems, log books and computerized systems. The information that should be noted on receipt of supplies includes date of receipt of supplies; supply batch number(s); order criteria pass or fail; and date the batch was opened for use or, if not suitable for use, date and method of disposal.

The bin card or stock logbook system is used to trail supplies on hand at any time. It is important to record the name and signature of the recipient of the supplies, date of receipt, date of expiry, quantity of the supply received, expected minimum stock to be held, and present balance in stock. Other important records are shelf position, number or name, and destination, for example, to –80°C freezer, to dryer. The bin card or stock logbook is best kept in the store, next to the relevant supplies or reagents.

Upon receipt in the lab, all supplies ought to be examined to rule out spoilage or expiries, to verify their conditions and to ensure they match the order specifications. Documentation should be developed and used for the receipt of supplies, and should bear the following records: recipient signature signifying receipt of supplies, date of receipt of each item and item expiry date. Besides, bin card/logbook/computer records should be updated or created for new items. For consumable supplies and reagents, storage on the shelf should be such that earlier expiry supplies are in front of later expiry supplies. Alternatively, if the expiry dates are the same, the new shipment should be stored behind existing shipment.

16.1.9 Good Storage Practices

Storage is a crucial part of inventory/supplies control. Good Storage Practices entail ensuring the stores are clean, organized, locked to protect the stocks; appropriately ventilated; and sheltered from direct sunlight. Supplies are stored as per the manufacturer's directions noting, especially, the temperature, safety and other requirements. Shelves for holding items should be sufficiently strong. Items should be diligently organized on the shelves to avoid motion or tumbles, while the shelves must be securely attached to prop up walls and prevent overturning. Items in the shelves should be within easy reach; strong step stools should be used for reaching upper shelves; bulkier/heavier items should be held on lower shelves. It is noteworthy that lab personnel should avoid lifting/carrying bulky or weighty items.

It is important to label shelves for supplies storage, as this rationalizes and orders the storage room. The process involves allocating numbers, letters or names to distinct areas of the shelves, cold rooms and other coolers, and noting in the log book which shelves are used for what items. This system helps to save personnel time that might be used in inordinately long searches through various shelves, and enables personnel unfamiliar with the stores to locate items more readily. Reagents should be labeled with the dates of their opening, ensuring the expiry dates are conspicuous.

16.1.10 Inventory Movement and Use

Inventory should be constantly monitored hence the importance of instituting and implementing systems for the same. To ensure this is effective, the following should be done:

- Assigning the task to a suitable person or persons to take charge
- Ensuring that all items supplied to the lab are always registered in the system and inventory records properly kept in all storage areas
- Carrying out weekly stock-takes of consumables to validate the system, and to monitor their movement and usage
- Ensuring that inventory records are always updated and maintained.

16.1.11 Automated Inventory Control

Most labs will do with simple automated/computerized systems that can be set up for inventory control. Use of computers proffers lots of advantages but also some disadvantages. The **main advantages** include the following:

- Close monitoring of equipment, apparatus and consumables at hand since it can be updated daily
- Enabling proper management of expiry dates by activating an alarm system to go off whenever it approaches the expiry dates of reagents. This way, reagents and other time-bound consumables will be optimally used
- Production of useful data for management, planning and procurement
- Use in coordination of the process of distribution of reagents and other consumables to satellite labs
- Reduction of the burden of inventory control.

Automated/computerized are not without flaws. Some of the disadvantages include the following:

- The fact that most computers are costly at the time of procurement and during use especially if an on-site computer is needed
- The need for personnel training, including on the use of the system.

16.2 LABORATORY WASTE MANAGEMENT

16.2.1 Introduction to Waste Management

This section presents the approaches and methods for the management and ultimate disposal of lab wastes. These wastes may contain chemical hazards as well as multi-hazardous compounds containing the combination of biological hazards, radioactive and chemical hazards (Ho & Chen, 2018). According to Wang et al. (2017), an effective waste management strategy aims at maximizing safety while lowering the environmental impacts. This effectiveness is considered from the time of purchase of raw materials utilized in the lab to the disposal of the resulting wastes from the lab processes. There are number of activities and processes for managing lab chemicals and wastes to ensure their probable adverse impacts on the environment are minimized, and maximum safety assured (Lee, Ellenbecker, & Moure-Ersaso, 2004). This section presents a discussion of the lab waste management strategies, starting with a description of the type of lab wastes followed with a discussion of the approaches that can be implemented in the management of the different types of chemical wastes. A discussion of the qualifications, competencies and role of personnel in the management of lab wastes is also presented.

16.2.2 Types of Laboratory Waste

Laboratory wastes are classified into two major categories, nonhazardous and hazardous wastes. Hazardous wastes, according to Walters, Lawrence, and Jalsa (2017), refer to the chemical, biological and radioactive waste that are dangerous to both humans and the environment. Handling of such wastes requires a high level of competency from the lab waste management personnel. Also, the management of such wastes requires the adoption of methods and strategies that will minimally impact the environment.

16.2.2.1 Hazardous Laboratory Waste

Hazardous lab waste is a waste that poses severe threat to the environment and human health if not disposed of properly. According to the Environmental Protection Agency (EPA), wastes that are considered hazardous has four major characteristics, ignitability, corrosivity, reactivity and toxicity (Oliveira et al., 2020). Having an ignitability characteristic means that the wastes have a flashpoint of less than 60°C, are nonliquid materials that cause fires, or are ignitable compressed oxidizers and gases. Wastes with corrosivity characteristics are acidic liquid wastes having a pH value of 2 or less and basic liquid wastes with a pH value of 12.5 and above; such wastes can corrode steel, thus labeled as corrosive hazardous wastes (Capoor & Bhowmik, 2017). Reactive hazardous wastes refer to wastes that are unstable under standard conditions and thus can react with water to produce toxic fumes. These kinds of wastes also have the capability to explode or detonate when heated. Lastly, toxic hazardous wastes include wastes that are dangerous to human health when inhaled, swallowed or absorbed. Such wastes can also leach through soil and cause serious contamination of the soil and groundwater.

16.2.2.1.1 Examples of Liquid Hazardous Laboratory Waste

- Liquid solvent waste
- Organic synthesis liquids
- Liquids associated with TLC and/or HPLC
- Sample vials that contain liquids for digestion, extraction or preservation
- Waste liquids associated with DNA extraction and cell lysis
- Specimen preservatives including alcohol, formalin, formaldehyde, paraformaldehyde, etc.
- Laboratory reagents which have been left behind or abandoned by previous users of the lab
- Unused portions of lab reagents which are no longer needed
- Photographic and X-ray-related darkroom chemicals

16.2.2.1.2 Examples of Solid Hazardous Wastes

- Adsorbent and absorbent materials used in chemical processes
- Paper used with chemical reagents and weighing boats
- Gloves used when handling hazardous chemicals
- Rags, paper towels, vermiculite used to clean up chemical spills
- Filters and ion exchange materials used during chemical processes
- Disposable pipette tips used for measuring and transferring chemicals
- Electrophoresis gels containing ethidium bromide

16.2.2.2 Nonhazardous Laboratory Waste

Nonhazardous waste refers to lab wastes that do not cause direct threat to the environment or human life. However, such wastes cannot be dumped into a sewer line or receptacles for the fear of causing blockages that will become environmental hazards.

16.2.2.2.1 Examples of Nonhazardous Wastes

- Empty containers
- Noncontaminated glass
- Organic chemicals such as sugars, starch, citric acid, lactic acid and naturally occurring amino acids
- Inorganic chemicals such as carbonates, oxides, sulfates, phosphates and chlorides
- Laboratory materials that are not contaminated with hazardous chemicals such as plastic and rubber protective clothing, filter paper, glassware and chromatographic absorbents

16.2.3 Management of Laboratory Wastes

The goal of the waste management strategies is to reduce the effects of the hazardous and nonhazardous compounds on the environment and humans. Different methods can be employed to deal with the problem. The waste management hierarchy as postulated by Nolasco et al. (2020) shows the methods of dealing with hazardous wastes in their order of preference. The most preferred option focus on reducing the amount of waste being released from the lab to prevent pollution in the first place. This approach is implemented through source reduction of the resulting wastes from various lab processes.

Nevertheless, not all wastes can be eliminated through source reduction or reuse, as such, other approaches have been proposed to deal with the generated wastes. The second option of managing such wastes include refining, recycling or recovering the wastes to be used in various processes (Ho & Chen, 2018). The second approach is also effective in lowering the rates of pollution since the wastes do not reach the environment and are acquired and utilized in other settings, before reaching the environment or landfills. Unfortunately, the second approach has also proven ineffective and impossible to achieve in the management of some kinds of wastes (Lee, Ellenbecker, & Moure-Ersaso, 2004). As a result, other strategies have been postulated that focus on treating the wastes to minimize their toxicity and potential for causing adverse effects on humans and the environment. The last option of managing the wastes is by disposing the hazardous and nonhazardous wastes in landfills or through incineration followed with a propped disposal of the residual ash.

As much as each of the identified methods may be vital in the management of wastes at certain times, it is important that the lab personnel always try to "move up the management hierarchy". Waste reduction should be the cornerstone of the efforts made toward the management of lab wastes, if their probable adverse effects on humans and the environment are to be curtailed.

16.2.3.1 Management of Hazardous Wastes

As illustrated above, there are four tiers to the management of hazardous lab waste. The four strategies entail source reduction and prevention of pollution, reuse of materials or their redistribution, treatment, reclamation or recycling of materials within the wastes, and disposal of the wastes through treatment, land burial or incineration. While the first two tiers are considered the most effective in the management of hazardous lab wastes, the last two tiers are also important in providing safety and environmental benefits. A discussion of each of the four tiers to lab hazardous waste management is advanced herein.

16.2.3.1.1 Pollution Prevention through Source Reduction

Hazard reduction is regarded as a significant part of pollution prevention. Such waste management approach is encouraged in labs that mostly deal with hazardous compounds. From a scientific point of view, it is highly feasible to lower the volume of the hazardous characteristics of chemicals used in various lab procedures and chemical reactions. According to Jiarong and Geng (2014), hazard reduction reactions can be included in the last steps in an experimental sequence, to minimize the amount of hazardous chemicals released as wastes. Significant economic benefits can be attained by performing a hazard reduction procedure as part of an experiment. Such benefits are attained when the necessity to accumulate, store, transport and treat hazardous wastes is eliminated.

Reduction of the hazardous wastes is achieved when the lab personnel keep up-to-date chemical inventories, order small quantities of hazardous materials and conduct experiments that lower the chance of waste release from the processes.

16.2.3.1.2 Chemical Inventory Management

Reduction of hazardous wastes begin with a good book-keeping of chemicals. Every lab is expected to establish a chemical inventory system within the lab. The system should include all the chemicals in the lab, detailing the name of the chemical, chemical abstracts services, procurement date, expiry date and the quantity. The lab manager has the responsibility of updating the required quantity of each chemical every three months and making the necessary adjustments. Comparing the quantity of stock and the volumes consumed provided a better guidance on the quantities that should be purchased annual. Procuring the right quantity is vital in ensuring that there are no unnecessary wastes from expired chemicals or reagents. Table 16.2 highlights how such comparisons can be made.

16.2.3.1.3 Chemical Exchange Program

After developing the computerized inventory, it is possible to initiate a chemical exchange program with the manufacturers. Such a program will enable the lab manager to request for new chemicals that have been depleted, while requesting for the manufacturers to uptake the unused chemicals. The arrangement is effective in reducing the amounts of unused chemicals from the lab as well as enabling to achieve maximum cost-savings.

TABLE 16.2
Average Volume of Chemicals Consumed Per Year

Different Chemicals Recorded	Stock (kg/year)	Consumption (kg/year)
Inorganic	6,150	1,867
Organic	3,604	1,323
Solvent	11,650	10,982
Total	21,404	14,172

TABLE 16.3
Substitution of Hazardous Chemicals with Nonhazardous Chemicals

Type of Experiment	Instead of These Chemicals	Use
Agar rose gel staining	Ethidium bromide	Fluorescent dyes or crystal violet
Glassware cleaning	Chromic acid solutions	Biodegradable surfactants, strong corrosive solutions, sonic baths
Kjeldahl digests	Mercury salt	Mercury-free catalysts (copper sulfate)
Storage of biological specimen	Formaldehyde	Ethanol

16.2.3.1.4 *Substitution of Chemicals*

This is another process that can be adopted in reducing the amount of hazardous chemicals in the lab. According to Lara et al. (2017), the process entails substituting hazardous chemicals with nonhazardous chemicals in certain lab programs and processes. Table 16.3 shows processes where hazardous chemicals can be substituted with nonhazardous chemicals.

16.2.3.1.5 *Material Recycling, Reuse or Reclamation*

Recycling and reclamation of materials from waste is another waste management method that can be adopted in the control and management of lab wastes. The recycling process is done to enable the lab manager utilized some of the materials reclaimed from the resulting wastes (Lara et al., 2017). Different methods can be adopted in recycling and reclaiming hazardous chemical wastes. Methods such as neutralization, distillation, digestion, encapsulation and other thermal treatment methods can be adopted in the recycling and reclamation of materials from lab wastes. The expense and the practicality of these processes are largely dependent on the nature of the material as well as its volume. Treatment and recycling of materials are more preferable than disposal of the wastes in the landfills or through incineration. As reported by Widyatmoko (2018), the recycling process ensures that minimal wastes are released into the environment, contributing to maximum savings on costs to the lab.

Examples of recycling processes includes treatment of high- and low-pH wastes with water to obtain various salts that can be used in other lab processes, and treatable wastewater that will have minimal adverse effects if released into the environment. Encapsulation of wastes containing mercury is another example of material reclamation since it can lead to the recovery of some mercury molecules and release of wastewater that have minimal adverse effects on the environment. Filtration of aqueous-based waste is also vital in lowering the volume of the wastes and releasing wastewaters that can be safely treated in a sewage treatment facility.

16.2.3.1.6 *Disposal and Incineration of Hazardous Wastes*

Disposal of chemical hazardous wastes can be done in sanitary sewers, landfills or incinerators for further treatment. While the disposal in the sewer system was initially considered an effective waste management method, its use in the current times has been prohibited (Lara et al., 2017). The environmental concerns associated with the disposal of the wastes in the sewer system have markedly changed the custom. In fact, most academic labs have eliminated sewer disposal as a waste management strategy since it causes major risks to the surrounding communities (Widyatmoko, 2018) based on the fact that most wastes being released into the environment are hazardous chemicals.

Chemical lab wastes that may be permissible for sewer disposal include wastes such as aqueous solutions that are less toxic and are biodegradable. Liquid lab wastes such as spent buffer solutions, caustics or neutralized mineral acids can be disposed of in the sanitary sewer (Lara et al., 2017). Other water-soluble nontoxic solids can also be disposed into the sewer. However, the disposable of water-miscible flammable liquids into the sewer system is severely prohibited.

As much as it may be permissible to dispose of safe lab wastes into the sewer system, the volume of wastes to be released should be based on the requirements set by the government. According to Lee, Ellenbecker, and Moure-Ersaso (2004), different governments and states have stipulated the effluent limits for various lab wastes being released into the sewer system. It is imperative that the lab manager complies with the set requirements on the effluent limits.

The release of waste in the form of vapor to the atmosphere is not an acceptable method of waste disposal. Apparatus for operations that are expected to release vapors should be equipped with appropriate trapping devices. Also, deliberate disposal of materials through evaporation of vapor should be avoided.

Chemical hoods have been devised to aid in the safe disposal of vapor wastes. As much as the hoods are capable of transporting vapor away from the lab personnel, such strategy should not be considered a routine method of vapor waste disposal, since it is not highly effective (Widyatmoko, 2018). The absorbent papers included in the hoods have limited absorbing capacity, lowering their effectiveness in absorbing the harmful chemicals from the vapor. Redirection of the vapor arising from the hood to a common trapping device, is effective in completely eliminating the harmful chemicals from the atmosphere.

Incineration of the lab wastes is a common method of waste management. According to Lee, Ellenbecker, and Moure-Ersaso (2004), incineration is done at temperatures of not less than 1,200°F in rotary kilns. The technology provides for the complete destruction of organic materials, reducing the volume of the wastes that must be disposed by landfill. While incineration is considered an effective waste disposal method in relation to landfill, it is an expensive option. The fuel needed to carry out the incineration enhances the overall cost of the waste management process, since the destruction of the wastes is done at higher temperatures (Widyatmoko, 2018). Also, not all wastes can be incinerated. Lab wastes containing mercury cannot be incinerated forcing the lab management to consider alternative methods of waste disposal other than incineration.

16.2.3.2 Management of Nonhazardous Wastes

Nonhazardous wastes are considered safe and can be disposed in the normal trash bin. The disposal of nonhazardous wastes via the normal bin or sewer can have significant reduction in the cost of waste management. However, many states and municipalities restrict the disposal of wastes in the landfills or sewer system. This is because a nonhazardous compound may be improperly labeled, exposing the people to serious harm.

It is, however, important to note that when any nonhazardous waste mixes with portions of hazardous material, then the resulting excess must be treated as a hazardous compound. Most labs have put in place policies that treats all wastes as hazardous. This ensures that proper waste management strategies are put in place to manage different kinds of wastes.

16.2.3.3 Management of Multi-hazardous Wastes

Multi-hazardous waste is a waste that presents any combination of chemical, radioactive or biological hazards. This kind of waste contains hazards that makes its management difficult and complex. For example, low-level mixed waste is a multi-hazardous waste that contains both hazardous wastes and low-level radioactive wastes (Widyatmoko, 2018). The hazardous characteristics, treatment methods and disposal requirements for these wastes are different and often incompatible. Other factors that further complicate the management of multi-hazardous wastes include a complex federal, state and local regulatory framework; limited disposal options; and high disposal costs.

Commercial treatment or disposal facilities for multi-hazardous waste from labs are scarce. There is little incentive for the development of a commercial market to treat and dispose of lab multi-hazardous waste because most of the waste that labs generate is unique to labs and small in volume. The management of multi-hazardous waste is particularly challenging for research labs where there are frequent changes in protocols, procedures, materials and waste generating processes.

These difficult and complex management issues can also make it difficult to promote and sustain prudent pollution prevention practices.

Medical, clinical, forensic and environmental labs, and biomedical, biochemical, radiological and other types of research labs generate multi-hazardous waste. Prudent management of these wastes is necessary to protect the health and safety of all lab personnel who handle, process and store the waste for disposal, and to minimize the potential of harm to public health and the environment. A further objective of prudent management of multi-hazardous waste is to promote excellence in environmental stewardship. The initiative can serve as a guide for developing prudent practices for managing multi-hazardous wastes as advanced below.

- *Source reduction:* Pollution should be prevented or reduced at the source whenever feasible.
- *Recycling:* Pollution that cannot be prevented should be recycled in an environmentally safe manner whenever feasible.
- *Treatment:* Pollution that cannot be prevented or recycled should be treated in an environmentally safe manner whenever feasible.
- *Disposal:* Disposal or other release into the environment should be employed only as a last resort and should be conducted in an environmentally safe manner.

Prudent waste management methods include a commitment by senior management to develop and support a waste minimization program. The program development should involve experienced lab personnel in planning waste minimization strategies and identifying source reduction options, such as incorporating pollution prevention goals into project proposals. Training of lab personnel to recognize opportunities for source reduction, reviewing research proposals to ensure adoption of available source reduction strategies, improving compliance with regulatory requirements, and institutional policy are among the new management initiatives at research institutions promoting pollution prevention. Multi-hazardous waste requires complex attention because of its combination of hazards and regulatory controls, as detailed in the following guidelines:

16.2.3.4 Assessment of the Risk Posed by Hazardous Wastes

A primary purpose of the risk assessment is to determine which hazardous constituent of the multi-hazardous waste presents the greatest risk. This knowledge can help identify source reduction and treatment possibilities to reduce the risk of the waste. An assessment that determines that a waste constituent does not present a significant risk may provide an opportunity for regulatory flexibility

16.2.3.5 Minimize the Hazardous Constituents in the Waste

Consider applying the waste minimization methods specific to each hazardous constituent of the waste. This strategy could result in reducing or eliminating one hazardous constituent from the waste stream and managing the waste as a single-hazard waste.

16.2.3.6 Determine Options to Manage the Multi-hazardous Waste

Waste management options include recycling, lab methods, management at institutional waste facilities and treatment and disposal at commercial sites. Options can vary considerably between labs depending upon institutional capabilities and state and local laws. It may be appropriate to manage the waste in order of risk priority, from high to low risk. Options must be compatible with all hazards, and combinations of waste management methods may be limited by their order of application. Reject any combination or sequence of methods that may create an unreasonable risk to waste handlers or the environment, or that might increase the overall risk.

Glossary of Terms

Adaptable lab configurations emphasize adaptability by incorporating elements such as modular furniture, adjustable partitions, and overhead utility distribution systems, enabling seamless reconfiguration in response to evolving research requirements.

Africa the second-largest and second-most populous continent, consisting of 54 nations. Bordered by the Atlantic, Mediterranean, Red, and Indian Oceans, it features diverse climates and geography, including the Sahara Desert and the Nile River.

Air-purifying respirator a respirator with an air-purifying filter or cartridge that removes airborne contaminants by drawing in ambient air and filtering out harmful substances.

Assessment the process of determining the quantity, worth, quality, or significance of a particular object or concept, which results in a conclusion or decision that reflects this evaluation.

Atmosphere-supplying respirator a respirator that supplies inhaled air independently from the surrounding atmosphere, including Supplied Air Respirators and Self-Contained Breathing Apparatus (SCBA).

Biological hazards (Biohazards) include agents like bacteria (*Salmonella*, *Escherichia coli*), viruses (COVID-19, influenza), parasites, molds, and toxins. Risks stem from blood, bodily fluids, and insects. Illness can occur via inhalation, ingestion, or skin contact with contaminated food, water, soil, or the environment.

Biological Weapons Convention (BWC) represents a significant multilateral agreement aimed at banning the development, production, and storage of biological and toxin weapons.

Biosafety the protocols to prevent the accidental spread of biological agents and toxins, aiming to protect human, animal, and plant health while preserving the environment.

Budget constraint restricts the quantity of goods a consumer can buy based on income and prices, influencing purchasing decisions and accessible product combinations.

Calibration see definition for **Testing and calibration**.

Chemical a substance with a specific composition, sourced naturally or synthesized. Some chemicals have distinct odors, while others are odorless. Toxic chemicals can pose health risks, so reducing exposure can significantly lower these risks.

Chemical facility an industrial plant that transforms raw materials into useful products using chemical or biological methods. These facilities are essential for producing chemicals like plastics, fertilizers, and pharmaceuticals.

Chemical insecurity the accidental or intentional release of harmful substances into the environment, mostly due to inadequate security measures, theft, and safety protocol failures. Poor handling of hazardous materials in high-risk facilities exacerbates the threats.

Chemical laboratory regulations focus on the reduction of accidents by prohibiting food, drinks, and cosmetics in the lab; enforcing strict safety measures like the use of personal protective equipment (PPEs) and fume hoods; and ensuring proper waste disposal.

Chemical security measures and strategies to prevent the misuse of chemicals through theft, sabotage, or terrorism. It focuses on protecting these substances from individuals with harmful intentions to prevent injury and environmental damage.

Chemical substitution exchanging a chemical entity within a molecular structure to reduce risks, enhance safety, or achieve a modified effect. It includes replacing hazardous chemicals in products and processes to protect human health and the environment. This concept also covers reactions where one atom or functional group in a molecule is replaced by another.

Chemical threat the potential or actual release of hazardous substances that could harm humans, animals, plants, or environment. These threats can arise from industrial accidents, terrorism, or the intentional use of chemical weapons.

Chemical Weapons Convention (CWC) is a legal framework that prohibits the development, production, possession, storage, and use of chemical weapons to protect human life.

Chinese herbal medicine a practice that uses plants, minerals, and some animal products to create remedies aimed at restoring balance in the body.

Constructivist theory posits that individuals actively shape their understanding of the world based on their prior knowledge and experiences. It highlights that learners find meaning through active engagement, social interaction, and reflection, rather than passively acquiring knowledge.

Cost-effective design provides advantages analogous to its counterparts while incurring a reduced total life cycle expense.

Cost-efficient materials provide economically viable, environmentally sustainable, and long-lasting options for construction and various applications by incorporating waste products, renewable resources, or easily accessible components. The optimal selection is contingent upon the particular requirements of the project.

Creativity the ability to generate innovative and useful ideas by using imagination and expertise. It can be an inherent talent or developed through time. Creativity involves finding original solutions, expressing oneself, and combining unrelated concepts.

Critical thinking skills the objective examination of information to arrive at reasoned judgments and decisions. This process includes identifying underlying assumptions, assessing the validity of arguments, and formulating logical conclusions based on the evidence presented.

Cyber-Physical Systems (CPS) integrate physical events with computational algorithms, linking users to the internet in smart energy grids and autonomous vehicles.

Dual-function laboratory bench an effective laboratory bench utilization approach considers a laboratory bench as comprising two distinct components, progressing from purely theoretical to operatively practical, each analyzable through dedicated utilization metrics.

Dual-use chemicals substances that have legitimate, peaceful applications in industry, agriculture, or medicine, but can also be misused for harmful purposes, such as creating chemical weapons or illegal drugs.

Efficacy and Safety whereas **efficacy** is a measure of how well a treatment works under controlled conditions, **safety** refers to the degree to which a treatment is free from harmful side effects.

Efficient material flow the systematic movement of raw materials, components, and finished products throughout the production or logistics processes. This organization is crucial for achieving prompt delivery and reducing waste.

Elastomeric a respirator facepiece constructed from either natural or synthetic elastic materials, including natural rubber, silicone, or ethylene propylene diene monomer (EPDM) rubber.

Emergency response the immediate actions taken before, during, and after an emergency to save lives, reduce distress, and lessen damage to people and the environment. It is crucial in the larger scope of emergency management, which also includes mitigation, preparedness, and recovery efforts.

Endocrine-disrupting chemicals (EDCs) adversely affect the endocrine system, which regulates hormones. They can mimic, inhibit, or interfere with hormonal functions, leading to health issues like reproductive problems, metabolic disturbances, and developmental anomalies.

Energy-efficient design minimizes energy usage, mitigates ecological effects, and decreases utility expenses. This is achieved through the implementation of both passive and active strategies in various elements, including architectural frameworks, material selection, and system functionality.

Environmental health examines the interplay between natural and built environments and their impact on human health. It analyzes determinants like air quality, water management, food safety, and housing conditions.

Equipment a wide variety of instruments, devices, and tools essential for performing scientific research activities.

Equipment density has two distinct interpretations. The first is power density, which quantifies the rate of power usage per unit of area, expressed in watts per square foot or per square meter. The second refers to physical density, which is characterized by the mass contained within a given volume.

Equipment maintenance is a systematic method for inspecting, servicing, and repairing machinery and tools to maintain operational efficacy, improve safety, and extend lifespan. It includes proactive measures like routine lubrication and adjustments, alongside reactive actions for addressing malfunctions and breakdowns.

Exhaust stacks (also called **smoke stacks**) function as conduits that direct exhaust gases through a system and into a specialized area, where these gases are expelled upward and away from the building or vehicle.

Experiential learning a method for acquiring knowledge, skills, and values through direct participation and reflection. It involves engaging in practical activities, problem-solving, and decision-making, followed by analyzing results to gain insights and understanding.

Extended warranty (also called **service agreement** or **service contract**) that individuals purchase for protection against repair and replacement costs for specific products after the initial manufacturer's warranty expires.

Failure Modes and Effects Analysis (FMEA) a methodical, forward-looking approach aimed at pinpointing potential failure modes within a product, process, or system. It involves a thorough examination of the consequences associated with these failures, alongside the development of strategies to avert or lessen their occurrence.

Financial reporting involves documenting, analyzing, and sharing a company's financial activities and performance with external stakeholders like investors and regulators. It also aids internal management in making informed decisions.

Flexible adaptability the ability to modify responses to changing conditions with minimal stress. It encompasses readiness to change plans and behaviors (flexibility) and effectively engaging with new environments (adaptability).

Flexible teaching methods adaptive pedagogical strategies used to adjust various components of the learning process (specifically the content, timing, methods, and environments) to accommodate the varied requirements of students, thereby improving accessibility and fostering inclusivity.

Flue Gas Desulfurization (FGD) a collection of technologies aimed at controlling pollution by extracting sulfur dioxide (SO_2) from flue gases. This process is mainly used in fossil fuel power generation and industrial operations to reduce acid rain and environmental damage from sulfur dioxide emissions.

Globally Harmonized System (GHS) Compliance a standardized framework for classifying and labeling chemicals to ensure consistent criteria and uniform Safety Data Sheets, for improving safety and clarity in chemical communication.

Good Laboratory Practices (GLP) a framework of principles, criteria, and procedures that ensures the quality and reliability of nonclinical health and environmental safety studies, supporting regulatory standards.

Green chemistry the development of chemical products and processes aimed at minimizing or eradicating hazardous materials. It relates to all phases of a chemical product's life cycle, encompassing its design through to its eventual disposal.

Harmonization the act of making different people, plans, situations, etc., suitable for each other, or the result of this: What is needed is a harmonization of regulations.

Hazard control systematic identification of threats in a work environment and implementing strategies to eliminate or reduce risks, protecting individuals and property.

Hazardous agents any solid, liquid, or gaseous materials that can cause physical harm or health problems, including those that are radioactive, toxic, explosive, flammable, corrosive, or biologically threatening.

Hazardous waste management a systematic approach to handling materials that pose threats to human health and the environment, and includes their collection, treatment, storage, transportation, and disposal.

Health risk the potential for negative effects on human health due to lifestyle choices, genetic factors, environmental exposures, or socioeconomic conditions, increasing the likelihood of specific diseases.

Heavy metals dense elements in the environment that can be hazardous in high concentrations, such as mercury, lead, arsenic, and cadmium. In contrast, iron and zinc are vital at low levels, showing a complex health–toxicity relationship.

Herbal Medicine Regulation varies greatly as some countries enforce strong quality, safety, and efficacy measures, while others lack effective oversight. Often, regulations mandate product registration, quality control (QC), and labeling. Manufacturers may demonstrate efficacy based on traditional usage instead of clinical trials.

Indian subcontinent constitutes a distinct physiographic area within Asia, situated beneath the Himalayas and extending into the Indian Ocean. This region is flanked by the Bay of Bengal to the east and the Arabian Sea to the west and constitutes the countries of Bangladesh, India, and Pakistan.

Industrial effluent wastewater from industrial activities containing organic and inorganic contaminants like heavy metals and hazardous materials, posing significant risks to public health and ecosystems.

Integrated Laboratory Network enhances diagnostic services related to patient care, thereby minimizing the necessity for patients to travel to different locations for a range of tests.

Interactive whiteboard a large, touch-sensitive display that serves as a computer interface, allowing users to write, draw, and access the internet, enhancing collaboration in educational and corporate settings.

Inventory control the process of managing the quantity of products or materials a business has in stock, from their arrival to their sale or use, to ensure that the right amount of goods is available at the right time and location.

ISO 45001 the first global standard for Occupational Health and Safety Management Systems, aimed at reducing workplace fatalities and injuries. It helps organizations create policies, identify hazards, and set controls to enhance occupational health and safety performance.

Key Performance Indicators measurable metrics that evaluate performance against an organization's strategic objectives over time. They establish targets, track progress, and guide data-driven decisions, aligning daily operations with the set goals.

Lab design elements encompass workflow, spatial organization, adaptable furniture, safety installations like fume hoods and eyewash stations, appropriate lighting, resilient utilities secure storage, digital technology integration, and sustainable methods, enhancing efficiency and safety based on research needs.

Lab Housekeeping effective laboratory cleanliness management that enhances safety and operational efficiency through a tidy environment. Key principles include "clean as you go," proper chemical storage, accessible emergency equipment, and regular chemical inventory assessments.

Lab layout design methodologies include Functional Layout, which organizes zones for tasks to improve access, and Process-Based Layout, which groups resources by research phases to enhance workflow.

Laboratory accreditation formally recognizes a laboratory's technical skills and impartiality in testing or calibrating. An authoritative body assesses the laboratory's quality management systems (QMS) and competencies. This evaluation guarantees reliable, accurate results aligned with international standards, namely ISO 17025 and ISO 15189.

Laboratory automation use of technology, robotics, and software to minimize human involvement in tasks, enhancing efficiency, precision, and productivity.

Laboratory design principles focus on Safety, Functionality, Flexibility, and Sustainability, ensuring effective development through safety measures, optimized workflows, adaptable spaces, and energy-efficient materials.

Laboratory ergonomics optimized lab designs that reduce injuries from repetitive actions and bad postures through adjustable equipment, neutral positions, organized spaces, and breaks.

Laboratory learning an educational method where learners gain knowledge and skills through practical, experimental engagement within a structured science lab setting.

Laboratory Policy Enforcement enhances safety, protects investments, reduces costs, and supports quality assurance. It may include Risk Management (security protocols, optimized policies, and access segmentation), insightful design, signage to aid navigation, and assessments to ensure compliance.

Laboratory safety the implementation of rigorous protocols accompanied by a robust safety culture. Key components include the application of safety practices, equipment, and PPE, and the ability to recognize potential hazards.

Laboratory safety culture a framework of values and practices that emphasizes safety, minimizes hazards, and reduces accidents. A commitment to safety at all levels, backed by management, communication, and training, is essential.

Laboratory Security protection of hazardous materials, sensitive data, and valuable equipment from theft or unauthorized access through barriers, electronic controls, inventory management, personnel vetting, and training.

Laboratory Security audit a methodical evaluation of a laboratory's security protocols that identifies weaknesses, checks regulatory adherence, and validates current measures.

Laboratory standards consist of regulations and protocols aimed at ensuring quality, safety, and reliability in laboratory practices. They cover equipment, procedures, and QMS. Examples include ISO standards like ISO 15189 and ISO 15190, and GLP, as well as guidelines from institutions like the Clinical and Laboratory Standards Institute (CLSI) and American Society for Testing and Materials (ASTM).

Laser safety the protection of eyes and skin from hazards by avoiding direct exposure to laser beams, using safety goggles, implementing confinement measures, and removing reflective objects and jewelry to reduce risk.

Life-span the length of time for which a person, animal, or thing exists.

Market variability analysis assesses current pricing requests against prices offered to similar clients, revealing insights about customers affected by pricing changes.

Material Safety Data Sheet (MSDS), or Safety Data Sheet (SDS) a detailed document from the manufacturer or importer of a chemical that provides vital information on the chemical's properties, risks, handling, storage, and emergency responses.

Methodology the framework of methods and principles used in a discipline like scientific inquiry or education. It is a strategy that justifies method selection and outlines their application to achieve a specific objective, such as a research question.

Modular furniture features standardized elements, which consist of autonomous modules that individuals have the flexibility to arrange in multiple configurations, thereby facilitating a personalized experience.

Monitoring the continuous, systematic collection and analysis of data to track progress toward goals, assess performance, and make informed decisions to ensure a project or program stays on track.

Open-plan laboratory design fosters collaborative efforts by featuring large communal spaces and minimal walls or physical barriers. Nonetheless, it requires meticulous design to effectively control noise levels and preserve individual privacy.

Operational efficiency is a vital business metric that measures an organization's ability to produce products or services using minimum resources, such as time, labor, materials, and costs. This aims to achieve the highest output and quality.

Optimal laboratory design enhances safety and efficiency by optimizing workflows, modular configurations, effective storage solutions, and using methodologies like bubble diagramming and 3D planning.

Optimal Laboratory Operations contributes to the reduction of downtime and expenses in laboratories by improving the efficiency of scheduled services such as preventive maintenance, while also increasing the effectiveness of instrument utilization and the quality of services provided.

Optimizing environment refers to two main aspects modifying physical spaces for better comfort and functionality, like designing homes or workplaces to improve productivity, and promoting planetary well-being by adopting eco-friendly behaviors and sustainable practices.

Personal exposure an individual's direct interactions with entities like people, places, concepts, or situations, and includes routine daily encounters and measurable exposures to environmental contaminants or radiation.

Petri Nets are mathematical modeling languages vital for outlining distributed systems, particularly in managing concurrency, synchronization, and resource sharing in CPS.

Pharmaceutical-Based Agents (PBAs) consist of chemical entities derived from pharmaceutical compounds. While these substances may possess valid medical applications, their misuse can lead to significant health hazards, including severe illness or mortality. These chemical agents can exist in various physical states (solid, liquid, or gas) and are susceptible to weaponization through a range of techniques.

Pharmaceutical security focuses on protecting products, data, and personnel through strategies like securing the supply chain against counterfeit medications, implementing cybersecurity for intellectual property and patient data, and enhancing physical security to reduce theft and contamination risks.

Pharmacopoeia a key reference that defines the quality, safety, and purity of medications and their ingredients. It includes details like chemical properties, identification methods, and assay limits for pharmaceutical compounds.

Physical hazards agents or conditions that can inflict harm through energy or mechanical means, often with minimal exposure, such as slips, electrical dangers, extreme temperatures, noise, vibrations, radiation, and fire risks.

Poisons harmful agents that adversely affect living organisms through various exposure routes, including ingestion, inhalation, dermal contact, or injection. Notable examples encompass certain botanical species, household cleaning agents, agricultural pesticides, pharmaceutical overdoses, and toxins derived from animals or fungi.

Procurement contract a formal agreement between a purchaser and a vendor that outlines the conditions for acquiring goods or services, including pricing, quality standards, and delivery timelines.

Quality control (QC) is a part of quality management focused on meeting quality specifications. It entails systematic testing, inspection, and validation of products or services to ensure adherence to standards and customer satisfaction before delivery.

Quality management system (QMS) an organized set of guidelines and processes that helps an organization deliver products or services that meet or exceed customer expectations, enhancing satisfaction.

Radiation safety the minimization of exposure to ionizing and nonionizing radiation using the principles of Time, Distance, and Shielding, along with the ALARA principle (As Low As Reasonably Achievable). Safety practices include reducing exposure time, increasing distance from sources, and using protective measures such as lead aprons and shielding walls.

Reflective testing enables lab specialists to add tests or notes based on their evaluations, delaying results for, say, physicians. Unclear findings often necessitate further testing for accurate diagnosis.

Regulation the oversight and administration of intricate systems based on a defined framework of guidelines and patterns. In the context of government, the term regulation predominantly refers to the secondary legislation enacted to implement and uphold primary legislation.

Regulatory frameworks comprise rules and guidelines that oversee industries and technologies to ensure fair, efficient, and secure operations while encouraging innovation. They are vital for risk management, protecting consumers and investors, and promoting public welfare.

Regulatory research or science establishes systematic methodologies and criteria for assessing the safety and effectiveness of various products within both the healthcare and environmental sectors. This research supports regulatory decisions, aids in translating scientific advancements into safe products, and highlights the essential role of regulatory bodies like the FDA in public health.

Remediation the act of remedying or correcting something that has been corrupted or that is deficient.

Reorder level (also called the reorder point) a specific inventory threshold established by an organization that serves as an indication for the organization to initiate a new order, thereby mitigating the risk of running out of stock.

Safety cabinet a storage system designed for the secure containment of hazardous substances, and reduction of risks from fire, biological agents, and accidental chemical releases.

Safety Equipment or PPE items that shield individuals from risks such as chemical exposure, falling objects, electrical hazards, and infections. Examples include hard hats, safety goggles, gloves, respirators, ear protection, protective footwear, and high-visibility clothing.

Safety guidelines established protocols designed to mitigate or eradicate risks associated with hazards or potential injuries.

Safety in affordability involves strategies for developing housing that is financially accessible while ensuring occupant security and livability. It aims to reduce risks from physical dangers, social unrest, and poor living conditions.

Safety officer or Health Safety and Environment (HSE) officer is essential for promoting workplace safety and health. They create and implement safety policies, conduct risk assessments to identify hazards, ensure compliance with regulations, and train employees to reduce accidents and injuries.

Safety protocols also called safety procedures systematic, sequential guidelines, processes, and criteria that direct employees in executing workplace tasks while ensuring their safety.

Satellite laboratory can refer to several concepts, including a mobile or smaller lab facility that supports a larger main laboratory, a dedicated research facility focused on satellite systems, a creative studio developing imaging technology, or even the term's use in a more figurative sense for supporting organizations or education in STEM fields.

Science education the formal and informal teaching aimed at imparting scientific concepts, methodologies, and skills. Its main goals are to improve understanding, develop critical skills, and inspire enthusiasm for exploring complex challenges.

Scientific skills abilities and methods essential for conducting inquiries, solving problems, and understanding phenomena. Key elements of the skills are observation, measurement, classification, communication, and prediction, and are vital for the scientific method, facilitating experiments, data analysis, and conclusions.

Shelf-life the period during which a product maintains its essential quality and safety for use, consumption, or sale under specific storage conditions. It is influenced by packaging and the product's composition. Research is done to determine this timeframe, ensuring adherence to quality standards and regulations.

Site security plan delineates systematic protocols designed to safeguard a site's personnel, assets, and financial resources against theft and unauthorized access, based on a thorough risk assessment.

Standard operating procedure (SOP) a formal framework of detailed, sequential instructions for organizations to guide employees in performing regular tasks. This approach is vital for enhancing operational efficiency, ensuring consistency, and maintaining quality standards.

Standardized workflows are predefined sequences of tasks designed to ensure consistency, reduce errors, and improve efficiency in repetitive processes.

Supplied Air Respirator (SAR) or airline respirator a respirator that provides breathable air from an external source, which is not intended to be portable or transported by the individual using the device.

Supplier qualification a methodical framework for assessing and choosing suppliers, aimed at guaranteeing that they can consistently adhere to the quality, reliability, financial stability, and ethical criteria of an organization.

Sustainable materials resources that exhibit minimal adverse effects on both the environment and social structures. Their key attributes include renewability, recyclability, biodegradability, and a low carbon footprint, all of which contribute to their overall sustainability.

Sustainable practices are actions and policies that promote long-term environmental, social, and economic health, ensuring resources are used responsibly without depleting them for future generations.

Sustainable remediation a methodical strategy for addressing the cleanup of polluted land and groundwater, with the objective of optimizing the cumulative environmental, social, and economic advantages associated with the initiative.

Terrorism and illicit trade are deeply linked, with terrorist groups using a variety of illegal activities, such as drug trafficking, arms smuggling, and the trade of natural resources, to generate revenue and fund their operations.

Testing and calibration are pivotal for QC, ensuring instrument reliability and measurement accuracy. Testing examines how well the product operates, while calibration ensures its output conforms to established norms. Calibration may necessitate modifications, whereas testing verifies adherence to these standards.

Threat analysis a systematic method for identifying, evaluating, and prioritizing threats to an organization, system, or asset. It involves assessing the likelihood and impact of threats, from cyberattacks to natural disasters and human errors. This process is crucial for creating effective strategies to manage risks.

Transformative education a conceptual framework that envisions future learning paradigms while simultaneously seeking to implement tangible strategies to enhance current educational practices. This approach emphasizes the need for innovation in learning methodologies to align contemporary education with prospective advancements.

Usage tracking the systematic collection and examination of data regarding user interactions with a system, software, or device. The primary objective of this process is to discern patterns in user behavior, enhance overall performance, and efficiently allocate resources.

User-Centered Design a cyclical methodology that prioritizes the end-user in product development, and emphasizes understanding user requirements and actions, ensuring these are incorporated into every design and development stage.

User satisfaction the degree to which individuals find contentment with a product, service, or experience when assessed against their preconceived expectations. Key elements affecting satisfaction are usability, reliability, functionality, and emotions, complicating evaluation.

Utility systems critical services that facilitate the operational capabilities of both buildings and communities. These systems include the infrastructure necessary for the provision of electricity, water, sewage, heating, cooling, ventilation, and telecommunications.

Vendor pre-qualification a systematic evaluation process that helps organizations assess potential suppliers' capabilities, competencies, and financial soundness before inviting them to submit tenders for goods, services, or works.

Vulnerability being susceptible to threats in social, cognitive, environmental, emotional, and military contexts. Methodically analyzing it involves thoroughly examining various risks and available resources that specifically cater to marginalized groups.

Vulnerability assessment identifies, evaluates, categorizes, and ranks security weaknesses in systems, networks, and applications to bolster defenses against cyber threats. It uses automated tools and manual techniques to detect vulnerabilities, assess severity, and recommend remediation.

Warranty claim a customer's request for a free repair, replacement, or compensation when a product fails to meet its promised quality standards within the warranty period.

Warranty voidance when a manufacturer or seller decides that a product's warranty is no longer valid, usually due to actions outside the warranty terms. Typical reasons include poor maintenance, unauthorized modifications, or misuse of the product.

Water pollution, or aquatic pollution the degradation of water bodies (oceans, seas, lakes, rivers, aquifers, reservoirs, and groundwater) from harmful substances introduced through various human activities. Contaminants compromise the quality of water in these ecosystems.

Workflow optimization a detailed analysis and refinement of business processes aimed at increasing efficiency, reducing expenditures, and enhancing overall productivity. This is achieved by removing unnecessary activities, automating routine tasks, and standardizing procedures.

Workplace safety protocol (usually called Safety Procedures) a framework for employees to safely execute workplace procedures, detailing both the procedural steps and the internal documentation created by the organization.

Bibliography

Academies, N. R. (2011). *Prudent Practices in the Laboratory: Handling and Management of Chemical Hazards: Updated Version*. National Academies Press, Washington, DC. Retrieved from: www.ncbi.nlm.nih.gov

Bailey, R. (2020). *Biology Lab Safety Rules*. Retrieved from: https://www.thoughtco.com/biology-lab-safety-rules-373321

Beheshti, H. M., Grgurich, D., & Gilbert, F. W. (2012). ABC inventory management support system with a clinical laboratory application. *Journal of Promotion Management*, 18(4), 414–435.

Buxton, P. (Ed.). (2021). *Metric Handbook: Planning and Design Data* (7th ed.). Routledge, New York. https://doi.org/10.4324/9781003052586

Consensus Study Report. (2016). *Chemical Lab Safety and Security. A Guide to Developing Standard Operating Procedures*. The National Academies Press, New York.

Diaconu, K., Chen, Y. F., Cummins, C., Moyao, G. J., Manaseki-Holland, S., & Lilford, R. (2017). Methods for medical device and equipment procurement and prioritization within low-and middle-income countries: Findings of a systematic literature review. *Globalization and Health*, 13(1), 1–16.

DiBerardinis, L. J., Baum, J. S., First, M. W., Gatwood, G. T., & Seth, A. K. (2013). *Guidelines for Laboratory Design: Health, Safety, and Environmental Considerations*. John Wiley & Sons, New Jersey.

Dittrich, E. (2015). *The Sustainable Laboratory Handbook: Design, Equipment, Operation*. Wiley-VCH, Weinheim.

Edward Allen, J. I. (2019). *Fundamentals of Building Construction: Materials and Methods*. John Wiley & Sons, Inc., New Jersey.

Emmanuel, J. C. (2010). Material & equipment, procurement & maintenance: Impact on blood safety. *Biologicals*, 38(1), 78–80.

Eva Kultermann, W. P. (2017). *Construction Materials, Methods & Techniques* (4th ed.). Cengage Learning, Boston, FL.

Farhadi, M. (2020). *What are the Biosafety Cabinet Classes?* Retrieved from: https://www.labconco.com/articles/whatare-the-biosafety-cabinet-classes

Guidelines for Establishing a Chemical Lab Safety and Security Program. National Research Council. (2010). *Promoting Chemical Lab Safety and Security in Developing Countries*. The National Academies Press, Washington, DC.

Hashim, N. M. Z., & Arifin, N. A. M. M. (2013). Laboratory inventory system. *International Journal of Science and Research (IJSR)*, 2, 261–264.

Health, T. N. (2013). *Laboratory Water, Its Importance and Applications*. Retrieved from: https://orf.od.nih.gov/TechnicalResources/Documents/DTR%20White%20Papers/Laboratory%20Water-Its%20Importance%20and%20ApplicationMarch-2013_508.pdf

Isin, A. (2011). GLP: Good Lab Practice, Modern Approaches to Quality Control, Dr. Ahmed Badr Eldin (Ed.). InTech, Available from: https://www.intechopen.com/books/modernapproaches-to-quality-control/glp-good-lab-practice

Littlefield, D. (2012). *Metric Handbook: Planning and Design Data* (4th ed.). Architectural Press, Oxford.

Neville Surti, C. N. (2018). Laboratories. In P. Buxton, *Metric Handbook: Planning and Design Data* Routledge, New York.

Omiko, A. (2015). Laboratory teaching: Implication on students' achievement in chemistry in secondary schools in Ebonyi State of Nigeria. *Journal of Education and Practice*, 6, 30.

Parmer, C. (2017). *Water Purification: A Guide to Methods of Water Treatment with a Focus on Distillation* Retrieved from: https://www.stuartequipment.com/adminimages/Water_Purification_focus_on_distillation_MAY18.pdf

Stubbins, K. (2010). *Sustainable Design of Research Laboratories: Planning, Design, and Operation*. John Wiley & Sons, Inc., New Jersey.

Watch, D. D. (2001). *Building Type Basics for Research Laboratories*. John Wiley & Sons, Inc., New York.

WHO. (2021). *Procurement Methods and Workflows*. Retrieved from: www.who.int/about/accountability/procurement/methods-and-workflows

World Health Organization. (2013). *Manual for Procurement of Diagnostics and Related Laboratory Items and Equipment*. WHO, Geneva, Switzerland.

World Health Organization. (2017). *Guidance for Procurement of In Vitro Diagnostics and Related Laboratory Items and Equipment.* WHO, Geneva, Switzerland.

Zahm, D. (2007). *Using Crime Prevention Through Environmental Design in Problem-Solving*. The National Institute for Occupational Safety and Health's Guidance for Protecting Building Environments from Airborne Chemical, Biological, or Radiological Attacks, Washington, DC.

References

Aarattuthodi, S., Kang, D., Gupta, S. K. et al. (2025). Cryopreservation of biological materials: Applications and economic perspectives. *In Vitro Cellular & Developmental Biology - Animal*, 25, 27. https://doi.org/10.1007/s11626-025-01027-0

Abalkhail, A. & Alslamah, T. (2022). Institutional factors associated with infection prevention and control practices globally during the infectious pandemics in resource-limited settings. *Vaccines*, 10(11), 1811. https://doi.org/10.3390/vaccines10111811

Abbate, E., Ragas, A. M. J., Caldeira, C. et al. (2025). Operationalization of the safe and sustainable by design framework for chemicals and materials: Challenges and proposed actions. *Integrated Environmental Assessment and Management*, 21(2), 245–262. https://doi.org/10.1093/inteam/vjae031

Abdul Wahab, N., Isa, N., & Ahmad Rahiza, F. (2022). Hazard identification, risk assessment and risk control (HIRARC) on laboratory waste disposal in academic laboratories. *Journal of Academia*, 10(2), 194–203.

Abed, S. F., Allain-Ioos, S., & Shindo, N. (2024). *Examining the Threat of Cyber-Attacks on Health Care during the COVID-19 Pandemic*. WHO, Geneva, Switzerland.

Abedi, E. A. (2023). "We first require basic technical training": Investigating formal professional development pathways and knowledge needs of teachers for technology integration. *Education and Information Technologies*, 29, 11793–11814. https://doi.org/10.1007/s10639-023-12344-w

Abedi, M., Alshybani, I., Shahadat, M. R. B., & Murillo, M. (2023). Beyond traditional teaching: The potential of large language models and chatbots in graduate engineering education. *Qeios*, 23, 59. https://doi.org/10.48550/arXiv.2309.13059

Abedsoltan, H. & Shiflett, M. B. (2024). Mitigation of potential risks in chemical laboratories: A focused review. *ACS Chemical Health & Safety*, 31(2), 104–120. https://doi.org/10.1021/acs.chas.3c00097

Abumalloh, R. A., Asadi, S., Nilashi, M. et al. (2021). The impact of coronavirus pandemic (COVID-19) on education: The role of virtual and remote laboratories in education. *Technology in Society*, 67, 101728.

Adamopoulos, I. P. & Syrou, N. F. (2022). Workplace safety and occupational health job risks hazards in public health sector in Greece. *European Journal of Environment and Public Health*, 6(2), em0118.

Advanced Life Support Group (ALSG), Gleeson, T., & Mackway-Jones, K. (2022). *Major Incident Medical Management and Support: The Practical Approach at the Scene* (4th ed). Wiley-Blackwell, New York.

Afrad, M. S. I., Monir, M. B., Haque, M. E., Barau, A. A., & Haque, M. M. (2020). Impact of industrial effluent on water, soil and rice production in Bangladesh: A case of Turag River Bank. *Journal of Environmental Health Science and Engineering*, 18(2), 825–834.

Agarwal, P. K. & Bain, P. M. (2019). *Powerful Teaching: Unleash the Science of Learning* (1st ed.). John Wiley & Sons, New York.

Agarwal, V. & Kumar, M. (2024). Comparative analysis of water management policies, guidelines, and regulations in Indian Subcontinent Rivers: A contrast between the developed and developing world. In M. Kumar, A. L. Ramanathan, S. Shrestha, K. Kuroda, & S. Mukherjee, *Ecohydrology from Catchment to Coast, River Basin Ecohydrology in the Indian Sub-Continent* (pp. 295–319). Elsevier, Amsterdam, Netherlands.

Agily, A., Khalawy, A., Homadi, A. Y., Shehri, A., Saleh, A., Harbi, A., … & Abdullah, H. (2022). Strategies for effective medical laboratory management: A comprehensive guide. *International Journal*, 10(6), 132–139.

Aguanno, R., Al Ariqi, L., Elkholy, A. A., Gardner, E., Grant, R., Pavade, G., … & Van Kerkhove, M. D. (2022). *Global Technical Meeting on MERS-CoV and Other Emerging Zoonotic Coronaviruses: Virtual Meeting, 15–16 November 2021 (Vol. 16)*. Food & Agriculture Org., Rome.

Agustian, H. Y., Finne, L. T., Jørgensen, J. T., Pedersen, M. I., Christiansen, F. V., Gammelgaard, B., & Nielsen, J. A. (2022). Learning outcomes of university chemistry teaching in laboratories: A systematic review of empirical literature. *Review of Education*, 10(2), e3360.

Ahmed, D. R., & Al Diab Al Azzawi, M. (2024). The current global threat of weapons of mass destruction to humanity and public health. *Discover Public Health*, 21, 37. https://doi.org/10.1186/s12982-024-00161-6

Ahmed, T. A. N. & Taha, S. (2016). Radiation exposure, the forgotten enemy: Toward implementation of national safety program. *The Egyptian Heart Journal*, 69(1), 55–62. https://doi.org/10.1016/j.ehj.2016.10.001

Aidonojie, P. A., Afolabi, M. T., Obieshi, E., Adeyemi-Balogun, M. J., & Wakili, S. A. (2024). Breaking legal and socio-economic challenges to plastic waste regulation in Nigeria: Lessons learned from Singapore. *Yustisia*, 13(1), 64–88.

Akharame, M., Oputu, O., Pereao, O., Olorunfemi, D., Fatoki, O., & Opeolu, B. (2022). Beta-FeOOH/polyamide nanocomposites for the remediation of 4-chlorophenol from contaminated waters. *Journal of Polymer Research*, 29(5), 172.

Akhtar, G. (2013). Indian pharmaceutical industry: An overview. *IOSR Journal of Humanities and Social Science*, 13(3), 51–66. https://doi.org/10.9790/0837-1335166

Akhtar, M. & Saeed, M. (2021). Applying activity based learning (ABL) in improving quality of teaching at secondary school level. *Pakistan Journal of Educational Research and Evaluation (PJERE)*, 3(2), 27–38.

Akyüz, K., Chassang, G., Goisauf, M., Kozera, Ł., Mezinska, S., Tzortzatou, O., & Mayrhofer, M. T. (2021). Biobanking and risk assessment: A comprehensive typology of risks for an adaptive risk governance. *Life Sciences, Society and Policy*, 17(1), 10.

Al Wahshi, F. S. S. (2019). *The Development of a Health, Safety and Environment Management System for an Integrated Gender Separated Campus in the Middle East Region – A Case Study*. Biology Theses, p. 8. https://scholarworks.uaeu.ac.ae/bio_theses/8

Alaamri, E. A. M. A., Hamzi, M. A. E., Baydi, H. A. M., Kariri, Y. I. Y., Ayashi, H. M. G., Majari, W. M. H., … & Otayfi, M. A. Q. (2024). The role of external quality assessments in clinical laboratories: Review of their impact on diagnostic accuracy and laboratory performance in molecular pathology. *Journal of Ecohumanism*, 3(8), 12799–12809.

Alali, M., Beheiry, S., & Atabay, S. (2024). Strategies for the design and construction of nature-inspired & living laboratory (NILL 1.0) buildings. *Biomimetics*, 9(7), 441. https://doi.org/10.3390/biomimetics9070441

Alavandar, Y. K. (2025). Transforming to a multinational cultural organization: Strategical change in the safety and environmental aspects and its regulation in hazardous chemicals handling industry. *Journal of Safety Education*, 3(1), 32–40.

Aldrighetti, R., Battini, D., Ivanov, D., & Zennaro, I. (2021). Costs of resilience and disruptions in supply chain network design models: A review and future research directions. *International Journal of Production Economics*, 235, 108103.

Alenezi, S., Alanazi, M., Aljazaeri, R., Almuzaini, M., Alrasheidi, S., Bin Shamlan, W., Aljohani, R., Alhawiti, G., Alqarni, M., Aljabri, E., Qmmash, M., & Kanan, M. (2023). Community pharmacies in the Asian countries of developing health system: Formation, regulation, and implication. *Pharmacy*, 11(4), 127. https://doi.org/10.3390/pharmacy11040127.

Alexander, F. K., Katsinas, S. G., Keeney, N. E., & Bray, N. J. (2023). The impact of federal funding on state budget priorities: Public higher education operating budgets and medicaid. *Journal of Education Finance*, 48(3), 234–250.

Al-Hussain, A. & Al-Ahmed, H. (2022). Role of lean principles and 5S programs in increasing maintenance productivity. In *International Petroleum Technology Conference*, Riyadh, Saudi Arabia (p. D012S121R004).

Ali, S., Rizwan, M., Farooq, M. A., Abbas, F., Saeed, R., Amir, M., ... & Kalaji, H. M. (2015). Effect of industrial wastewater on wheat germination, growth, yield, nutrients and bioaccumulation of lead.

Ali, N. M., Khan, M. K., Mazhar, B., & Mustafa, M. (2025). Impact of water pollution on waterborne infections: Emphasizing microbial contamination and associated health hazards in humans. *Discover Water*, 5(1), 19.

Alkhatib, R. & Gaede, K. I. (2024). Data management in biobanking: Strategies, challenges, and future directions. *BioTech*, 13(3), 34. https://doi.org/10.3390/biotech13030034

Alquier, L. (2021). Regulatory Harmonization in a Resource-Limited Setting: The World Health Organization Collaborative Procedure for Accelerated Registration. Aug. 2021.

AlShammari, W., Alhussain, H., & Rizk, M. N. (2021). Risk management assessments and recommendations among students, staffs, and health care workers in educational biomedical laboratories. *Risk Management and Healthcare Policy*, 14, 185–198. https://doi.org/10.2147/RMHP.S278162

Alshehry, T. H. A., Alshahri, R. I. M., & Alzahrani, H. S. (2022). Aerosol generating procedures and associated control/mitigation measures: A comprehensive review. *Journal of Namibian Studies: History Politics Culture*, 31, 310–332.

Altıntaş, F. & Koluman, A. (2024). Enhancing biosafety and biosecurity: Quality management in high-containment laboratories. *Etlik Veteriner Mikrobiyoloji Dergisi*, 35(1), 594. https://doi.org/10.35864/evmd.1366594

Altmeyer, K., Kapp, S., Thees, M., Malone, S., Kuhn, J., & Brünken, R. (2020). The use of augmented reality to foster conceptual knowledge acquisition in STEM laboratory courses—Theoretical background and empirical results. *British Journal of Educational Technology*, 51(3), 611–628.

Al-Zoubi, A., San Cristobal, E., Shahroury, F. R., & Castro, M. (2024). The Middle East higher education experience: Implementing remote labs to improve the acquisition of skills in industry 4.0. *IEEE Transactions on Learning Technologies*, 17, 982–991. https://doi.org/10.1109/TLT.2023.3341490

Amadi, L. (2020). Globalization and the changing liberal international order: A review of the literature. *Research in Globalization*, 2, 100015. https://doi.org/10.1016/j.resglo.2020.100015

Amanat, I., Kanwal, S., & Akhter, N. (2024). Exploring the female students and teachers problems in conducting laboratory practicals. *Journal of Excellence in Social Sciences*, 3(2), 188–202.

Amaresh, K., Ganachari, M. S., & Devarinti, R. (2024). Assessing views on using surplus human bio specimens for future research among pharmaceutical sponsors: A prospective interventional study. *Journal of Young Pharmacists*, 16(3), 506–519.

American Chemical Society and Committee on Chemical Safety. (2016). *Guidelines for Chemical Laboratory Safety in Secondary Schools*. American Chemical Society, Washington, DC.

Amin, M. T., Khan, F., Halim, S. Z., & Pistikopoulos, S. (2022). A holistic framework for process safety and security analysis. *Computers & Chemical Engineering*, 165, 107963. https://doi.org/10.1016/j.compchemeng.2022.107963

Amusat, A. I., Okewole, S. A., Rafiu, R. A., & Amusat, M. A. (2023). Overview of pesticide usage, misuse and its impact on environmental degradation in south-western states and some part of northern states in Nigeria. *Journal of Pesticide Science and Pest Control*, 2(1), 13. https://doi.org/10.58489/2833-0943/013

Anaduaka, E. G., Uchendu, N. O., Asomadu, R. O., Ezugwu, A. L., Okeke, E. S., & Chidike, E. T. P. (2023). Widespread use of toxic agrochemicals and pesticides for agricultural products storage in Africa and developing countries: Possible panacea for ecotoxicology and health implications. *Heliyon*, 9(4), e15173. https://doi.org/10.1016/j.heliyon.2023.e15173

Ananda, L. R., Rahmawati, Y., & Khairi, F. (2023). Critical thinking skills of chemistry students by integrating design thinking with STEAM-PjBL. *Journal of Technology and Science Education*, 13(1), 352–367.

Andersen, B. M. (2018). Laboratories. *Prevention and Control of Infections in Hospitals*, 25, 959–967. https://doi.org/10.1007/978-3-319-99921-0_73

Aneck-Hahn, N., Bornman, M., & de Jager, C. (2008). Preliminary assessment of oestrogenic activity in water sources in Rietvlei Nature Reserve, Gauteng, South Africa. *African Journal of Aquatic Science*, 33(3), 249–254.

Aneck-Hahn, N., Schulenburg, G., Bornman, M., Farias, P., & De Jager, C. (2007). Impaired semen quality associated with environmental DDT exposure in young men living in a malaria area in the Limpopo Province, South Africa. *Journal of Andrology*, 28(3), 423–434.

Aneck-Hahn, N., Van Zijl, M., Swart, P., Truebody, B., Genthe, B., Charmier, J., & De Jager, C. (2018). Estrogenic activity, selected plasticizers and potential health risks associated with bottled water in South Africa. *Journal of Water and Health*, 16(2), 253–262.

Angon, P. B., Islam, Md. S., Shreejana, K. C., Das, A., Anjum, N., Poudel, A., & Suchi, S. A. (2024). Sources, effects and present perspectives of heavy metals contamination: Soil, plants and human food chain. *Heliyon*, 10(7), e28357.

Anjum, M., Siddique, N., Younis, H., Faiz, Y., Shafique, M. A., Mahnoor, A., Sajid, A., & Altaf, M. (2024). Heavy metals and radionuclides in Islamabad's industrial area: A comprehensive analysis of soil and water pollution, source apportionment and health effects using statistical and geospatial tools. *Journal of Trace Elements and Minerals*, 8, 100127.

Annaratone, L., De Palma, G., Bonizzi, G., Sapino, A., Botti, G., Berrino, E., … & Alleanza Contro il Cancro (ACC) Pathology and Biobanking Working Group. (2021). Basic principles of biobanking: From biological samples to precision medicine for patients. *Virchows Archiv: European Journal of Pathology*, 479, 233–246.

Ansah, F. O., Assem, H. D., Ossei-Anto, T. A., Acheampong, A., & Owusu, M. (2024). Evaluating the availability and effectiveness of safety equipment in chemistry laboratories at public colleges of education in Ghana. *Creative Education*, 15(10), 2224–2257.

Anto, I. J. C., Buagas, I. R. A., Ong, P. M. V. J., Naparan, G. B., & Villaver, A. V. (2023). Challenges and coping strategies of science teachers. *Canadian Journal of Educational and Social Studies*, 3(4), 148–166.

Anwar, G. & Abdullah, N. N. (2021). Inspiring future entrepreneurs: The effect of experiential learning on the entrepreneurial intention at higher education. *International Journal of English Literature and Social Sciences*, 6(2), 183–184. https://doi.org/10.22161/ijels.62.26

Anwar, M., Iqbal, Q., & Saleem, F. (2020). Improper disposal of unused antibiotics: An often overlooked driver of antimicrobial resistance. *Expert Review of Anti-Infective Therapy*, 18(8), 697–699. https://doi.org/10.1080/14787210.2020.1754797

Anza, M., Bibiso, M., Kuma, B., & Osuman, K. (2016). Investigation of laboratory and chemical safety in Wolaita Sodo University, Ethiopia. *Chemistry and Materials Research*, 8, 11.

Archer, E., Petrie, B., Kasprzyk-Hordern, B., & Wolfaardt, G. (2017). The fate of pharmaceuticals and personal care products (PPCPs), endocrine disrupting contaminants (EDCs), metabolites and illicit drugs in a WWTW and environmental waters. *Chemosphere*, 174, 437–446.

Archer, E., Wolfaardt, G., & van Wyk, J. (2017). Pharmaceutical and personal care products (PPCPs) as endocrine disrupting contaminants (EDCs) in South African surface waters. *Water SA*, 43(4), 684–706.

Archer, E., Wolfaardt, G., van Wyk, J., & van Blerk, N. (2020). Investigating (anti)estrogenic activities within South African wastewater and receiving surface waters: Implication for reliable monitoring. *Environmental Pollution*, 263, 114424. https://doi.org/10.1016/j.envpol.2020.114424

Ardiyanti, R., Azrague, K., Medema, G., Hallé, C. (2024). Navigating quality, health and environmental risk: A novel framework for wastewater resource recovery products. *Journal of Cleaner Production*, 434, 140260. https://doi.org/10.1016/j.jclepro.2023.140260

Arnesdotter, E., Rogiers, V., Vanhaecke, T., & Vinken, M. (2021). An overview of current practices for regulatory risk assessment with lessons learnt from cosmetics in the European Union. *Critical Reviews in Toxicology*, 51(5), 395–417.

Arora, D. R. (2004). Quality assurance in microbiology. *Indian Journal of Medical Microbiology*, 22(2), 81–86.

Arote, K. S., Salade, D. A., & Patil, N. V. (2023). A brief review on regulatory affairs: Ensuring compliance, safety, and market access. *International Journal of Pharmaceutical Sciences*, 1(10), 22–30.

Arthur, C. (2019). Making It: Industry for Development (Number 27; pp. 1–46). United Nations Industrial Development Organization, Geneva, Switzerland. https://www.unido.org/sites/default/files/files/2019-11/MakingIt-27-46pp.pdf

Arukwe, A., Eggen, T., & Moder, M. (2012). Solid waste deposits as a significant source of contaminants of emerging concern to the aquatic and terrestrial environments—A developing country case study from Owerri, Nigeria. *Science of the Total Environment*, 438, 94–102.

Asem-Hiablie, S., Church, C., Elliott, H., Shappell, N., Schoenfuss, H., Drechsel, P., Williams, C., Knopf, A., & Dabie, M. (2013). Serum estrogenicity and biological responses in African catfish raised in wastewater ponds in Ghana. *Science of the Total Environment*, 463, 1182–1191.

Aslan, D. (2018). Which skills are needed and how they should be gained by laboratory medicine professionals for successful ISO 15189 accreditation. *EJIFCC*, 29(4), 264–273.

Asrani, P., Hussain, A., Nasreen, K., AlAjmi, M. F., Amir, S., Sohal, S. S., & Hassan, M. I. (2021). Guidelines and safety considerations in the laboratory diagnosis of SARS-CoV-2 infection: A prerequisite study for health professionals. *Risk Management and Healthcare Policy*, 2021, 379–389.

Atinkut Asmare, B., Freyer, B., & Bingen, J. (2022). Women in agriculture: Pathways of pesticide exposure, potential health risks and vulnerability in sub-Saharan Africa. *Environmental Sciences Europe*, 34, 89. https://doi.org/10.1186/s12302-022-00638-8

Australian Prescriber. (2003). Chinese herbal medicines – A cautionary tale. *Australian Prescriber*, 26(4), 88–90.

Authority, E. F. S. (2024). The European Union one health 2023 Zoonoses report. *EFSA Journal*, 21(12). 8442. https://doi.org/10.2903/j.efsa.2023.8442

Avivar, C. (2012). Strategies for the successful implementation of viral laboratory automation. *The Open Virology Journal*, 6, 115–121. https://doi.org/10.2174/1874357901206010115

Awofolu, R. & Fatoki, O. (2003). Persistent organochlorine pesticide residues in freshwater systems and sediments from the Eastern Cape, South Africa. *Water SA*, 29(3), 323–330.

Azanaw, A., Birlie, B., Teshome, B., & Jemberie, M. (2022). Textile effluent treatment methods and eco-friendly resolution of textile wastewater. *Case Studies in Chemical and Environmental Engineering*, 6, 100230.

Azanaw, J., Dagne, H., Andualem, Z., & Adane, T. (2021). Food safety knowledge, attitude, and practice of college students, Ethiopia, 2019: A cross-sectional study. *BioMed Research International*, 1, 6686392.

Aziz, M. & Ojumu, T. (2020). Exclusion of estrogenic and androgenic steroid hormones from municipal membrane bioreactor wastewater using UF/NF/RO membranes for water reuse application. *Membranes*, 10(3), 37.

Azizullah, A., Khattak, M. N. K., Richter, P., & Häder, D.-P. (2011). Water pollution in Pakistan and its impact on public health—A review. *Environment International*, 37(2), 479–497.

Babalola, O. & van Wyk, J. (2021). Exposure impacts of Imazapyr formulation on larval development and thyroid histology of Xenopus laevis. *Environmental Science and Pollution Research*, 28(37), 50967–50974.

Bae, S. & Martin, C. S. (2023). The impact of measurable findings from pre- and post-occupancy evaluations of indoor environmental quality in the primary workspace. *Buildings*, 13(10), 2486. https://doi.org/10 3390/buildings13102486

Baek, K., Park, J.-T., & Kwak, K. (2022). Systematic review and meta-analysis of cancer risks in relation to environmental waste incinerator emissions: A meta-analysis of case-control and cohort studies. *Epidemiology and Health*, 44, e2022070.

Bai, M., Liu, Y., Qi, M., Roy, N., Shu, C. M., Khan, F., & Zhao, D. (2022). Current status, challenges, and future directions of university laboratory safety in China. *Journal of Loss Prevention in the Process Industries*, 74, 104671.

Balakrishnan, K., Dey, S., Gupta, T. et al. (2019). The impact of air pollution on deaths, disease burden, and life expectancy across the states of India: The global burden of disease study 2017. *The Lancet Planetary Health*, 3(1), e26–e39.

Balakrishnan, S. M., Deepika, S. R., & Srinivasan, P. M. (2024). Innovative teaching strategies in higher education and their impact on overall educational effectiveness: A theoretical model and instrument development. In E. A. Railean, *Implementing Interactive Learning Strategies in Higher Education* (pp. 214–240). IGI Global, New York.

Balaram, V. & Satyanarayanan, M. (2022). Data quality in geochemical elemental and isotopic analysis. *Minerals*, *12*(8), 999. https://doi.org/10.3390/min12080999

Balkrishna, A., Sharma, N., Srivastava, D. et al. (2024). Exploring the safety, efficacy, and bioactivity of herbal medicines: bridging traditional wisdom and modern science in healthcare. *Future Integrative Medicine*, 3(1), 35–49. https://doi.org/10.14218/FIM.2023.00086

Ball, N., Bars, R., Botham, P. A., Cuciureanu, A., Cronin, M. T., Doe, J. E., … & van Ravenzwaay, B. (2022). A framework for chemical safety assessment incorporating new approach methodologies within REACH. *Archives of Toxicology*, 96(3), 743–766.

Bamakan, S. M. H., Malekinejad, P., & Ziaeian, M. (2022). Towards blockchain-based hospital waste management systems: applications and future trends. *Journal of Cleaner Production*, 349, 131440.

Barbulescu, A., Barbes, L., & Dumitriu, C. S. (2021). Assessing the water pollution of the Brahmaputra River using water quality indexes. *Toxics*, 9(11), 297.

Bărbulescu, A., & Hosen, K. (2025). Cement industry pollution and its impact on the environment and population health: a review. *Toxics*, 13(7), 587. https://doi.org/10.3390/toxics13070587

Barling, J. (2023). *Brave New Workplace: Designing Productive, Healthy, and Safe Organizations*. Oxford University Press, Oxford.

Barnhoorn, I., van Dyk, J., Pieterse, G., & Bornman, M. (2010). Intersex in feral indigenous freshwater *Oreochromis mossambicus*, from various parts in the Luvuvhu River, Limpopo Province, South Africa. *Ecotoxicology and Environmental Safety*, 73(7), 1537–1542.

Baron, N. (2021). *How We Read Now: Strategic Choices for Print, Screen, and Audio*. Oxford University Press, Oxford.

Baroudi, S. & Helder, M. R. (2019): Behind the scenes: Teachers' perspectives on factors affecting the implementation of inquiry-based science instruction. *Research in Science & Technological Education*, 39(1), 68–89. https://doi.org/10.1080/02635143.2019.1651259

Barrasso, A. P. & Spilios, K. E. (2021). A scoping review of literature assessing the impact of the learning assistant model. *International Journal of STEM Education*, 8, 12. https://doi.org/10.1186/s40594-020-00267-8

Barron, H. A., Brown, J. C., & Cotner, S. (2021). The culturally responsive science teaching practices of undergraduate biology teaching assistants. *Journal of Research in Science Teaching*, 58(9), 1320–1358.

Bartolo, N. S., Azzopardi, L. M., & Serracino-Inglott, A. (2021). Pharmaceuticals and the environment. *Early Human Development*, 155, 105218.

Barton-Maclaren, T. S., Wade, M., Basu, N. et al. (2022). Innovation in regulatory approaches for endocrine disrupting chemicals: The journey to risk assessment modernization in Canada. *Environmental Research*, 204, 112225.

Baskin, C. R., & Zelicoff, A. P. (Eds.). (2016). *Ensuring National Biosecurity: Institutional Biosafety Committees*. Academic Press.

Basu, S. (2021). *The History of Forensic Science in India* (1st ed.). Routledge, London.

Batty, L. & Reilly, K. (2023). Understanding barriers to participation within undergraduate STEM laboratories Towards development of an inclusive curriculum. *Journal of Biological Education*, 57(5), 1147–1169.

Bauchner, H., Fontanarosa, P. B., & Livingston, E. H. (2020). Conserving supply of personal protective equipment - a call for ideas. *JAMA*, 323(19), 1911. https://doi.org/10.1001/jama.2020.4770

Bauer, J. H. & Pickels, E. G. (1936). A high speed vacuum centrifuge suitable for the study of filterable viruses *Journal of Experimental Medicine*, 64(4), 503–528. https://doi.org/10.1084/jem.64.4.503

Bayot, M. L. & King, K. C. (2025). *Biohazard Levels*. StatPearls Publishing, Treasure Island, FL. https://www.ncbi.nlm.nih.gov/books/NBK535351/

Bayrak, T. (2025). Patient safety in healthcare: A proposal for ensuring the use of regulation-compliant safety devices. *Therapeutic Innovation & Regulatory Science*, 2, 863. https://doi.org/10.1007/s43441-025-00863-2

Begum, N. F., Ramalingam, K., Ramani, P., & Begum III, N. F. (2024). Storage, retention, and use of leftover pathology specimens: The underestimated treasures. *Cureus*, 16(1), e53025. https://doi.org/10.7759/cureus.53025

Bekmirzayevich, A. N. & Bakhodirovich, K. D. (2024). A process approach to ISO/IEC 17025 in the implementation of a quality management system in testing laboratories. *Web of Discoveries: Journal of Analysis and Inventions*, 2(12), 62–70.

Bell, R. & Bell, H. (2020). Applying educational theory to develop a framework to support the delivery of experiential entrepreneurship education. *Journal of Small Business and Enterprise Development*, 27(6), 987–1004.

Bell, T. M., Facemire, P., Bearss, J. J., Raymond, J. L., Chapman, J., Zeng, X., … & Mucker, E. M. (2024). Smallpox lesion characterization in placebo-treated and tecovirimat-treated macaques using traditional and novel methods. *PLoS Pathogens*, 20(2), e1012007.

Bello, U., Madekurozwa, M., Groenewald, H., Aire, T., & Arukwe, A. (2014). The effects on steroidogenesis and histopathology of adult male Japanese quails (*Coturnix coturnix japonica*) testis following pre-pubertal exposure to di(*n*-butyl) phthalate (DBP). *Comparative Biochemistry and Physiology C-Toxicology & Pharmacology*, 166, 24–33.

Bengu, T., du Plessis, J., Modley, L., & van Dyk, J. (2017). Health effects in fish from the polluted Orlando Dam and Klipspruit wetland system, Soweto, South Africa. *African Journal of Aquatic Science*, 42(2), 131–141.

Benishek, L. E., Radonovich, L. J., Blackley, B. H., & Weissman, D. N. (2023). Healthcare workers' infection risk perceptions of aerosol-generating procedures and affective response. *Antimicrobial Stewardship & Healthcare Epidemiology*, 3(1), e29.

Bennari, H., Omari, S. E., Habsaoui, A., Chaair, H., & Belouafa, S. (2025). Toward ISO/IEC 17025 compliance: Challenges and strategies for a university research laboratory. *Accreditation and Quality Assurance*, 30, 55–66.

Bennett, E. (2024). Case studies in green chemistry: Successful industrial applications. *Chemistry Journal*, 11(2), 1–2. https://www.purkh.com/index.php/tocomp

Bennett, S. A., Conn, C. M., Gill, H. E., Holmen, B. K., McDevitt, T. M., Miliander, C. L., … & Hanson, C. A. (2025). Regulatory requirements for laboratory developed tests in the United States. *Journal of Immunological Methods*, 537, 113813.

Berezin, C. T., Aguilera, L. U., Billerbeck, S. et al. (2023). Ten simple rules for managing laboratory information. *PLoS Computational Biology*, 19(12), e1011652. https://doi.org/10.1371/journal.pcbi.1011652

Berggren, E. & Worth, A. P. (2023). Towards a future regulatory framework for chemicals in the European Union–Chemicals 2.0. *Regulatory Toxicology and Pharmacology: RTP*, 142, 105431. https://doi.org/10.1016/j.yrtph.2023.105431

Bergmann, M., Schäpke, N., Marg, O., Stelzer, F., Lang, D. J., Bossert, M., … & Sußmann, N. (2021). Transdisciplinary sustainability research in real-world labs: Success factors and methods for change. *Sustainability Science*, 16, 541–564.

Bertrand, P. G., Ahmed, H. A. M., A., Ngwafor, R., & Frazzoli, C. (2016). Toxicovigilance systems and practices in Africa. *Toxics*, 4(3), 13. https://doi.org/10.3390/toxics4030013

Besenyő, J. & Sinkó, G. (2024). Terrorist organizations' activities against crucial installations: Al-Shabaab's attacks on critical infrastructure in Kenya. In J. Besenyő, *Terrorism and Counter-Terrorism in Modern Sub-Saharan Africa* (pp. 169–193). Springer Nat, Cham.

Bhat, M. A. (2024). Airborne microplastic contamination across diverse university indoor environments: A comprehensive ambient analysis. *Air Quality, Atmosphere and Health*, 17, 1851–1866. https://doi.org/10.1007/s11869-024-01548-9

Bhattarai, N., Lobell, D. B., Balwinder, S., Fishman, R., Kustas, W. P., Pokhrel, Y., & Jain, M. (2023). Warming temperatures exacerbate groundwater depletion rates in India. *Science Advances*, 9(35), eadi1401. https://doi.org/10.1126/sciadv.adi1401

Bhusnure, O. G., Dongare, R. B., Gholve, S. B., & Giram, P. S. (2018). Chemical hazards and safety management in pharmaceutical industry. *Journal of Pharmacy Research*, 12(3), 13.

Bisong, N. M. & Kemeji, M. J. (2024). A case of drug-facilitated crime by the use of benzodiazepines. *Journal of Medicine in Africa*, 2(1), 28–30.

Bizami, N. A., Tasir, Z., & Kew, S. N. (2023). Innovative pedagogical principles and technological tools capabilities for immersive blended learning: A systematic literature review. *Education and Information Technologies*, *28*(2), 1373–1425.

Blacksell, S. D., Dhawan, S., Kusumoto, M. et al. (2023). The biosafety research road map: The search for evidence to support practices in human and veterinary laboratories. *Applied Biosafety*, 28(2), 64–71.

Blakey, D. H., Lafontaine, M., Lavigne, J., Sokolowski, D., Philippe, J. M., Sapori, J. M. ... & Global Health Security Initiative. (2013). A screening tool to prioritize public health risk associated with accidental or deliberate release of chemicals into the atmosphere. *BMC Public Health*, 13, 253. https://doi.org/10.1186/1471-2458-13-253

Blobner, J., Dengler, L., Blobner, S., Eberle, C., Weller, J., Teske, N., ... & von Baumgarten, L. (2023). Significance of molecular diagnostics for therapeutic decision-making in recurrent glioma. *Neuro-Oncology Advances*, 5(1), vdad060.

Blok, J. & Balk, E. (2020). Environmental regulation in the European community. In G. M. Rand, *Fundamentals of Aquatic Toxicology* (2nd ed.). CRC Press, Boca Raton, FL. https://doi.org/10.1201/9781003075363

Blot, S., Ruppé, E., Harbarth, S. et al. (2022). Healthcare-associated infections in adult intensive care unit patients: Changes in epidemiology, diagnosis, prevention and contributions of new technologies. *Intensive and Critical Care Nursing*, 70, 103227.

Bolujoko, N., Ogunlaja, O., Alfred, M., Okewole, D., Ogunlaja, A., Olukanni, O., Msagati, T., & Unuabonah, E. (2022). Occurrence and human exposure assessment of parabens in water sources in Osun State, Nigeria. *Science of the Total Environment*, 814, 152448. https://doi.org/10.1016/j.scitotenv.2021.152448

Boom, F. A., Le Brun, P. P. H., Ris, J. M., Veenbaas, T., & Touw, D. (2023). Reducing the risk of non-sterility of aseptic handling in hospital pharmacies, part C: Applying risk assessment and risk control in practice. *European Journal of Hospital Pharmacy*, 30(3), 160–166. https://doi.org/10.1136/ejhpharm-2021-002747.

Bornman, M., Barnhoorn, I., de Jager, C., & Veeramachaneni, D. (2010). Testicular microlithiasis and neoplastic lesions in wild eland (Tragelaphus oryx): Possible effects of exposure to environmental pollutants? *Environmental Research*, 110(4), 327–333.

Boronow, K. E., Cohn, B., Havas, L., Plumb, M., & Brody, J. G. (2023). The effect of individual or study-wide report-back on knowledge, concern, and exposure-reducing behaviors related to endocrine-disrupting chemicals. *Environmental Health Perspectives*, 131(9), 97005. https://doi.org/10.1289/EHP12565

Borovečki, A., Mlinarić, A., Horvat, M., & Šupak Smolčić, V. (2018). Informed consent and ethics committee approval in laboratory medicine. *Biochemia Medica (Zagreb)*, 28(3), 030201. https://doi.org/10.11613/BM.2018.030201

Borup, R. & Traulsen, J. (2016). Falsified medicines-bridging the gap between business and public health. *Pharmacy*, 4(2), 16. https://doi.org/10.3390/pharmacy4020016.

Bossi, E., Limo, E., Pagani, L., Monza, N., Serrao, S., Denti, V., ... & Paglia, G. (2024). Revolutionizing blood collection: Innovations, applications, and the potential of microsampling technologies for monitoring metabolites and lipids. *Metabolites*, 14(1), 46.

Boudehane, A., Lounas, A., Moussaoui, Y., Balducci, C., & Cecinato, A. (2016). Levels of organic compounds in interiors (school, home, university and hospital) of Ouargla city, Algeria. *Atmospheric Environment*, 144, 266–273.

Braathen, M., Mdegela, R., Correia, D., Rundberget, T., Myburgh, J., Botha, C., Skaare, J., & Sandvik, M. (2009). Vitellogenin in African sharptooth catfish (*Clarias gariepinus*): Purification, characterization, and ELISA development. *Journal of Toxicology and Environmental Health-Part A-Current Issues*, 72(3–4), 173–183.

Braga, F., Infusino, I., & Panteghini, M. (2015). Role and responsibilities of laboratory medicine specialists in the verification of metrological traceability of in vitro medical diagnostics. *Journal of Medical Biochemistry*, 34(3), 282–287. https://doi.org/10.1515/jomb-2015-0004

Brink, K., van Vuren, J., & Bornman, R. (2012). The lack of endocrine disrupting effects in catfish (*Clarias gariepinus*) from a DDT sprayed area. *Ecotoxicology and Environmental Safety*, 79, 256–263.

Brizee, S., van Passel, M. W. J., van den Berg, L. M., Feakes, D., Izar, A., Lin, K. T. B., Podin, Y., Yunus, Z., & Bleijs, D. A. (2019). Development of a biosecurity checklist for laboratory assessment and monitoring. *Applied Biosafety*, 24(2), 83–89. https://doi.org/10.1177/1535676019838077.

Broughton, E. (2005). The Bhopal disaster and its aftermath: A review. *Environmental Health*, 4(1), 6. https://doi.org/10.1186/1476-069X-4-6.

Bruinen de Bruin, Y., Franco, A., Ahrens, A., Morris, A., Verhagen, H., Kephalopoulos, S., ... & Fantke, P. (2022). Enhancing the use of exposure science across EU chemical policies as part of the European exposure science strategy 2020–2030. *Journal of Exposure Science and Environmental Epidemiology*, 32(4), 513–525. https://doi.org/10.1038/s41370-021-00388-4

Bugingo, J. B., Yadav, L. L., Mugisha, I. S., & Mashood, K. K. (2024). Improving teachers' and students' views on nature of science through active instructional approaches: A review of the literature. *Science & Education*, 33, 29–71. https://doi.org/10.1007/s11191-022-00382-8

Burch, T. R., Stokdyk, J. P., Rice, N., Anderson, A. C., Walsh, J. F., Spencer, S. K., … & Borchardt, M. A. (2022). Statewide quantitative microbial risk assessment for waterborne viruses, bacteria, and protozoa in public water supply wells in Minnesota. *Environmental Science & Technology*, 56(10), 6315–6324.

Burden, N., Embry, M. R., Hutchinson, T. H. et al. (2021). Investigating endocrine-disrupting properties of chemicals in fish and amphibians: Opportunities to apply the 3Rs. *Integrated Environmental Assessment and Management*, 18(2), 442–458.

Bureau of Indian Standards. (n.d.). Origin of BIS. Accessed on 02-08-2025 from: https://www.bis.gov.in/the-bureau/origin-of-bis/

Cahyaningrum, D., Tegar Muktiana Sari, H., & Iswandari, D. (2019). Faktor-Faktor yang Berhubungan dengan Kejadian Kecelakaan Kerja di Laboratorium Pendidikan. *Jurnal Pengelolaan Laboratorium Pendidikan*, 1(2), 41–47. https://doi.org/10.14710/jplp.1.2.41-47

Cai, S., Ni, J., & Li, H. (2019). The toxicity and safety of traditional Chinese medicines: Please treat with rationality. *Current Drug Metabolism*, *20*(12), 939–947.

Calabrese, M., Portarapillo, M., Di Nardo, A., Venezia, V., Turco, M., Luciani, G., & Di Benedetto, A. (2024). Hydrogen safety challenges: A comprehensive review on production, storage, transport, utilization, and CFD-based consequence and risk assessment. *Energies*, 17(6), 1350.

Callihan, D. R., Downing, M., Meyer, E., Ochoa, L. A., Petuch, B., Tranchell, P., & White, D. (2021). Considerations for laboratory biosafety and biosecurity during the coronavirus disease 2019 pandemic: Applying the ISO 35001: 2019 standard and high-reliability organizations principles. *Applied Biosafety*, 26(3), 113–122.

Cameron, E. E., Kuo, C. J., & Ho, P. (2010). The chemical security engagement program. *Journal of Chemical Health and Safety*, 17(5), 5–12.

Capoor, M. R. & Bhowmik, K. T. (2017). Current perspectives on biomedical waste management: Rules, conventions and treatment technologies. *Indian Journal of Medical Microbiology*, 35(2), 157–164.

Casillas, R. P., Tewari-Singh, N., & Gray, J. P. (2021). Special issue: Emerging chemical terrorism threats. *Toxicology Mechanisms and Methods*, 31(4), 239–241. https://doi.org/10.1080/15376516.2021.1904472

Castegnaro, M. & Sansone, E. B. (2012). *Chemical Carcinogens: Some Guidelines for Handling and Disposal in the Laboratory*. Springer Science & Business Media, New York.

Castro, G., Fourie, A., Marlin, D., Venkatraman, V., Gonzalez, S., & Asimakopoulos, A. (2022). Occurrence of bisphenols and benzophenone UV filters in wild brown mussels (*Perna perna*) from Algoa Bay in South Africa. *Science of the Total Environment*, 813, 571. https://doi.org/10.1016/j.scitotenv.2021.152571

Caves Jr, J. P. & Carus, W. S. (2022). Controlling chemical weapons in the new international order. *WMD Proceedings*, 1, 8. Accessed on 06-07-2025 from: https://digitalcommons.ndu.edu/wmd-proceedings/1

Chakraborty, K., Aravind, S., Hashim, M., & Bils, S. (2024). Ensuring laboratory safety: Key elements of hazard assessment, risk evaluation, and experiment planning. In: *Course Manual: ICAR Winter School on Harnessing Recent Advances in High-Value Compound Development and Seaweed Biomass Utilization for Human Well-being: Propelling Atmanirbhar Swastha Bharat and Empowering Farmers 15th February to 06th March 2024. CMFRI Training Manual Series (38)* (pp. 70–95). ICAR- Central Marine Fisheries Research Institute, Kochi.

Chan, P., Van Gerven, T., Dubois, J. L., & Bernaerts, K. (2021). Virtual chemical laboratories: A systematic literature review of research, technologies and instructional design. *Computers and Education Open*, 2, 100053.

Charlton, C. L., Hull, N., Sloma, C. R. et al. (2021). How to prepare for the unexpected: A public health laboratory response. *Clinical Microbiology Reviews*, 34(3), 10–11.

Chatha, S. A. S., Asgher, M., & Iqbal, H. M. N. (2017). Enzyme-based solutions for textile processing and dye contaminant biodegradation—A review. *Environmental Science and Pollution Research*, 24(16), 14005–14018.

Chavez, S., Koyfman, A., Gottlieb, M., Brady, W. J., Carius, B. M., Liang, S. Y., & Long, B. (2023). Ebola virus disease: A review for the emergency medicine clinician. *The American Journal of Emergency Medicine*, 70, 30–40.

Che Huei, L., Ya-Wen, L., Chiu Ming, Y. et al. (2020). Occupational health and safety hazards faced by healthcare professionals in Taiwan: A systematic review of risk factors and control strategies. *SAGE Open Medicine*, 8, 2050312120918999.

Chen, G., Wu, Y., Men, J., Zhou, L., & Li, X. (2024a). A novel approach for failure and consequence assessment considering multi-hazard coupling scenario in chemical industrial parks. *Engineering Failure Analysis*, 156, 107867. https://doi.org/10.1016/j.engfailanal.2023.107867

Chen, S. H., Garcia, C. B., Lazear, E., Lentz, T. B., Robertson, S. D., Li, Z., & Goller, C. C. (2024b). Implementation of electronic lab notebooks (ELNs) in science laboratory classes: Student response and lessons learned. *Discover Education*, 3(1), 70.

Chen, S., Xu, J., Liu, C., Zhu, Y., Nelson, D. R., Zhou, S., & Xiao, P. (2021a). Good agricultural and collection practices for medicinal plants in China: A review. *Journal of Ethnopharmacology*, 268, 113610.

Chen, S., Xu, J., Liu, C., Zhu, Y., Nelson, D. R., Zhou, S., & Xiao, P. (2021b). Globalisation of traditional Chinese medicine: Challenges and opportunities. *Chinese Herbal Medicines*, **13**(2), 161–167.

Chen, X., Lei, Z., Chang, J.-S., & Lee, D.-J. (2025). Navigating future wastewater treatment plants with artificial intelligence: Applications, challenges, and innovations. *Journal of Cleaner Production*, 504, 145467.

Chen, Y. G., Huang, J. H., Luo, R., Ge, H. Z., Wołowicz, A., Wawrzkiewicz, M., … & Chen, S. H. (2021c). Impacts of heavy metals and medicinal crops on ecological systems, environmental pollution, cultivation, and production processes in China. *Ecotoxicology and Environmental Safety*, 219, 112336. https://doi.org/10.1016/j.ecoenv.2021.112336

Chen, L., Wu, J., & Zhang, Q. (2022). Historical context and modern innovations in Chinese herbal medicine: A systematic review. *Journal of Traditional and Complementary Medicine*, *12*(4), 315–327.

Cheng, S., Zhou, J., Xu, G., Xin, M., ……….. & Liu, M. (2020). Acupuncture and moxibustion for pain relief and quality of life improvement in patients with knee osteoarthritis: A protocol for systematic review and meta-analysis. *Medicine*, *99*(22), e20171. https://doi.org/10.1097/MD.0000000000020171

Cheng, W., Du, J., Chen, C. et al. (2024). Exploring the influencing factors of adverse drug reaction reporting among medical personnel: A COM-B model-based study. *Journal of Pharmaceutical Policy and Practice*, 17(1), 2417410.

Chew, S. L. & Cerbin, W. J. (2021). The cognitive challenges of effective teaching. *The Journal of Economic Education*, 52(1), 17–40.

Chipangura, J. K., Abdussamad, A. M., & Lewis, D. I. (2024). The veterinarian's role in biocontainment research animal facilities and prevention of spread of pathogens: A case of Nigeria and South Africa. *Laboratories*, 1, 103–115.

Chokwe, T., Okonkwo, J., Sibali, L., & Ncube, E. (2012). Optimization and simultaneous determination of alkyl phenol ethoxylates and brominated flame retardants in water after SPE and heptafluorobutyric anhydride derivatization followed by GC/MS. *Chromatographia*, 75(19–20), 1165–1176.

Chopra Committee. (1931). *Report of the Drugs Enquiry Committee, 1930* – 31. Central Publication Branch, Calcutta. https://indianculture.gov.in/report-drugs-enquiry-committee-1930-31

Choudhury, A., Singh, P. A., Bajwa, N., Dash, S., & Bisht, P. (2023). Pharmacovigilance of herbal medicines: Concerns and future prospects. *Journal of Ethnopharmacology*, 309, 116383. https://doi.org/10.1016/j.jep.2023.116383

Choudhury, P., Bera, T., Pratihar, S., Das, S., & Chatterjee, A. (2025). Mitigation of pathogenic microorganisms in infectious biomedical waste: Biomedical waste management. In M. Tsakona, *Hospital Waste Management and Toxicity Evaluation* (pp. 163–190). IGI Global Scientific Publishing, Hershey, PA.

Christensen, R., Fisher, D., Salmon, S. et al. (2021). Training for outbreak response through the global outbreak alert and response network. *BMC Medicine*, 19, 123. https://doi.org/10.1186/s12916-021-01996-5

Christiansen, S., Axelstad, M., Scholze, M. et al. (2020). Grouping of endocrine disrupting chemicals for mixture risk assessment–Evidence from a rat study. *Environment International*, 142, 105870.

Chu, S. K. W., Reynolds, R. B., Tavares, N. J., Notari, M., & Lee, C. W. Y. (2021). *21st Century Skills Development through Inquiry-Based Learning from Theory to Practice*. Springer International Publishing, New York.

Çilekrenkli, A. & Kaya, E. (2023). Learning science in context: Integrating a holistic approach to nature of science in the lower secondary classroom. *Science & Education*, 32, 1435–1469.

CITES. (2022). Convention on International Trade in Endangered Species of Wild Fauna and Flora. Retrieved January 13 2025, from https://cites.org/eng

Clapp, P. W., Sickbert-Bennett, E. E., Samet, J. M. et al. (2021). Evaluation of cloth masks and modified procedure masks as personal protective equipment for the public during the COVID-19 pandemic. *JAMA Internal Medicine*, 181(4), 463–469.

Clements, B. W. & Casani, J. A. P. (2016). Emerging and reemerging infectious disease threats. *Disasters and Public Health*, 6, 245–265. https://doi.org/10.1016/B978-0-12-801980-1.00010-6

Coker, A. C. & Kayode, C. A., eds. (2001). *Chapter Twelve - Safety in Chemical Reaction Engineering, Modeling of Chemical Kinetics and Reactor Design* (pp. 910–1033). Gulf Professional Publishing, Oxford, UK. https://doi.org/10.1016/B978-088415481-5/50014-0

Collett, G., Ladyman, M., Hazael, R., & Temple, T. (2021). The use of a predictive threat analysis to propose revisions to existing risk assessments for precursor chemicals used in the manufacture of home-made explosives (HME). *Heliyon*, 7(12), e08343. https://doi.org/10.1016/j.heliyon.2021.e08343

Cordaillat-Simmons, M., Rouanet, A., & Pot, B. (2020). Live biotherapeutic products: The importance of a defined regulatory framework. *Experimental & Molecular Medicine*, 52(9), 1397–1406.

Cornish, N. E., Anderson, N. L., Arambula, D. G. et al. (2021). Clinical laboratory biosafety gaps: Lessons learned from past outbreaks reveal a path to a safer future. *Clinical Microbiology Reviews*, 34(3), 10–11.

Cortijo-Alfonso, M. E., Yuste, S., Piñol-Felis, C., Romero, M. P., Macià, A., & Rubió-Piqué, L. (2025). Finger-prick blood sampling using volumetric absorptive microsampling (VAMS) method for monitoring the main (poly) phenolic metabolites in human blood after barley biscuit intake. *Journal of Chromatography B*, 1256, 124527.

Costantino, R. C. (2021). The U.S. medicine chest: Understanding the U.S. pharmaceutical supply chain and the role of the pharmacist. *Journal of the American Pharmacists Association*, 61(1), e87–e92. https://doi.org/10.1016/j.japh.2020.07.018.

Costello, J., Kaur, M., Reformat, M. Z., & Bolduc, F. V. (2023). Leveraging knowledge graphs and natural language processing for automated web resource labeling and knowledge mobilization in neurodevelopmental disorders: Development and usability study. *Journal of Medical Internet Research*, 25, e45268.

Covarrubias, C. E., Rivera, T. A., Soto, C. A., Deeks, T., & Kalergis, A. M. (2022). Current GMP standards for the production of vaccines and antibodies: An overview. *Frontiers in Public Health*, 10, 1021905. https://doi.org/10.3389/fpubh.2022.1021905.

CPCB. (2025a). *Hazardous and other Wastes (Management & Transboundary Movement) Rules, 2016*. CPCB, Delhi. https://cpcb.nic.in/rules/

CPCB. (2025b). *CPCB Guidelines on Environmental Monitoring*. CPCB, Delhi. https://www.ppsthane.com/blog/cpcb-guidelines-for-environmental-monitoring

Crowley, M. & Dando, M. (2024). Regulation of toxins and bioregulators under the chemical weapons convention and the biological and toxin weapons convention. *Journal of Biosafety and Biosecurity*, 6(2), 99–112. https://doi.org/10.1016/j.jobb.2024.03.003

Crowley, M., Shang, L., & Dando, M. (2018). Preventing chemical weapons as sciences converge. *Science*, 362(6416), 753–755. https://doi.org/10.1126/science.aav5129

Czigle, S., Nagy, M., Mladěnka, P., Tóth, J., & OEMONOM. (2023). Pharmacokinetic and pharmacodynamic herb-drug interactions-part I. Herbal medicines of the central nervous system. *PeerJ*, 11, e16149. https://doi.org/10.7717/peerj.16149

D'Angelo, J. G. (2023). Project time! A course-based undergraduate research experience, a CURE for the traditional organic lab. *Journal of Chemical Education*, 100(8), 2904–2916. https://doi.org/10.1021/acs.jchemed.2c00694

Dagher, G. (2022). Quality matters: International standards for biobanking. *Cell Proliferation*, 55(8), e13282. https://doi.org/10.1111/cpr.13282

Dalvie, M. & London, L. (2009). Risk assessment of pesticide residues in South African raw wheat. *Crop Protection*, 28(10), 864–869.

Das Neves, J., Barnhoorn, I., & Wagenaar, G. (2018). The effects of environmentally relevant concentrations of aldrin and methoxychlor on the testes and sperm of male *Clarias gariepinus* (Burchell, 1822) after short-term exposure. *Fish Physiology and Biochemistry*, 44(6), 1421–1434.

de Bienassis, K., Slawomirski, L., & Klazinga, N. S. (2021). The economics of patient safety Part IV: Safety in the workplace: Occupational safety as the bedrock of resilient health systems. OECD Health Working Papers, No. 130, OECD Publishing, Paris. https://doi.org/10.1787/b25b8c39-en

de Bruin, W., van der Merwe, C., Kritzinger, Q., Bornman, R., & Korsten, L. (2017). Morphological characterisation of lettuce plasma membrane ultrastructure and vesicle formation caused by nonylphenol: A scanning electron microscopy study. *South African Journal of Botany*, 111, 176–181.

de Vries, B. (2023). Recent developments in the national implementation of biological weapons convention: What happened since resolution 1540? *Journal of Conflict and Security Law*, 28(3), 579.

De, M. (2017). Review of common hazardous waste generated from educational institutions: Case study from plant DNA isolation protocol in undergraduate college laboratory. *Coherence. Seminar Edition*, 2, 2.

DeCoito, I. & Estaiteyeh, M. (2022). Transitioning to online teaching during the COVID-19 pandemic: An exploration of STEM teachers' views, successes, and challenges. *Journal of Science Education and Technology*, 31(3), 340–356. https://doi.org/10.1007/s10956-022-09958-z

de Jager, C., Aneck-Hahn, N., Van Zijl, M., Hayward, S., Swart, P., & Genthe, B. (2019). Endocrine disrupting chemicals in commercially available cling film brands in South Africa. *Human and Ecological Risk Assessment*, 25(6), 1633–1644. https://doi.org/10.1080/10807039.2018.1471659

Dehal, A., Vaidya, A. N., & Kumar, A. R. (2022). Biomedical waste generation and management during COVID-19 pandemic in India: Challenges and possible management strategies. *Environmental Science and Pollution Research*, 29(10), 14830–14845. https://doi.org/10.1007/s11356-021-16736-8

Dehkordi, M. M., Nodeh, Z. P., Dehkordi, K. S., Salmanvandi, H., Khorjestan, R. R., & Ghaffarzadeh, M. (2024). Soil, air, and water pollution from mining and industrial activities: Sources of pollution, environmental impacts, and prevention and control methods. *Results in Engineering*, 23, 102729.

Demirci, C. (2010). Constructing a philosophy: Prospective teachers' opinions about constructivism. *Procedia - Social and Behavioral Sciences*, 9, 278–285.

Deshmukh, P., Amolic, U. V., & Mutkule, U. (2025). Impact of industrialization on accumulation of heavy metals in irrigation water, soil and vegetables. *International Journal of Advanced Biochemistry Research*, 9(3), 472–476.

Destura, R. V., Estrella, F. P., & Bello, P. R. (2025). Introduction to biosafety and biosecurity in the Philippines. In J. J. G. Guerrero, *Biosafety and Biosecurity* (1st ed.). CRC Press, Boca Raton, FL.

DiBerardinis, L. J., Baum, J. S., First, M. W., Gatwood, G. T., & Seth, A. K. (2013). *Guidelines for Laboratory Design: Health, Safety, and Environmental Considerations*. John Wiley & Sons, New Jersey.

Di Pietro, G., Forcucci, F., & Chiarelli, F. (2023). Endocrine disruptor chemicals and children's health. *International Journal of Molecular Sciences*, 24(3), 2671.

Di Renzo, G. C., Conry, J. A., Blake, J. et al. (2015). International federation of gynecology and obstetrics opinion on reproductive health impacts of exposure to toxic environmental chemicals. *International Journal of Gynecology & Obstetrics*, 131(3), 219–225.

DiBerardinis, L. J., Baum, J. S., First, M. W., Gatwood, G. T., & Seth, A. K. (2013). *Guidelines for Laboratory Design: Health, Safety, and Environmental Considerations*. John Wiley & Sons, New Jersy.

Dickmann, P., Sheeley, H., & Lightfoot, N. (2015). Biosafety and biosecurity: A relative risk-based framework for safer, more secure, and sustainable laboratory capacity building. *Frontiers in Public Health*, 3, 241. https://doi.org/10.3389/fpubh.2015.00241

Differding, E. (2017). The drug discovery and development industry in India—Two decades of proprietary small-molecule R&D. *ChemMedChem*, 12(11), 786–818.

Dixon, H., Armstrong, G., Barton, M. et al. (2019). Discovery of common chemical exposures across three continents using silicone wristbands. *Royal Society Open Science*, 6(2), 1818136.

Domański, J., Janczura, A., Wanat, M. et al. (2023). Preservation fluids of heritage anatomical specimens—a challenge for modern science. Studies of the origin, composition and microbiological contamination of old museum collections. *Journal of Anatomy*, 243(1), 148–166.

Doshi, A. & Clay, C. (2017). Rethink space: (Re)designing a workspace using human-centered design to support flexibility, collaboration, and engagement among clinical and translational research support services. *Journal of Clinical and Translational Science*, 1(3), 160–166. https://doi.org/10.1017/cts.2017.5

du Plessis, M., Fourie, C., Stone, W., & Engelbrecht, A.-M. (2023). The impact of endocrine disrupting compounds and carcinogens in wastewater: Implications for breast cancer. *Biochimie*, 209, 103–115.

Du, P., Lam, W. C., Leung, C. et al. (2024). Efficacy and safety of Chinese herbal medicine to prevent and treat COVID-19 household close contacts in Hong Kong: An open-label, randomized controlled trial. *Frontiers in Immunology*, 15, 1359331

Dua, H. S., Ting, D. S. J., Al Saadi, A., & Said, D. G. (2020). Chemical eye injury: Pathophysiology, assessment and management. *Eye*, 34(11), 2001–2019.

Dubale, S., Usure, R. E., Mekasha, M. T. et al. (2025). Traditional herbal medicine legislative and regulatory framework: A cross-sectional quantitative study and archival review perspectives. *Frontiers in Pharmacology*, 16, 1475297. https://doi.org/10.3389/fphar.2025.1475297

Dyer, O. (2024). Covid-19 originated in Wuhan lab, alleges Republican congressional report. *BMJ (Clinical Research ed.)*, 387, q2765. https://doi.org/10.1136/bmj.q2765

Eccher, A., Dei Tos, A. P., Scarpa, A., L'Imperio, V., Munari, E., Troncone, G., … & Pagni, F. (2023). Cost analysis of archives in the pathology laboratories: From safety to management. *Journal of Clinical Pathology*, 76(10), 659–663.

Edgerly, A., Gillespie, G. L., Bhattacharya, A., & Hittle, B. M. (2024). Summarizing recommendations for the prevention of occupational heat-related illness in outdoor workers: a scoping review. *Workplace Health & Safety*, 2024, 21650799241281998.

Edo, G. I., Obasohan, P., Makia, R. S. et al. (2024). The use of quality control parameters in the evaluation of herbal drugs. A review. *Discover Medicine*, 1, 168.

Edward Steward, J. (2018). The Status of Safety Culture at Louisiana State University. *LSU Doctoral Dissertations*, p. 4498. https://doi.org/10.31390/gradschool_dissertations.4498 https://repository.lsu.edu/gradschool_dissertations/4498

Edwards, B., Novossiolova, T., Crowley, M., Whitby, S., Dando, M., & Shang, L. (2022). Meeting the challenges of chemical and biological weapons: Strengthening the chemical and biological disarmament and non-proliferation regimes. *Frontiers in Political Science, 4*, 805426. doi.org/10.3389/fpos.2022.805426

Eggert, S., Mieszczanek, P., Meinert, C., & W Hutmacher, D. (2020). OpenWorkstation: A modular open-source technology for automated in vitro workflows. *HardwareX*, 8, e00152. https://doi.org/10.1016/j.ohx.2020.e00152.

Ekici, G. (2011). Analysis of variables that express perception of high school students towards the class environment of biology laboratories. *Procedia–Social and Behavioral Sciences*, 15, 1901–1905.

Ekwempu, A. I. & Anderson, D. (2019). Knowledge and attitudes of safe agrochemical handling by users in Plateau State, Nigeria. *International Research Journal of Public and Environmental Health*, 6(7), 162–169. Accessed on 20-07-2025 from: https://www.journalissues.org/IRJPEH/ https://doi.org/10.15739/irjpeh.19.020

ELABSS. (2025). Comprehensive Guide on Laboratory Compliance Requirement in India. ELABSS. https://elabss.ai/comprehensive-guide-on-laboratory-compliance-requirement-in-india/?utm

El-Hajj, M. & Mirza, Z. A. (2024). Protecting small and medium enterprises: A specialized cybersecurity risk assessment framework and tool. *Electronics*, 13(19), 3910. https://doi.org/10.3390/electronics13193910

Elhakim, M., Ghazy, R. M., & Samhouri, D. (2024). How the IHR (2005) shaped the COVID-19 pandemic response in the eastern mediterranean region: What went well and what did not. *Health Secure*, 22(6), 437–444. https://doi.org/10.1089/hs.2024.0011

Elkhalifa, S. M., Altilmisani, N. M., Albishri, F. et al. (2024). Overview of quality control and safety in public health pest laboratory in Jeddah, Saudi Arabia. *International Journal of Agriculture and Biosciences*, 13(2), 92–100.

Elzagheid, M. (2023). *Chemical Technicians: Good Laboratory Practice and Laboratory Information Management Systems*. De Gruyter, Berlin, Boston. https://doi.org/10.1515/9783111191492

Emanuel, R. E. (2021). *A Cybersecurity Model for the Health Sector: A Case Study of Hospitals in Nairobi, Kenya*. United States International University (USIU), Nairobi, Africa.

Emeka, A., Margaret, A. F., Jacob, O. N., & Olatunde-Aiyedun, T. G. (2021). Problems facing science teachers in public secondary schools in Nigeria and way forward. *International Journal of Discoveries and Innovations in Applied Sciences*, 1, 118–129.

Emkani, M., Yazdi, M., Zarei, E., Klockner, K., Alimohammadlou, M., & Kamalinia, M. (2024). Advancing understanding of vulnerability assessment in process industries: A systematic review of methods and approaches. *International Journal of Disaster Risk Reduction*, 107, 104479. https://doi.org/10.1016/j.ijdrr.2024.104479

Emmert, E. A. & ASM Task Committee on Laboratory Biosafety (2013). Biosafety guidelines for handling microorganisms in the teaching laboratory: Development and rationale. *Journal of Microbiology & Biology Education*, 14(1), 78–83. https://doi.org/10.1128/jmbe.v14i1.531

Erasmus, P. M. (2012). Informed consent and the secondary use of biospecimens in oncology research legal and bioethics perspectives. Thesis, University of Cape Town. https://hdl.handle.net/11427/4693

Erkaboev, A., Obidov, J., Madmarova, U., & Alikhonov, E. (2023). Analysis of the ISO 9001 standard model of risk management in analytical testing laboratories. *E3S Web of Conferences*, 452, 06009.

Esterhuyse, M., Venter, M., Veldhoen, N., Helbing, C., & van Wyk, J. (2009). Characterization of vtg-1 mRNA expression during ontogeny in Oreochromis mossambicus (PETERS). *Journal of Steroid Biochemistry and Molecular Biology*, 117(1–3), 42–49.

Etteieb, S., Kawachi, A., Han, J., Tarhouni, J., & Isoda, H. (2015). Estrogenic activity assessment in environmental matrices by means of in vitro cell-based bioassays coupled with molecular investigation. *Clean-Soil Air Water*, 43(9), 1279–1288.

Eze, C. U. (2006) Ensuring Effective utilization of the laboratory in science pedagogy: A challenge to science education in the 21st century Nigeria. Ebonyi State University. *Journal of Education*, 4(1), 84–88.

Ezeigweneme, C. A., Umoh, A. A., Ilojianya, V. I., & Adegbite, A. O. (2024). Review of telecommunication regulation and policy: Comparative analysis USA and Africa. *Computer Science & IT Research Journal*, 5(1), 81–99.

Ezenwa, S., Talpade, A. D., Ghanekar, P., Joshi, R., Devaraj, J., Ribeiro, F. H., & Mentzer, R. (2022). Toward improved safety culture in academic and industrial chemical laboratories: An assessment and recommendation of best practices. *ACS Chemical Health & Safety*, 29(2), 202–213.

Ezzelle, J., Rodriguez-Chavez, I. R., Darden, J. M., Stirewalt, M., Kunwar, N., Hitchcock, R., … & D'souza, M. P. (2008). Guidelines on good clinical laboratory practice: Bridging operations between research and clinical research laboratories. *Journal of Pharmaceutical and Biomedical Analysis*, 46(1), 18–29.

Fabia, I., Fabia, U., & Piotrowski, S. (2024). Cross-border cooperation as a key element of internal security–selected aspects. *Scientific Journal of Bielsko-Biala School of Finance and Law*, 28(3), 16. https://doi.org/10.19192/wsfip.sj3.2024.16

Fadilah, K., Rosada, K. K., Affandy, M. A., Ali, M., & Sunardi, S. (2023). Effect of leather tanning effluent on chromium contamination in paddy field. *Plant Science Today (Early Access)*, 19, 2188. https://doi.org/10.14719/pst.2188

Fagre, A. C., Cohen, L. E., Eskew, E. A., Farrell, M., Glennon, E., Joseph, M. B., … & Albery, G. F. (2022). Assessing the risk of human-to-wildlife pathogen transmission for conservation and public health. *Ecology Letters*, 25(6), 1534–1549.

Fang, W. Y., Ravindar, L., Rakesh, K. P., Manukumar, H. M., Shantharam, C. S., Alharbi, N. S., & Qin, H. L. (2019). Synthetic approaches and pharmaceutical applications of chloro-containing molecules for drug discovery: A critical review. *European Journal of Medicinal Chemistry*, 173, 117–153.

Farahi-Ashtiani, I., Jahangiri, K., Sohrabizadeh, S., Gholamnia, R., & Rasouli, M. R. (2023). Explaining of the experts' points of view about the types and importance of man-made hazards in Iran's process industries: A qualitative study. *Heliyon*, 9(2), e13472.

Farhat, H., Alinier, G., Chaabna, K. et al. (2024). Preparedness and emergency response strategies for chemical, biological, radiological and nuclear emergencies in disaster management: A qualitative systematic review. *Journal of Contingencies and Crisis Management*, 32(3), e12592.

Farounbi, A. & Ngqwala, N. (2020). Occurrence of selected endocrine disrupting compounds in the Eastern Cape province of South Africa. *Environmental Science and Pollution Research*, 27(14), 17268–17279.

Fatemi, F., Dehdashti, A., & Jannati, M. (2022). Implementation of chemical health, safety, and environmental risk assessment in laboratories: A case-series study. *Frontiers in Public Health*, 10, 898826. https://doi.org/10.3389/fpubh.2022.898826

Fatoki, O. & Awofolu, R. (2003). Methods for selective determination of persistent organochlorine pesticide residues in water and sediments by capillary gas chromatography and electron-capture detection. *Journal of Chromatography A*, 983(1–2), 225–236.

Faul, A., Julies, E., & Pool, E. (2013). Oestrogen, testosterone, cytotoxin and cholinesterase inhibitor removal during reclamation of sewage to drinking water. *Water SA*, 39(4), 499–506.

Faul, A., Julies, E., & Pool, E. (2014). Steroid hormone concentrations and physiological toxicity of water from selected dams in Namibia. *African Journal of Aquatic Science*, 39(2), 189–198.

Federal Bureau of Investigation (FBI). (2017). Pharmaceutical Theft. Retrieved March 2025, from https://www.fbi.gov/news/stories/pharmaceutical-theft

Feeney, A. J., Goad, J. A., & Flaherty, G. T. (2024). Global perspective of the risks of falsified and counterfeit medicines: A critical review of the literature. *Travel Medicine and Infectious Disease*, 61,102758.

Felder, R. M. & Brent, R. (2024). *Teaching and Learning STEM: A Practical Guide* (2nd ed.). John Wiley & Sons, New York.

Fernando, I., Fei, J., Cahoon, S., & Close, D. C. (2024). A review of the emerging technologies and systems to mitigate food fraud in supply chains. *Critical Reviews in Food Science and Nutrition*, 65(26), 5108–5135. https://doi.org/10.1080/10408398.2024.2405840

FICCI. (2023). *Indian Chemical Industry: Growth and Future Prospects.* Federation of Indian Chambers of Commerce & Industry. Accessed on 03-10-2025 from: https://ficci.in/public/storage/SPDocument/23338/Indian-Specialty-Chemical-Industry.pdf

Finkler, S. A., Calabrese, T. D., & Smith, D. L. (2025). *Financial Management for Public, Health, and Not-for-Profit Organizations* (8th ed.). SAGE Publications, Washington DC.

Flaws, B. (2019). *Arisal of the Clear and Turbid: An Essay on the History of Traditional Chinese Medicine.* Blue Poppy Press, Portland, Oregon.

Fleming, F. F. (2010). Nitrile-containing pharmaceuticals: Efficacious roles of the nitrile pharmacophore. *Journal of Medicinal Chemistry*, 53(22), 7902–7917.

Fouzi, N. F. R., Aziz, H. A., & Yaakub, N. (2024). Systematic review of chemical safety and chemical security risk management approach. *Process Safety and Environmental Protection*, 183, 676–686.

Franić, Z., Bituh, T., Godec, R., Čačković, M., Meštrović, T., & Šiško, J. (2020). Experiences with the accreditation of the institute for medical research and occupational health, Zagreb, Croatia. *Archives of Industrial Hygiene and Toxicology*, 71(4), 312–319. https://doi.org/10.2478/aiht-2020-71-3449

Franson, J. C., Friend, M., Toweill, E. D., Brownell, L. R., Nettles, F. V., Davis, S. D., & Foreyt, J. W. (1999). Introduction to general field procedures (field manual of wildlife diseases). In M. Friend, J. Christian Franson, *Field Manual of Wildlife Diseases: General Field Procedures and Diseases of Birds.* USGS-National Wildlife Health Center, Honolulu, HI. https://www.nwhc.usgs.gov/publications/field_manual/

Freese, T., Elzinga, N., Heinemann, M., Lerch, M. M., & Feringa, B. L. (2024). The relevance of sustainable laboratory practices. *RSC Sustainability*, 2(5), 1300–1336. https://doi.org/10.1039/d4su00056k

Fromm, J., Radianti, J., Wehking, C., & Stieglitz, S. (2021). More than experience? - On the unique opportunities of virtual reality to afford a holistic experiential learning cycle. *The Internet and Higher Education*, 50, 100804. https://doi.org/10.1016/j.iheduc.2021.100804

Fudvoye, J., Bourguignon, J. P., & Parent, A. S. (2014). Endocrine-disrupting chemicals and human growth and maturation: A focus on early critical windows of exposure. *Vitamins and Hormones*, 94, 1–25. https://doi.org/10.1016/B978-0-12-800095-3.00001-8

Furlan, M. C. R., Ferreira, A. M., da Silva Barcelos, L., Rigotti, M. A., de Sousa, A. F. L., Dos Santos Junior, A. G., de Andrade, D., de Almeida, M. T. G., & da Silva Barreto, M. (2019). Evaluation of disinfection of surfaces at an outpatient unit before and after an intervention program. *BMC Infectious Diseases*, 19(1), 355. https://doi.org/10.1186/s12879-019-3977-4

Furxhi, I., Costa, A., Vázquez-Campos, S., Fito-López, C., Hristozov, D., Ramos, J. A. T., … & Farcal, L. (2023). Status, implications and challenges of European safe and sustainable by design paradigms applicable to nanomaterials and advanced materials. *RSC Sustainability*, 1(2), 234–250.

Gabel, J. & Rubin, E. (2019). Chemical terrorism: A global threat. *Journal of Homeland Security*, 23(3), 12–19.

Gajdzik, B. & Wycislik, A. (2012). Planning and implementation of investments in the chemical laboratory. Guidelines for the management of complex projects for example the purchase of equipment and laboratory devices. *Przemysl Chemiczny*, 91(10), 1965–1972.

Gamaralalage, P. J. D., Yagasa, R., Onogawa, K., Basir, A., Alexender, P., & Ghosh, S. K. (2019). *A Roadmap for Sustainable Waste Management and Resource Circulation in South Asia, 2019–2030.* Institute for Global Environmental Strategies, Hayama, Kanagawa, Japan. https://www.iges.or.jp/en/pub/roadmap-south-asia/en

Gao, W, Wu, Z., Zuo, K., Xiang, Q., Chen, X., Zhang, L., & Liu, H. (2025). Biosafety concept: Origins, evolution, and prospects. *Biosafety and Health*, 7(4), 209–217. https://doi.org/10.1016/j.bsheal.2025.07.003

Gao, W., Wu, Z., Zuo, K., Xiang, Q., Zhang, L., Chen, X., … & Liu, H. (2024). From biosafety to national security: The evolution and challenges of biosafety laboratories. *Laboratories*, 1(3), 158–173.

Garg, S., Reinhart, K., Couture, A., Kniss, K., Davis, C. T., Kirby, M. K., … & Olsen, S. J. (2025). Highly pathogenic avian influenza A (H5N1) virus infections in humans. *New England Journal of Medicine*, 392(9), 843–854.

Garzón, J., Baldiris, S., Gutiérrez, J., & Pavón, J. (2020). How do pedagogical approaches affect the impact of augmented reality on education? A meta-analysis and research synthesis. *Educational Research Review*, 31, 100334. https://doi.org/10.1016/j.edurev.2020.100334

Gastineau, T. (2020). What can vaccines learn from aviation? *Vaccine*, 38(33), 5082–5084. https://doi.org/10.1016/j.vaccine.2020.06.027.

Gautam, P. K., Chandra, S., & Henry, P. K. (2024). Monitoring of the groundwater level using GRACE with GLDAS satellite data in Ganga Plain, India to understand the challenges of groundwater, depletion, problems, and strategies for mitigation. *Environmental Challenges*, 15, 100874. https://doi.org/10.1016/j.envc.2024.100874

Gavrila, A.-M., Diacon, A., Iordache, T.-V., Rotariu, T., Ionita, M., & Toader, G. (2024). Hazardous materials from threats to safety: Molecularly imprinted polymers as versatile safeguarding platforms. *Polymers*, 16(19), 2699.

Gemler, B. T., Mukherjee, C., Howland, C. A., Huk, D., Shank, Z., Harbo, L. J., … & Bartling, C. M. (2022). Function-based classification of hazardous biological sequences: Demonstration of a new paradigm for biohazard assessments. *Frontiers in Bioengineering and Biotechnology*, 10, 979497.

Genthe, B., Le Roux, W., Schachtschneider, K., Oberholster, P., Aneck-Hahn, N., & Chamier, J. (2013). Health risk implications from simultaneous exposure to multiple environmental contaminants. *Ecotoxicology and Environmental Safety*, 93, 171–179.

George, M., Sichilongo, K., Ramabulana, T., Madala, N., & Dubery, I. (2019). Comparison of Soxhlet and reflux techniques for extraction and characterisation of potential endocrine-disrupting compounds from solid waste dumpsite soil. *Environmental Monitoring and Assessment*, 191(3), 149.

Gericke, N., Högström, P., & Wallin, J. (2022). A systematic review of research on laboratory work in secondary school. *Studies in Science Education*, 59(2), 245–285.

Ghai, B., Liao, Q. V., Zhang, Y., Bellamy, R., & Mueller, K. (2020). Explainable active learning (XAL): An empirical study of how local explanations impact annotator experience. https://doi.org/10.48550/arXiv.2001.09219

Gholamizadeh, K., Zarei, E., Yazdi, M., Rodrigues, M. A., Shirmohammadi-Khorram, N., & Mohammadfam, I. (2023). An integration of intelligent approaches and economic criteria for predictive analytics of occupational accidents. *Decision Analytics Journal*, 9, 100357.

Giljum, S., Maus, V., Kuschnig, N., Luckeneder, S., Tost, M., Sonter, L. J., & Bebbington, A. J. (2022). A pantropical assessment of deforestation caused by industrial mining. *Proceedings of the National Academy of Sciences*, 119(38), e2118273119.

Gillum, D. R. (2024). Balancing innovation and safety: Frameworks and considerations for the governance of dual-use research of concern and potential pandemic pathogens. *Applied Biosafety*, 30(2), 69–78. https://doi.org/10.1089/apb.2024.0033

Giovanni, C., Marques, F. L. N., & Günther, W. M. R. (2021). Laboratory chemical waste: Hazard classification by GHS and transport risk. *Revista de Saúde Pública*, 55, 102.

Godman, B., Egwuenu, A., Wesangula, E. et al. (2022). Tackling antimicrobial resistance across sub-Saharan Africa: Current challenges and implications for the future. *Expert Opinion on Drug Safety*, 21(8), 1089–1111.

Gomez-del Rio, T. & Rodríguez, J. (2022). Design and assessment of a project-based learning in a laboratory for integrating knowledge and improving engineering design skills. *Education for Chemical Engineers*, 40, 17–28.

Gonyora, M. & Ventura-Medina, E. (2024). Investigating the relationship between human and organisational factors, maintenance, and accidents. The case of chemical process industry in South Africa. *Safety Science*, 176, 106530.

Gorıa, C. & Konstantınıdıs, A. (2023). A participatory pedagogical model for online distance learning: Ideation and implementation. *Turkish Online Journal of Distance Education*, 24(1), 145–161.

Gough, W. W. (2021). Chemical Facility Anti-Terrorism Standards in Washington. Walden Dissertations and Doctoral Studies, p. 11076. Walden University. https://scholarworks.waldenu.edu/dissertations/11076

Government of Kenya, Ministry of Health, Pharmacy and Poisons Board. (2023). Guideline for scheduling of Health Products and Technologies, Version 1, 2022. https://web.pharmacyboardkenya.org/download/guidelines-for-scheduling-of-health-products-and-technologies/

Grandhi, S. P., Dagwar, P. P., & Dutta, D. (2024). Policy pathways to sustainable E-waste management: A global review. *Journal of Hazardous Materials Advances*, 16, 100473.

Grillo, R., Fraceto, L. F., Amorim, M. J., Scott-Fordsmand, J. J., Schoonjans, R., & Chaudhry, Q. (2021). Ecotoxicological and regulatory aspects of environmental sustainability of nanopesticides. *Journal of Hazardous Materials*, 404, 124148.

Grobelak, A., Całus-Makowska, K., Jasińska, A., Klimasz, M., Wypart-Pawul, A., Augustajtys, D., … & Kowalska, A. (2024). Environmental impacts and contaminants management in sewage sludge-to-energy and fertilizer technologies: Current trends and future directions. *Energies*, 17(19), 4983.

Groh, K. J., Geueke, B., Martin, O., Maffini, M., & Muncke, J. (2021). Overview of intentionally used food contact chemicals and their hazards. *Environment International*, 150, 106225.

Guenther, M. H. (2022). The essential or critical use of chemical substances in the United States and the European Union. *International Chemical Regulatory and Law Review*, 5(1), 3–12.

Guide, E. P., Route, E., Placard, B., & Authorizations, R. U. (2001). *Safety Standards. Series NS-G-2.3*. IAEA, Vienna.

Guo, J. & Luo, C. (2022). Risk assessment of hazardous materials transportation: A review of research progress in the last thirty years. *Journal of Traffic and Transportation Engineering (English Edition)*, 9(4), 571–590.

Gupta, A. K., Nair, S. S. & Singh, S. (2013). Environmental Legislation for Disaster Risk Management, National Institute of Disaster Management & Deutsche Gesellschaft für internationale Zusammenarbeit GmbH (GIZ), 108 p

Gupta, D. K., Tiwari, A., Yadav, Y., Soni, P., & Joshi, M. (2024). Ensuring safety and efficacy in combination products: Regulatory challenges and best practices. *Frontiers in Medical Technology*, 6, 1377443.

Gupta, P. P., Bankar, N. J., Mishra, V. H., Sanghavi, S., & Badge, A. K. (2023). The efficient disposal of biomedical waste is critical to public health: Insights from the central pollution control board guidelines in India. *Cureus*, 15(10), e47303.

Gyu Kim, M., Huh, S., Han, N., Hyun Kim, J., Kim, K., Lee, E., Kim, I. W., & Mi Oh, J. (2019). Analysis of issues and future trends impacting drug safety in South Korea. *International Journal of Environmental Research and Public Health*, 16(18), 3368. https://doi.org/10.3390/ijerph16183368.

Halasa, S., Abusalim, N., Rayyan, M. et al. (2020). Comparing student achievement in traditional learning with a combination of blended and flipped learning. *Nursing Open*, 7(4), 1129–1138.

Hale, S. E., Neumann, M., Schliebner, I., Schulze, J., Averbeck, F. S., Castell-Exner, C., … & Arp, H. P. H. (2022). Getting in control of persistent, mobile and toxic (PMT) and very persistent and very mobile (vPvM) substances to protect water resources: Strategies from diverse perspectives. *Environmental Sciences Europe*, 34(1), 22.

Hall, S. K. (2018). *Chemical Safety in the Laboratory*. CRC Press, Boca Raton, FL.

Hamed, G. & Aljanazrah, A. (2020). The effectiveness of using virtual experiments on students' learning in the general physics lab. *Journal of Information Technology Education: Research*, 19, 977–996. https://doi.org/10.28945/4668

Hansen, C.H. & Hansen, K.L. (2021). *Noise Control: From Concept to Application* (2nd ed.). CRC Press, Boca Raton, FL.

Hantoko, D., Li, X., Pariatamby, A., Yoshikawa, K., Horttanainen, M., & Yan, M. (2021). Challenges and practices on waste management and disposal during COVID-19 pandemic. *Journal of Environmental Management*, 286, 112140.

Harefa, D., Sarumaha, M., Telaumbanua, K., Telaumbanua, T., Laia, B., & Hulu, F. (2023). Relationship student learning interest to the learning outcomes of natural sciences. *International Journal of Educational Research & Social Sciences*, 4(2), 240–246.

Hassim, M. H., Zakaria, Z., & Mahmud, H. (2024). Process safety teaching in universities-current practices and way forward. *Process Safety and Environmental Protection*, 105, 1431–1443. https://doi.org/10.1016/j.psep.2024.10.105

Hawkins, J. W. (2010). Explosives Recognition and Awareness Training: A Psychological Approach to Pre-Blast Mitigation. Masters Theses, p. 4855. Accesses on 27-07-2025 from: https://scholarsmine.mst.edu/masters_theses/4855

Hawthorn, A., Berzins, I. K., Dennis, M. M., Kiupel, M., Newton, A. L., Peters, E. C., … & Work, T. M. (2023). An introduction to lesions and histology of scleractinian corals. *Veterinary Pathology*, 60(5), 529–546.

Hayes, L. H., Hopkins, S. E., Liu, S., Pardo, C. A., Garcia-Dominguez, M. A., Oleszek, J., … & Benson, L. A. (2023). Challenges in the clinical recognition of acute flaccid myelitis and its implications. *The Journal of Pediatrics*, 253, 55–62.

He, J., Lewis, B., & Guffey, S. (2017). Experimental study on centerline velocities of a rectangular capture hood under realistic conditions. *Journal of Occupational and Environmental Hygiene*, 15(267), 957. https://doi.org/10.1080/15459624.2017.1395957

Heslop, D. J. & Blain, P. G. (2020). Threat potential of pharmaceutical based agents. *Intelligence and National Security*, 35(4), 539–555.

Hewitt, N. J., Troutman, J., Przibilla, J. et al. (2022). Use of in vitro metabolism and biokinetics assays to refine predicted in vivo and in vitro internal exposure to the cosmetic ingredient, phenoxyethanol, for use in risk assessment. *Regulatory Toxicology and Pharmacology*, 131, 105132.

Hlekelele, L., Franklyn, P., Dziike, F., & Durbach, S. (2018). TiO_2 composited with carbon nanofibers or nitrogen-doped carbon nanotubes synthesized using coal fly ash as a catalyst: Bisphenol-A photodegradation efficiency evaluation. *New Journal of Chemistry*, 42(6), 4531–4542.

Hlekelele, L., Nomadolo, N. E., Setshedi, K. Z., Mofokeng, L. E., Chetty, A., & Chauke, V. P. (2019). Synthesis and characterization of polyaniline, polypyrrole and zero-valent iron-based materials for the adsorptive and oxidative removal of bisphenol-A from aqueous solution. *RSC advances*, 9(25), 14531–14543. https://doi.org/10.1039/c9ra01666j

Ho, C. C. & Chen, M. S. (2018). Risk assessment and quality improvement of liquid waste management in Taiwan University chemical laboratories. *Waste Management*, 71, 578–588.

Ho, K. & Tenkate, T. (2024). Safety data sheets as a hazard communication tool: An assessment of suitability and readability. *Safety and Health at Work*, 15(2), 192–199.

Hoang, T., Moskwa, N., & Halvorsen, K. (2018). A 'smart' tube holder enables real-time sample monitoring in a standard lab centrifuge. *PLoS One*, 13(4), e0195907. https://doi.org/10.1371/journal.pone.0195907

Hock, S. C., Pheh, A., Sachdeva, V., & Wah, C. L. (2022). Challenges in the manufacture, storage, distribution and regulation of traditional and novel vaccines. *Generics and Biosimilars Initiative Journal (GaBI Journal)*, 11(1), 13–24. https://doi.org/10.5639/gabij.2022.1101.004

Hock, S. C., Wah, C. L., & Kian, S. M. (2020). Global challenges in the manufacture, regulation and international harmonization of GMP and quality standards for biopharmaceuticals. *Generics and Biosimilars Initiative Journal (GaBI Journal)*, 9(2), 52–63. https://doi.org/10.5639/gabij.2020.0902.010

Hoffman, K. C. & Dicks, A. P. (2023). Incorporating the United Nations sustainable development goals and green chemistry principles into high school curricula. *Green Chemistry Letters and Reviews*, 16(1), 108. https://doi.org/10.1080/17518253.2023.2185108

Hoffmann, F. & Kloas, W. (2010). An environmentally relevant endocrine-disrupting antiandrogen, vinclozolin, affects calling behavior of male Xenopus laevis. *Hormones and Behavior*, 58(4), 653–659.

Hofstein A. (2017). The role of laboratory in science teaching and learning. In K. S. Taber, & B. Akpan. *Science Education. New Directions in Mathematics and Science Education.* Sense Publishers, Rotterdam.

Hogan, D. & O'Flaherty, J. (2022). Exploring the nature and culture of science as an academic discipline: Implications for the integration of education for sustainable development. *International Journal of Sustainability in Higher Education*, 23(8), 120–147.

Holland, I. & Davies, J. A. (2020). Automation in the life science research laboratory. *Frontiers in Bioengineering and Biotechnology*, 8, 571777.

Holmes, J. & von Winterfeldt, D. (2025). Evaluating compliance with UN security council resolution 1540 on nonproliferation of weapons of mass destruction. *Risk Analysis*, 45(7), 1766–1797. https://doi.org/10.1111/risa.17697

Horak, I., Horn, S., & Pieters, R. (2021). Agrochemicals in freshwater systems and their potential as endocrine disrupting chemicals: A South African context. *Environmental Pollution (Barking, Essex: 1987)*, 268(Pt A), 115718.

Horn, S., Pieters, R., & Bohn, T. (2020). May agricultural water sources containing mixtures of agrochemicals cause hormonal disturbances? *Science of the Total Environment*, 711, 134862.

Horowitz, G. L., Zaman, Z., Blanckaert, N. J. et al. (2005). Modular analytics: A new approach to automation in the clinical laboratory. *Journal of Automated Methods & Management in Chemistry*, 2005, 8–25. https://doi.org/10.1155/jammc.2005.8

Hou, J., Gai, W., Cheng, W., & Deng, Y. (2021). Hazardous chemical leakage accidents and emergency evacuation response from 2009 to 2018 in China: A review. *Safety Science*, 135(3), 105101. https://doi.org/10.1016/j.ssci.2020.105101

Housni, H., Bendahhou, K., Tahiri, M., & Tahiri Jouti, N. (2023). Evaluation of chemical risks management during handling operations and determination of storage conditions in scientific laboratories. *Chemistry Africa*, 6(12), 1. https://doi.org/10.1007/s42250-022-00530-1

Hu, X., Chen, H., & Yu, M. (2020). Exploring the non-technical competencies for on-scene public health responders in chemical, biological, radiological, and nuclear emergencies: a qualitative study. *Public Health*, 183, 23–29. https://doi.org/10.1016/j.puhe.2020.04.015

Hu, Z., Yang, X., Ho, P. C., Chan, S. Y., Heng, P. W., Chan, E., ... & Zhou, S. (2021). Herb-drug interactions in clinical practice: A literature review. *Drugs*, *65*(9), 1239–1282.

Hua, W., Chen, J., Qin, Q., Wan, Z., & Song, L. (2021). Causation analysis and governance strategy for hazardous cargo accidents at ports: Case study of Tianjin Port's hazardous cargo explosion accident. *Marine Pollution Bulletin*, 173(Pt B), 113053. https://doi.org/10.1016/j.marpolbul.2021.113053

Huang, K., Chen, G., & Khan, F. (2022a). Vulnerability assessment method for domino effects analysis in chemical clusters. *Process Safety and Environmental Protection*, 164(9), 539–554. https://doi.org/10.1016/j.psep.2022.06.021

Huang, T., Liu, P., Lien, A. S. Y., Yang, S., Chang, H., & Yen, H. (2014). A nationwide population-based study of traditional Chinese medicine usage in children in Taiwan, *Complementary Therapies in Medicine*, Volume 22, Issue 3, 2014, pp 500–510. https://doi.org/10.1016/j.ctim.2014.04.002.

Huang, Y., Li, X., & Zhang, J. (2022b). Integrated pest management for stored traditional Chinese medicinal materials: A review of natural repellants and controlled atmosphere storage. *Frontiers in Pharmacology*, 13, 945123.

Huang, S. J., Amendola, L. M., & Sternen, D. L. (2022d). Variation among DNA banking consent forms: Points for clinicians to bank on. *Journal of Community Genetics*, 13(4), 389–397. https://doi.org/10.1007/s12687-022-00601-3

Huang, L., Wang, Y., & Li, X. (2023). Integrating artificial intelligence into the modernization of traditional Chinese medicine industry: A review. *Frontiers in Pharmacology*, *14*, 1181183.

Humphries, M., Myburgh, J., Campbell, R., Buah-Kwofie, A., & Combrink, X. (2021). Organochlorine pesticide bioaccumulation in wild Nile crocodile (Crocodylus niloticus) fat tissues: Environmental influences on changing residue levels and contaminant profiles. *Science of the Total Environment*, 753, 142068. https://doi.org/10.1016/j.scitotenv.2020.142068

Hungerbühler, K., Boucher, J. M., Pereira, C., Roiss, T., Scheringer, M., Hungerbühler, K., … & Scheringer, M. (2021). The role of legislation. In K. Hungerbühler, *Chemical Products and Processes: Foundations of Environmentally Oriented Design* (pp. 27–53). Springer, Cham.

Hunter, P. (2024). Security challenges by AI-assisted protein design: The ability to design proteins in silico could pose a new threat for biosecurity and biosafety. *EMBO Reports*, 25(5), 2168–2171. https://doi.org/10.1038/s44319-024-00124-7

Huyser, C. (2014). Prevention of infections in an ART laboratory: A reflection on simplistic methods. *Facts, Views and Vision in ObGyn*, 6(4), 231–234.

Iaiani, M., Moreno, V. C., Reniers, G., Tugnoli, A., & Cozzani, V. (2021). Analysis of events involving the intentional release of hazardous substances from industrial facilities. *Reliability Engineering and System Safety*, 212, 107593. https://doi.org/10.1016/j.ress.2021.107593

IBEF. (2025a). *Indian Pharmaceuticals Industry Report*. Ministry of Commerce and Industry, Government of India. https://www.ibef.org/industry/pharmaceutical-india

IBEF. (2025b). *Biotechnology Industry Report*. Ministry of Commerce and Industry, Government of India. https://www.ibef.org/industry/biotechnology-india

Ibrahim, A. & Harabawy, A. (2014). Sublethal toxicity of carbofuran on the African catfish *Clarias gariepinus*: Hormonal, enzymatic and antioxidant responses. *Ecotoxicology and Environmental Safety*, 106, 33–39.

Idris, N., Talib, O., & Razali, F. (2022). Strategies in mastering science process skills in science experiments: A systematic literature review. *Jurnal Pendidikan IPA Indonesia*, 11(1), 155–170. https://doi.org/10.15294/jpii.v11i1.32969

Igarashi, Y., Tateishi, S., Sawajima, T., Kikuchi, K., Kawasumi, M., Matsuoka, J., … & Mori, K. (2022). What is the role of occupational physicians in the workplace during the COVID-19 pandemic in Japan? A qualitative interview study. *BMC Health Services Research*, 22(1), 1294.

Igbashio, M. D., Idemudia, O. E., & Ochoyama, H. (2022). A review of hazards and precautionary measures in science and technology laboratories. *NIPES-Journal of Science and Technology Research*, 4(4), 551.

Igwama, G. T., Olaboye, J. A., Maha, C. C., Ajegbile, M. D., & Abdul, S. (2024). Integrating electronic health records systems across borders: Technical challenges and policy solutions. *International Medical Science Research Journal*, 4(7), 788–796.

Ikegami, K., Ando, H., Kurogi, K., & Ogami, A. (2023). Perceived workplace health support and severe psychological distress among Japanese workers: A prospective cohort study. *Journal of Occupational and Environmental Medicine*, 65(12), 992–997.

Ikranbegiin, R., Schmid, G., Hoos, D., Young, A., Della-Latta, P., Spearman, P., Ramos, A., Alemayehu, B., Achmetova, B., Nauryzova, G., & Albetkova, A. (2019). Challenges and solutions for instituting an efficient maintenance program for laboratory equipment in Central Asian, and developing world, countries. *BMC Public Health*, 19(Suppl 3), 476. https://doi.org/10.1186/s12889-019-6782-5

Ilinca, R., Chiriac, I. A., Luțescu, D. A., Ganea, I., Hristodorescu-Grigore, S., & Dănciulescu-Miulescu, R. E. (2023). Understanding the key differences between ISO 15189: 2022 and ISO 15189: 2012 for an improved medical laboratory quality of service. *Revista Romana de Medicina de Laborator*, 31(2), 77–82.

Imran, M., Najmi, A. K., Rashid, M. F., Tabrez, S., & Shah, M. A. (2013). Clinical research regulation in India-history, development, initiatives, challenges and controversies: Still long way to go. *Journal of Pharmacy and Bioallied Sciences*, 5(1), 2–9. https://doi.org/10.4103/0975-7406.106553

Imran, S. B. M., Mohanty, B. S., & Behera, A. R. (2025). Building a sustainable circular economy through waste management in India: A review. In M. Imran, S. B. M., Mohanty, & A. R. Behera, *Multidisciplinary Explorations: A Scholarly Compilation*. CIRS Publication, Patna, Bihar. https://doi.org/10.5281/ZENODO.15151671

Inam, E., Nwoke, I., Udosen, E., & Offiong, N. (2019). Ecological risks of phenolic endocrine disrupting compounds in an urban tropical river. *Environmental Science and Pollution Research*, 26(21), 21589–21597.

Institute of Medicine. (2002). *Biological Threats and Terrorism: Assessing the Science and Response Capabilities: Workshop Summary*. The National Academies Press, Washington, DC.

International Criminal Police Organization (INTERPOL). (2025). Pharmaceutical crime operations. Retrieved March 2025, from INTERPOL: https://www.interpol.int/en/Crimes/Illicit-goods/Pharmaceutical-crime-operations

International Organization for Standardization (ISO). (2021). Traditional Chinese medicine - Vocabulary - Part 1: General terms (ISO/TS 22316-1:2021).

International Union for Conservation of Nature (IUCN). (2020). The IUCN Red List of Threatened Species. Retrieved January 11 2025, from https://www.iucnredlist.org

Interpol. (2023). Global illicit medicines targeted by INTERPOL operation. Retrieved March 2025, from Interpol: https://www.interpol.int/en/News-and-Events/News/2023/Global-illicit-medicines-targeted-by-INTERPOL-operation

Iqbal, H. H., Siddique, A., Qadir, A., Ahmad, S. R., Liess, M., & Shahid, N. (2024). Human health and ecology at risk: A case study of metal pollution in Lahore, Pakistan. *Environmental Sciences Europe*, 36(1), 9.

Irima, F. K. (2023). Institutional Factors Influencing Implementation of Adult and Continuing Education in Adult Education Centers in Imenti North Sub-county, Meru, Kenya. Thesis, University of Nairobi. https://erepository.uonbi.ac.ke/handle/11295/164701

Isabel, J. M., Ellinger, P. J., & Griffin, D. M. (2024). *Laboratory Operations and Management: Foundations of Practice* (1st ed.). F.A. Davis Company, Philadelphia, PE.

Isere, E. E. & Omorogbe, N. E. (2024). Quality management in clinical and public health research: A panacea for minimising and eliminating protocol deviations in research operations. *Nigerian Medical Journal*, 65(3), 367–375.

Issahaku, G. R., Fischer, H. T., Appiah-Brempong, E., Opoku, D., & Hanefeld, J. (2024). Strategies to foster community engagement for epidemic and pandemic preparedness and response in sub-Saharan Africa: A scoping review protocol. *BMJ Open*, 14(6), e082757.

Iwen, P. C., Alter, R., Herrera, L. V., Sambol, A. R., Stiles, K. L., & Hinrichs, S. H. (2018). Laboratory processing of specimens. In A. Hewlett & A. K. Murthy, *Bioemergency Planning*. Springer, Cham. https://doi.org/10.1007/978-3-319-77032-1_6

Jafari, Y., Yin, M., Lim, C. et al. (2022). Effectiveness of infection prevention and control interventions, excluding personal protective equipment, to prevent nosocomial transmission of SARS-CoV-2: A systematic review and call for action. *Infection Prevention in Practice*, 4(1), 100192.

Jagadeeswaran, I. & Sriram, H. (2022). EU 1272/2008–Classification, labelling and packaging of substances and mixtures. In P. S. T. Shanmugam, *Medical Device Guidelines and Regulations Handbook* (pp. 261–295). Springer International Publishing, Cham.

Jagun, C. (2018). Strategies for Compliance with Government Regulations in a Pharmaceutical Company. Accessed on 06-07-2025 from: https://scholarworks.waldenu.edu/cgi/viewcontent.cgi?article=6631&context=dissertations

Jahan, S. & Singh, A. (2023). Causes and impact of industrial effluents on receiving water bodies: A review. *Malaysian Journal of Science and Advanced Technology*, 3(2), 111–121. https://doi.org/10.56532/mjsat.v3i2.144

Jayamani, J., Janardan, C. C., Appan, S. V., Kathamuthu, K., & Ahmed, M. E. (2022). A practical tool for risk management in clinical laboratories. *Cureus*, 14(12), e32774. https://doi.org/10.7759/cureus.32774

Jepson, P. C., Guzy, M., Blaustein, K., Sow, M., Sarr, M., Mineau, P., & Kegley, S. (2014). Measuring pesticide ecological and health risks in West African agriculture to establish an enabling environment for sustainable intensification. *Philosophical Transactions of the Royal Society B. Biological Sciences*, 369(1639), 20130491. https://doi.org/10.1098/rstb.2013.0491

Jernigan, D. B., Raghunathan, P. L., Bell, B. P. et al. (2002). Investigation of bioterrorism-related Anthrax, United States, 2001: Epidemiologic findings. *Emerging Infectious Diseases*, 8(10), 1019–1028. https://doi.org/10.3201/eid0810.020353

Jia, C., Hew, K. F., Bai, S., & Huang, W. (2022). Adaptation of a conventional flipped course to an online flipped format during the Covid-19 pandemic: Student learning performance and engagement. *Journal of Research on Technology in Education*, 54(2), 281–301.

Jia, Z., Nourian, P., Luscuere, P., & Wagenaar, C. (2023). Spatial decision support systems for hospital layout design: A review. *Journal of Building Engineering*, 67, 106042.

Jiang, Y., Nan, Z., & Yang, S. (2013). Risk assessment of water quality using Monte Carlo simulation and artificial neural network method. *Journal of Environmental Management*, 122, 130–136. https://doi.org/10.1016/j.jenvman.2013.03.015.

Jiao, Y., Yeophantong, P., & Lee, T. M. (2021). Strengthening international legal cooperation to combat the illegal wildlife trade between Southeast Asia and China. *Frontiers in Ecology and Evolution*, 9, 2021. https://doi.org/10.3389/fevo.2021.645427

Jiarong, F. & Geng, L. (2014). The control and treatment of wastes from college biology laboratories. *Journal of Huangshan University*, 2014, 5.

Johnson, J., Nkomozepi, P., Opute, P., & Mbajiorgu, E. (2021). Cardiac and cerebellar histomorphology and inositol 1,4,5-trisphosphate (IP3R) perturbations in adult xenopus laevis following atrazine exposure. *Applied Sciences-Basel*, 11(21), 10006.

Jonai, H., Ono, M., Hirachi, K., Tanaka, M., Sohara, H., & Umemura, Y. (2021). Transition of chemical management in Japan-Shift to self-regulation and measures for small businesses. *Industrial Health*, 59(5), 298–307.

Jonnalagadda, R., Singh, P., Gogineni, A., Reddy, R. R. S., & Reddy, H. B. (2022). Developing, implementing and evaluating training for online graduate teaching assistants based on Addie Model. *Asian Journal of Education and Social Studies*, 28(1), 1–10.

Joseph, T. (2021). Management system approach for addressing biosafety and biosecurity of emerging pathogens in a biosafety level-3 core facility. *Applied Biosafety*, 26(4), 210–220. https://doi.org/10.1089/apb.2021.0007

Joubert, B. R., Mantooth, S. N., & McAllister, K. A. (2020). Environmental health research in Africa: Important progress and promising opportunities. *Frontiers in Genetics*, 10, 1166. https://doi.org/10.3389/fgene.2019.01166.

Kamau, E. M. (2013). Navigating laboratory services quality in challenging environments: A perspective for implementation in small, low-income countries and post-conflict settings. *African Journal of Laboratory Medicine*, 2(1), 48. https://doi.org/10.4102/ajlm.v2i1.48

Kang, E. & Hwang, H. J. (2023). The importance of anonymity and confidentiality for conducting survey research. *Journal of Research and Publication Ethics*, 4(1), 1–7. https://doi.org/10.15722/jrpe.4.1.202303.1

Kanwar, A. S., Carr, A., Ortlieb, K., & Mohee, R. (2020). Opportunities and challenges for campus-based universities in Africa to translate into dual-mode delivery. *Exploring Dual and Mixed Mode Provision of Distance Education*, 39(1), 8–26.

Kapoor, A. & Mehendale, S. (2021). Challenges in implementation of green chemistry in Indian pharmaceutical sector. *Journal of Pharmaceutical Research International*, 33, 160–168.

Karthikeyan, L., Suresh, V. M., Krishnan, V., Tudor, T., & Varshini, V. (2018). The management of hazardous solid waste in India: An overview. *Environments*, 5(9), 9.

Karungamye, P., Rugaika, A., Mtei, K., & Machunda, R. (2022). The pharmaceutical disposal practices and environmental contamination: A review in East African countries. *HydroResearch*, 5, 99–107.

Kasambala, H., Rwiza, M., & Mdegela, R. (2019). Levels and distribution of progesterone in receiving waters and wastewaters of a growing urban area. *Water Science and Technology*, 80(6), 1107–1117.

Kasonga, T., Coetzee, M., Kamika, I., & Momba, M. (2022). Assessing a co-culture fungal granule ability to remove pharmaceuticals in a sequencing batch reactor. *Environmental Technology*, 43(11), 1684–1699.

Kasonga, T., Coetzee, M., Van Zijl, C., & Momba, M. (2019). Removal of pharmaceutical' estrogenic activity of sequencing batch reactor effluents assessed in the T47D-KBluc reporter gene assay. *Journal of Environmental Management*, 240, 209–218.

Kassotis, C. D., Vandenberg, L. N., Demeneix, B. A., Porta, M., Slama, R., & Trasande, L. (2020). Endocrine-disrupting chemicals: Economic, regulatory, and policy implications. *The Lancet Diabetes & Endocrinology*, 8(8), 719–730.

Kato, S. & Kansha, Y. (2024). Comprehensive review of industrial wastewater treatment techniques. *Environmental Science and Pollution Research*, 31(39), 51064–51097.

Kättström, D., Beronius, A., Boije af Gennäs, U., Rudén, C., & Ågerstrand, M. (2024). Fulfilling the criteria for CLP classification: the implications for substances under the EU chemicals legislation. *Frontiers in Toxicology*, 6, 1452065.

Kaur, R., Downey, C., & Choate, R. (2018). *Promoting Laboratory Quality Management Programs in Cuenca.* Worcester Polytechnic Institute, Ecuador.

Kawamoto, T. & Kawamoto, K. (2021). Preparation of thin frozen sections from nonfixed and undecalcified hard tissues using Kawamoto's film method (2020). *Skeletal Development and Repair: Methods and Protocols*, 1130, 259–281.

Kelle, A. (2021). The CWC at 25: From verification of chemical-weapons destruction to attribution of their use. *The Nonproliferation Review*, 28(4–6), 319–336.

Kelley, E. W. (2021). LAB theory, HLAB pedagogy, and review of laboratory learning in chemistry during the COVID-19 pandemic. *Journal of Chemical Education*, 98(8), 2496–2517.

Kennepohl, D. (Ed.). (2023). *Teaching Science Online: Practical Guidance for Effective Instruction and Lab Work* (1st ed.). Routledge, New York.

Kenya Chemical Society (KCS) and Sandia National Laboratories (SNL). (2023). Securing Highly Toxic Agrochemicals and Rudimentary Chemical Materials in Somalia Through a Train-the-Trainer Engagement and Outreach Campaigns. Phase 2 - Micro-Grants for Mini-Workshops and Outreach Campaign. Sponsored by Chemical Security Program (CSP). Internal Report, Somalia.

Kenya Chemical Society (KCS) and Sandia National Laboratories (SNL). (2024). Chemical Inventory Management Training for Kenya Chemical Suppliers/Procedural Documentation Development and Chemical Inventory Management for Kenya Chemical Suppliers. Sponsored by Chemical Security Program. Internal Report.

Kenya Chemical Society (KCS) and Sandia National Laboratories (SNL). (2022a). Chemical Security Risk Assessment for Somalia Academia and Small-to-Medium Industry and Physical Security Upgrades Grants (PSUG). Sponsored by Chemical Security Program (CSP). Internal Report.

Kenya Chemical Society (KCS) and Sandia National Laboratories (SNL). (2022b). Mitigating Risk of Pharmaceutical-Based Agents in Somalia Through Sustainable Partnerships with Pharmaceutical Industry Stakeholders. Internal Report.

Kenya Chemical Society (KCS). (2022a). Follow-On Workshop on Chemical Risk Assessment for Pharmaceutical Manufacturers, Distributors, and Suppliers in Kenya. Sponsored by Chemical Security Program (CSP). Internal Report.

Kenya Chemical Society (KCS). (2022b). Workshop on Curriculum Development for Pharmaceutical and Agricultural Training Institutions to Mainstream Chemical Security. Sponsored by Chemical Security Program (CSP). Internal Report.

Kenya Chemical Society (KCS). (2023). Unregulated Chemical Industries and Informal Markets Workshops and Door to Door Campaigns for Mombasa. Kenya Chemical Society, Nairobi.

Kenyan Pharmacy and Poisons Board (PPB). (2017). Seizure of Counterfeit Drugs. Retrieved March 2025, from https://www.ppb.go.ke.

Key, T., Mathai, N. J., Venkatesan, A. S., Farnell, D., & Mohanty, K. (2020). Personal protective equipment during the COVID-19 crisis: A snapshot and recommendations from the frontline of a university teaching hospital. *Bone & Joint Open*, 1(5), 131–136.

Keyter, A., Banoo, S., Salek, S., & Walker, S. (2018). The South African regulatory system: Past, present, and future. *Frontiers in Pharmacology*, 9, 1407. https://doi.org/10.3389/fphar.2018.01407.

Khan, S. & Malik, A. (2014). Environmental and health effects of textile industry wastewater. In A. Malik, E. Grohmann, & R. Akhtar, *Environmental Deterioration and Human Health* (pp. 55–71). Springer, Netherlands.

Khan, W. U., Ahmed, S., Dhoble, Y., & Madhav, S. (2023). A critical review of hazardous waste generation from textile industries and associated ecological impacts. *Journal of the Indian Chemical Society*, 100(1), 100829.

Khan, R., Sanyal, D., & Ahmed, M. Z. (2024). Seasonal variation, spatial distribution, and health risk assessment of potentially toxic elements in surface water and sediment of the highly polluted Old Brahmaputra River, Bangladesh. *Science of The Total Environment*, *958*, 169974.

Khanizade, A., Moslehi, S., Dowlati, M., Moradimajd, P., & Moradian, M. J. (2025). Preparedness dimensions and components of emergency medical services in chemical hazards: A systematic review. *BMC Emergency Medicine*, 25(1), 24.

Khankhoje, R. (2023). An in-depth review of test automation frameworks: Types and trade-offs. *International Journal of Advanced Research in Science, Communication and Technology (IJARSCT)*, 3(1), 55–64.

Khatri, R. B., Endalamaw, A., Erku, D., Wolka, E., Nigatu, F., Zewdie, A., & Assefa, Y. (2023). Preparedness, impacts, and responses of public health emergencies towards health security: Qualitative synthesis of evidence. *Archives of Public Health*, 81(1), 208.

Khattar, R. B. & Nehme, M. E. (2024). Emergence and evolution of standardization systems in medical biology laboratories. *Advances in Laboratory Medicine/Avances en Medicina de Laboratorio*, 5(3), 261–267.

Khosroniya, M., Hosnavi, R., & Zahedi, M. R. (2024). Enhancing operational performance in industry 4.0: The mediating role of total quality management and total productive maintenance at Zarharan industrial complex. *International Journal of Industrial Engineering and Operational Research*, 6(1), 96–122.

Kibambe, M., Momba, M., Daso, A., Van Zijl, M., & Coetzee, M. (2020). Efficiency of selected wastewater treatment processes in removing estrogen compounds and reducing estrogenic activity using the T47D-KBLUC reporter gene assay. *Journal of Environmental Management*, 260, 110135. https://doi.org/10.1016/j.jenvman.2020.110135

Kiechle, F. L. & Arcenas, R. C. (2016). Utilization management in a large community hospital. *Utilization Management in the Clinical Laboratory and Other Ancillary Services*, 1, 151–170. https://doi.org/10.1007/978-3-319-34199-6_14

Kiige, M. J. & Atina, J. O. (2016). The effectiveness of SMASSE teacher training programme on KCSE performance in mathematics and chemistry subjects in Kikuyu District, Kenya. *Journal of Education and Practice*, 7(6), 1735.

Kilian, E., Delport, R., Bornman, M., & de Jager, C. (2007). Simultaneous exposure to low concentrations of dichloro-diphenyltrichloroethane, deltamethrin, nonylphenol and phytoestrogens has negative effects on the reproductive parameters in male Spraque-Dawley rats. *Andrologia*, 39(4), 128–135.

Kim, J., Yang, S., Moodie, E., Obida, M., Bornman, R., Eskenazi, B., & Chevrier, J. (2022). Prenatal exposure to insecticides and weight trajectories among South African children in the VHEMBE birth cohort. *Epidemiology*, 33(4), 505–513.

Kirechu, P. (2020). Al-Shabaab's improvised explosive device supply chain gambit in Somalia. Retrieved March 2025, from https://warontherocks.com/2020/09/al-shabaabs-improvised-explosive-device-supply-chain-gambit-in-somalia/

Kirillova, V., Ahmad, R., & Wilden, R. (2024). Assuring safe use of chemicals: An overview of the regulatory developments, challenges, and outlook for The EU. *International Chemical Regulatory and Law Review*, 7, 2.

Kleine-Wechelmann, S., Bastiaanse, K., Freundel, M., & Becker-Asano, C. (2023). Designing the mobile robot Kevin for a life science laboratory. In *Conference: 2022 31st IEEE International Conference on Robot and Human Interactive Communication (RO-MAN)*, Naples, Italy. https://doi.org/10.1109/RO-MAN53752.2022.9900786

Knezevic, C. E., Das, B., El-Khoury, J. M., Jannetto, P. J., Lacbawan, F., & Winter, W. E. (2022). Rising to the challenge: Shortages in laboratory medicine. *Clinical Chemistry*, 68(12), 1486–1492.

Kocak, N. (2019). The level of knowledge and opinions of science teacher candidates on safety measurements in general chemistry laboratory studies. *Journal of Education and Training Studies*, 7(11), 4378. https://doi.org/10.11114/jets.v7i11.4378

Kolil, V. K. & Achuthan, K. (2024). Virtual labs in chemistry education: A novel approach for increasing student's laboratory educational consciousness and skills. *Education and Information Technologies*, 29(18), 1–25.

Korde, A., Patt, M., Selivanova, S. V. et al. (2024). Position paper to facilitate patient access to radiopharmaceuticals: Considerations for a suitable pharmaceutical regulatory framework. *EJNMMI Radiopharmacy and Chemistry*, 9(1), 2. https://doi.org/10.1186/s41181-023-00230-2

Kortenkamp, A., Martin, O. V., Faust, M. et al. (2011). State of the art assessment of endocrine disrupters. Final report. European Commission, Directorate-General for the Environment (Project Contract No. 070307/2009/550687/SER/D3).

Korucu-Kış, S. (2021). Preparing student teachers for real classrooms through virtual vicarious experiences of critical incidents during remote practicum: A meaningful-experiential learning perspective. *Education and Information Technologies*, 26(6), 6949–6971.

Kothinti, R. R. (2024). Deep learning in healthcare: Transforming disease diagnosis, personalized treatment, and clinical decision-making through AI-driven innovations. *World Journal of Advanced Research and Reviews*, 24(02), 2841–2856.

Krausmann, E., Girgin, S., & Necci, A. (2019). Natural hazard impacts on industry and critical infrastructure: Natech risk drivers and risk management performance indicators. *International Journal of Disaster Risk Reduction*, 40, 101163.

Kravtsova, A., Zinicovscaia, I., Peshkova, A., Nekhoroshkov, P., Cepoi, L., Chiriac, T., & Rudi, L. (2024). Impact of metal-containing industrial effluents on leafy vegetables and associated human health risk. *Foods*, 13(21), 3420.

Krishan, U. (2021). Laboratory safety and prevention of laboratory associated infections. *Prevention of Healthcare Associated Infections: Infection Prevention and Control*, 2021, 335.

Kruger, T., Barnhoorn, I., van Vuren, J., & Bornman, R. (2013). The use of the urogenital papillae of male feral African sharptooth catfish (*Clarias gariepinus*) as indicator of exposure to estrogenic chemicals in two polluted dams in an urban nature reserve, Gauteng, South Africa. *Ecotoxicology and Environmental Safety*, 87, 98–107.

Kruid, J., Fogel, R., & Limson, J. (2018). Unsubstituted metallophthalocyanine catalysts for the removal of endocrine disrupting compounds using H_2O_2 as oxidant. *Environmental Science and Pollution Research*, 25(32), 32346–32357.

Krutzsch, W. & Trapp, R. (2023). *A Commentary on the Chemical Weapons Convention*. Brill - Nijhoff, Leiden, Netherlands.

Kumar Badjatya, J., Bodla, R., & Musyuni, P. (2013). Export registration of pharmaceuticals in rest of world countries (row). *Journal of* Drug Delivery and Therapeutics, 3(1), 391. https://jddtonline.info/index.php/jddt/article/view/391

Kumar, A., Thakur, A. K., Gaurav, G. K., Klemeš, J. J., Sandhwar, V. K., Pant, K. K., & Kumar, R. (2023). A critical review on sustainable hazardous waste management strategies: A step towards a circular economy. *Environmental Science and Pollution Research International*, 30(48), 105030–105055.

Kumar, K. S. R., Mufti, S. S., Sarathy, V., Hazarika, D., & Naik, R. (2021a). An update on advances in COVID-19 laboratory diagnosis and testing guidelines in India. *Frontiers in Public Health*, 9, 568603. https://doi.org/10.3389/fpubh.2021.568603

Kumar, P., Kausar, M. A., Singh, A. B., & Singh, R. (2021b). Biological contaminants in the indoor air environment and their impacts on human health. *Air Quality, Atmosphere & Health*, 14(11), 1723–1736.

Kumar, P., Maurya, P., & Henly, S. (2021). Regulatory framework on chemicals in India: Latest developments and challenges. *International Chemical Regulatory and Law Review*, 4(3), 101–106.

Kumar, S., Chhabra, G., Sehrawat, K. S., & Singh, M. (2024). Developing a competency assessment framework for medical laboratory technologists in primary healthcare settings in India. *PLoS One*, 19(4), e0294939. https://doi.org/10.1371/journal.pone.0294939

Kumar, S., Smith, S. R., Fowler, G., Velis, C., Kumar, S. J., Arya, S., Kumar, R., & Cheeseman, C. (2017). Challenges and opportunities associated with waste management in India. *Royal Society Open Science*, 4(3), 160764. https://doi.org/10.1098/rsos.160764

Kumra, S. (2025). Need for technical standards in India's chemical sector: Reflections on ICMSR and the role of GHS. Indian Chemical News. https://www.indianchemicalnews.com/

Lahori, A. H., Tunio, M., Ahmed, S. R., Mierzwa-Hersztek, M., Vambol, V., Afzal, A., Kausar, A., Vambol, S., Umar, A., & Muhammad, A. (2024). Role of pressmud compost for reducing toxic metals availability and improving plant growth in polluted soil: Challenges and recommendations. *Science of the Total Environment*, 951, 175493.

Lamare, R. E., & Singh, O. P. (2016). Limestone mining and its environmental implications in Meghalaya, India. *ENVIS Bulletin on Himalayan Ecology*, 24, 87–100.

Lameras, P., Arnab, S., De Freitas, S., Petridis, P., & Dunwell, I. (2021). Science teachers' experiences of inquiry-based learning through a serious game: A phenomenographic perspective. *Smart Learning Environments*, 8(1), 7.

Landucci, G., Tonelli, J., Tortolini, A., Marroni, G., Pepe, E., & Tamburini, F. (2022). Attractiveness assessment of chemical facilities operating in critical areas: Development of a semi-quantitative methodology tailored to the Maghreb context. *Chemical Engineering Transactions*, 90, 421–426.

Lara, E. R., De la Rosa, J. R., Castillo, A. I. R., de Jesús Cerino-Córdova, F., Chuken, U. J. L., Delgadillo, S. S. F., & Rivas-García, P. (2017). A comprehensive hazardous waste management program in a Chemistry School at a Mexican university. *Journal of Cleaner Production*, 142, 1486–1491.

Le, H. T., Quoc, K. L., Nguyen, T. A., Dang, K. T., Vo, H. K., Luong, H. H., ... & Duong-Trung, N. (2022). Medical-waste chain: A medical waste collection, classification and treatment management by blockchain technology. *Computers*, 11(7), 113.

Lee, B. K., Ellenbecker, M. J., & Moure-Ersaso, R. (2004). Alternatives for treatment and disposal cost reduction of regulated medical wastes. *Waste Management*, 24(2), 143–151.

Lee, H. E., Kim, M.-G., Yoon, S. J., Huh, D.-A., & Moon, K.-W. (2021). Changes in risk in medium business plating and paint manufacturing plants following the revision of the Korean chemical accident prevention system. *International Journal of Environmental Research and Public Health*, 18(22), 11982. https://doi.org/10.3390/ijerph182211982

Lekei, E. E., Ngowi, A. V., & London, L. (2014). Pesticide retailers' knowledge and handling practices in selected towns of Tanzania. *Environmental Health*, 13, 79. https://doi.org/10.1186/1476-069X-13-79

Lellis, B., Fávaro-Polonio, C. Z., Pamphile, J. A., & Polonio, J. C. (2019). Effects of textile dyes on health and the environment and bioremediation potential of living organisms. *Biotechnology Research and Innovation*, 3(2), 275–290.

Lerch, M., Kenerson, H., Theiss, A., Chafin, D., Westerhoff, M., Otter, M., Yeung, R., & Baird, G. (2019). Rapid tissue processing using a temperature-controlled collection device to preserve tumor biomarkers. *Cell Tissue Bank*, 21(1), 89–97. https://doi.org/10.1007/s10561-019-09800-8

Lestari, R. R. & Pratama, A. P. (2023). Hubungan Pengetahuan, Pendidikan Dan Pelatihan Dengan Tingkat Kepatuhan Penggunaan Alat Pelindung Diri (Apd) Pada Petugas Penyapu Jalan di Kecamatan Bangkinang Kota Tahun 2018. *Prepotif: Jurnal Kesehatan Masyarakat*, 3(1), 1–10. https://doi.org/10.31004/prepotif.v3i1.387

Levine, C. & Marshall, K. (2024). Planning for the Next Infectious Disease Threat: Considerations to Ensure Every Facility is Prepared. (Report No. IHS/CR-2024-1020). The Sam Houston State University Institute for Homeland Security. https://hdl.handle.net/20.500.11875/4880

Li, C., Dos Reis, A. A., Ansari, A., Bertelli, L., Carr, Z., Dainiak, N., ... & Zhang, J. (2022a). Public health response and medical management of internal contamination in past radiological or nuclear incidents: A narrative review. *Environment International*, 163, 107222. https://doi.org/10.1016/j.envint.2022.107222

Li, J., Zhang, N., & Qian, W. (2020). Clinical efficacy and safety of traditional Chinese medicine for chronic diseases: A systematic review. *Evidence-Based Complementary and Alternative Medicine*, 2020, 8860376.

Li, J., Zhang, N., & Qian, W. (2023). Integrating traditional knowledge and modern science for the safety assessment of Chinese herbal medicine. *Phytomedicine*, 107, 154485.

Li, S. (1578/2021). *Bencao Gangmu: Compendium of Materia Medica.* People's Medical Publishing House, China.

Li, S. L., Lin, G., & Han, T. (2019). Herbalome: A new perspective for Chinese herbal medicine research. *Chinese Medicine*, 14(1), 1–14.

Li, H., Wang, Y., & Zhang, T. (2022b). Modernization of traditional Chinese medicine: Quality control of cultivation and processing. *Frontiers in Pharmacology*, *13*, **956264.**

Li, X., Shen, X., Jiang, W., Xi, Y., & Li, S. (2024). Comprehensive review of emerging contaminants: Detection technologies, environmental impact, and management strategies. *Ecotoxicology and Environmental Safety*, 278, 116420. https://doi.org/10.1016/j.ecoenv.2024.116420

Liang, H. Y., Yan, T., & Zhao, W. W. (2024). Comprehensive assessment of recent major chemical accidents in China and path to sustainable solutions. *Smart Construction and Sustainable Cities*, 2(1), 24. https://doi.org/10.1007/s44268-023-00024-0

Liang, Z., Lai, Y., Li, M., Shi, J., Lei, C. I., Hu, H., & Ung, C. O. L. (2023). Applying regulatory science in traditional Chinese medicines for improving public safety and facilitating innovation in China: A scoping review and regulatory implications. *Chinese Medical Journal*, 16(1), 23. https://doi.org/10.1186/s13020-021-00433-2.

Liao, B., Zeng, X., Ling, Z., Zhao, S., Li, B., & Han, X. (2025). Recent advances in zero discharge treatment technologies for desulfurization wastewater in coal-fired power plants: A mini-review. *Processes*, 13(4), 982.

Liao, M. & Helen, H. (2021). An analysis of the relationship between WHMIS certification and holder's knowledge after being certified. *BCIT Environmental Public Health Journal*, 2021, 181. https://doi.org/10.47339/ephj.2021.181

Lieberman, N. A., Mathias, P. C., Bradley, B. T., & Greninger, A. L. (2022). Clinical performance and trends during the first two months of monkeypox virus PCR testing at two United States reference labs. *Journal of Clinical Microbiology*, 60(12), e01371–22.

Lippi, G. & Plebani, M. (2020). The critical role of laboratory medicine during coronavirus disease 2019 (COVID-19) and other viral outbreaks. *Clinical Chemistry and Laboratory Medicine (CCLM)*, 58(7), 1063–1069.

Lippi, G., Adeli, K., Ferrari, M. et al. (2020). Biosafety measures for preventing infection from COVID-19 in clinical laboratories: IFCC taskforce recommendations. *Clinical Chemistry and Laboratory Medicine (CCLM)*, 58(7), 1053–1062.

Liu, Y., Li, F., & Zuo, J. (2020). Historical evolution and modern development of traditional Chinese medicine. *Frontiers in Pharmacology*, 11, 584.

Lo, Y. C., Dooyema, C. A., Neri, A., Durant, J., Jefferies, T., Medina-Marino, A., & Brown, M. J. (2010). Childhood lead poisoning associated with gold ore processing: A village-level investigation—Zamfara State, Nigeria. *Environmental Health Perspectives*, 120(10), 1450–1455. https://doi.org/10.1289/ehp.1104793

Lombardi, D., Shipley, T. F., & Astronomy Team, Biology Team, Chemistry Team, Engineering Team, Geography Team, Geoscience Team, and Physics Team. (2021). The curious construct of active learning. *Psychological Science in the Public Interest*, 22(1), 8–43.

Lu, Y. & Ahmad, Y. (2023). Heritage protection perspective of sustainable development of traditional villages in Guangxi, China. *Sustainability*, 15(4), 3387. https://doi.org/10.3390/su15043387

Luardo, F. Q. M., Alvarado, K. B. C., Alcontin, W. N. et al. (2024). Musculoskeletal disorders in construction labourers: Health risks and mitigation strategies using ergonomic cement trolleys. *European Journal of Modern Medicine and Practice*, 4(6), 487–507.

Luqman, A., Khan, M., & Rahman, A. (2024). Top 10 public health challenges for 2024: Charting a new direction for global health security. *Public Health Challenges*, *3*(1), e70022.

Luken, R. A. (2009). Equivocating on the polluter-pays principle: The consequences for Pakistan. *Journal of Environmental Management*, 90(11), 3479–3484.

Lwande, C., Muchemi, L., & Oboko, R. (2021). Identifying learning styles and cognitive traits in a learning management system. *Heliyon*, 7(8), e07701.

Ma, L., Ma, X., Zhang, J., Yang, Q., & Wei, K. (2021). A methodology for dynamic assessment of laboratory safety by SEM-SD. *International Journal of Environmental Research and Public Health*, 18(12), 6545. https://doi.org/10.3390/ijerph18126545

Maciocia, G. (2019). *The Foundations of Chinese Medicine: A Comprehensive Text for Acupuncturists and Herbalists.* Elsevier Health Sciences, Amsterdam, Netherlands.

Mackey, T. K., Liang, B. A., York, P., & Kubic, T. (2015). Counterfeit drug penetration into global legitimate medicine supply chains: A global assessment. *American Journal of Tropical Medicine and Hygiene*, 92(6 Suppl), 59–67. https://doi.org/10.4269/ajtmh.14-0389. Epub 2015 Apr 20.

Madhav, N., Oppenheim, B., Gallivan, M. et al. (2017). Pandemics: Risks, impacts, and mitigation. In D. T. Jamison, H. Gelband, S. Horton, et al., *Disease Control Priorities: Improving Health and Reducing Poverty* (3rd ed.). The International Bank for Reconstruction and Development/The World Bank, Washington, DC. https://doi.org/10.1596/978-1-4648-0527-1_ch17. https://www.ncbi.nlm.nih.gov/books/NBK525302/

Madigan, D. J., Kim, L. E., & Glandorf, H. L. (2024). Interventions to reduce burnout in students: A systematic review and meta-analysis. *European Journal of Psychology of Education*, 39(2), 931–957.

Madjar, R. M., Vasile Scăețeanu, G., & Sandu, M. A. (2024). Nutrient water pollution from unsustainable patterns of agricultural systems, effects and measures of integrated farming. *Water*, 16(21), 3146.

Mahdavi, V., Eslami, Z., Molaee-Aghaee, E., Peivasteh-Roudsari, L., Sadighara, P., Thai, V. N., ... & Ravanlou, A. A. (2022). Evaluation of pesticide residues and risk assessment in apple and grape from western Azerbaijan Province of Iran. *Environmental Research*, 203, 111882. https://doi.org/10.1016/j.envres.2021.111882

Mahmood, S. U., Crimbly, F., Khan, S., Choudry, E., & Mehwish, S. (2020). Strategies for rational use of personal protective equipment (PPE) among healthcare providers during the COVID-19 crisis. *Cureus*, 12(5), e8248.

Mahmud, A. F. (2022). *Development of Safety Self-Audit Mechanism for Chemical Laboratories in Higher Learning Institute: A Case Study in Universiti Malaya*. Universiti Malaya, Malaya.

Mahomed, S., Voyi, K., Aneck-Hahn, N., & de Jager, C. (2008). Oestrogenicity and chemical target analysis of water from small-sized industries in Pretoria, South Africa. *Water SA*, 34(3), 357–363.

Majlesi, M., Khaloei, A., & Rabouri, M. M. (2015). Evaluation of safety standards in medical diagnostic laboratory in Kerman in 2014. *Safety Promotion and Injury Prevention (Tehran)*, 3(2), 83–90.

Makene, V. & Pool, E. (2019). The effects of endocrine disrupting chemicals on biomarkers of inflammation produced by lipopolysaccharide stimulated RAW264.7 macrophages. *International Journal of Environmental Research and Public Health*, 16(16), 2914.

Makgato, M. (2012). Identifying constructivist methodologies and pedagogic content knowledge in the teaching and learning of technology. *Procedia - Social and Behavioral Sciences*, 47, 1398–1402.

Mallawaarachchi, H., Karunasena, G., Sandanayake, Y., & Liu, C. (2023). Conceptualising a model to assess the optimum water flow of industrial symbiosis (IS). *Sustainability*, 15(11), 8627.

Mallonee, S., Shariat, S., Stennies, G., Waxweiler, R., Hogan, D., & Jordan, F. (1996). Physical injuries and fatalities resulting from the Oklahoma City bombing. *JAMA*, 276(5), 382–387. https://doi.org/10.1001/jama.1996.03540050042021

Mandal, G., Ansari, M., & Aswal, D. K. (2021). Quality management system at NPLI: Transition of ISO/IEC 17025 from 2005 to 2017 and implementation of ISO 17034: 2016. *Mapan*, 36(3), 657–668. https://doi.org/10.1007/s12647-021-00490-w

Mani, V., Prakash, M., & Lai, W. C. (2022). Cloud-based blockchain technology to identify counterfeits. *Journal of Cloud Computing*, 11(1), 67. https://doi.org/10.1186/s13677-022-00341-2.

Mani, Z., Plummer, V., Kuhn, L., Khorram-Manesh, A., Tin, D., & Goniewicz, K. (2024). Public health responses to CBRN terrorism in the Middle East and North Africa. *Disaster Medicine and Public Health Preparedness*, 18, e87.

Manickum, T. & John, W. (2014). Occurrence, fate and environmental risk assessment of endocrine disrupting compounds at the wastewater treatment works in Pietermaritzburg (South Africa). *Science of the Total Environment*, 468, 584–597.

Mantle, J. L., Rammohan, J., Romantseva, E. F., Welch, J. T., Kauffman, L. R., McCarthy, J., Schiel, J., Baker, J. C., Strychalski, E. A., Rogers, K. C., & Lee, K. H. (2019). Cyberbiosecurity for biopharmaceutical products. *Frontiers in Bioengineering and Biotechnology*, 7, 116. https://doi.org/10.3389/fbioe.2019.00116.

Manyilizu, M. C. (2023). Effectiveness of virtual laboratory vs. paper-based experiences to the hands-on chemistry practical in Tanzanian secondary schools. *Education and Information Technologies*, 28, 4831–4848. https://doi.org/10.1007/s10639-022-11327-7

Marcelino, S., Hamdane, S., Gaspar, P. D., & Paço, A. (2023). Sustainable agricultural practices for the production of medicinal and aromatic plants: Evidence and recommendations. *Sustainability*, 15(19), 14095. https://doi.org/10.3390/su151914095

Marchand, M., Pieterse, G., & Barnhoorn, I. (2008). Preliminary results on sperm motility and testicular histology of two feral fish species, *Oreochromis mossambicus* and *Clarias gariepinus*, from a currently DDT-sprayed area, South Africa. *Journal of Applied Ichthyology*, 24(4), 423–429.

Marchand, M., Pieterse, G., & Barnhoorn, I. (2010). Sperm motility and testicular histology as reproductive indicators of fish health of two feral fish species from a currently DDT sprayed area, South Africa. *Journal of Applied Ichthyology*, 26(5), 707–714.

Marchand, M., van Dyk, J., Pieterse, G., Barnhoorn, I., & Bornman, M. (2009). Histopathological alterations in the liver of the sharptooth catfish *Clarias gariepinus* from polluted aquatic systems in South Africa. *Environmental Toxicology*, 24(2), 133–147.

Marescotti, D., Narayanamoorthy, C., Bonjour, F., Kuwae, K., Graber, L., Calvino-Martin, F., Ghosh, S., & Hoeng, J. (2021). AI-driven laboratory workflows enable operation in the age of social distancing. *SLAS Technology*, 27(3), 195–203. https://doi.org/10.1016/j.slast.2021.12.001

Maritano, V., Barge, P., Biglia, A. et al. (2024). Anticounterfeiting and fraud mitigation solutions for high-value food products. *Journal of Food Protection*, 87(4), 100251. https://doi.org/10.1016/j.jfp.2024.100251

Marks, B. & Thomas, J. (2022). Adoption of virtual reality technology in higher education: An evaluation of five teaching semesters in a purpose-designed laboratory. *Education and Information Technologies*, 27(1), 1287–1305. https://doi.org/10.1007/s10639-021-10653-6

Marron, J. (2024), *Implementing the Health Insurance Portability and Accountability Act (HIPAA) Security Rule A Cybersecurity Resource Guide, Special Publication (NIST SP)*. National Institute of Standards and Technology, Gaithersburg, MD. https://doi.org/10.6028/NIST.SP.800-66r2

Martin, J., Rodriguez-Gomez, R., Zafra-Gomez, A., Alonso, E., Vilchez, J., & Navalon, A. (2016). Polar stir bars for isolation and preconcentration of perfluoroalkyl substances from human milk samples prior to UHPLC-MS/MS analysis. *Bioanalysis*, 8(7), 633–647.

Martins, F. & Ribeiro, M. H. L. (2025). Quality and regulatory requirements for the manufacture of master cell banks of clinical grade iPSCs: The EU and USA perspectives. *Stem Cell Reviews and Reports*, 21(3), 645–679. https://doi.org/10.1007/s12015-024-10838-9

Marwaha, J. S., Landman, A. B., Brat, G. A., Dunn, T., & Gordon, W. J. (2022). Deploying digital health tools within large, complex health systems: Key considerations for adoption and implementation. *NPJ Digital Medicine*, 5(1), 13.

Marx-Stoelting, P., Rivière, G., Luijten, M., Aiello-Holden, K., Bandow, N., Baken, K., … & Sanders, P. (2023). A walk in the PARC: Developing and implementing 21st century chemical risk assessment in Europe. *Archives of Toxicology*, 97(3), 893–908.

Mary Sadler, P. E., Stephanie Ishii, D., Tim Devine, P. E., & Sawyer, H. (2022). https://www.hazenandsawyer.com/articles/non-traditional-approaches-to-tds-local-limits-development-and-npdes-permit.

Masgoret, M., Botha, C., Myburgh, J., Naude, T., Prozesky, L., Naidoo, V., Van Wyk, J., Pool, E., & Swan, G. (2009). Molasses as a possible cause of an "endocrine disruptive syndrome" in calves. *Onderstepoort Journal of Veterinary Research*, 76(2), 209–225.

Mason, A. M., Wright, K. B., & Lambert, L. (2021). Understanding the impacts of emerging infectious disease (EID) media exposure: The case of US Ebola virus disease (EVD). *Journal of Communication in Healthcare*, 14(1), 31–40.

McCarthy, R., Gino, B., d'Entremont, P., Barari, A., & Renouf, T. S. (2020). The importance of personal protective equipment design and donning and doffing technique in mitigating infectious disease spread: A technical report. *Cureus*, 12(12), e12084. https://doi.org/10.7759/cureus.12084

McCowan, T., Omingo, M., Schendel, R., Adu-Yeboah, C., & Tabulawa, R. (2022). Enablers of pedagogical change within universities: Evidence from Kenya, Ghana and Botswana. *International Journal of Educational Development*, 90, 102558.

McHugh, K., Smit, N., Van Vuren, J., Van Dyk, J., Bervoets, L., Covaci, A., & Wepener, V. (2011). A histology-based fish health assessment of the tigerfish, Hydrocynus vittatus from a DDT-affected area. *Physics and Chemistry of the Earth*, 36(14–15), 895–904.

McLeod, R. W. (2022). Approaches to understanding human behavior when investigating incidents in academic chemical laboratories. *ACS Chemical Health & Safety*, 29(3), 263–279. https://doi.org/10.1021/acs.chas.2c00020

McManus, A. L. (2023). The Other Chemists' War: The Uses, Dual Uses, and Abuses of Chemical Weapons in World War II. PhD Thesis. Princeton University.

McMurtry, A. (2024). Learning theory and interdisciplinarity: Reframing learning in terms of evolving coherences rather than "true" representations or correspondences. In T. Marauhn, *Handbook of Interdisciplinary Teaching and Administration* (pp. 13–24). Edward Elgar Publishing, Cheltenham.

Meawad, A. S., Bojinova, D. Y., & Pelovski, Y. G. (2010). An overview of metals recovery from thermal power plant solid wastes. *Waste Management*, 30(12), 2548–2559.

Ménard, A. D. & Trant, J. F. (2020). A review and critique of academic lab safety research. *Nature Chemistry*, 12(1), 17–25.

Mendonça, A. O. (2024). Enhancing biosafety management and governance: a comprehensive assessment of high-containment biological laboratories in Brazil. 398f. Tese (Doutorado em Bioquímica Aplicada) - Universidade Federal de Viçosa, Viçosa. https://locus.ufv.br/handle/123456789/32466

Metcalfe, C. D., Bayen, S., Desrosiers, M., Munoz, G., Sauv´e, S., & Yargeau, V. (2022). An introduction to the sources, fate, occurrence and effects of endocrine disrupting chemicals released into the environment. *Environmental Research*, 207, 112658. https://doi.org/10.1016/j.envres.2021.112658

Meulenbroeks, R., van Rijn, R., & Reijerkerk, M. (2024). Fostering secondary school science students' intrinsic motivation by inquiry-based learning. *Research in Science Education*, 54(3), 339–358.

Meuten, D. J., Moore, F. M., Donovan, T. A., Bertram, C. A., Klopfleisch, R., Foster, R. A., … & Whitley, D. (2021). International guidelines for veterinary tumor pathology: A call to action. *Veterinary Pathology*, 58(5), 766–794.

Mhlongo, S. A., Sibali, L. L., & Ndibewu, P. (2023). Occurrence, quantification and removal of triclosan in wastewater of Umbogintwini Industrial Complex in KwaMakhutha, South Africa. *South African Journal of Science*, 119(11–12), 1–9.

Mia, R., Selim, M., Shamim, A. M., Chowdhury, M., Sultana, S., Armin, M., Hossain, M., Akter, R., Dey, S., & Naznin, H. (2019). Review on various types of pollution problem in textile dyeing & printing industries of Bangladesh and recommendation for mitigation. *Journal of Textile Engineering & Fashion Technology*, 5(4), 220–226.

Mihaila, C. O. (2018). Counterfeiting and illicit drug sales. Effects on intellectual property rights. *Juridical Tribune - Review of Comparative and International Law, Bucharest Academy of Economic Studies*, 8(1), 39–53.

Miller, S. (2018). *Dual Use Science and Technology, Ethics and Weapons of Mass Destruction*. Springer, Cham.

Ministry of Environment and Mineral Resources. (2011). *Kenya National Profile to Assess the Chemicals Management*. Ministry of Environment and Mineral Resources, Kenya.

Ministry of Environment, Forest & Climate Change, Government of India. (2016). *Guidelines on Implementing Liabilities for Environmental Damages Due to Handling & Disposal of Hazardous Waste and Penalty*. Central Pollution Control Board, Delhi.

Ministry of Health & Family Welfare. (2021). Guidelines on Integrated Public Health Laboratories. Government of India. Accessed on 03-10-2025 from: https://nhsrcindia.org/sites/default/files/Guidelines%20on%20integrated%20public%20health%20laboratories_dt%2007%20jan.pdf

Mnguni, S., Schoeman, C., Marais, S., Cukrowska, E., & Chimuka, L. (2018). Determination of oestrogen hormones in raw and treated water samples by reverse phase ultra-fast liquid chromatography mass spectrometry—A case study in Johannesburg South, South Africa. *Water SA*, 44(1), 111–117.

Mocerino, M., Yeo, S., & Zadnik, M. (2009). Preparing demonstrators for first year science laboratories. In *Proceedings of The Australian Conference on Science and Mathematics Education*, held 30 September–2 October, in Sydney, Australia.

Moedritzer, M. J. (2023). Biotech patent eligibility: Why we care about conventionality. *Business, Entrepreneurship, and Tax Law Review*, 112, 8. https://scholarship.law.missouri.edu/betr/vol7/iss1/8

Mohamadkhani, A. & Poustchi, H. (2015). Repository of human blood derivative biospecimens in biobank: Technical implications. *Middle East Journal of Digestive Diseases*, 7(2), 61–68.

Mohammed, M. (2024). The Devil's Breath: The Most Perilous Drug in The World? Retrieved from NACADA: https://nacada.go.ke/devils-breath-most-perilous-drug-world

Mohammed, S. M. & Amponsah, K. D. (2021). Teachers' and educational administrators' conceptions of inquiry: Do they promote or constrain inquiry-based science teaching in junior high schools? *Journal of Curriculum and Teaching*, 10(3), 58. https://doi.org/10.5430/jct.v10n3p58

Mohzana, M., Murcahyanto, H., Fahrurrozi, M., & Supriadi, Y. N. (2023). Optimization of management of laboratory facilities in the process of learning science at high school. *Jurnal Penelitian Pendidikan IPA*, 9(10), 8226–8234.

Morgan, S., Stewart, M., & Bennett, T. (2020). Simplifying COSHH and improving chemical safety. *Process Safety and Environmental Protection*, 137, 66–72.

Morris, T. H. (2020). Experiential learning–a systematic review and revision of Kolb's model. *Interactive Learning Environments*, 28(8), 1064–1077. https://doi.org/10.1080/10494820.2019.1570279

Moshtari, M. & Safarpour, A. (2024). Challenges and strategies for the internationalization of higher education in low-income East African countries. *Higher Education*, 87, 89–109. https://doi.org/10.1007/s10734-023-00994-1

Mossmann, H., Nicklas, W., & Hedrich, H. (2008). Management of immunocompromised and infected animals. *Methods in Microbiology*, 32, 183–231. https://doi.org/10.1016/S0580-9517(02)32093-2

Mosteiro-Romero, M., Miller, C., Chong, A., & Stouffs, R. (2023). Elastic buildings: Calibrated district-scale simulation of occupant-flexible campus operation for hybrid work optimization. *Building and Environment*, 237, 110318. https://doi.org/10.1016/j.buildenv.2023.110318

Moura, M. J., Pereira, A. D., Santos, D. J., Silva, A. G., Paiva, C. C., & Duarte, B. P. (2025). Cleaning validation in pharmaceutical quality control laboratories: A structured protocol for contamination risk mitigation. *DARU Journal of Pharmaceutical Sciences*, 33(2), 1–13.

Mourya, D. T., Yadav, P. D., Khare, A., & Khan, A. H. (2017). Certification & validation of biosafety level-2 & biosafety level-3 laboratories in Indian settings & common issues. *Indian Journal of Medical Research*, 146(4), 459–467. https://doi.org/10.4103/ijmr.IJMR_974_16

Muigua, K. (2022). Strengthening the environmental liability regime in Kenya for sustainable development. *Journal of CMSD*, 8(3), 1–45. Accessed on 20-07-2025 from: https://journalofcmsd.net/wp-content/uploads/2022/05/Strengthening-the-Environmental-Liability-Regime-in-Kenya-for-Sustainable-Development.pdf

Mukherjee, A., Saha, D., Harvey, C. F., Taylor, R. G., Ahmed, K. M., & Bhanja, S. N. (2015). Groundwater systems of the Indian Sub-Continent. *Journal of Hydrology: Regional Studies*, 4, 1–14.

Munson, E., Bowles, E. J., Dern, R. et al. (2018). Laboratory focus on improving the culture of biosafety: Statewide risk assessment of clinical laboratories that process specimens for microbiologic analysis. *Journal of Clinical Microbiology*, 56(1), e01569–17. https://doi.org/10.1128/JCM.01569-17

Murch, S. H. (2022). Biotechnology and biosecurity: An overview. *Biosecurity and Bioterrorism: Biodefense Strategy, Practice, and Science*, 20(4), 208–221.

Murtaza, G., Ghafoor, A., Qadir, M., Owens, G., Aziz, M. A., Zia, M. H., & SAIFULLAH. (2010). Disposal and use of sewage on agricultural lands in Pakistan: A review. *Pedosphere*, 20(1), 23–34.

Mustafa, M., Hatah, E., & Makmor-Bakry, M. (2023). Perspectives of pharmacists on medication reviews-exploring implementation research in service establishment in community settings. *Research in Social and Administrative Pharmacy*, 19(4), 673–680.

Mutengwe, M., Aneck-Hahn, N., Korsten, L., Van Zijl, M., & De Jager, C. (2016). Pesticide residues and estrogenic activity in fruit and vegetables sampled from major fresh produce markets in South Africa. *Food Additives and Contaminants Part A-Chemistry Analysis Control Exposure & Risk Assessment*, 33(1), 95–104.

Mutlu, N. G., Altuntas, S., & Dereli, T. (2023). The evaluation of occupational accident with sequential pattern mining. *Safety Science*, 166, 106212. https://doi.org/10.1016/j.ssci.2023.106212

Muzammil, K., Rayyani, S., Abbas Sahib, A., Gholizadeh, O., Naji Sameer, H., Jwad Kazem, T., … & Yasamineh, S. (2024). Recent advances in Crimean-Congo hemorrhagic fever virus detection, treatment, and vaccination: Overview of current status and challenges. *Biological Procedures Online*, 26(1), 20.

Nabavi-Pelesaraei, A., Mohammadkashi, N., Naderloo, L., Abbasi, M., & Chau, K. W. (2022). Principal of environmental life cycle assessment for medical waste during COVID-19 outbreak to support sustainable development goals. *Science of the Total Environment*, 827, 154416.

Najafabadi, M. M., Mirzaei, A., Laskookalayeh, S. S., & Azarm, H. (2022). An investigation of the relationship among economic growth, agricultural expansion and chemical pollution in Iran through decoupling index analysis. *Environmental Science and Pollution Research*, 29(50), 76101–76118.

Nasrallah, I. M., El Kak, A. K., Ismaiil, L. A., Nasr, R. R., & Bawab, W. T. (2022). Prevalence of accident occurrence among scientific laboratory workers of the public university in Lebanon and the impact of safety measures. *Safety and Health at Work*, 13(2), 155–162.

Nation Africa. (2022). The 'mchele' they use to spike your drinks. (N. Africa) Retrieved March 2025, from https://nation.africa/kenya/healthy-nation/the-mchele-they-use-to-spike-your-drinks-3664576

National Academies of Sciences, Engineering, and Medicine. (2018). *Reducing the Threat of Improvised Explosive Device Attacks by Restricting Access to Explosive Precursor Chemicals*. The National Academies Press, Washington, DC. https://doi.org/10.17226/24862

National People's Congress of China. (2016). *Wildlife Protection Law of the People's Republic of China*. National People's Congress of China, China.

National People's Congress of China. (2017). *Traditional Chinese Medicine Law of the People's Republic of China*. National People's Congress of China, China.

National People's Congress of China. (2019). *Drug Administration Law of the People's Republic of China*. National People's Congress of China, China.

National Research Council (US) Committee on Prudent Practices in the Laboratory. (2011). *Prudent Practices in the Laboratory: Handling and Management of Chemical Hazards: Updated Version*. National Academies Press, Washington, DC. https://www.ncbi.nlm.nih.gov/books/NBK55880/

Ndangili, P., Jijana, A., Baker, P., & Iwuoha, E. (2011). 3-Mercaptopropionic acid capped ZnSe quantum dot-cytochrome P450 3A4 enzyme biotransducer for 17 beta-estradiol. *Journal of Electroanalytical Chemistry*, 653(1–2), 67–74.

Newman, D. & Cragg, G. (2020). Natural products as sources of new drugs over the nearly four decades from 01/1981 to 09/2019. *Journal of Natural Products*, 83(3), 770–803.

Newmaster, S., Grguric, M., Dhivya, S. et al. (2013). DNA barcoding detects substitutions and contamination in herbal products. *BMC Medicine*, 11(1), 1–11

Ney, M., Kloss, G. K., & Schreiber, A. (2011). Using provenance to support good laboratory practice in grid environments. *Studies in Computational Intelligence*, 426, 7. https://doi.org/10.1007/978-3-642-29931-5-7

Nicolás, J. P. (2023). Research with human biological samples: Systematization of the phases and affected rights. In E. V. J. A. Lecaros, *Handbook of Bioethical Decisions. Volume I: Decisions at the Bench* (pp. 279–294). Springer International Publishing, Cham.

Niede, R. & Benbi, D. K. (2022). Integrated review of the nexus between toxic elements in the environment and human health. *AIMS Public Health*, 9(4), 758–789. https://doi.org/10.3934/publichealth.2022052

NIH. (2024). NIH Guidelines for Research Involving Recombinant or Synthetic Nucleic Acid molecules (NIH GUIDELINES) April 2024 Department of Health and Human Services, National Institutes of Health. Accessed on 28-08-2025 from: https://osp.od.nih.gov/wp-content/uploads/NIH_Guidelines.pdf

Nikolaou, N., Papadakis, A., Psychogyios, K., & Zahariadis, T. (2023). Vulnerability identification and assessment for critical infrastructures in the energy sector. *Electronics*, 12(14), 3185. https://doi.org/10.3390/electronics12143185

NIOSH. (2019). A Guide to Atmosphere-Supplying Respirators. By Cichowicz, J., Cofey, C., and Fries, M. Pittsburgh, PA: U.S. Department of Health and Human Services, Centers for Disease Control and Prevention, National Institute for Occupational Safety and Health. DHHS (NIOSH) Publication No. 2019-174, https://doi.org/10.26616/NIOSHPUB2019174

Nirgude, N. T., Shukla, S., & Venkatachalam, A. (2013). Physico-chemical analysis of some industrial effluents from Vapi industrial area, Gujarat, India. *RASAYAN Journal of Chemistry*, 6, 68–72.

Nkomo, P., Richter, L., Kagura, J., Mathee, A., Naicker, N., & Norris, S. (2018). Environmental lead exposure and pubertal trajectory classes in South African adolescent males and females. *Science of the Total Environment*, 628–629, 1437–1445.

Nolasco, E., Duraes, P. H. V., Gonçalves, J. P., de Oliveira, M. C., de Abreu, L. M., & de Almeida, A. N. (2020). Characterization of solid wastes as a tool to implement waste management strategies in a university campus. *International Journal of Sustainability in Higher Education*, 1(1), 1–20.

Novossiolova, T. (2024). Role of the life science community in strengthening the web of prevention for biosafety and biosecurity. In L. Shang, *Essentials of Biological Security: A Global Perspective* (pp. 107–120). Wiley, Berlin.

Nugroho, R. L. & Pangeran, P. (2021). Application of ISO 31000-based risk assessment to improve balanced scorecard performance at Shofa pharmacy. *International Journal of Multicultural and Multireligious Understanding*, 8(2), 251–263. https://doi.org/10.18415/ijmmu.v8i2.2321

Nweke, O. & Sanders, W. H. (2024). Modern environmental health hazards: A public health issue of increasing significance in Africa. *Environmental Health Perspectives*, 117(6), 863–70. https://doi.org/10.1289/ehp.0800126

Nyansa, M. M., Burrows, N. L., Galerneau, A. J., Bekkala, A. P., & Hungwe, K. N. (2024). Beyond intentions: Understanding the gap between safety education and student behaviors. *Journal of Chemical Education*, 101(3), 798–806.

O'Dwyer, M., Filieri, R., & O'Malley, L. (2023). Establishing successful university–industry collaborations: Barriers and enablers deconstructed. *Journal of Technology Transfer*, 48, 900–931. https://doi.org/10.1007/s10961-022-09932-2

Oakenfull, D. & Sidhu, G. S. (2023). *Saponins IN Toxicants of Plant Origin* (1st ed.). CRC Press, Boca Raton, FL.

Obayori, O. S. & Salam, L. B. (2023). Control of bioaerosols, biosafety, and biosecurity. In M. O. Ilori, *Aeromicrobiology*. Academic Press, New York. https://doi.org/10.1016/B978-0-323-96122-6.00002-5

Odunayo, A. (2024). Leveraging artificial intelligence (AI) for the maintenance of science laboratory equipment. *African Journal of Advances in Science and Technology Research*, 16(1), 131–148.

Ogunode, N.J. & Aiyedun, T.G. (2020). Administration of science programme in Nigerian higher institutions: Issues, challenges and way forward. *Middle European Scientific Bulletin*, 6, 94–99. https://doi org/10.47494/mesb.v6i.121

Ohoro, C., Adeniji, A., Okoh, A., & Okoh, O. (2021). Spatial and seasonal variations of endocrine disrupting compounds in water and sediment samples of Markman Canal and Swartkops River Estuary, South Africa and their ecological risk assessment. *Marine Pollution Bulletin*, 173, 113012. https://doi.org/10.1016/j.marpolbul.2021.113012

Ojemaye, C. & Petrik, L. (2019). Occurrences, levels and risk assessment studies of emerging pollutants (pharmaceuticals, perfluoroalkyl and endocrine disrupting compounds) in fish samples from Kalk Bay harbour, South Africa. *Environmental Pollution*, 252, 562–572.

Okaiyeto, K. & Oguntibeju, O. O. (2021). African herbal medicines: Adverse effects and cytotoxic potentials with different therapeutic applications. *International Journal of Environmental Research and Public Health*, 18(11), 5988.

Okillan Asianut, G. (2021). Enhancing learners' hands-on experiences in the acquisition of integrated production skills at Shimoni Core primary teachers' college, Wakiso district in Uganda. Thesis, Kyambogo University. https://kyuspace.kyu.ac.ug/xmlui/handle/20.500.12504/459

Okolie-Osemene, J. & Romaniuk, S. N. (2024). Jihad in the Horn of Africa: Somalia's Al-Shabaab Islamist Insurgency. In S. N. Romaniuk, A. Roul, A. P. Fabe, & J. Besenyő, *Handbook of Terrorist and Insurgent Groups* (pp. 261–270). CRC Press, Boca Raton, FL.

Okoro, H. K., Orosun, M. M., Oriade, F. A., Momoh-Salami, T. M., Ogunkunle, C. O., Adeniyi, A. G., … & Ngila, J. C. (2023). Potentially toxic elements in pharmaceutical industrial effluents: A review on risk assessment, treatment, and management for human health. *Sustainability*, 15(8), 6974.

Okumura, T., Takasu, N., Ishimatsu, S., Miyanoki, S., Mitsuhashi, A., Kumada, K., Tanaka, K., & Hinohara, S. (1996). Report on 640 victims of the Tokyo subway sarin attack. *Annals of Emergency Medicine*, 28(2), 129–135. https://doi.org/10.1016/S0196-0644(96)70052-5

Oladimeji, T. E., Oyedemi, M., Emetere, M. E., Agboola, O., Adeoye, J. B., & Odunlami, O. A. (2024). Review on the impact of heavy metals from industrial wastewater effluent and removal technologies. *Heliyon*, 10(23), e40370.

Oliveira, A. C. R. D., Braga, A. M. C. B., Villardi, J. R. W., & Krauss, T. M. (2020). Waste management in laboratories of a Brazilian public university: A challenge for environmental health and occupational health. *Saúde em Debate*, 43, 63–77.

Oloruntoba, K., Sindiku, O., Osibanjo, O., Balan, S., & Weber, R. (2019). Polybrominated diphenyl ethers (PBDEs) in chicken eggs and cow milk around municipal dumpsites in Abuja, Nigeria. *Ecotoxicology and Environmental Safety*, 179, 282–289.

Olowu, R., Arotiba, O., Mailu, S., Waryo, T., Baker, P., & Iwuoha, E. (2010). Electrochemical aptasensor for endocrine disrupting 17 beta-estradiol based on a poly(3,4-ethylenedioxylthiopene)-gold nanocomposite platform. *Sensors*, 10(11), 9872–9890.

Olujimi, O., Aroyeun, O., Akinhanmi, T., & Arowolo, T. (2017). Occurrence, removal and health risk assessment of phthalate esters in the process streams of two different wastewater treatment plants in Lagos and Ogun States, Nigeria. *Environmental Monitoring and Assessment*, 189(7), 345.

Olver, P., Bohn, M. K., & Adeli, K. (2023). Central role of laboratory medicine in public health and patient care. *Clinical Chemistry and Laboratory Medicine (CCLM)*, 61(4), 666–673.

Omer, M. A. (2024). Assessing the safety of chemical management practices in academic laboratories in Hargeisa, Somaliland. *Cogent Education*, 11(1), 2372188.

Omiko, A. (2007). *Job Orientation and Placement: The Role of Science Education in a Developing Economy*. Larry and Caleb Publishing House, Ebonyi –Nigeria, Abakaliki.

Omiko, A. (2015a). Chemistry teachers' attitude and knowledge of the use of information communication technology (ICT) in chemistry instruction delivery at the secondary school level in Ebonyi State of Nigeria. *Journal of Curriculum Organization of Nigeria (CON)*, 2015, 8.

Omiko, A. (2015b). Laboratory teaching: Implication on students' achievement in chemistry in secondary schools in Ebonyi State of Nigeria. *Journal of Education and Practice*, 6, 30.

Onyekwere, O., Okonkwo, C., Okoroafor, A., & Okonkwo, C. (2019). Occurrence and risk assessment of phenolic endocrine disrupting chemicals in shallow groundwater resource from selected Nigerian rural settlements. *Ovidius University Annals of Chemistry*, 30(2), 101–107.

Organisation for the Prohibition of Chemical Weapons (OPCW). (2021). *Indicative Guidelines for Chemical Safety and Security in Small and Medium-sized Enterprises to Foster the Peaceful Uses of Chemistry* (L. Z. N. Kojevnikov, Ed.) The Hague, Netherlands. Retrieved March 7 2025, from https://www.opcw.org/resources/capacity-building/international-cooperation-programmes/chemical-safety-and-security

Organisation for the Prohibition of Chemical Weapons (OPCW). (2022). Most Traded Scheduled Chemicals. Retrieved from https://www.opcw.org/resources/declarations/most-traded-scheduled-chemicals-2022

Organization for the Prohibition of Chemical Weapons (OPCW). (2025). Capacity Building: Associate Programme. Retrieved March 2025, from https://www.opcw.org/resources/capacity-building/international-cooperation-programmes/associate-programme

Orosz, G., Németh, V., Kovács, L., Somogyi, Z., & Korom, E. (2023). Guided inquiry-based learning in secondary-school chemistry classes: A case study. *Chemistry Education Research and Practice*, 24(1), 50–70.

Osei, D. M. & Zubair, M. (2025). *Packaging and Shipping Infectious Materials*. StatPearls Publishing, Treasure Island, FL.

Ostad-Ali-Askari, K. (2022). Management of risks substances and sustainable development. *Applied Water Science*, 12, 7. https://doi.org/10.1007/s13201-021-01562-7

Otitoju, O., Onwurah, I., Otitoju, G., Ndem, J., & Uwah, A. (2011). Effect of Bonny light crude oil contaminated diet on serum cholesterol and testosterone concentrations in Wistar albino rats. *African Journal of Biotechnology*, 10(53), 10986–10990.

Panagiotidou, E., Chountalas, P. T., Magoutas, A. I., & Kitsios, F. C. (2025). The multifaceted impact of ISO/IEC 17025 accreditation: A sector-specific analysis in civil engineering testing and calibration laboratories. *The TQM Journal*, 37(4), 998–1035.

Pandey, S., Joshi, N., & Kumar, M. (2020). Agrochemicals and human well-being: A review in context of Indian agriculture. *International Journal of Chemical Studies*, 8(1), 1539–1543. https://doi.org/10.22271/chemi.2020.v8.i1v.8477

Papadopoli, R., Nobile, C. G. A., Trovato, A., Pileggi, C., & Pavia, M. (2020). Chemical risk and safety awareness, perception, and practices among research laboratories workers in Italy. *Journal of Occupational Medicine and Toxicology*, 15, 1–11.

Pappas, G. (2011). Pakistan and water: New pressures on global security and human health. *American Journal of Public Health*, 101(5), 786–788.

Parida, V. K., Sikarwar, D., Majumder, A., & Gupta, A. K. (2022). An assessment of hospital wastewater and biomedical waste generation, existing legislations, risk assessment, treatment processes, and scenario during COVID-19. *Journal of Environmental Management*, 308, 114609.

Parmar, A., Narasimha, V. L., & Nath, S. (2024). National drug laws, policies, and programs in India: A narrative review. *Indian Journal of Psychological Medicine*, 46(1), 5–13.

Parra-Arroyo, L., González-González, R. B., Castillo-Zacarías, C., Martínez, E. M. M., Sosa-Hernández, J. E., Bilal, M., … & Parra-Saldívar, R. (2022). Highly hazardous pesticides and related pollutants: Toxicological, regulatory, and analytical aspects. *Science of the Total Environment*, 807, 151879.

Pascarella, G., Rossi, M., Montella, E., Capasso, A., De Feo, G., Botti, G., … & Morabito, A. (2021). Risk analysis in healthcare organizations: Methodological framework and critical variables. *Risk Management and Healthcare Policy*, 14, 2897–2911.

Patel, R. R. & George, M. (2021). Updating the international health regulations: Revising IHR decision instruments and the binary PHEIC declaration system. *Washington University Global Studies Law Review*, 20(3), 661–685.

Patel, V., Chesmore, A., Legner, C. M., & Pandey, S. (2022). Trends in workplace wearable technologies and connected-worker solutions for next-generation occupational safety, health, and productivity. *Advanced Intelligent Systems*, 4(1), 2100099.

Patrick, S., Aneck-Hahn, N., Van Wyk, S., Van Zijl, M., Huma, M., & de Jager, C. (2020). Veterinary growth promoters in cattle feedlot runoff: Estrogenic activity and potential effects on the rat male reproductive system. *Environmental Science and Pollution Research*, 27(12), 13939–13948.

Patrick, S., Bornman, M., Joubert, A., Pitts, N., Naidoo, V., & de Jager, C. (2016). Effects of environmental endocrine disruptors, including insecticides used for malaria vector control on reproductive parameters of male rats. *Reproductive Toxicology*, 61, 19–27.

Paudyal, S. (2025). Chemical regulations and the economy in India. Enhesa. https://www.enhesa.com/resources/article/chemical-regulations-and-the-economy-in-india

Paul, G. (2021). Emerging and re-emerging infectious diseases: Role of forensic pathology services in detection and biosafe practices. *Recent Advances in Forensic Medicine & Toxicology*, 3, 191.

Pawar, K. N., Gore, R. T., & Palekar, S. R. (2023). A review on approval process and regulation of medical devices as per US FDAand CDSCO. *International Journal of Drug Regulatory Affairs*, 11(1), 61–70. https://ijdra.com/index.php/journal/article/view/586

Payne, M. K., Nelson, A. W., Humphrey, W. R., & Straut, C. M. (2020). The chemical management system (CMS): A useful tool for inventory management. *Journal of Chemical Education*, 97(7), 1795–1798.

Pereira, P. (2023). *Trends in the Accreditation of Medical Laboratories by ISO 15189*. IntechOpen, New York. https://doi.org/10.5772/intechopen.113829

Perkins, D. & Bakanidze, L. (2024). Examples of biorisk management national regulatory frameworks. In L. Shang, *Essentials of Biological Security: A Global Perspective* (pp. 173–187). Wiley, Berlin.

Persily, A. (2015). Challenges in developing ventilation and indoor air quality standards: The story of ASHRAE standard 62. *Building and Environment*, 91, 61–69. https://doi.org/10.1016/j.buildenv.2015.02.026

Pharmacy and Poisons Board (PPB). (2022a). Annual Report 2021/2022. Pharmacy and Poisons Board. Retrieved March 2025, from https://web.pharmacyboardkenya.org/download/annual-report-2021-2022/

Pharmacy and Poisons Board (PPB). (2022b). Pharmacy and Poisons Board. Retrieved from https://web.pharmacyboardkenya.org/download/health-products-worth-24-million-seized-in-ppb-led-operation/

Pieterse, G., Marchand, M., van Dyk, J., & Barnhoorn, I. (2010). Histological alterations in the testes and ovaries of the sharptooth catfish (*Clarias gariepinus*) from an urban nature reserve in South Africa. *Journal of Applied Ichthyology*, 26(5), 789–793.

Pillai, S. P. & Fox, E. (2025). Laboratory quality management system fundamentals. *Frontiers in Bioengineeri ng and Biotechnology*, 13, 1578654. https://doi.org/10.3389/fbioe.2025.1578654

Pistollato, F., Madia, F., Corvi, R., Munn, S., Grignard, E., Paini, A., … & Zuang, V. (2021). Current EU regulatory requirements for the assessment of chemicals and cosmetic products: Challenges and opportunities for introducing new approach methodologies. *Archives of Toxicology*, 95, 1867–1897.

Pisu, A., Elia, N., Pompianu, L., Barchi, F., Acquaviva, A., & Carta, S. (2024). Enhancing workplace safety: A flexible approach for personal protective equipment monitoring. *Expert Systems with Applications*, 238, 122285.

Poller, B., Tunbridge, A., Hall, S., Beadsworth, M., Jacobs, M., Evans, C., & Crook, B. (2018). A unified personal protective equipment ensemble for clinical response to possible high consequence infectious diseases: A consensus document on behalf of the HCID programme. *Journal of Infection*, 77(6), 496–502. https://doi.org/10.1016/j.jinf.2018.08.016

Polshettiwar, S. A., Sawant, D. H., Abhale, N. B., Chavan, N. B., Baheti, A. M., Wani, M. S., Tagalpallewar, A. A., Deshmukh, C. D., & Polshettiwar, A. P. (2022). Review on regulation of herbal products used as a medicine across the globe: A case study on turmeric - golden medicine. *Biomedical and Pharmacology Journal*, 15(3), 2458.

Popov, O., Iatsyshyn, A., Pecheny, V., Kovach, V., & Kovalenko, V. (2023). Approaches to assessing consequences of accidents during transportation of hazardous substances by road. In A. Zaporozhets, O. Popov, *Systems, Decision and Control in Energy IV: Volume II. Nuclear and Environmental Safety* (pp. 327–342). Springer Nature Switzerland, Cham.

Powell, J. A. (2023). Importance of maintaining laboratory emergency equipment—showers and eyewashes. *ACS Chemical Health and Safety*, 30(4), 179–181. https://doi.org/10.1021/acs.chas.3c00028

Predieri, B., Alves, C. A. D., & Iughetti, L. (2022). New insights on the effects of endocrine-disrupting chemicals on children. *Jornal de Pediatria*, 98(Suppl 1), S73–S85. https://doi.org/10.1016/j.jped.2021.11.003

Preston, C., Valdez, M. L., & Bond, K. (2012). Strengthening medical product regulation in low- and middle-income countries. *PLOS Medicine*, 9(10), e1001327. https://doi.org/10.1371/journal.pmed.1001327.

Puri, M., Gandhi, K., & Kumar, M. S. (2023). Emerging environmental contaminants: A global perspective on policies and regulations. *Journal of Environmental Management*, 332, 117344. https://doi.org/10.1016/j.jenvman.2023.117344

Qhanya, L., Mthakathi, N., Boucher, C., Mashele, S., Theron, C., & Syed, K. (2017). Isolation and characterisation of endocrine disruptor nonylphenol-using bacteria from South Africa. *South African Journal of Science*, 113(5–6), 39–45.

Quainoo, E. A., Quansah, F., Adams, F., & Opoku, E. (2020). Comparative review of educational reforms, policies, and systems: A case of China and Ghana. *Journal of Comparative Studies and International Education*, 2, 5–21.

Qusairy, Z. & Rada, M. (2023). International Biospecimen Sourcing Platform: A Key Player in Advancing Health Research and Drug Discovery. Preprints. https://doi.org/10.20944/preprints202312.1256.v1

Rabbi, H. R. & Rahman, A. (2017). Ship breaking and recycling industry of Bangladesh; issues and challenges. *Procedia Engineering*, 194, 254–259.

Rachkowski, D. (2024). Strengthening environmental protection for a healthier Canada Act: The right to a healthy environment & accountability in chemicals management. University of British Columbia. Retrieved 03 January 2025, from https://open.library.ubc.ca/collections/ubctheses/24/items/1.0445216

Ragavan, G. K., Selvaraj, P., Murugesan, D., Krishnaswamy Gopalan, T., Chinnaiyan, S., Gopal, D. R., … & Palavesam, A. (2023). Management practices and technologies for efficient biological sample collection from domestic animals with special reference to Indian field conditions. *Animal Diseases*, 3(1), 34.

Rahim, M. S. A., Reniers, G., Yang, M., & Bajpai, S. (2024). Risk assessment methods for process safety, process security and resilience in the chemical process industry: A thorough literature review. *Journal of Loss Prevention in the Process Industries*, 88, 105274. https://doi.org/10.1016/j.jlp.2024.105274

Rahman, M. T., Sobur, M. A., Islam, M. S., Ievy, S., Hossain, M. J., El Zowalaty, M. E., Rahman, A. T., & Ashour, H. M. (2020). Zoonotic diseases: Etiology, impact, and control. *Microorganisms*, 8(9), 1405. https://doi.org/10.3390/microorganisms8091405.

Raj, A., Yadav, T., Patil, S. et al. (2025). Counterfeit medicine: A major public health concern and effective remedies for combatting the crisis. *Discover Pharmaceutical Sciences*, 1, 4. https://doi.org/10.1007/s44395-025-00004-6

Rampley, C. P. N., Whitehead, P. G., Softley, L., Hossain, M. A., Jin, L., David, J., Shawal, S., Das, P., Thompson, I. P., Huang, W. E., Peters, R., Holdship, P., Hope, R., & Alabaster, G. (2020). River toxicity assessment using molecular biosensors: Heavy metal contamination in the Turag-Balu-Buriganga river systems, Dhaka, Bangladesh. *Science of the Total Environment*, 703, 134760.

Rank, A., Ramos, R. S., da Silva, R. S., Soares, J. R. S., Picanço, M. C., & Fidelis, E. G. (2020). Risk of the introduction of *Lobesia botrana* in suitable areas for Vitis vinifera. *Journal of Pest Science*, 93, 1167–1179.

Rashid, A., Schutte, B. J., Ulery, A., Deyholos, M. K., Sanogo, S., Lehnhoff, E. A., & Beck, L. (2023). Heavy metal contamination in agricultural soil: Environmental pollutants affecting crop health. *Agronomy*, 13(6), 1521.

Rashidi, M. A., Khazaei, S. R., Samimi, K., Khodakarim, S., Khatabakhsh, A., & Pouyakian, M. (2023). Application of social media in chemical safety training: A case study of training GHS standards to students and laboratory staff at a university. *Journal of Chemical Education*, 100(2), 517–527.

Raveh, Y., Livingstone, J., Vianna, R., & Nicolau-Raducu, R. (2022). Personal protective equipment for liver transplant in SARS-CoV-2 polymerase chain reaction-positive convalescing recipients. *Transplantation Proceedings*, 54(6), 1528–1533. https://doi.org/10.1016/j.transproceed.2022.05.014

Ravinetto, R., Pinxten, W., & Rägo, L. (2018). Quality of medicines in resource-limited settings: Need for ethical guidance. *Global Bioethics*, 29(1), 81–94. https://doi.org/10.1080/11287462.2018.1522991.

Redinger, C. F. (2024). *Organizational Risk Management: An Integrated Framework for Environmental, Health, Safety, and Sustainability Professionals, and their C-Suites* (1st ed.). Wiley, Berlin.

Reeves, S. M. & Crippen, K. J. (2021). Virtual laboratories in undergraduate science and engineering courses: A systematic review, 2009–2019. *Journal of Science Education and Technology*, 30(6), 16–30. https://doi.org/10.1007/s10956-020-09866-0

Regasa, M. B. & Nyokong, T. (2022). Design and fabrication of electrochemical sensor based on molecularly imprinted polymer loaded onto silver nanoparticles for the detection of 17-β-estradiol. *Journal of Molecular Recognition*, 35(10), e2978.

Reniers, G. & Khakzad, N. (2017). Revolutionizing safety and security in the chemical and process industry: Applying the CHESS concept. *Journal of Integrated Security and Safety Science*, 1(1), 1547. https://doi.org/10.18757/jiss.2017.1.1547

Rethorst, D. N. (2015). Animal health equipment management. *Veterinary Clinics of North America: Food Animal Practice*, 31(2), 259–267. https://doi.org/10.1016/j.cvfa.2015.03.009

Ricci, F., Moreno, V. C., & Cozzani, V. (2021). A comprehensive analysis of the occurrence of Natech events in the process industry. *Process Safety and Environmental Protection*, 147, 703–713.

Ripanda, A. S., Rwiza, M. J., Nyanza, E. C., Machunda, R. L., & Vuai, S. H. (2022). Contribution of illicit drug use to pharmaceutical load in the environment: A focus on Sub-Saharan Africa. *Journal of Environmental and Public Health*, 2022(1), 9056476.

Risi, G. F., Bloom, M. E., Hoe, N. P., Arminio, T., Carlson, P., Powers, T., Feldmann, H., & Wilson, D. (2010). Preparing a community hospital to manage work-related exposures to infectious agents in biosafety level 3 and 4 laboratories. *Emerging Infectious Diseases*, 6(3), 373–378. https://doi.org/10.3201/eid1603.091485

Roberson, J., Wrenn, A., Poole, J., Jaeger, A., & Eltoum, I. A. (2013). Constructing a modern cytology laboratory: A toolkit for planning and design. *Cytojournal*, 10, 3. https://doi.org/10.4103/1742-6413.107983

Rodell, M., Velicogna, I., & Famiglietti, J. S. (2009). Satellite-based estimates of groundwater depletion in India. *Nature*, 460(7258), 999–1002.

Rogers, B. J. (2013). Collaboration: The effects of joint agency training for both national guard and department of homeland security. College of technology directed projects. Paper 47. https://docs.lib.purdue.edu/techdirproj/47

Roshni, N. A., Hasan, M. K., Wadud, M. A., Hasan, A. K., Akter, R., Khan, R. N. A., & Hemel, S. A. K. (2023). Toxic metal levels in forest soils caused by industrialization and impacts on tree biomass carbon stock. *Environmental Processes*, 10(3), 45.

Rossete, C. A. & Ribeiro, M. G. (2021). Laboratory technicians' use and interpretation of hazard communication elements on chemical labels. *ACS Chemical Health & Safety*, 28(3), 211–223. https://doi.org/10.1021/acs.chas.0c00125

Roy, S., Mandal, B., & Das, B. K. (2025). Quality assurance in laboratory practices and equipment maintenance: Ensuring precision, reliability, and compliance. In B. K. Das, V. Kumar, *Laboratory Techniques for Fish Disease Diagnosis*. Springer, Singapore. https://doi.org/10.1007/978-981-96-4620-3_5

Rozsypal, T., Pejchal, J., Opravil, J., Haralampiev, M., Bocos-Bintintan, V., & Kobliha, Z. (2025). Destruction of chemical weapons stockpiles in the Russian federation: A review. *Environmental Chemistry Letters*, 23, 393–418. https://doi.org/10.1007/s10311-024-01812-5

Ruba, U. B., Chakma, K., Senthi, J. Y., & Rahman, S. (2021). Impact of industrial waste on natural resources: A review in the context of Bangladesh. *Current World Environment*, 16(2), 348–361.

Rush, A., Byrne, J. A., & Watson, P. H. (2024). Applying findable, accessible, interoperable, and reusable principles to biospecimens and biobanks. *Biopreservation and Biobanking*, 22(6), 550–556. https://doi.org/10.1089/bio.2023.0110

Russell, T. & Martin, A. K. (2023). Learning to teach science. In N. G. Lederman, *Handbook of Research on Science Education* (pp. 1162–1196). Routledge, London.

Rutebemberwa, E., Aku, F. Y., Zein, E. I. K. A., & Bellali, H. (2020). Reasons for and barriers to biosafety and biosecurity training in health-related organizations in Africa, Middle East and Central Asia: Findings from GIBACHT training needs assessments 2018–2019. *Pan African Medical Journal*, 37, 64. https://doi.org/10.11604/pamj.2020.37.64.23390

Ryan, J. R. (2016). Category C diseases and agents. *Biosecurity and Bioterrorism*, 2016, 113–134. https://doi.org/10.1016/B978-0-12-802029-6.00005-0

Safdar, M., Ullah, M., Bibi, A., Khan, M. A., Rehman, M., Fatima, Z., … & Naeem, M. (2023). The evolving landscape of biosafety and biosecurity: A review of international guidelines and best practices. *Journal of Women Medical and Dental College*, 2(2), 73.

Saha, A. K., Jahin, M. A., Rafiquzzaman, M., & Mridha, M. F. (2024). Ergonomic design of computer laboratory furniture: Mismatch analysis utilizing anthropometric data of university students. *Heliyon*, 10(14), e34063. https://doi.org/10.1016/j.heliyon.2024.e34063

Saha, A., Engelhardt, G. R., Macdonald, D. D., & Khan, R. A. (2023). Industrial pollution and its effect in the context of Bangladesh. *World Journal of Advanced Research and Reviews*, 18(2), 936–945.

Saina, E. J., Odimu, K. N., & Otara, A. M. (2017). Levels of awareness on safety and health in use of agro-chemicals among large scale flower farm workers in Uasin Gishu County, Kenya. *Research on Humanities and Social Sciences*, 7(13), 2224–5766. https://www.iiste.org/Journals/index.php/RHSS/article/viewFile/38124/39202

Salaudeen, T., Okoh, O., & Okoh, A. (2019). Performance assessment of wastewater treatment plants with special reference to phenol removal. *International Journal of Environmental Science and Technology*, 16(1), 401–412.

Salaudeen, T., Okoh, O., Agunbiade, F., & Okoh, A. (2018). Fate and impact of phthalates in activated sludge treated municipal wastewater on the water bodies in the Eastern Cape, South Africa. *Chemosphere*, 203, 336–344.

Salonga, E. A. G. (2025). Philippine biosafety and biosecurity regulations: From global response to domestic realities. In J. J. G. Guerrero, *Biosafety and Biosecurity* (1st ed.). CRC Press, Boca Raton, FL.

Samaranayake, A. I., Nishadya, S., & Jayasundara, U. K. (2022). Analyzing safety culture in Sri Lankan industrial chemical laboratories. *Safety and Health at Work*, 13(1), 86–92. https://doi.org/10.1016/j.shaw.2021.11.001

Santella, N. & Steinberg, L. J. (2011). Accidental releases of hazardous materials and relevance to terrorist threats at industrial facilities. *Journal of Homeland Security and Emergency Management*, 8(1), 53. https://doi.org/10.2202/1547-7355.1809. https://www.bepress.com/jhsem/vol8/iss1/53

Saravanakumar, K., De Silva, S., Santosh, S. S., Sathiyaseelan, A., Ganeshalingam, A., Jamla, M., … & Wang, M. H. (2022). Impact of industrial effluents on the environment and human health and their remediation using MOFs-based hybrid membrane filtration techniques. *Chemosphere*, 307, 135593.

Sarı, U., Duygu, E., Şen, F., & Kırındı, T. (2020). The effects of STEM education on scientific process skills and STEM awareness in simulation based ınquiry learning environment. *Journal of Turkish Science Education*, 17(3), 387–405. https://doi.org/10.36681/tused.2020.34

Sato, R., Jakobsson, U., & Midlöv, P. (2024). A proposed medical system change in Japan inspired by Swedish primary health care: Important role of general practitioners and specialist nurses at primary health care centers. *Journal of General and Family Medicine*, 25(6), 295–304.

Saxena, R. R. (2024). Intelligent Approaches to Predictive Analytics in Occupational Health and Safety in India (Version 3). arXiv. https://doi.org/10.48550/ARXIV.2412.16038

Schaetz, D. (2012). Conceptions of Crisis Management: The Analysis of Pharmaceutical Companies in Germany, Japan, and the United States. Global Business Leadership Student Work, p. 1. https://digitalcommons.csbsju.edu/gbl_students/1

Schellenberg, T., Subramanian, V., Ganeshan, G., Tompkins, D., & Pradeep, R. (2020). Wastewater discharge standards in the evolving context of urban sustainability–The case of India. *Frontiers in Environmental Science*, 8, 30. https://doi.org/10.3389/fenvs.2020.00030

Schwarz, C. V., Braaten, M., Haverly, C., & de los Santos, E. X. (2021). Using sense-making moments to understand how elementary teachers' interactions expand, maintain, or shut down sense-making in science. *Cognition and Instruction*, 39(2), 113–148.

Scott, N. B. & Pocock, N. S. (2021). The health impacts of hazardous chemical exposures among child labourers in low- and middle-income countries. *International Journal of Environmental Research and Public Health*, 18(10), 5496. https://doi.org/10.3390/ijerph18105496

Seery, M. K., Agustian, H. Y., & Zhang, X. (2019). A framework for learning in the chemistry laboratory. *Israel Journal of Chemistry*, 59(6–7), 546–553.

Segarra, J. D., Bensi, M., & Modarres, M. (2021). A Bayesian network approach for modeling dependent seismic failures in a nuclear power plant probabilistic risk assessment. *Reliability Engineering and System Safety*, 213, 107678.

Seifrid, M., Pollice, R., Aguilar-Granda, A., Morgan Chan, Z., Hotta, K., Ser, C. T., … & Aspuru-Guzik, A. (2022). Autonomous chemical experiments: Challenges and perspectives on establishing a self-driving lab. *Accounts of Chemical Research*, 55(17), 2454–2466.

Seka, A. M., Zhang, J., Zhang, D., Ayele, E. G., Han, J., Prodhan, F. A., Zhang, G., & Liu, Q. (2022). Hydrological drought evaluation using GRACE satellite-based drought index over the lake basins, East Africa. *Science of the Total Environment*, 852, 158425.

Selvam, V., Ashok, D., Rajamanoharan, I. D., Rajalakshmi, V., & Vidhya, K. (2022). Impact of Ujjwala Yojana scheme and its effect on behavioural changes among rural women. *International Journal of Asian Business and Information Management*, 13(1), 1–14.

Serrano-Perez, J. J., González-García, L., Flacco, N., Taberner-Cortés, A., García-Arnandis, I., Pérez-López, G., … & Romá-Mateo, C. (2023). Traditional vs. virtual laboratories in health sciences education. *Journal of Biological Education*, 57(1), 36–50.

Setioningsih, E. D. & Setiawan, S. Y. (2025). Enhancing laboratory efficiency: A community service program for centrifuge maintenance at Randegansari Husada Hospital. *Frontiers in Community Service and Empowerment*, 4(1):16–20. https://doi.org/10.35882/ficse.v4i1.90

Shah, S. H., Shah, N. U., & Jbeen, A. (2024). Exploration of LIS professionals efforts in Pakistan towards the improvements of technological competencies in 21st century. *Global Knowledge, Memory and Communication*, 73(4/5), 596–613.

Shaji, E., Santosh, M., Sarath, K. V., Prakash, P., Deepchand, V., & Divya, B. V. (2021). Arsenic contamination of groundwater: A global synopsis with focus on the Indian Peninsula. *Geoscience Frontiers*, 12(3), 101079.

Shalu, K., Singh, K., Kaur, S., & Sikka, P. (2023). Petroleum export performance in India: Analysis of growth and structural dynamics. *Indian Journal of Economics and Development*, 19(4), 776–785. https://doi.org/10.35716/IJED-23070

Shamsuddin, N. & Kaur, J. (2020). Students' learning style and its effect on blended learning, does it matter? *International Journal of Evaluation and Research in Education*, 9(1), 195–202.

Shana, Z. & Abulibdeh, E. S. (2020). Science practical work and its impact on high students' academic achievement. *JOTSE*, 10(2), 888. https://doi.org/10.3926/jotse.888

Shang, W. J., Jing, W. Z., Jue, L. I. U., & Min, L. I. U. (2023). Global epidemic of Ebola virus disease and the importation risk into China: An assessment based on the risk matrix method. *Biomedical and Environmental Sciences*, 36(1), 86–93.

Shankar, S., Shanker, U., & Shikha, S. (2014). Arsenic contamination of groundwater: A review of sources, prevalence, health risks, and strategies for mitigation. *The Scientific World Journal*, 2014, 304524.

Shapira, M., Cohen, B., Friemann, S., Tal, Y., Teper, Z., Dudkiewicz, M., … & Shriki, D. D. (2025). The impact of clinical sample transportation by unmanned aerial systems on the results of laboratory tests. *Drones*, 9(3), 179.

Sharma, A. K., Ramacharyulu, D. A., Lakhanpal, S., Shakier, H. G., Reddy, U., & Kumari, V. (2024a). Retracted: Exploring the future of advanced materials processing: Innovations and challenges ahead: A review. *E3S Web of Conferences*, 505, 01021. https://doi.org/10.1051/e3sconf/202450501021

Sharma, A., Kosasih, E., Zhang, J., Brintrup, A., & Calinescu, A. (2022a). Digital twins: State of the art theory and practice, challenges, and open research questions. *Journal of Industrial Information Integration*, 30, 100383.

Sharma, B. M., Scheringer, M., Chakraborty, P. et al. (2022b). Unlocking India's potential in managing endocrine-disrupting chemicals (EDCs): Importance, challenges, and opportunities. *Exposure and Health*, 15(4), 8. https://doi.org/10.1007/s12403-022-00519-8

Sharma, M., Rawat, S., Kumar, D., Awasthi, A., Sarkar, A., Sidola, A., Choudhury, T., & Kotecha, K. (2024b). The state of the Yamuna River: A detailed review of water quality assessment across the entire course in India. *Applied Water Science*, 14(8), 175.

Sharma, P., Bihari, V., Agarwal, S. K., Verma, V., Kesavachandran, C. N., Pangtey, B. S., Mathur, N., Singh, K. P., Srivastava, M., & Goel, S. K. (2012). Groundwater contaminated with hexavalent chromium [Cr (VI)]: A health survey and clinical examination of community inhabitants (Kanpur, India). *PLoS One*, 7(10), e47877.

Sharma, P., Iqbal, H. M. N., & Chandra, R. (2022c). Evaluation of pollution parameters and toxic elements in wastewater of pulp and paper industries in India: A case study. *Case Studies in Chemical and Environmental Engineering*, 5, 100163.

Sharma, R. & Hossain, M. M. (2019). Strengthening public health partnerships in India: Envisioning the role of law enforcement during public health emergencies. *Indian Journal of Forensic and Community Medicine*, 44(3), 188–192. https://doi.org/10.4103/ijcm.IJCM_110_19

Sharma, S. K., Mishra, M., & Mudgal, S. K. (2020). Efficacy of cloth face mask in prevention of novel coronavirus infection transmission: A systematic review and meta-analysis. *Journal of Education and Health Promotion*, 9, 192.

Sharma, S., Mohler, J., Mahajan, S. D., Schwartz, S. A., Bruggemann, L., & Aalinkeel, R. (2023). Microbial biofilm: A review on formation, infection, antibiotic resistance, control measures, and innovative treatment. *Microorganisms*, 11(6), 1614.

Sharma, V., Gidwani, B. D., Sharma, V., & Meena, M. L. (2019). Implementation model for cellular manufacturing system using AHP and ANP approach. *Benchmarking: An International Journal*, 26(5), 1605–1630.

Shekarloo, M. V., Sadat, S. M., Moradpour, Z., & Zendehdel, R. (2024). Evaluation of control of substances hazard to health (COSHH) essentials model by measuring occupational exposure in a chemical production industry. *Journal of Chemical Health Risks*, 14(2), 1614.

Shekhar, C. & Rai, P. (2025). Indian Pharmaceutical Industry: Past, Present, and Future Outlook (2024–2030). https://doi.org/10.13140/RG.2.2.32592.01284. https://ssrn.com/abstract=5162548 or https://doi.org/10.2139/ssrn.5162548

Shem, P. M., Otieno, G., Onyango, J. O., Owuor, J. J., Mbugua, P. W., & Aluoch, A. O. (2021). Chemical security in Kenya: Efforts, challenges, and opportunities in academic institutions and the chemical industry. *ACS Chemical Health & Safety*, 28(2), 112–118.

Shishodia, V., Jindal, D., Sinha, S., & Singh, M. (2023). Identification and removal of biological contamination in the media and cell suspensions. In V. Shishodia, *Animal Cell Culture: Principles and Practice* (pp. 235–249). Springer International Publishing, Cham.

Shu, Q., Li, Y., & Gao, W. (2023). Emergency treatment mechanism of laboratory safety accidents in university based on IoT and context aware computing. *Heliyon*, 9(9), e19406. https://doi.org/10.1016/j.heliyon.2023.e19406

Sidwaba, U., Ntshongontshi, N., Feleni, U., Wilson, L., Waryo, T., & Iwuoha, E. (2019). Manganese peroxidase-based electro-oxidation of bisphenol A at hydrogellic polyaniline-titania nanocomposite-modified glassy carbon electrode. *Electrocatalysis*, 10(4), 323–331.

Signe, L. (2018). The potential of manufacturing and industrialization in Africa: Trends, opportunities, and strategies. Brookings Institution. https://www.brookings.edu/wp-content/uploads/2018/09/Manufacturing-and-Industrialization-in-Africa-Signe-20180921.pdf

Sikolia, F. M., Toili, W. W., Owiti, D., & Fedha, S. S. (2017). Limiting factors influencing implementation of SMASSE (ASEI and PDSI) instructional strategies in teaching biology in secondary schools in Kakamega North Sub County, Kenya. *IOSR Journal of Research & Method in Education (IOSR-JRME)*, 7(1), 69–82. https://doi.org/10.9790/7388-0701026982

Sile, I., Teterovska, R., Oskars Onzevs, O., & Ardava, E. (2023). Safety concerns related to the simultaneous use of prescription or over-the-counter medications and herbal medicinal products: Survey results among Latvian Citizens. *International Journal of Environmental Research and Public Health*, 20(16), 6551. https://doi.org/10.3390/ijerph20166551

Singh, A. K., Kumar, A., & Chandra, R. (2022). Environmental pollutants of paper industry wastewater and their toxic effects on human health and ecosystem. *Bioresource Technology Reports*, 20(1–2), 101250. https://doi.org/10.1016/j.biteb.2022.101250

Singh, B. P., Samanta, P., Choudhury, M., Gupta, P., Chadha, U., & Tolera Kuma Eticha, T. K. (2024a). Physicochemical and heavy metal pollution level in Hindon River ecosystem: An implication to public health risk assessment. *Environmental Quality Management*, 34(2), 22263. https://doi.org/10.1002/tqem.22263

Singh, H., Chawla, A. S., & Kapoor, V. K. (1985). Medicinal chemistry research in India. *Progress in Medicinal Chemistry*, 22, 243–266. https://doi.org/10.1016/S0079-6468(08)70232-7

Singh, R. & Frank, A. L. (2025). Work-related migration to the Alang shipbreaking industry from other parts of India: An overview of health-related issues. *Annals of Global Health*, 91(1), 32, 1–8.

Singh, R., Singh, A., Balomajumder, C., & Vidyarthi, A. K. (2025). Assessment of industrial effluent discharges contributing to Ganga water pollution through a multivariate statistical framework: Investigating the context of Indian industries. *Environmental Science and Pollution Research*, 2025, 6. https://doi.org/10.1007/s11356-024-35823-0

Singh, S., Sharma, P., Pal, N., Sarma, D. K., Tiwari, R., & Kumar, M. (2024b). Holistic one health surveillance framework: Synergizing environmental, animal, and human determinants for enhanced infectious disease management. *ACS Infectious Diseases*, 10(3), 808–826.

Smallwood, N., Harrex, W., Rees, M., Willis, K., & Bennett, C. M. (2022). COVID-19 infection and the broader impacts of the pandemic on healthcare workers. *Respirology*, 27(6), 411–426.

Smith, M. Y. & Benattia, I. (2016). The patient's voice in pharmacovigilance: Pragmatic approaches to building a patient-centric drug safety organization. *Drug Safety*, 39(9), 779–785. https://doi.org/10.1007/s40264-016-0426-9.

Smith, T. D., DeJoy, D. M., & Dyal, M. A. (2020). Safety specific transformational leadership, safety motivation and personal protective equipment use among firefighters. *Safety Science*, 131, 104930. https://doi.org/10.1016/j.ssci.2020.104930

Solecki, R., Kortenkamp, A., Bergman, Å., & Chahoud, I. (2017). Scientific principles for the identification of endocrine disrupting chemicals: A consensus statement. *Archives of Toxicology*, 91(2), 1001–1006. https://doi.org/10.1007/s00204-016-1866-9

Soltanzadeh, A., Zarei, E., Mahdinia, M., & Sadeghi-Yarandi, M. (2024). An integrated approach to assess safety and security risks in chemical process industries. *Journal of Loss Prevention in the Process Industries*, 90, 105344. https://doi.org/10.1016/j.jlp.2024.105344

Sorathia, K. & Servidio, R. (2012). Learning and experience: Teaching tangible interaction and edutainment. *Procedia–Social and Behavioral Sciences*, 64, 265–274.

Sorensen, G., Dennerlein, J. T., Peters, S. E., Sabbath, E. L., Kelly, E. L., & Wagner, G. R. (2021). The future of research on work, safety, health and wellbeing: A guiding conceptual framework. *Social Science & Medicine*, 269, 113593.

Spath, J., Arumugam, P., Lindberg, R., Abafe, O., Jansson, S., Fick, J., & Buckley, C. (2021). Biochar for the removal of detected micropollutants in South African domestic wastewater: A case study from a demonstration-scale decentralised wastewater treatment system in eThekwini. *Water SA*, 47(4), 396–416.

Sperry, J. B., Azuma, M., & Stone, S. (2021). Explosive hazard identification in pharmaceutical process development: A novel screening method and workflow for shipping potentially explosive materials. *Organic Process Research & Development*, 25(2), 212–224.

Srinivasulu, M., Chandra, M. S., Gooty, J. M., & Madhavi, A. (2022). Personal care products—fragrances, cosmetics, and sunscreens—in the environment. In M. Z. Hashmi, *Environmental Micropollutants* (pp. 131–149). Elsevier, Amsterdam, Netherlands.

Ssebugere, P., Sillanpaa, M., Wang, P., Li, Y., Kiremire, B., Kasozi, G., Zhu, C., Ren, D., Shang, H., Zhang, Q., & Jiang, G. (2014). Polychlorinated dibenzo-p-dioxins, polychlorinated dibenzofurans and polybrominated diphenyl ethers in sediments and fish species from the Murchison Bay of Lake Victoria, Uganda. *Science of the Total Environment*, 500, 1–10.

State Administration for Market Regulation (SAMR). (2022). *Regulations on the Supervision and Administration of Online Trading of Drugs*. SAMR, China.

Steere, N. V. (1973). Research needed for laboratory standards-Part II. *Journal of Chemical Education*, 50(12), A579.

Stein, S. & Luna, J. (2021). Toxic sensorium: Agrochemicals in the African Anthropocene. *Environment and Society*, 12(1), 87–107. Retrieved June 8 2025, from https://doi.org/10.3167/ares.2021.120106

Steward, J. E. (2018). The Status of Safety Culture at Louisiana State University. LSU Doctoral Dissertations, p. 4498. https://repository.lsu.edu/gradschool_dissertations/4498

Stitt-Fisher, M. (2015). Biological hazards and select agents. *Research Regulatory Compliance*, 2015, 79–111. https://doi.org/10.1016/B978-0-12-420058-6.00004-6

Stopforth, A., Burger, B., Crouch, A., & Sandra, P. (2007). The analysis of estrone and 17 beta-estradiol by stir bar sorptive extraction-thermal desorption-gas chromatography/mass spectrometry: Application to urine samples after oral administration of conjugated equine estrogens. *Journal of Chromatography B-Analytical Technologies in the Biomedical and Life Sciences*, 856(1–2), 156–164.

Street, M. E., Angelini, S., Bernasconi, S. et al. (2018). Current knowledge on endocrine disrupting chemicals (EDCs) from animal biology to humans, from pregnancy to adulthood: Highlights from a National Italian meeting. *International Journal of Molecular Sciences*, 19(6), 1647. https://doi.org/10.3390/ijms19061647

Stroupe, D. & Carlone, H. B. (2022). Leaving the laboratory: Using field science to disrupt and expand historically enduring narratives of science teaching and learning. *Science & Education*, 31(8), 893–921. https://doi.org/10.1007/s11191-021-00296-x

Stucki, A. O., Barton-Maclaren, T. S., Bhuller, Y., Henriquez, J. E., Henry, T. R., Hirn, C., … & Clippinger, A. J. (2022). Use of new approach methodologies (NAMs) to meet regulatory requirements for the assessment of industrial chemicals and pesticides for effects on human health. *Frontiers in Toxicology*, 4, 964553. https://doi.org/10.3389/ftox.2022.964553

Subedi, D., Dhakal, A., Jyoti, S. et al. (2025). Zoonotic diseases awareness and food safety practices among livestock farmers in Nepal. *Frontiers in Veterinary Science*, 11, 1514953. https://doi.org/10.3389/fvets.2024.1514953

Sun, D. A., Wu, L., & Fan, G. (2021). Laboratory information management system for biosafety laboratory: Safety and efficiency. *Journal of Biosafety and Biosecurity*, 3(1), 28–34. https://doi.org/10.1016/j.jobb.2021.03.001

Supena, I., Darmuki, A., & Hariyadi, A. (2021). The influence of 4C (constructive, critical, creativity, collaborative) learning model on students' learning outcomes. *International Journal of Instruction*, 14(3), 873–892. https://doi.org/10.29333/iji.2021.14351a

Suroowan, S., Abdallah, H. H., & Mahomoodally, M. F. (2021). Herb-drug interactions and toxicity: Underscoring potential mechanisms and forecasting clinically relevant interactions induced by common phytoconstituents via data mining and computational approaches. *Food and Chemical Toxicology*, 156, 112432. https://doi.org/10.1016/j.fct.2021.112432

Surujlal-Naicker, S. & Bux, F. (2013). Application of radio-immunoassays to assess the fate of estrogen EDCs in full scale wastewater treatment plants. *Journal of Environmental Science and Health Part A-Toxic/Hazardous Substances & Environmental Engineering*, 48(1), 37–47.

Syed, M., Saleem, T., Shuja-ur-Rehman, I. M. A., Javed, F., Khan, M. B. S., & Sadiq, K. (2010). Effects of leather industry on health and recommendations for improving the situation in Pakistan. *Archives of Environmental & Occupational Health*, 65(3), 163–172.

Sylverken, A. A., Owusu, M., Agbavor, B., Kwarteng, A., Ayisi-Boateng, N. K., Ofori, P., … & Phillips, R. O. (2022). Using drones to transport suspected COVID-19 samples; Experiences from the second largest testing centre in Ghana, West Africa. *PLoS One*, 17(11), e0277057.

Szklarski, Ł. (2023). CBRN threats–advancing national security through interdisciplinary innovations: An analytical framework for chemical hazard detection technologies. *Zeszyty Naukowe SGSP*, 88(2), 93–118.

Szklarski, Ł. (2024). The threat of CBRN terrorism: An overview and improvised use of chemical, biological, radiological and nuclear materials. *Scientific Reports of Fire University*, 2(91), 39–62.

Szymkowiak, A., Melović, B., Dabić, M., Jeganathan, K., & Kundi, G. S. (2021). Information technology and Gen Z: The role of teachers, the internet, and technology in the education of young people. *Technology in Society*, 65, 101565.

Ta, G. C., Amir-Sultan, M. M., Lee, K. E., Mokhtar, M., Tan, L. L., Omar, A. S., & Sulkafle, N. H. (2020). A proposed integrated framework for chemical safety and chemical security. *Journal of Chemical Education*, 97(7), 1769–1774.

Ta, G. C., bin Soud, N., bin Nasir, K., bin Abdullah, F. A., Rosli, M. I., Nordin, D., … & Takriff, M. S. (2022). Prevention of technological disasters: Adoption of indicative criteria associated with GHS in regulating major accident hazards. *Process Safety and Environmental Protection*, 162, 583–594.

Ta, L., Gosa, L., & Nathanson, D. A. (2019). Biosafety and biohazards: Understanding biosafety levels and meeting safety requirements of a biobank. *Methods in Molecular Biology*, 1897, 213–225. https://doi.org/10.1007/978-1-4939-8935-5_19

Taloor, A. K., Sharma, S., Suryakiran, S., Sharma, R., & Sharma, M. (2024). Groundwater contamination and health risk assessment in Indian subcontinent: A geospatial approach. *Current Opinion in Environmental Science & Health*, 39, 100555.

Tan, C. & Ng, C. S. L. (2021). *Constructivism in Education*. Oxford Research Encyclopedia of Education, Oxford. https://doi.org/10.1093/acrefore/9780190264093.013.92

Tang, Q., Yan, F., Yuan, L., Tang, Y., Chen, H., Sun, Y., … & Song, G. (2024). Enhancing laboratory biosafety management: A comprehensive strategy from theory to practice. *Frontiers in Public Health*, 12, 1439051.

Tanutomo, N. M. & Octavia, T. (2016). Designing an integrated product and process layout using a simulation: The case of plastic bag company. *International Journal of Industrial Research and Applied Engineering*, 1(1), 24. https://doi.org/10.9744/JIRAE.1.1.15-24

Tariq, A. & Mushtaq, A. (2023). Untreated wastewater reasons and causes: A review of most affected areas and cities. *International Journal of Chemical and Biochemical Science*, 23(1), 121–143.

Tejada-Pérez, J. J., Herrera-Burgos, M. R., Parrón-Carreño, T., & Alarcón-Rodríguez, R. (2022). Biohazard accidents, harmful elements to the wellness of healthcare workers, and their risk factors. *International Journal of Environmental Research and Public Health*, 19(20), 13214. https://doi.org/10.3390/ijerph192013214

Tenkate, T., Kramer, D. M., Strahlendorf, P., & Szymanski, T. (2021). Setting priorities: Testing a tool to assess and prioritize workplace chemical hazards. *Work*, 70(1), 85–98. https://doi.org/10.3233/WOR-205270

Teotia, A. K., Chauhan, A. S., Thota, P. et al. (2024). Engineering advances in safe laboratory design for pharmaceutical microbiology and assessment of their biohazard risk measures. *Safety Science*, 174, 106458.

Thacker, S. B., Dannenberg, A. L., & Hamilton, D. H. (2001). Epidemic intelligence service of the centers for disease control and prevention: 50 years of training and service in applied epidemiology. *American Journal of Epidemiology*, 154(11), 985–992. https://doi.org/10.1093/aje/154.11.985

Thacker, S. B., Sencer, D. J., & Jaffe, H. W. (2017). Centers for disease control. *International Encyclopedia of Public Health*, 2017, 448–454. https://doi.org/10.1016/B978-0-12-803678-5.00057-6

Thobela, M. S., Maseme, M. R., & Duma, B. M. (2022). An overview of the national biobank of the national health laboratory service: A South African national treasure for biological resources. *Journal of Laboratory Medicine*, 46(2), 83–97.

Thompson, L., Ikenaka, Y., Darwish, W., Yohannes, Y., van Vuren, J., Wepener, V., Smit, N., Assefa, A., Tharwat, A., Eldin, W., Nakayama, S., Mizukawa, H., & Ishizuka, M. (2018). Investigation of mRNA expression changes associated with field exposure to DDTs in chickens from KwaZulu-Natal, South Africa. *PLoS One*, 13(10), e0204400.

Tian, L., Zheng, J., Pineda, M., Yargeau, V., Furlong, D., Chevrier, J., Bornman, R., Obida, M., Goodyer, C., & Bayen, S. (2022). Targeted screening of 11 bisphenols and 7 plasticizers in food composites from Canada and South Africa. *Food Chemistry*, 385, 675. https://doi.org/10.1016/j.foodchem.2022.132675

Toews, M. E. (2024). Making the Case for a Privacy Law Approach to Human Biomaterial Regulation. Thesis. University of Alberta. https://doi.org/10.7939/r3-yat2-8216

Tongu, S., Sha'Ato, R., Okonkwo, J., Olukunle, O., Tor-Anyiin, T., & Eneji, I. (2021). Levels of polybrominated diphenyl ethers (PBDEs) in water and sediment from open city drains in Makurdi Metropolitan Area, North Central Nigeria. *Environmental Monitoring and Assessment*, 193(5), 244.

Tourapi, C., Christoforou, E., Gaudêncio, S. P., & Vasquez, M. I. (2024). Aquatic biomaterial repositories: Comprehensive guidelines, recommendations, and best practices for their development, establishment, and sustainable operation. *Marine Drugs*, 22(9), 427.

Trapp, R. (2022). The chemical weapons convention. In T. Marauhn, *Research Handbook on International Arms Control Law* (pp. 279–296). Edward Elgar Publishing, Cheltenham.

Tricco, A. C., Lillie, E., Zarin, W., O'Brien, K. K., Colquhoun, H., Levac, D., Moher, D., Peters, M. D. J., Horsley, T., Weeks, L., Hempel, S., Akl, E. A., Chang, C., McGowan, J., Stewart, L., Hartling, L., Aldcroft, A., Wilson, M. G., Garritty, C., & Straus, S. E. (2018). PRISMA extension for scoping reviews (PRISMA-ScR): Checklist and explanation. *Annals of Internal Medicine*, 169(7), 467–473.

Trinidad & Tobago, Government. (2018). National Environmental Policy of Trinidad & Tobago, 2018. Accessed on 12-07-2025 from: https://www.planning.gov.tt/sites/default/files/National%20Environmental%20Policy%20%28NEP%29%20T%26T%202018.pdf

Tripathi, D., Singh, L. K., Tripathi, A. K., & Chaturvedi, A. (2021). Model based security verification of cyber-physical system based on Petrinet: A case study of nuclear power plant. *Annals of Nuclear Energy*, 159, 108306.

Tripathy, A. P., Dixit, P. K., & Panigrahi, A. K. (2022). Impact of effluent of pulp & paper industry on the flora of river basin at Jaykaypur, Odisha, India and its ecological implications. *Environmental Research*, 204(Pt A), 111769.

Trump, B. D., Florin, M. V, Perkins, E. et al. (2021). Biosecurity for synthetic biology and emerging biotechnologies: Critical challenges for governance. In B. D. Trump, M. V. Florin, E. Perkins, *Emerging Threats of Synthetic Biology and Biotechnology: Addressing Security and Resilience Issues*. Springer, Dordrecht, DE. https://www.ncbi.nlm.nih.gov/books/NBK584259/#_ncbi_dlg_citbx_NBK584259

Truter, J., van Wyk, J., Oberholster, P., Botha, A., & de Klerk, A. (2016). An in vitro and in vivo assessment of endocrine disruptive activity in a major South African river. *Water Air and Soil Pollution*, 227(2), 54.

Truter, J., van Wyk, J., Oberholster, P., Botha, A., & Mokwena, L. (2017). An evaluation of the endocrine disruptive potential of crude oil water accommodated fractions and crude oil contaminated surface water to freshwater organisms using in vitro and in vivo approaches. *Environmental Toxicology and Chemistry*, 36(5), 1330–1342.

Tucker, J. B. (2012). *Innovation, Dual Use, and Security: Managing the Risks of Emerging Biological and Chemical Technologies*. MIT Press, New York. https://doi.org/10.7551/mitpress/9147.003.0005

Tun, T. & Sim, X. (2021). Uncertified facility (BSL 2 plus): Its journey through life for preparations and setting up, compliance with biosafety regulations, implementation, and registration of the facility with the ministry of health, Singapore. *Korean Journal of Clinical Laboratory Science*, 53(1):68–80. https://doi.org/10.15324/kjcls.2021.53.1.68

Ubani, E., Madu, C. N., & Umeifekwem, U. (2024). Assessment of the impact of chemical security measures on domestic terrorism: A study of Boko Haram's use of precursor chemicals in Nigeria. *SN Social Sciences*, 4(10). https://doi.org/10.1007/s43545-024-00932-7

Uddin, Md. J., & Jeong, Y.-K. (2021). Urban river pollution in Bangladesh during last 40 years: Potential public health and ecological risk, present policy, and future prospects toward smart water management. *Heliyon*, 7(2), e06107.

Ugrina, M., & Milojković, J. (2024). Advances in wastewater treatment. *Energies*, 17(6), 1400.

Ugwu, C. N., Ugwu, O. P. C., Alum, E. U., Eze, V. H. U., Basajja, M., Ugwu, J. N., … & Uti, D. E. (2025). Medical preparedness for bioterrorism and chemical warfare: A public health integration review. *Medicine*, 104(18), e42289.

UNDESA. (2019). Millenium Development Goals Report. United Nations Department of Economic and Social Affairs. https://unstats.un.org/sdgs/report/2019/The-Sustainable-Development-Goals-Report-2019.pdf

UNDP/World Bank/WHO Special Programme for Research and Training in Tropical Diseases. (2001) Handbook: Good Laboratory Practice (GLP) /UNDP/World Bank/WHO Special Programme for Research and Training in Tropical Diseases. World Health Organization, Geneva, Switzerland. https://iris.who.int/handle/10665/66894

UNEP. (2013). Cost of Inaction on the Sound Management of Chemicals. https://wedocs.unep.org/bitstream/handle/20.500.11822/8412/-Costs%20of%20inaction%20on%20the%20sound%20management%20of%20chemicals 2013Report_Cost_of_Inaction_Feb2013.pdf?sequence=3&amp%3BisAllowed=

UNEP. (2017). Endocrine Disrupting Chemicals. United Nations Environment Programme. https://wedocs.unep.org/bitstream/handle/20.500.11822/34376/EDC_brochure.pdf?sequence=1&isAllowed=y

UNEP/WHO. (2012). *State of the Science of Endocrine Disrupting Chemicals—2012*. United Nations Environment Programme/World Health Organization, Geneva, Switzerland.

UNIDO. (2020). *How Industrial Development Matters to the Well-Being of the Population* (pp. 1–67). United Nations Industrial Development Organization, Geneva, Switzerland. https://www.unido.org/sites/default/files/files/2020-02/wellbeing_final_report.pdf

United Nations Office on Drugs and Crime (UNODC). (2024). *World Drug Report 2024: Special Points of Interest*. UNODC, Geneva, Switzerland. Retrieved March 2025, from https://www.unodc.org/documents/data-and-analysis/WDR_2024/WDR_2024_SPI.pdf

United Nations Office on Drugs and Crime (UNODC). (2025). *United Nations Office on Drugs and Crime*. UNODC, Geneva, Switzerland. Retrieved March 2025, from United Nations Office on Drugs and Crime: https://www.unodc.org/easternafrica/en/illicit-drugs/drug-trafficking-patterns.html

United Nations Security Council (UNSC). (2016). Report of the Monitoring Group on Somalia and Eritrea pursuant to Security Council resolution 2244 (2015): Somalia. https://www.un.org

United Nations Security Council (UNSC). (2020). Al-Shabaab may look to acquire explosive materials through illicit smuggling, Somalia Sanctions Committee Chair warns Security Council (SC/14209). Al-Shabaab may look to acquire explosive materials through illicit smuggling, Somalia Sanctions Committee Chair warns Security Council (SC/14209). https://press.un.org/en/2020/sc14209.doc.htm

United Nations Security Council (UNSC). (2023). Panel of Experts Report on Somalia addressed to the Chair of the Security Council Committee pursuant to resolution 751 (1992) concerning Al-Shabaab.

United **States Chemical Safety and Hazard Investigation Board (CSB) investigation report**.(2023). Fatal Chemical Decomposition Reaction and Explosion at Optima Belle LLC.

Unnikrishnan, S. & Hegde, D. S. (2007). Environmental training and cleaner production in Indian industry—A micro-level study. *Resources, Conservation and Recycling*, 50(4), 427–441.

Vaka, D. K. (2024). Integrating inventory management and distribution: A holistic supply chain strategy. *International Journal of Managing Value and Supply Chains*, 15(2), 13–23.

van Dijk, J., Gustavsson, M., Dekker, S. C., & van Wezel, A. P. (2021). Towards 'one substance–one assessment': An analysis of EU chemical registration and aquatic risk assessment frameworks. *Journal of Environmental Management*, 280, 111692.

Van Ham, P., van der Meer, S., & Ellahi, M. (2017). Chemical Weapons Challenges Ahead: The Past and Future of the OPCW. Netherlands Institute of International Relations Clingendael. The Hague: Clingendael Institute. https://www.clingendael.org/sites/default/files/2017- 10/Clingendael_Report_Chemical_Weapons_Challenges_Ahead_2017.pdf

van Wyk, J., Pool, E., & Leslie, A. (2003). The effects of anti-androgenic and estrogenic disrupting contaminants on breeding gland (nuptial pad) morphology, plasma testosterone levels, and plasma vitellogenin levels in male Xenopus laevis (African clawed frog). *Archives of Environmental Contamination and Toxicology*, 44(2), 247–256.

Vandendorpe, J., Adam, B., Wilbrandt, J., Lindstädt, B., & Förstner, K. U. (2024). Ten simple rules for implementing electronic lab notebooks (ELNs). *PLoS Computational Biology*, 20(6), e1012170.

Vandy, J. F., Mohanty, S., & Mohapatra, S. (2025). Biotechnology industry as a catalyst for circular economy in India. *Discover Sustainability*, 6(1), 728. https://doi.org/10.1007/s43621-025-01473-6

Varadarajan, D. (2014). Trade Secret Fair Use, 83 Fordham L. Rev. 1401 (2014). https://ir.lawnet.fordham.edu/flr/vol83/iss3/9

Varshney, R., Singh, P., & Yadav, D. (2021). Hazardous wastes treatment, storage, and disposal facilities. In D. Yadav, P. Kumar, P. Singh, D. A. Vallero, *Hazardous Waste Management—An Overview of Advanced and Cost-Effective Solutions*. Elsevier, Amsterdam, Netherlands. https://doi.org/10.1016/B978-0-12-824344-2.00009-4

Veleva, V. R., Cue, B. W., Todorova, S., Thakor, H., Mehta, N. H., & Padia, K. B. (2018). Benchmarking green chemistry adoption by the Indian pharmaceutical supply chain. *Green Chemistry Letters and Reviews*, 11(4), 439–456. https://doi.org/10.1080/17518253.2018.1530802

Vierendeels, G., Reniers, G. L. L., & Ale, B. J. M. (2011). Modeling the major accident prevention legislation change process within Europe. *Safety Science*, 49(3), 513–521. https://doi.org/10.1016/j.ssci.2010.11.011

Vinceti, S. R., Docea, A. O., Tsitsimpikou, C., & Filippini, T. (2021). Updating the European Union's regulation on classification, labelling and packaging of substances and mixtures (CLP): A key opportunity for consumers, workers and stakeholders with interests in the legislation and toxicology of hazardous chemicals. *Toxicology Reports*, 8, 1865–1868. https://doi.org/10.1016/j.toxrep.2021.11.011

Volonnino, G., La Russa, R., Di Fazio, N., Ottaviani, M., Zamponi, M. V., Spadazzi, F., & Umani-Ronchi, F. (2023). Z-drugs and their use in drug-facilitated crimes: A review of the literature. *Clinical Therapeutics*, 74(5), 451–468.

Vounba, P., Loul, S., Tamadea, L. F., & Siawaya, J. F. (2022). Microbiology laboratories involved in disease and antimicrobial resistance surveillance: Strengths and challenges of the central African states. *African Journal of Laboratory Medicine*, 11(1), 1570.

Vourtsis, D., Papageorgiou, E., Kriebardis, A., Karikas, G. A., van Willigen, G., Kotrokois, K., … & Karkalousos, P. (2025). Managing biological risks in biomedical laboratories of public hospitals in Athens, Greece, based on the biosafety requirements. *ESI Preprints*, 21(37), 122–122.

Wagenaar, G., Botha, T., & Barnhoorn, I. (2012). Sperm motility and testicular histology as reproductive indicators in *Clarias gariepinus* from an eutrophic impoundment, South Africa. *Journal of Applied Ichthyology*, 28(6), 990–997.

Wagner, J. K., Tanniru, J. K., Chane, C. A., & Meyer, M. N. (2022). Exploring access to genomic risk information and the contours of the HIPAA public health exception. *Journal of Law and the Biosciences*, 9(2), lsac034.

Wahab, N. A. A., Aqila, N. A., Isa, N., Husin, N. I., Zin, A. M., Mokhtar, M., & Mukhtar, N. M. A. (2021). A systematic review on hazard identification, risk assessment and risk control in academic laboratory. *Journal of Advanced Research in Applied Sciences and Engineering Technology*, 24(1), 47–62.

Wahidul, A. Md. (2023). Land-based marine pollution: An emerging threat to Bangladesh. In M. Mancuso, M. H. H. Abbas, T. Bottari, & A. A. Abdelhafez, *Environmental Sciences* (Vol. 5). IntechOpen, New York.

Wahono, B., Lin, P. L., & Chang, C. Y. (2020). Evidence of STEM enactment effectiveness in Asian student learning outcomes. *IJ Stem Ed*, 7, 36. https://doi.org/10.1186/s40594-020-00236-1

Walker, J. R. (2003). Strengthening the BTWC. The role of biological and toxin weapons convention in combating natural and deliberate disease outbreaks. *EMBO Reports*, 4(Suppl 1), S61–S64. https://doi.org/10.1038/sj.embor.embor852.

Walters, A. U., Lawrence, W., & Jalsa, N. K. (2017). Chemical lab safety awareness, attitudes and practices of tertiary students. *Safety Science*, 96, 161–171.

Walters, D. B. (2015). Safety, security, and dual-use chemicals. *Journal of Chemical Health and Safety*, 22(5), 3–16.

Wan, M., Li, T., Chen, H., Mao, C., & Shen, J. (2021). Biosafety, functionalities, and applications of biomedical micro/nanomotors. *Angewandte Chemie International Edition*, 60(24), 13158–13176.

Wang, W., Russell, A., & Yan, Y. (2014). *Traditional Chinese medicine and new concepts of predictive, preventive and personalized medicine in diagnosis and treatment of suboptimal health.* The EPMA Journal2014 5:4. doi:10.1186/1878-5085-5-4. Available from: https://www.researchgate.net/publication/260167397_Traditional_Chinese_medicine_and_new_concepts_of_predictive_preventive_and_personalized_medicine_in_diagnosis_and_treatment_of_suboptimal_health [accessed Feb 25 2026].

Wang, D., Yang, G., Han, J., Duo, Y., Zhou, X., & Tong, R. (2024a). Quantitative assessment of human error of emergency behavior for hazardous chemical spills in chemical parks. *Process Safety and Environmental Protection*, 189, 930–949.

Wang, H., Chen, Y., Wang, L., Liu, Q., Yang, S., & Wang, C. (2023). Advancing herbal medicine: Enhancing product quality and safety through robust quality control practices. *Frontiers in Pharmacology*, 14, 1265178. https://doi.org/10.3389/fphar.2023.1265178

Wang, J., Fan, Y., & Niu, Y. (2022). Routes to failure: Analysis of chemical accidents using the HFACS. *Journal of Loss Prevention in the Process Industries*, 75, 104695. https://doi.org/10.1016/j.jlp.2021.104695

Wang, J., Zhou, W., Li, Y., Yu, M., & Zhu, Z. (2024b). The laboratory skills contest: An activity to improve and evaluate students' laboratory practice through the determination of water hardness. *Journal of Chemical Education*, 101(4), 1758–1765.

Wang, L. K., Wang, M. H. S., Hung, Y. T., Shammas, N. K., & Chen, J. P. (Eds.). (2017). *Handbook of Advanced Industrial and Hazardous Wastes Management* (Vol. 7). CRC Press, Boca Raton, FL.

Wang, S., Ding, L., Sui, H., & Gu, Z. (2021a). Cybersecurity risk assessment method of ICS based on attack-defense tree model. *Journal of Intelligent & Fuzzy Systems*, 40(6), 10475–10488.

Wang, J., Wang, L., & Li, J. (2021b). *Traditional Chinese medicine in the 21st century: Accessibility and affordability in rural healthcare*. Beijing, China: Science Press.

Wang, M., Li, Y., & Chen, R. (2021d). Processing of traditional Chinese medicine: A review of history, theory, and current standards. *Journal of Ethnopharmacology*, *264*, **113264.**

Weber, D. J., Rutala, W. A., & Sickbert-Bennett, E. (2023). Emerging infectious diseases, focus on infection prevention, environmental survival and germicide susceptibility: SARS-CoV-2, mpox, and Candida auris. *American Journal of Infection Control*, 51(11), A22–A34.

Weldon, D. A. (2012). Planning and designing spaces for the sciences. *Journal of Undergraduate Neuroscience Education*, 11(1), A114–A118.

Weng, S. T., Li, Q. W., Gao, Y. D., & Qiu, Y. F. (2024). Biosafety risk control strategies in laboratory animal research. *Safety and Health at Work*, 15(1), 118–122. https://doi.org/10.1016/j.shaw.2023.11.005

Westwood, N., Londesbrough, D., Ford, S. J., & Halbert, G. W. (2016). Explosivity: An unusual challenge in drug development. In: *10th World Meeting on Pharmaceutics, Biopharmaceutics and Pharmaceutical Technology*, 2016-04-04–2016-04-07, SECC.

White, T. E., Wong, W. B., Janowiak, D., & Hilborne, L. H. (2021). Strategies for laboratory professionals to drive laboratory stewardship. *Practical Laboratory Medicine*, 26, e00249.

WHO. (2013). *WHO Traditional Medicine Strategy: 2014–2023*. World Health Organization, Geneva.

Widyatmoko, H. (2018). Management of hazardous waste in Indonesia. *IOP Conference Series: Earth and Environmental Science*, 106(1), 012032.

Wiesner, L., Araya, S., & Lovsin Barle, E. (2022). Identifying nonhazardous substances in pharmaceutical manufacturing and setting default health-based exposure limits (HBELs). *Journal of Applied Toxicology*, 42(9), 1443–1457.

Williams, M., Armstrong, L., & Sizemore, D. C. (2025). *Biologic, Chemical, and Radiation Terrorism Review*. StatPearls Publishing, Treasure Island, FL. https://www.ncbi.nlm.nih.gov/books/NBK493217/

Williamson, S. M. & Prybutok, V. (2024). Balancing privacy and progress: A review of privacy challenges, systemic oversight, and patient perceptions in AI-driven healthcare. *Applied Sciences*, 14(2), 675. https://doi.org/10.3390/app14020675

Win, A. M. (2024). A Study on Awareness of Personal Protective Equipment in Workplace (Case Study: Zaykabar Industrial Zone) (Aung Myo Win).

Winkelmann, K., Keeney-Kennicutt, W., Fowler, D., Macik, M. L., Guarda, P. P., & Ahlborn, C. J. (2023). Learning gains and attitudes of students performing chemistry experiments in an immersive virtual world. *Interactive Learning Environments*, 28(5), 82–96.

Wohlleben, W., Mehling, A., & Landsiedel, R. (2023). Lessons learned from the grouping of chemicals to assess risks to human health. *Angewandte Chemie International Edition*, 62(22), e202210651.

Wooding, M., Rohwer, E., & Naude, Y. (2017a). Comparison of a disposable sorptive sampler with thermal desorption in a gas chromatographic inlet, or in a dedicated thermal desorber, to conventional stir bar sorptive extraction-thermal desorption for the determination of micropollutants in water. *Analytica Chimica Acta*, 984, 107–115.

Wooding, M., Rohwer, E., & Naude, Y. (2017b). Determination of endocrine disrupting chemicals and antiretroviral compounds in surface water: A disposable sorptive sampler with comprehensive gas chromatography—Time-of-flight mass spectrometry and large volume injection with ultra-high performance liquid chromatography-tandem mass spectrometry. *Journal of Chromatography A*, 1496, 122–132.

World Health Organization (2020a). *The WHO Recommended Classification of Pesticides by Hazard and Guidelines to Classification 2019*. WHO, Geneva, Switzerland.

World Health Organization (WHO). (2016). *International Health Regulations* (3rd ed.). WHO, Geneva, Switzerland. https://www.who.int/publications/i/item/9789241580496

World Health Organization (WHO). (2020b). *Cybersecurity in the Pharmaceutical Industry*. WHO, Geneva, Switzerland. https://www.who.int

World Health Organization (WHO). (2020c). *Laboratory Biosafety Manual, Fourth Edition and Associated Monographs* (pp. 103–137). WHO, Geneva, Switzerland.

World Health Organization. (2024). *Guidance on Regulations for the Transport of Infectious Substances 2023–2024*. WHO, Geneva, Switzerland.

World Intellectual Property Organization (WIPO). (2021). *Intellectual Property and Traditional Knowledge*. WHO, Geneva, Switzerland.

World Trade Organization (WTO). (2019). *Trade-Related Aspects of Intellectual Property Rights (TRIPS) Agreement*. WHO, Geneva, Switzerland.

Wu, X., Ni, J., & Li, H. (2020). Safety assessment of traditional Chinese medicine: A systematic review of herb-drug interactions and adverse effects. *Frontiers in Pharmacology*, *11*, 575.

Wu, P., Lin, S., Cao, G. et al. (2022). Absorption, distribution, metabolism, excretion and toxicity of microplastics in the human body and health implications. *Journal of Hazardous Materials*, 437, 129361.

Wunsch, C. R. (1980). Trade secret confidentiality and toxic substances regulation: A non-tariff trade barrier in the chemical trade. *Cornell International Law Journal*, 14(1), 7. https://scholarship.law.cornell.edu/cilj/vol14/iss1/7

Xie, P. (2011). Toxicity of Chinese herbal medicines. *Clinical Pharmacology & Therapeutics*, 89(6), 821–826.

Xin, B., Mu, S., Tan, T. et al. (2020).Belief in and use of traditional Chinese medicine in Shanghai older adults: a cross-sectional study. *BMC Complement Med Ther* **20**, 128. https://doi.org/10.1186/s12906-020-02910-x

Xu, W., Boehm, G., & Zheng, Q. (2016). Factors to consider in developing individual pharmaceutical product quality risk profiles useful to government procurement agencies. *Acta Pharmaceutica Sinica B*, 6(1), 79–92. https://doi.org/10.1016/j.apsb.2015.09.009

Xu, Y. & Wang, Y. (2023). *IP Protection of Traditional Knowledge: A Dilemma Faced by the Chinese Traditional Medicine in Global Competition*. IntechOpen, New York. https://doi.org/10.5772/intechopen.1001820

Yakar, Z. & Baykara, H. (2014). Inquiry-based laboratory practices in a science teacher training program. *Eurasia Journal of Mathematics, Science and Technology Education*, 10(2), 173–183.

Yang, J., Jia, Z., Zhang, J., Wang, H., Dai, Y., Zhu, G., ... & Chen, S. (2023). Unblocking high-value botanical value chains: Is there a role for blockchain technology? *Frontiers in Pharmacology*, *14*, 1179099.

Yazdani, M. & Haghani, M. (2023). Logistics of patient evacuation in response to disease outbreaks: Critical considerations for transportation planning. *Transportation Research Interdisciplinary Perspectives*, 22, 100975.

Yazdi, M. (2024). Maintenance strategies and optimization techniques. In M. Yazdi, *Advances in Computational Mathematics for Industrial System Reliability and Maintainability* (pp. 43–58). Springer Nature Switzerland, Cham.

Yazid, M. F. H. A., Ta, G. C., & Mokhtar, M. (2020). Classified chemicals in accordance with the globally harmonized system of classification and labeling of chemicals: Comparison of lists of the European Union, Japan, Malaysia and New Zealand. *Safety and Health at Work*, 11(2), 152–158. https://doi.org/10.1016/j.shaw.2020.03.002

Yoo, J. H. (2018). Review of disinfection and sterilization – Back to the basics. *Infection & Chemotherapy*, 50(2):101–109. https://doi.org/10.3947/ic.2018.50.2.101

Yu, D. G., Li, Q., Song, W., Xu, L., Zhang, K., & Zhou, T. (2023). Advanced technique-based combination of innovation education and safety education in higher education. *Journal of Chemical Education*, 100(2), 507–516.

Yuan, S., Reniers, G., & Yang, M. (2022). The necessity of integrating safety and security barriers in the chemical process industries and its potential framework. *Chemical Engineering Transactions*, 91, 13–18. https://doi.org/10.3303/CET2291003

Yuan, S., Yang, M., & Reniers, G. (2024). Integrated process safety and process security risk assessment of industrial cyber-physical systems in chemical plants. *Computers in Industry*, 155, 104056. https://doi.org/10.1016/j.compind.2023.104056

Zaboli, P., Hashemi-Meshkini, A., Varmaghani, M., Gholami, H., Vazirian, I., Zekri, H. S., Eslamitabar, S., & Kebriaeezadeh, A. (2016). Pharmaceutical laws and regulations in Iran: An overview. *Journal of Research in Pharmacy Practice*, 5(3), 155–161. https://doi.org/10.4103/2279-042X.185709

Zerizghi, T., Guo, Q., Tian, L., Wei, R., & Zhao, C. (2022). An integrated approach to quantify ecological and human health risks of soil heavy metal contamination around coal mining area. *Science of the Total Environment*, 814, 152653. https://doi.org/10.1016/j.scitotenv.2021.152653

Zhang, Z. Y., Wang, Y. J., Yan, H., Chang, X. W., ………, & Duan, J. A. (2021). Rapid Geographical Origin Identification and Quality Assessment of Angelicae Sinensis Radix by FT-NIR Spectroscopy. *Journal of analytical methods in chemistry*, *2021*, 8875876. https://doi.org/10.1155/2021/8875876

Zhang, S., Han, Y., Peng, J., Chen, Y., Zhan, L., & Li, J. (2023). Human health risk assessment for contaminated sites: A retrospective review. *Environment International*, 171, 107700. https://doi.org/10.1016/j.envint.2022.107700

Zhang, X., Zhang, Y., & Li, G. (2022). Safety and risk assessment of traditional Chinese medicine in pregnancy and breastfeeding: A systematic review. *Journal of Traditional Chinese Medicine***,** *42*(3), 426–434.

Zhang, L., Jin, S., Qin, C., Ma, D., Ye, J. and Liu, Q. (2025). Bibliometric analysis of traditional Chinese medicine for viral infections through immune modulation (2015–2025). *Front. Immunol.* 16:1647900. doi: 10.3389/fimmu.2025.1647900

Zhao, G., Vazquez-Noguerol, M., Liu, S., & Prado-Prado, J. C. (2024a). Agri-food supply chain resilience strategies for preparing, responding, recovering, and adapting in relation to unexpected crisis: A cross-country comparative analysis from the COVID-19 pandemic. *Journal of Business Logistics*, 45(1), e12361.

Zhao, H., Liu, H., Wei, G. et al. (2021). Comparative life cycle assessment of emergency disposal scenarios for medical waste during the COVID-19 pandemic in China. *Waste Management*, 126, 388–399.

Zhao, X., Bai, Z., Zhan, X., Wang, J., Cheng, Y., & Xiao, X. (2024b). Safety evaluation of traditional Chinese medicine: New era, new strategy. *Acupuncture and Herbal Medicine*, 4(2), 171–175. https://doi.org/10.1097/hm9.0000000000000119

Zhi, J. & Lien, J. M. (2024). Improving human-robot collaboration via computational design. *IEEE Robotics and Automation Letters*, 99, 1–8. https://doi.org/10.1109/lra.2024.3519863

Zhong, B., Pan, X., Love, P. E., Sun, J., & Tao, C. (2020). Hazard analysis: A deep learning and text mining framework for accident prevention. *Advanced Engineering Informatics*, 46, 101152.

Zhou, S., Pan, S. Q., & Huang, M. (2021). Pharmacokinetic and pharmacodynamic herb-drug interactions in older adults with polypharmacy. *Frontiers in Pharmacology***,** *14*, 115623.

Zhu, R., Hu, X., Li, X., Ye, H., & Jia, N. (2020). Modeling and risk analysis of chemical terrorist attacks: A Bayesian network method. *International Journal of Environmental Research and Public Health*, 17(6), 2051. https://doi.org/10.3390/ijerph17062051

Zodrow, J., Vedagiri, U., Sorell, T., McIntosh, L., Larson, E., Hall, L., … & Anderson, J. (2022). PFAS Experts symposium 2: PFAS toxicology and risk assessment in 2021—Contemporary issues in human and ecological risk assessment of PFAS. *Remediation Journal*, 32(1–2), 29–44.

Zubanov, P. S., Tregub, P. P., Goldberg, A. S., Godkov, M. A., & Akimkin, V. G. (2025). Comprehensive assessment of medical laboratory performance: A 4D model of quality, economics, velocity, and productivity indicators. *Clinical Chemistry and Laboratory Medicine (CCLM)*, 63(10), 1928–1940.

Zulfiqar, N. (2025). *Industrial and agricultural contributions to water pollution in Pakistan: Policy interventions for sustainable water management.* [SSRN]. https://ssrn.com/abstract=5294649.

Zytoon, M. A. (2020). A decision support model for prioritization of regulated safety inspections using integrated Delphi, AHP and double-hierarchical TOPSIS approach. *IEEE Access*, 8, 83444–83464. https://doi.org/10.1109/access.2020.2991179

Index

Note: **Bold** page numbers refer to tables and *italic* page numbers refer to figures.

D

G

H

I

J

K

M

Q

R

S

W

Y

For Product Safety Concerns and Information please contact our EU representative GPSR@taylorandfrancis.com
Taylor & Francis Verlag GmbH, Kaufingerstraße 24, 80331 München, Germany

www.ingramcontent.com/pod-product-compliance
Ingram Content Group UK Ltd.
Pitfield, Milton Keynes, MK11 3LW, UK
UKHW020345170826
474444UK00003B/28

* 9 7 8 1 0 4 1 0 4 8 5 2 7 *